THE ULTIMATE GUIDE TO

HOMESTEADING

Introduction

WHAT THIS BOOK IS ABOUT

Be content with what you have; rejoice in the way things are. When you realize there is nothing lacking, the whole world belongs to you.

—Lao Tzu

This book is a guide for those who are seeking useful knowledge on how to live a life that is real, with the freedom to focus on what is truly important. There is no skill more important than self-reliance and being prepared for any situation.

The information contained in this book is meant as a starting guide. Some of the activities described may be dangerous, or even possibly illegal where you live (for example: producing ethanol or growing hemp). I can't accept any responsibility for what might happen when you try some of them. The information provided is as accurate as possible and I hope that it will do some good.

My goal for this book was this: If you had nothing, and suddenly had to survive in the world by doing everything yourself, what would you need to know? What if you lived in an apartment and suddenly the grocery store was empty? Or you had no gasoline and couldn't drive to work and had to survive without any money? What if you lived in a rural area and wanted to "get off the grid"? Or you wanted to live with as little impact on the earth as possible? Preparedness is not just about having enough food to eat for a year; it's about living as well as you can in less-than-wonderful circumstances.

I have included simple instructions from a variety of resources, including pioneer techniques and nineteenth-century expertise, as well as military survival, and organic agricultural methods. The most efficient, long-term, and cheapest method of survival is a vegetarian lifestyle that uses animals to work the ground. A homestead, as I define it, is any location in which all the necessities of life are being produced on site. Your suburban house, your apartment, or your houseboat could be a homestead if you utilize your space. If you grow most of your food, create the things you need to live, and use the resourcefulness of the early homesteaders that built North America, then you're a homesteader too.

I don't guarantee that anything in here will help you during Armageddon, but it might make the quality of your life a bit better. The simple skills that I describe here are also organic and earth-friendly, so even if you aren't in bad circumstances, this book might help make you and your world a bit healthier. Human breast milk has been found to contain over 350 man-made chemicals that are ingested by a mom and then passed on to her baby. An apple you buy at the grocery store may have been treated up to forty times with any of 100 chemicals in use by industrial farms today. And that's just your food; it doesn't count the air you breathe, the chemicals you clean with, or the synthetic materials in your clothes.

The purpose of this book is to provide the knowledge you need to live well with absolutely nothing.... although having a pocketknife to get started might help. When following the instructions, it is important to remember that unlike most homesteading guides, this one is meant to be an instruction manual. Please read the instructions through carefully, then do each step one at a time. Sometimes you may have to figure some stuff out for yourself, and I guarantee that everything in here takes practice. Gardening, pottery, weaving, woodcarving.... these are all art forms that some people take a lifetime to learn. Be patient with yourself and take the time to do things right. Don't forget to use the handy dictionary in the back of the book if you run into a word you don't know. I would also suggest investing in some good plant guides (both wild and garden varieties), because this is a book about taking action, and not about plant identification. So get going! Good luck, my friend, and may you find peace on your journey.

—N.F., 2011

Success is relative. It is what we can make of the mess we have made of things.

—T. S. Eliot

1 | What Is Homesteading?

THE HISTORY OF HOMESTEADING

Man—despite his artistic pretensions, his sophistication, and his many accomplishments—owes his existence to a six-inch layer of topsoil and the fact that it rains.

—Author Unknown

Congress passed the Homestead Act in 1862. It was the height of the Civil War, and the American government had been scooping up as much land as possible all the way to the Pacific Ocean. Previous to that, it had been very difficult and expensive to buy land to the west, and hardly anyone did. But suddenly land became incredibly affordable—free! Any person 21 years or older, male or female, could file a patent for 160 acres, a quarter mile on each side, of unclaimed public land. All you had to do was build a house, dig a well, plow a field, and live there for the next five years and the government would give you the title. As time passed they made it even easier. By 1873 if you planted trees you could have an extra 160 acres of land, because they believed that trees produced rainfall (they don't). If you moved to the desert, you could get 640 acres.

Of course it really wasn't that easy. If you actually made it to your land in a covered wagon without serious injury or death, you still had to survive on it through drought, disease, attacks, and other hazards. The right to call yourself a "homesteader" was earned not by getting the land, but by being tough enough and passionate enough to make your dream successful. There are numerous stories of pioneer women working in the fields nine months pregnant, who after delivering the baby in the middle of cutting down hay, just proceeded to wipe the newborn off and strap it to her back so she could keep on working. Fathers worked sunup to sundown, their young sons right alongside them. Their only pleasures were the odd fishing trip, evening singing around the stove, or a potluck dinner.

People continued to get free land up into the 1900s, mostly in Alaska, until 1973 when the Act was repealed and the government stopped accepting applications. Many of the people today who want to be homesteaders feel a bit jealous that they missed out by only one generation. We idealize the homestead, the art of self-sufficiency. Do homesteaders exist today? I believe they do, and you can call yourself a homesteader if some of the following applies to you:

1. You have invested all your time and money into a piece of property that most people wouldn't want.
2. You don't care about a career or success, and are mostly interested in the size of your tomatoes or the health of your baby goats.

3. You want your kids to learn to feed chickens and feel that it is an important life skill that teaches them to appreciate hard work and the value of their food.
4. You sometimes feel that your animals do things to annoy you and then laugh about it behind your back.
5. You manage to grow so much food in a small area that you have trouble canning it and giving it away to neighbors.
6. You have very little money but you know if you had it you would only be spending it on your homestead.
7. You purposely try to do things in a less convenient way because you feel that doing things yourself makes you a better person.
8. When you talk about your "family," it means your animals, children, birds in your backyard, the worms you raise, and the bees you are trying to take care of.
9. You have really tried to check off everything on the list of skills at the back of Carla Emery's book.
10. You feel out of place in this time period—you sometimes feel as if you should have been born a century ago.
11. You appreciate natural beauty a thousand times more than anything that could be printed in a fashion magazine. Not to mention you haven't looked at one of those in a long time.
12. You can't wait to retire so you can work even harder on your homestead.

Walden Pond

▲ Viking farm

13. You feel so passionate about a dream that you can't really explain that you are willing to sell everything you have to make it happen.

Today a homestead is a place that grows all the essentials of life, food and heat and water, and teaches the basic skills of living to the inhabitants. Here is the quote that inspired me, written by the great Thoreau. It is the ultimate summary of what homesteading really is:

> "I went to the woods because I wished to live deliberately, to front only the essential facts of life, and see if I could not learn what it had to teach, and not, when I came to die, discover that I had not lived. I did not wish to live what was not life, living is so dear; nor did I wish to practice resignation, unless it was quite necessary. I wanted to live deep and suck out all the marrow of life, to live so sturdily and Spartan-like as to put to rout all that was not life, to cut a broad swath and shave close, to drive life into a corner, and reduce it to its lowest terms, and, if it proved to be mean, why then to get the whole and genuine meanness of it, and publish its meanness to the world; or if it were sublime, to know it by experience, and be able to give a true account of it in my next excursion. For most men, it appears to me, are in a strange uncertainty about it, whether it is of the devil or of God, and have somewhat hastily concluded that it is the chief end of man here to 'glorify God and enjoy him forever.'"

LEARNING FROM THE PAST

The great eventful Present hides the Past; but through the din of its loud life hints and echoes from the life behind steal in.

~John Greenleaf Whittier

The Viking Farm

Vikings generally grew flax for making their clothes, as well as barley, rye, and wheat for bread. Their green vegetables

▲ Medieval farm

were turnips, beans, cabbage, and other staples. They used horses and wooden plows for breaking the earth. They had a live storage in a storage house near to the garden. They also kept cows, sheep, and pigs in barns and fenced enclosures, and hunted whale, deer, moose, seals, rabbits, walrus, polar bear, wild boar, geese, and seagulls. Their houses had thatched roofs, and they were also good blacksmiths. They used slaves to supplement their work force.

The Medieval European Farm

Medieval farmers were serfs who worked for a lord, and so farming was basically communal. An acre was the amount of land that a peasant could plow during one day's work, and they used three or four crop rotations; for example: wheat the first year, barley the second year, and then the third year it would rest (or lie fallow). Cattle and other animals such as sheep were grazed in separate fields, and kept in the barn or sometimes kept in the house. These animals usually did not last through winter, however. They also had thatched roofs.

The 1800s European Farm

Because farmers now had their own small individual farms, conditions were somewhat improved. They continued crop rotation but added in complimentary plants; for example: wheat the first year, turnips the second, barley the third year and clover the fourth year. This eliminated the fallow year. They gathered the manure from cattle and horses and spread it over the fields, which increased food production enough to keep the cattle all year round.

1800s North American Farm

After the Civil War agriculture could no longer depend on slave labor.

▲ Nineteenth-century European farm

Pioneering families moved farther and farther westward and had many children to supplement their work force. Trees had to be cleared and so fields were often planted around tree stumps, and sometimes they did not plant any grain, focusing only on vegetables in order to get the highest yield possible. Horses or oxen were used to do this heavy labor. Main crops were potatoes, turnips, and if they grew grain it was limited to corn, wheat, and barley. Cattle became a major

▼ Nineteenth-century North American farm

necessity not only for milk and meat but also for their draft ability. The common housing was cabins with wood shingles.

Not much has changed through time but you can learn a few things from the early homesteader's mistakes. Knowing your soil and using plants wisely can mean the difference between life and death. Working smart is much better than owning slaves or having fifty children, so planning ahead can save you time and effort. These people spent every daylight hour working for their survival, but with some ingenuity you can be self-sufficient and have an evening to relax now and then. They also had severe health problems and short life spans. Take care of yourself and your family, and find balance.

▲ Urban community garden

A FEW NOTES ON INTENTIONAL COMMUNITY

I am of the opinion that my life belongs to the whole community and as long as I live, it is my privilege to do for it whatever I can. I want to be thoroughly used up when I die, for the harder I work the more I live.

~George Bernard Shaw

What is an intentional community ?

Spontaneous communities are those that just happen, usually because of a geographic resource. An intentional community is planned out by a group of people who decide they want to live together for a common goal or purpose. People have been doing this for a long time, ranging from communities in which personal property doesn't exist, to modern cohousing where people buy shares just like a corporation.

What is the benefit of community?

Community is a support system. In an emergency people will come together for a common goal—to survive. If your community already exists, you are more likely to succeed because you are organized. Community provides emotional support as well as financial support and resources. People are a valuable commodity.

What types of communities are there?

Neighborhood: Some neighborhoods have a good sense of community, others don't. They are not intentional. A neighborhood can develop community by making a community garden, removing fences, and having neighborhood activities.

Commune: Communes are groups of people who have no individual ownership of property. For thousands of years, idealistic people in search of a life without greed have started this type of community. But, it is the natural instinct of people to have ownership, and these have always failed—they are not long-lived, although many have been temporarily beneficial to those who started them.

Eco-village: A community united with the goal of living a sustainable lifestyle by pooling resources and studying alternative, ecological methods of living.

Cohousing: Cohousing is usually started in an urban area in order to lower housing costs, or at least it used to be. You buy a unit in a carefully designed living space just like you would a townhouse. However, modern cohousing actually usually costs more than a regular house because they often use sophisticated ecological building techniques and materials.

Other intentional communities: In other communities the people live separately or in a group house, and have a religious or political reason for living together. Each community is different.

What are the rules of starting a successful community?

Write down the common goal of the community, and make sure everyone agrees with it.

Recognize the different personalities and individuals in the group and respect opinions.

Have a set way to conduct your meetings, so that everyone presents concerns in a calm manner.

If you will be purchasing land or other legal actions, consult a lawyer to prevent problems.

When you choose a leader, uphold that leader, and make sure the leader upholds the group.

Have rituals and celebrations to bring the group together.

HOMESTEADING EQUIPMENT YOU WILL NEED

There is just one life for each of us: our own.

~Euripides

What things do I need before moving to a primitive homestead?

The following items are for those making the big move to the land and who plan to live simply, with wood heat, using hand tools, etc. To buy all of this equipment new would cost you somewhere around $40,000, so please start collecting these things early from garage sales and secondhand stores. Also, you may not want everything on this list

Washtub and board

Flat iron ▸

depending on what you are doing and what kind of house you will live in. Some questions to ask yourself are:

If you have no animals and do not plan to hunt, do you still want to have the rifle for protection?

Are you going to need a rifle to butcher cows or goats?

What kind of heat are you going to cook over and heat your house with?

What kind of house are you building?

How will you get water to your garden? Can you use a garden hose?

Are you going to use a tractor or horse or manpower to turn the soil?

What animals do you want to keep and what is their purpose?

How much food storage do you really need? (The food storage is vital . . . that first year will be the toughest and knowing you've got some put away already is a lifesaver.)

What climate are you going to?

What safety equipment will you need? Are you going to be chopping down trees?

For the house:

Battery-powered smoke alarm and batteries

Wood cookstove

Wood stove for heat

Galvanized wash tub and scrub board

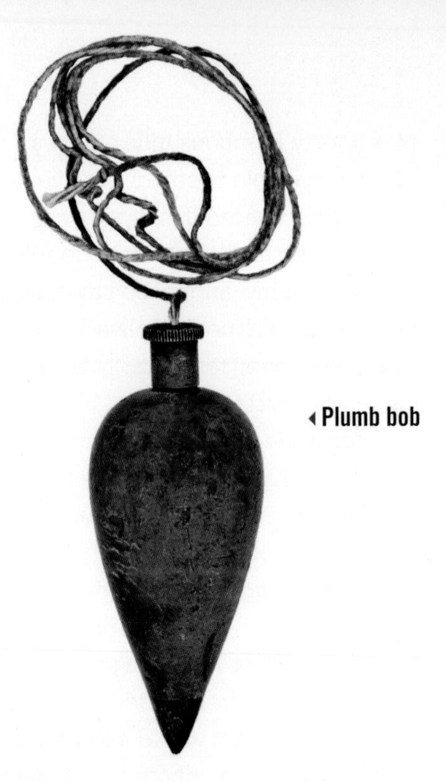
◂ Plumb bob

Clothesline
Clothespins
Dustpans
Straw broom
Flat iron and trivet
Wood ironing board
Treadle sewing machine
Baskets

Building equipment:

12-inch hammer
Large ax
Hatchet with steel head and sheath
2" x 8" clamps
4" x 16" clamps
Assorted chisels of forged steel
5-pound wood splitting wedge
Low angle hand plane
Sledgehammer
24" level
Brass plumb bob [chapter1–10]
Chalk line
Bottle of blue chalk
Pick
100-foot tape measure

Supply of nails of different sizes
45-pound anvil
Blacksmithing kit
Crosscut saw
Carpenter saw
Keyhole saw
Saw sharpening kit
Sandpapers and sanding block
Linseed oil
Emery paper
Phillips and Robertson's screwdrivers
Ratchet set
Sawhorse
Square
Pipe wrench

Hunting/fishing equipment:

.22 rifle
Gun license
Rifle ammunition
Fishing reel and rod
Fishing line
Assorted lures and hooks
Large hunting knife
Fishing knife

Farm equipment:

Pots for plants
Spade shovels
Dung shovels
Snow shovels
Leather and cotton gloves
Pitchforks
Posthole digger
Garden hoses
Disc plow
Cultivator
Plow
Spring-tooth harrow
Mower
Tether
Lots of seeds
Rope
Several sizes of pulleys
Chains and padlocks
Fence stretcher
Scythe and sharpening stone
Machete
Push lawn mower
Garden hoe
Dutch flat hoe
Yard rake
Garden rake
Tote baskets
Wheelbarrow
Dung forks
Hay rake

Animals:

Team of draft horses
Chickens
Rooster
Adopted dog
Assorted chicken feeders
Chicken water dishes
Chicken feed
Hay for horses
Dog food
Chicken wire
Sheep fencing
Barbed wire
Iodine
Lamb's nipples and bottles
Old blankets
Vapor rub

Dutch oven

Worming medicine
Stainless steel milking container
Horse harness
Horse cart
Farrier tools
Horseshoes
Grooming tools
Horse blanket
Leadropes, halters, and bridles
Chisel, file, and hoof pick
Dustpans
Straw broom

Energy and water:

Hand water pump
Pipes

Cooking equipment:

Cooking thermometer
Ceramic crocks
Hand grain mill
Butcher knife
Steel bowls
Unbleached muslin
Dutch oven
Enamel canner
Pressure canner
Pressure gage
Jar rack
Canning jars
Canning lids
Hydrometer (for water content)

Metal ice chest
Large iron frying pan
1-and 2-quart iron pots
Roasting pan
Bread pans
Muffin pans
Pizza pans
Steel brush
Scouring pads
Steel wool
Cutting board
Rolling pin
Wire racks
Dishpan
Wash pan
Drain rack
Large bowls
Stainless steel teakettle
Teapot
Cooling rack
Breadboard
Whisk
Pie and cookie tins
Wood mixing spoons
Pot holders and dish cloths
Strainers
Butter churn and paddle

Food storage for 2 adults and 3 kids for one year:

25 lbs—Wheat
60 lbs—Enriched white flour
109 lbs—Cornmeal
150 lbs—Rolled oats
260 lbs—Enriched white rice
13 lbs—Pearled barley
150 lbs—Pasta
180 lbs—Dry beans
8 lbs—Dry lima beans
8 lbs—Dry soybeans
8 lbs—Dry split peas
8 lbs—Dry lentils
28 lbs—Dry soup mix
19 qt—Cooking oil
10 qt—Shortening
4 qt—Mayonnaise
4 qt—Salad dressing
15 qt—Peanut butter
56 lbs—Nonfat dry milk
50 (12 oz) cans—Evaporated milk
160 lbs—Granulated sugar
12 lbs—Brown sugar
4 lbs—Molasses
20 lbs—Honey
12 lbs—Corn syrup
25 lbs—Jam
24 lbs—Powdered fruit drink
4 lbs—Jell-O
24 lbs—Salt
2 lbs—Dry yeast
10 lbs—Baking soda
7 lbs—Baking powder
56 gal—Water
Supply of canned fruits and vegetables

Personal items:

Supply of toilet paper
Personal hygiene items
First aid kit

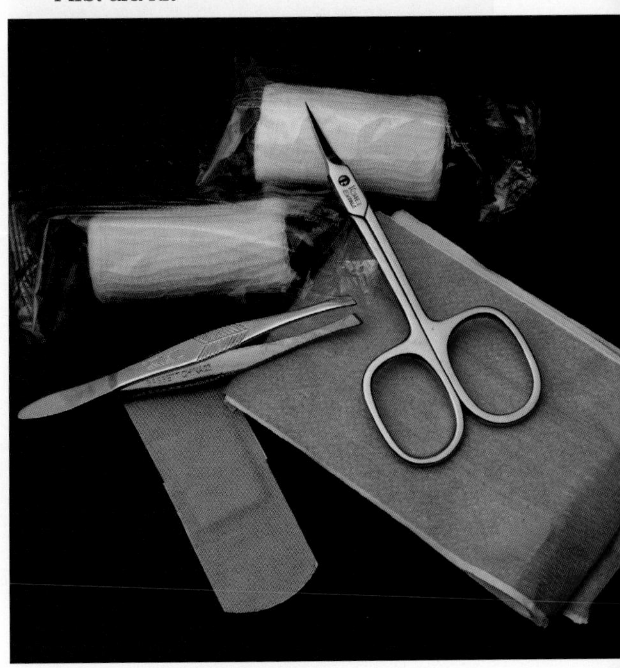

▲ Kerosene lantern

Miscellaneous items:

Board games
Flashlights and a battery supply
Pocketknives
Rope
Weather thermometer
Humidity gage
Assorted storage bins and buckets
Kerosene/oil lamps and kerosene/oil supply
Supply of gasoline
Sleeping bags
Pillows
Long underwear
Socks
Steel-toed boots
Tents
Supply of tarps
Hard hats and safety glasses
Vaseline
Books for home schooling

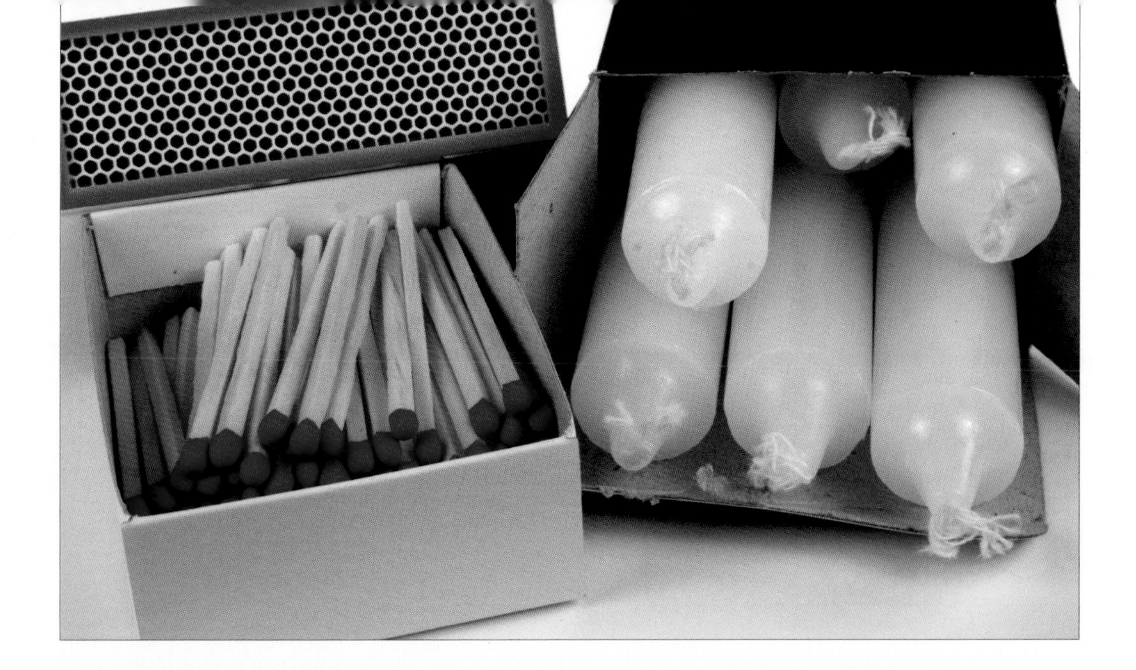

Sewing kit
Candles and candleholders
Waterproof matches
Wind-up and solar radio
Wool blankets
Epsom salts

Money: It is a good idea to have a year of savings in the bank to pay your bills, plus some for an emergency.

◂ **24" chainsaw**

If you are going to want to have electricity and some more modern conveniences, then add the following items to your list. It includes some very handy lumber-making tools:

14-cubic-foot chest freezer
Refrigerator
Crew cab 4x4 truck
85cc 24-inch chainsaw
33cc 14-inch chainsaw
9-foot chainsaw mill
Chainsaw sharpening kit
Circular saw
16-inch saw and extra blades
12-½-inch portable planer
Debarker
Drills and drill bits
Tiller

EMERGENCY KIT

Life belongs to the living, and he who lives must be prepared for changes.

~Goethe

week. Don't forget candles (three per day or more), candleholders, a big box of matches, and oil or electric lamps and flashlights, extra batteries and lamp oil, and a battery-powered fire detector. It is also useful to have a woodstove and winter's supply of wood.

What is a primitive emergency kit?

For your house you could have a herbal first aid kit, clean cloths for sanitation, pocket knife, ½ gallon water per person per day stored in glass jugs, cloth diapers, paper and cloth scraps for toilet paper, candles or lamps, flint, and canned and dried food. If you already live this way then an emergency may not be a big deal.

What should I keep in the car?

You should have blankets, a warm coat, walking shoes, winter hiking boots, matches and candles, caps, mittens, and overshoes, especially if you live in a colder climate. If you live in the desert you should have sun-protection clothes like sun hats and sun block, and you should also have gallon jugs of water for your car and yourself. You should also have a compass and road maps, a knife, hatchet, high-calorie non-perishable food, two tow chains, transistor radio, flashlight, extra batteries, and jumper cables.

What is a modern emergency kit?

A more modern kit for your house includes a first aid kit, acetaminophen, your prescription medicine, and personal sanitation supplies. Also you should have a dried or canned food storage for one year, disposable dishes and utensils, and plant identification books for food foraging. Store ½ gallon of water per person per day stored in glass jars or jugs. Store a supply of disposable diapers if necessary, and one roll of toilet paper per person per

What should a very well stocked first aid have in it?

Adhesive tape
Antiseptic ointment
Band-Aids
Batteries (for flashlight)
Blanket
Cold pack
Disposable gloves
Sterile gauze pads
Hand cleaner
Plastic bags
Roller gauze
Scissors
Small flashlight
Knife
Steri-Strip wound closure strips
Butterfly bandages
Spenco 2nd Skin
Polysporin antibiotic ointment

Splints (including finger size)
Eyewash
Bandage scissors
Standard First Aid (a book by the American Red Cross)
Syrup of Ipecac
Triangular bandages
Tweezers
Hydrogen peroxide
Betadine (diluted)
Acetaminophen
Ibuprofen
Dental floss
Soft dental or orthodontic wax
Cotton balls
Tempanol or Cavit filling material
Small tweezers
Toothache medicine (such as Red Cross, Dent's, or Orajel) or clove oil
Catgut sutures
Cotton swabs

What medications should I have on hand?

Pain medication
Fever (acetaminophen works for both pain and fever)
Anti-diarrhea
Anti-nausea
Poisoning: Syrup of Ipecac
Rash, itching
Cough medicine
Earache drops
Eye drops

What should aspirin be used for?

Aspirin should ***only*** be used for pain after an injury, and even then only give it to adults, never to children. Never give aspirin for pain from a virus or other illness, because aspirin has been linked to Reye's syndrome, a mysterious illness that develops after aspirin is given for chicken pox or the flu. Reye's Syndrome can result in the victim falling into a coma. In fact, acetaminophen or ibuprofen is just as effective as a pain reliever and you should avoid using aspirin unless you take it to prevent a heart attack.

What can I use to clean a cut or wound instead of store-bought stuff?

Iodine
Garlic
Salt water
Honey
Sphagnum moss (a natural source of iodine)

FOOD STORAGE

Hunger is the best pickle.

~Benjamin Franklin

Note: The following is a general guide—use your own estimates based on these figures and consider the eating habits of your own family.

Food Type	Amount for Adult Per Year	Amount for Child (1–5 years) Per Year
Baking powder	1 pound	1 pound
Baking soda	1 pound	1 pound
Brown sugar	3 pounds	3 pounds
Canned fruits	5 #10 cans (more if possible)	3 #10 cans (more if possible)
Cooking oil	2 gallons	1 gallon
Cornmeal	30 pounds	12 pounds
Dry beans	45 pounds	15 pounds

Food Type	Amount for Adult Per Year	Amount for Child (1–5 years) Per Year
Dry lentils	5 pounds	2 pounds
Dry lima beans	5 pounds	2 pounds
Dry pasta	40 pounds	22 pounds
Dry potatoes	5 #10 cans (65 cups)	3 #10 cans (39 cups)
Dry soup mix	7 pounds	5 pounds
Dry soy beans	40 pounds	20 pounds
Dry split peas	5 pounds	2 pounds
Dry yeast	½ pound	½ pound
Enriched white flour	25 pounds	12 pounds
Enriched white rice	80 pounds	30 pounds
Evaporated milk	12 12-ounce cans (2 pounds)	6 12-ounce cans
Honey	3 pounds	1 pound
Jam	5 pounds	2 pounds
Jell-O	1 pound	1 pound
Mayonnaise	2 quarts	1 quart
Molasses	1 pound	1 pound
Peanut butter	4 pounds	2 pounds
Pearled barley	80 pounds	5 pounds
Powdered eggs	3 #10 cans (39 cups)	2 #10 cans (26 cups)
Powdered fruit drink	6 pounds	3 pounds
Powdered milk	60 pounds	30 pounds
Rolled oats	50 pounds	12 pounds

Food Type	Amount for Adult Per Year	Amount for Child (1–5 years) Per Year
Salad dressing	1 quart	1 quart
Salt	10 pounds	5 pounds
Shortening	4 pounds	2 pounds
Sugar	40 pounds	20 pounds
Tomatoes	5 #10 cans sauce or slices	3 #10 cans sauce or slices
Tuna or canned meat	10 cans or more	5 cans or more
Vinegar	½ gallon	½ gallon
Wheat	200 pounds	75 pounds

What other supplies should I store?

Supply of spices: garlic powder, onion powder, basil, oregano, pepper, chili powder, and cinnamon.

Packages of dried fruit and raisins

Beef and chicken bouillon

Baking cocoa

Ketchup and mustard

Salsa

Vanilla

Seeds for sprouting alfalfa, mung, radish, peas, lentil sprouts

Pancake mix

Pickles

Boxes of cereal

Crackers, graham crackers, and baby crackers

Cans of nuts

Ground, vacuum-sealed coffee

Tea

Children's and adult's vitamins

Baby cereal and formula

▲ Home-canned food storage

Meals in a box (such as macaroni and cheese)

How do I build a food storage?

The simplest, cheapest and easiest way to build up a food storage is to buy bulk at a price club or when the grocery store is having a sale. If you grow your own, slowly grow more and can and store the increase. There are suppliers who will get you all you need in easy-to-store containers for a price. If you really need to live off your food storage and all you have is the basics you will very quickly get tired of beans and rice and mayonnaise. The extra supplies such as pickles and cocoa at the bottom of the list may be the most important because they will alleviate the drudgery of the same meals all the time—and you should also buy a variety of pasta, beans, soups, etc. It is also good to keep some things on hand in the freezer.

How do I care for a food storage?

Keep the storage in a cool (70°F or less), dry, dark place in containers that are rodent-proof. It is important to rotate the food if it is stored for a long period of time. Don't leave the food in sacks and bags—always store in airtight containers made especially for food, or in plastic buckets lined with food-packaging plastic liners. Moisture and rodents are the biggest enemies of food storage so it is important that you follow the rules and make sure the food is extremely dry before sealing the container.

How often should I rotate the food?

1–2 years: instant potatoes
2–3 years: powdered milk
3–4 years: hot cocoa, rice

4–5 years: rolled oats, vanilla pudding, white flour, and soup mix

6–8 years: dry pinto beans, dry pink beans, dry white beans, apple slices, spaghetti, macaroni, chopped dry onions

8–10 years: carrots, fruit drink mix

20 years: wheat, sugar

What kinds of containers are appropriate for food storage?

#10 aluminum cans

Foil pouches

Glass canning jars

PETE (polyethylene terephthalate) plastic bottles

Plastic buckets with food liners

How much water should I store?

Each person should have 14 gallons for a two-week supply. The purpose of a water storage is to keep some on hand if there isn't any water available, anywhere. It is obviously very cumbersome. If you live in an area with lots of water, investing in a very good water purifier can save space.

BUYING LAND AND EVALUATING YOUR LOCATION

No man but feels more of a man in the world if he have a bit of ground that he can call his own. However small it is on the surface, it is four thousand miles deep; and that is a very handsome property.

~Charles Dudley Warner

Some factors to consider when finding a location:

What is the climate like?

What is the cost of living in the local area? Is food expensive and what are the utility costs?

▲ Who are your neighbors?

What are the taxes?

Is it beautiful to you?

Are there employment opportunities?

What are the schools like or what are the home school laws?

Where is the nearest hospital?

Is there a low crime rate?

Is there a good sense of community? Do they accept new people?

Is there culture and education?

Are there churches and other community places?

What is the transportation like? Are roads good?

Is there a farmer's market or grocery nearby?

Steps for buying property:

1. *Decide what you want.*

Where do you want to live?

What are you going to do with the land, and what do you need to do it with?

How much money do you have?

Can you guarantee that you can make a living?

2. *When you find a property you like, investigate.*

Does every detail of the property fit all your priorities?

If it doesn't, can it be made to?

3. *Consider aspects of the land for country living.*

Does it have a woodlot with hardwood?

Does the property flood at any time?

Can you live there right away?

Does it have good soil, and are there any signs of erosion?

Is there a busy road right by it?

Who lives nearby?

Does the land have good drainage for a septic or outhouse?

If it has a septic tank, how old is it and when was it last cleaned?

4. *Test the water.*

Is there an existing water system and does it work?

Has the water been tested for minerals and salt?

Has the water been tested for purity?

If there's aboveground water, is it safe?

Is there a year-round creek or stream?

How do I inspect an existing house?

Are there any torn shingles, patches, unevenness, and curled shingles?

How old is the roof and how many layers of roofing are there?

Does it sag or dip on the surface? (A sign of broken rafters or rotting boards.)

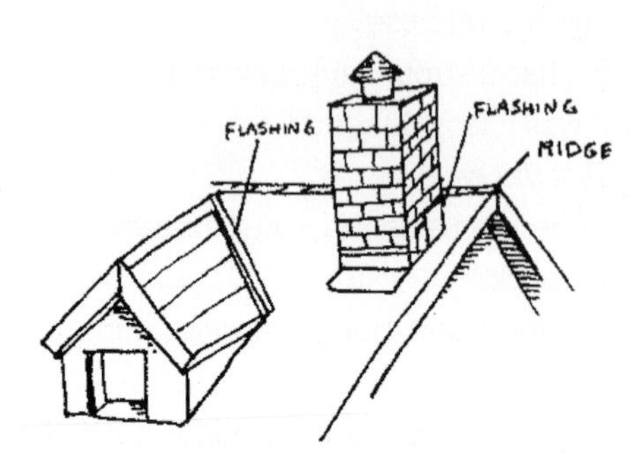

▲ **Flashing and ridgeline**

Does the ridge of the roof sag or dip? (A sign of a weak foundation, termites, rot, or bowed interior walls.)

Does the eave line sag? (A sign that the eave is rotten or broken. If both the eave and the ridge sag at one end, it could be the foundation settling, termites, a heavy chimney or drainage problems.)

Is the flashing (sheet metal) around the chimney and around dormers where a vertical wall connects to the roof good?

Are the gutters and downspouts deteriorating?

Do the sills (the wood that rests on the foundation), main supporting beams, and posts show any sign of weakness, rot or termites?

Do the sewer lines (3–4 inch cast iron, copper or plastic pipes under the house) have tape around them, are dirty, or are cracked?

What type is the wiring under the house? Is it knob-and-tube or wrapped in cloth? (If it is, it is too old and would have to be replaced.)

Are the wires in the junction boxes overcrowded and the insulation crumbling?

Is the wiring good?

How old is the siding and wood on the outside? Are there any bad materials that need replacing?

Is there any paint peeling? (Could be a sign that water is seeping in—poke the wall with a knife to look for rotting wood.)

Are any windowsills rotting?

If it is a brick house, are there any really huge cracks, bowed walls, or fallen bricks?

If it is a stucco house, are there any dark spots? (Could be a sign of rotted wood underneath.)

Are the pipes inside the walls, or on the surface?

Does the water run out fast and is there any rust in it? (Low pressure could mean there are mineral deposits, and rust can mean rusty pipes.)

Do the drainpipes show any sign of leaking, with green or white crystals around the joints?

Are there any missing ceramic tiles or broken tiles?

Is there any mold?

Check for radiator leaks at the valves; do the valves turn on and off?

Is the kitchen modern, big, and convenient?

Are kitchen cabinets, appliances, and countertops workable?

What is the total amperage of the house? (A 200 amp service is enough for a large family in a big house, while 60 is common in older homes.)

Are there any loose switches or outlets, and are there enough?

When you tap lightly on the ceiling does it sound hollow? (A sign that the plaster or lath has pulled away from the joists.)

Does the ceiling sag, and can you move it ½ inch or more?

Are there any stains on the ceiling that indicate a leak?

Are there any rough or broken wall surfaces?

Are there any broken windows or rotted window sashes?

Do the windows open and close properly?

Do the floors look good or will you need to sand or cover with carpeting?

Does the floor sag, dip or slope? (Check the baseboards where it slopes to see if there is a crack.)

Does the room shake when you walk or jump on it?

Does the basement leak or have lots of moisture? Has it ever flooded?

How old is the hot-water heater? Is it gas or electric?

Does the hot-water heater leak or have rust on it?

In the attic, is the house insulated, and how deep is it?

Is anything living in the attic?

Is there asbestos in the attic?

How do I go through all the legal procedures?

1. Before signing a contract, make sure you can secure adequate financing. The contract should include a description of all the encumbrances on the property.

2. If the land has no direct access, obtain an easement (a legal right-of-way), to go across a neighbor's property.
3. Do not accept a land contract or a mortgage. The agency then holds the title on your property and can take it away from you. People do it but it's bad.
4. Make sure there are no mineral, timber, water, or other encumbrances. This is when a former owner sold or reserved the rights to mine, cut trees, or take the water. If you bought the place and tried to get water you would be stealing it.
5. Check how much taxes you would be paying on the property, and make sure there are no unpaid taxes.
6. Have a title search done to make sure the property is free of liens (a legal claim) and encumbrances. Buy title insurance for this to guarantee its accuracy.
7. Your contract should include a clause giving you a minimum amount of water supply. Make sure this is legal also.
8. Check zoning laws and building and health codes to make sure you can use the property the way you want.

How can I get the best deal?

1. Sometimes it's cheaper to by a place that is "For sale by owner."
2. Walk all over the property so you know what you're getting.
3. Find at least three similar properties to compare prices, unless you're sure.
4. Buy quickly when you know what you want.
5. Don't offer ***all*** your cash—keep some for other things.
6. Offer the smallest down and total price you think could have a chance.
7. The owner may offer a counteroffer, and you can accept or give your own counteroffer.
8. When making a counteroffer, don't offer more interest.
9. Calculate how much interest you'll pay.
10. How high can you make the monthly payments?
11. You should have a down payment saved up beforehand.

DESIGNING AN EFFICIENT HOMESTEAD

I never had any other desire so strong, and so like to covetousness, as that one which I have had always, that I might be master at last of a small house and a large garden.

~Abraham Cowley, The Garden, *1666*

How do I plan an efficient homestead?

When designing a layout, it is important to keep in mind what you are going to do, and how efficient it can be. Make a list of the different elements of your homestead and what they need and produce. Then you can overlap the different needs of the different elements; for instance, plants need water, and you can get water from a pond, so put your fish in the pond and put the pond near the garden. The fish will provide fertilizer and food for you also. An example is your garden: a garden needs light, water, organic material, manure, cultivation, and protection from cold. A garden makes organic material, acts as a windbreak, feeds animals and humans, and provides fuel, mulch, and erosion control.

Apartment garden

What elements are needed for an indoor homestead?

Container garden for food production
Light source for the garden
Rainwater collection for the garden and drinking
Living area for humans
Solar or electric heat
Solar or electric light
Waste disposal
Manufacturing living necessities
Food preservation

What are the elements of a backyard homestead?

Intensive garden
Small grain production
Small orchard
Greenhouse
Rainwater collection for the garden and drinking
Living area for humans
Solar, wood, or electric heat
Solar, fire, or electric light
Waste disposal
Duck, chicken, and goose area
Goat shelter and grazing
Necessity manufacturing
Food preservation
Beehives

What are the elements of a large homestead?

Intensive or row garden
Large grain production
Big orchard
Greenhouse
Rainwater collection for the garden and drinking
Stream, fishpond, or river
Living area for humans

Urban garden

Solar, wood, or electric heat
Solar, fire, or electric light
Waste disposal
Duck, chicken, and goose area
Goat, cow, and horse shelter, grazing
Necessity manufacturing
Food preservation
Beehives
Storage facility

▾ **Homestead life cycles**

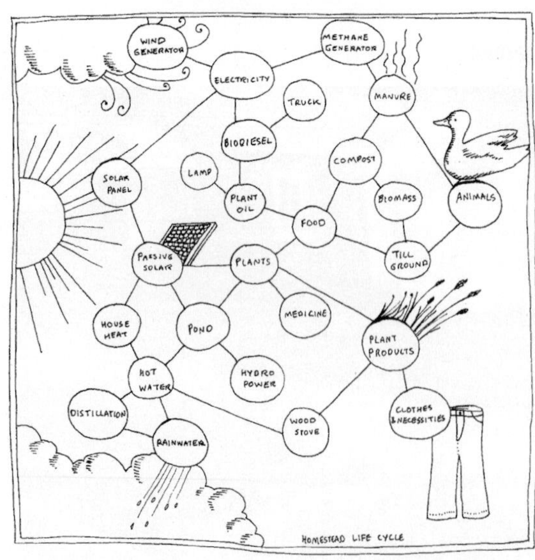

What kinds of gardens are there?

Human food garden—this is pretty obvious. A vegetable garden usually does best in rows or raised beds near a water source and with lots of sunlight.

Cooking herb garden—culinary herbs taste much better when they come directly from the garden. Basil, oregano, cilantro, garlic, and dill by your kitchen door are convenient, and can grow in pots very successfully.

Medicinal herb garden—if you look through the herbal remedy part of this book you will notice some herbs can be applied to a wide variety of ailments. A small herb garden provides you with a constant supply of these and gives you extra to dry for use later.

Animal food garden, or forage—rather than letting a field lie fallow, it can be a space-saving and cheap solution to plant a cover crop such as alfalfa and allow the animals to forage. Alternatively you could plant a garden that not only feeds chickens, but attracts insects away from your vegetable garden so that they can eat your pests.

Soil fertilizing, green manure—a ground cover like alfalfa not only saves topsoil, it also adds nutrients to the soil.

Windbreak, erosion control—besides ground cover, a healthy stand of trees that can grip the soil is the best way to combat erosion.

Building material—you could grow your house if you are building something with bamboo, reeds, wool, straw or other material that doesn't take as long to grow.

Cloth production—flax, nettles, hemp, cotton, hemp, and wool need space to produce in enough quantity for cloth production, and a relatively flat area.

Fuel production—a woodlot is a very valuable thing to have, and if managed properly it can provide fuel for generations to come. Preserving a woodlot is also a very ecologically responsible thing to do.

Insect control—there are a variety of plants that deter pests. Rimming your gardens with them is a vital method of increasing the yield of your vegetable garden.

Where should I put the different elements of my homestead?

Animals: Birds should be located near to a vegetable garden because they can be let in to forage for pests. However, they will munch on your vegetables too so if you use this method plant extra. Goats and horses should be located away from gardens you eat from, but near to grain sources that they can forage on. Fish go in the pond, and bees

The necessities of life: plants, water, fish, shelter, transportation, etc.

go near gardens that flower, such as the orchard.

You: Your herb and vegetable gardens should be near to the house, so you can pick as you eat them. You should also be near water, and close enough to get to animals easily. Necessity manufacturing and food preservation will be either in the house or next to it.

Water: Rainwater will most likely be harvested off shelters and living areas, and so will be near to where it is needed. You cannot predict where a stream will be, but build your house near it but still far enough away to dam up the stream for a pond. Water can also be used for passive solar reflection.

Grain and orchard: Grain should be near to a storage facility such as a barn, and not close to horses and goats that like to get out. Orchards and building material should be utilized as windbreaks, to prevent soil erosion.

Waste: Manure from animals should be covered; human waste should be kept away from water supplies and must stay out of groundwater. You can have a composting toilet above ground anywhere, but for a regular outhouse you must have it at least 100 feet from the house (privies have special health codes).

THE CYCLES OF WORK

From breakfast, or noon at the latest, to dinner, I am mostly on horseback, attending to my farm or other concerns, which I find healthful to my body, mind, and affairs.

~Thomas Jefferson

What work needs to be done every day on a sustainable homestead?

These tasks need to be done no matter what the season:

Weed and water the garden
Rinse out the sprout jar
Care for greenhouse plants
Water fruit trees if they need it
Irrigate the fields
Check on the sourdough starter
Bake bread
Mend the clothes
Fill up the wood box
Care for the compost pile
Move the animals to new pasture
Feed and water the animals
Milk the cows and goats twice a day
Strain, pasteurize, cool, and store the milk
Skim yesterday's milk and store the cream
Groom the horses
Clean and muck out the stable

What is on the calendar?

If you did everything possible on a homestead, your calendar would look something like this . . . but it's not all work. Homemade holidays and seasonal fun are included here also. Some homesteaders have a different breeding schedule—this is a suggestion.

Note: Vaccinations on this calendar are optional. Many organic small farms do not vaccinate, and have perfectly healthy animals.

January

Fun: New Year's Day. If you are in a cold climate, put cod liver oil in the bird feeder.

Personal: Make this year's clothes and mend your tools.

Fowl: Watch for broody geese and collect eggs.

Rabbit: Check water twice a day for freezing.

Goats: Trim the goat's hooves.

Sheep: Worm ewes if they were last treated in November. Examine them for ticks and trim hooves. Vaccinate all sheep. Tag ewes before lambing. Give ewes a pregnancy check at the end of the month.

Horses: Shoe and worm the horses. Clean tack once a week.

Cattle: Finish drying up pregnant milk cows.

February

Fun: Valentine's Day.

Trees: Chop down a maple tree in preparation to make syrup. Cut fruit tree scions for grafting.

Bees: Order new bees.

Rabbit: Check water twice a day for freezing.

Goats: Trim the goat's hooves. Young goats should be eating solid food well. Begin separating them from their mothers.

Sheep: Watch pregnant ewes for lambing, dock newborn lambs, and castrate non-breeding males at 2 weeks old.

Horses: Shoe and worm the horses. Clean tack once a week.

March

Fun: St. Patrick's Day. Hold a sugaring-off party. The 21st is the first day of spring.

Trees: When the time is right, tap maple trees for making syrup. Plant new trees (depends on variety). Prune older trees. Start grafting trees when buds start to swell.

Personal: Take down storm windows and winter sealing.

Garden: Start transplants for your early garden (March 1st). Plant bee garden.

Field: Plow the fields when the soil is dry enough. Plant grain amaranth and wheat when frost is past. Plant sweet sorghum 2 weeks after frost.

Bees: Watch for swarming bees. If you want more hives, leave them alone. If not, pinch out the queen cells.

Fowl: Get goose pasture ready. Rooster mating season starts.

Rabbit: Check water twice a day for freezing.

Goats: Check the goat's hooves. Buy a young buck goat or use your own. Worm goats and give shots before turning out to pasture.

Sheep: Buy weaned lambs. Sheer sheep and clean the wool. Treat ewes and lambs for ticks and lice. Vaccinate lambs.

Horses: Shoe and worm horses. Check pasture for weeds. Clean tack once a week. Clip horses for plowing.

Cattle: Watch for bred dairy cows calving from the 1st to the middle of the month.

April

Fun: April Fools', Easter, Arbor Day, and Passover.

Personal: Store the winter quilts and clothes. Weave cloth in spare time. Beginning of danger from Lyme disease if it is in your region.

Farm: Clean out the manure from the barns.

Field: Turn over the virgin soil you turned last fall. Disc old pasture and broadcast grass seeds. Check sheep pasture for weeds with burrs.

Garden: Plant your early garden transplants or plant seeds in the garden (April 1st & 15th). Start spring garden transplants (April 1st).

Fowl: End of rooster fertility.

Goats: Check goat hooves. Goats bred in November should be due April 1st.

Sheep: Give hay to sheep, vaccinate, cull out old ewes, and then turn them out. Treat ewes for lice and wean lambs one week before turning out.

Horses: Shoe and worm horses and turn out to pasture. Clean tack once a week.

May

Fun: May Day, Mother's Day, Memorial Day, and Victoria Day (Canada).

Garden: Harvest your earliest garden vegetables. Plant spring garden transplants (May 15th) or sow seeds in garden (May 1st & 15th). Start the latest spring garden transplants (May 1st).

Field: Harvest wheat, then plant grain sorghum. Plant alfalfa hay for goats and sheep.

Pond: Blanket weed will cover the pond and you can remove it.

Bees: Catalog bees should arrive if you ordered them.

Goats: Check goat hooves.

Sheep: Watch for footrot and isolate infected animals. Weigh lambs and choose replacements.

Horses: Shoe and worm horses. Clean tack once a week.

Cattle: Breed the cows that are in heat from the 15th to the end of the month.

June

Fun: Father's Day, Midsummer. The 21st is the beginning of summer.

Personal: Gather flowers for dyes, and material for baskets.

Garden: Harvest your early garden and spring garden. Plant your latest spring garden transplants (June 15th) or sow seeds in garden (June 1st & 15th). Start summer garden transplants (June 1st).

Field: Harvest sweet sorghum on a clear, chilly day. Harrow pasture and irrigate.

Goats: Check goat hooves and worm goats.

Sheep: Check lambs for diseases, and treat all lambs for worms.

Horses: Shoe and worm horses. Clean tack once a week.

July

Fun: Independence Day (U.S.) and Canada Day.

Personal: Gather flowers and berries for dyes.

Garden: Harvest spring garden and your latest spring garden. Plant summer garden transplants (July 15th) or sow seeds in garden (July 15th). Start latest summer garden transplants (July 1st). Harvest fruit.

Field: Harvest grain amaranth and alfalfa hay. Plant broomcorn 10–14 days after amaranth harvest.

Goats: Check goat hooves.

Sheep: Rotate pasture. Purchase ram if necessary, and shear all rams and replacement ewe lambs early in month.

Horses: Shoe and worm horses. Clean tack once a week.

August

Fun: County Fair.

Personal: Gather berries for dyes. Gather cornhusks for baskets or beds.

Garden: Harvest spring garden and latest spring garden. Sow late summer garden seeds if you didn't transplant (August 1st & 15th). Start fall garden transplants.

Trees: Harvest nuts and apples. Cut trees for firewood.

Goats: Check goat hooves. Worm goats and butcher meat goats.

Bees: Collect honey.

Sheep: Put fat ewes on a diet.

Horses: Shoe and worm horses. Clean tack once a week.

September

Fun: Labor Day, school starts, and Grandparents' Day. The 23rd is the beginning of autumn.

Personal: Gather cattail stalks and husks and other basket material. Gather bayberry for bayberry candles. Clean out barns.

Field: Test the soil of all pastures and fields. Turn over virgin soil and plow the fields. Plant winter grain.

Garden: Compost the garden (add material). Harvest latest spring garden, summer garden, and late summer garden. Plant fall garden transplants (September 15th), or sow seeds (September 1st & 15th).

Trees: Harvest nuts.

Bees: Start feeding bees.

Fowl: Chickens may start to molt.

Goats: Check goat's hooves. Dry up goats that are kidding in November.

Sheep: Flush, and tag and check hooves on ewes (September 1st). Check rams for injury or illness, shear on September 1st.

October

Fun: Halloween, Thanksgiving (Canada).

Personal: End of Lyme disease season. Prepare to butcher larger animals when temperature reaches 40 degrees during the day. Change the feathers in your feather beds. Store up quilts, clothes, and candles. Make soap. Collect berry vines for baskets

Garden: Harvest summer garden, late summer garden, and fall garden. Plant winter garden.

Fowl: Goose molting season starts; collect feathers.

Goats: Check goat hooves. Worm goats, but only worm does 6–8 weeks before kidding and start giving them grain.

Sheep: Finish breeding season by October 20th. Breed ewe lambs by mid-October.

Horses: Shoe and worm horses. Clean tack once a month.

Cattle: Butcher the male calves that are over 15 months old.

November

Fun: Thanksgiving, Remembrance Day (Canada).

Personal: Chop wood, put up storm windows, and seal cracks.

Garden: Harvest late summer garden and fall garden.

Trees: Plant new trees (check variety), and prune older trees.

Goats: Check goat hooves. Stop milking goats, and breed goats November 1st.

Fowl: Cull old, non-laying geese.

Sheep: Flush, eye, tag, and check hooves of ewes (November 1st). Breed ewes (November 14th). Cull older ewes.

Horses: Shoe and worm horses. Clean tack once a week.

Rabbit: Check water twice a day for freezing.

December

Fun: Advent, St. Nicholas Day, St. Lucia Day, Christmas, Chanukah, New Year's Eve, Boxing Day (Canada). The 22nd is the beginning of winter.

Personal: Put cod liver oil in bird feed in cold climate. Plan next year's farming.

Goats: Check goat hooves and trim them. Taper off doe's grain around the 15th.

Sheep: Plan for lambing and watch for health problems.

Horses: Shoe and worm horses. Clean tack once a week.

Rabbit: Check water twice a day for freezing.

Cattle: Start drying up pregnant milk cows.

How important are the seasons?

The earth rotates at a tilt around the sun. This tilt causes the seasons to be different lengths and causes opposite seasons in the Northern and Southern hemispheres. We track the seasons by the tilt of the North and South poles. As you live on the land you will become more and more in tune with the cycles of the seasons. All of the work outlined in the calendar is really dictated by the seasons and natural cycles of life, and you will find yourself fitting into that rhythm.

Beginning dates of seasons (for the Northern Hemisphere—they are opposite in the Southern):

Summer solstice: On June 21st, the North Pole leans most toward the sun, and it is the longest day of the year.

Winter solstice: On December 22nd, the South Pole leans most toward the sun, and it is the shortest day of the year.

Vernal (ring) equinox: On March 21st, the earth is most sideways to the sun; day and night are the same length.

Autumnal equinox: On September 23, the earth is most sideways to the sun; day and night are the same length.

Lengths (for Northern Hemisphere):

Spring: 92 days and 20 hours.
Summer: 93 days and 14 hours.
Autumn: 89 days and 19 hours.
Winter: 89 days and 1 hours.

What natural indicators can tell me what the weather will do?

Good weather signs: Red sunset, flying beetles, busy spiders, high-flying

birds, heavy dew, gray morning, soft–edged clouds, pinecones with scales open.

Bad weather signs: Pale yellow sky at sunset, red sunrise, red eastern sky, dogs sniff the air a lot, birds ruffle feathers and huddle together, hard–edged clouds, pinecones with scales shut.

How do I know how far away the lightning is?

Count the seconds between the flash of lightning and the thunder. The lighting is about 1 mile away for every 5 seconds you count.

How do I measure wind speed?

1–3 mph: Smoke rises straight up.

4–7 mph: Leaves rustle slightly.

8–12 mph: Loose paper scraps lift, wind felt on face.

13–18 mph: Loose dust is blown, small branches move.

19–24 mph: Small trees sway.

25–31 mph: Large branches move constantly.

32–38 mph: Entire trees in motion, walking affected.

39–46 mph: Walking difficult.

47–54 mph: Gale wind, slight building damage.

55–63 mph: Trees uprooted, heavy building damage.

64–75 mph: Very violent storm.

How do I determine the wind chill factor?

The following index can help you determine how cold it ***really*** is outside. Although it might not help you if you don't have a thermometer, if you can remember some of these numbers and guess the temperature it may useful if you are stuck outside. If the wind chill factor is -18°F or lower, frostbite can occur in 15 minutes or less. That means if you know that it's about -5°F outside but there is a 10 mph breeze blowing, it doesn't seem that bad but really it is very cold. The wind takes your body heat away from

Wind Speed	30°	25°	20°	15°	10°	5°	0°	-5°	-10°	-15°
5 mph	25	19	13	7	1	-5	-11	-16	-22	-28
10 mph	21	15	9	3	-4	-10	-16	-22	-28	-35
15 mph	19	13	6	0	-7	-13	-19	-26	-32	-39
20 mph	17	11	4	-2	-9	-15	-22	-29	-35	-42
25 mph	16	9	3	-4	-11	-17	-24	-31	-37	-44
30 mph	15	8	1	-5	-12	-19	-26	-33	-39	-46
40 mph	13	6	-1	-8	-15	-22	-29	-36	-43	-50
50 mph	12	4	-3	-10	-17	-24	-31	-38	-45	-52
60 mph	10	3	-4	-11	-19	-26	-33	-40	-48	-55

you, so it should warn you to dress very warm.

IMPORTANT INFO ON CHEMICALS AND TOXINS

The longer I live the less confidence I have in drugs and the greater is my confidence in the regulation and administration of diet and regimen.

~John Redman Coxe, 1800

What are product ingredients really made of?

Some products list ingredients that sound awful but really aren't, while other products contain all kinds of carcinogens. The following is a list of ingredients; useful not only to know what's in there but, if you really wanted, you could replicate some store-bought things yourself.

Gum arabic: acacia vera

Acetic acid: 3–5 percent solution of vinegar

Alum: in recipes a spice not frequently used anymore except in pickling, as a chemical it is aluminum powder, sulfate, carbonate, etc.

Ammonium carbonate: baker's ammonia or smelling salts

Amylacetate: banana oil.

Arabic gum powder: acacia vera.

Arrowroot: herb, powder substitute for cornstarch, tapioca starch, and rice starch or flour.

Ascorbic acid: vitamin C.

Bicarbonate of soda: baking soda.

Calcium carbonate: chalk or agricultural lime

Calcium hydroxide: slaked or slacked lime

Calcium oxide: unslaked quicklime

Calcium sulfate: plaster of Paris

Citric acid: derived acidic fruits

Furfuraldehyde: bran oil

Glucose: corn syrup

Glycerin: by-product of the saponification of vegetable oil or animal fats

Graphite: pencil lead

Hydrogen peroxide: peroxide

Iodine: tincture of iodine (4%)

Isopropyl alcohol: rubbing alcohol 70–90 percent

Lye: made from ashes

Magnesium hydroxide: milk of magnesia

Magnesium silicate: talc

Magnesium sulfate: Epsom salt

Methyl salicylate: wintergreen oil, sweet birch oil, or teaberry oil

Potassium bitartrate: cream of tartar, pearl ash, salt of wormwood

Potassium carbonate: potash

Potassium chloride: potash muriate

Silica/Silicon dioxide: sand

Sodium chloride: table salt

Sodium hypochlorite: bleach

Sucrose: cane sugar

Talc: talcum powder, alternative is arrowroot powder

Tincture of iodine: 47 percent alcohol, 4percent iodine

Whiting: chalk mixed with linseed oil to form putty, add water and other things to make whitewash.

There have been 85,000 new chemicals invented in the world since World War II. Of these, about 3,000 are produced in quantities over 1 million pounds per year. Of those largely produced chemicals that are used in foods, cleaning supplies, pesticides, water, and air, only 43 percent have been tested for toxicity. Only 10 percent have ever been tested with fetuses and children in mind. The second leading cause of death (after injury) in children is cancer. Only fifty years ago the leading causes were diseases like malaria, measles, and TB that are still prevalent in thirdworld countries. The leading cause of hospitalization is asthma. Learning and developmental disabilities affect one in six children, and are ***increasing***. ("Children's Health at Risk" by Elizabeth Hauge, *totalHealth* Vol. 26 No. 3)

Sandra Steingraber wrote, in *Having Faith: An Ecologists Journey to Motherhood*: "If the world's environment is contaminated, so too is the ecosystem of a mother's body. If a mother's body is contaminated, so too is the child who inhabits it. These truths should inspire us all—mothers, fathers, grandparents, doctors, midwives and everyone concerned about future generations—to action." It seems clear that our society's dependence and use of chemicals can be directly linked to the health problems of our children. Theoretically, if the chemicals were no longer used, would they have any health problems at all, since we no longer have trouble with malaria and TB?

While technology is a great time saver, and medical advancement saves lives, why are we shooting ourselves in the foot with our contaminants? The first belief that I have as a homesteader is in purity. I want my air clean, my water clear, my home free from toxic chemicals, and my clothes and vegetables pesticide free, not only for myself but also for my children. While we possibly can live longer through modern medicine, we probably will die young and suffer because of modern chemical "miracles." Production, use and disposal of these materials poisons us three times—people are harmed, the environment is harmed, and generations from now our great-grandchildren are harmed because our waste won't decompose. Meanwhile, psychologically and genetically they will be dealing with irreversible problems. Once genes are altered, they never go back. I believe the future of humanity rests on our choice to forgo the convenience of chemicals and disposable products, and choose responsibility.

What kinds of toxins exist in the average home?

The major toxins in the average North American home are plastics, chemicals in products we use, and pesticides and herbicides in food and clothing. These toxins not only are breathed into your lungs in the form of tiny particles, they are absorbed through your skin and ingested as well. Most of these are known to cause cancer when exposed to them over time (which most of us are). Here are a few of the known carcinogens that poison us every day:

PVC: Found in most plastic products, including shower curtains, grocery bags, bottle caps, children's toys, vinyl siding, carpet, car upholstery, plastic piping, three-ring binders, medical gloves, thermal blankets, garden hose, inflatable pools, raincoats, Venetian blinds, telephone cables, etc. Vinyl chloride (PVC) causes cancer by exposure usually by breathing it in the air or drinking water contaminated with it. Workers are at more risk, but any kind of contact is harmful and even just having it on your skin can cause irritation. It does produce toxic fumes, especially when newly manufactured.

Aluminum chlorhydrate and zirconium: Found in all antiperspirants, they work by blocking the pores from breathing. They are forms of aluminum, and while you are much more likely to absorb aluminum in foods and by breathing it in, any exposure to aluminum can't be good because it can cause cancer and brain damage.

Formaldehyde: Found in non-iron clothing, shampoo, toothpaste, mascara, and air fragrance. Many homes have a few pressed-wood and fiberboard furniture pieces, all of which contain formaldehyde. These products release fumes constantly, and the levels get higher in lots of moisture and humidity. It can cause allergy-like symptoms and over time can cause cancer. There is no safe level of exposure to formaldehyde.

Makeup and beauty products: Makeup and other beauty products contain so many different toxic materials, including coal-tar dye, benzyl violet 4B or violet 2, lead, formaldehyde, lead acetate, progesterone (a drug regulated by the FDA), selenium sulfide, ethanol, shellac, acetone, limonene, and sand or quartz forms of silica. These chemicals can be absorbed through the skin and breathed into the lungs. They cause cancer and permanent physical damage to humans, and some of them cause brain damage to babies and young children even through very little exposure. Check the ingredients of everything you buy.

Fragrances: Not only put into cars, petroleum is also found in most perfumes and fragrances. Toluene, ethanol, acetone, formaldehyde, benzene, and methylene chloride are all in fragrances and are known to cause cancer, birth defects, infertility, and nervous system damage. Even fabric softener contains chloroform, and many products contain fragrances, including dryer sheets, soaps, disinfectants, hair products, beauty products, lotions, and incense. Watch out because they can even say "unscented" on them.

Cleaners: Soap and water clean just as well as cleaners, because antibacterial cleansers need to be left on the surface for at least two minutes in order to kill the bacteria. They also kill weaker strains of bacteria and allow stronger germs to flourish, and possibly weaken your immune system since you no longer build up resistance to germs. Not to mention the dangerous fumes and skin contact that you will have with toxic chemicals that cause cancer and other horrible problems. These products should only be used if you are cleaning something very hazardous, such as a hospital or a very contaminated mess.

2 The Basics

MATH AND MONEY

If I can acquire money and also keep myself modest and faithful and magnanimous, point out the way, and I will acquire it.

—*Epictetus*

What are common abbreviations for measurements?

There are many different measurements for a variety of things. This is an attempt to standardize some of the abbreviations used in this book:

cal - calories
c - Cup
deg/° - degrees
EHP - electric horsepower
ft/' - foot
g - gram
hr - hour
Hz - hertz
in/" - inch
kg - kilogram
l - liter
lb - pound
m - meter
mi - mile
ml - milliliter
mph - miles per hour
oz - ounce
pg. - page
psi - pounds per square inch
pt - pint
qt - quart
Tbsp/T - tablespoon
tsp/t - teaspoon

What are the home equivalents of can sizes?

Cans that are bought at stores have different equivalencies. Some recipes call for a "can" of something but you might only have a home-canned jar, which are usually in pints. This table shows the amounts for different can sizes.

8 ounces = 1 cup
Picnic = 10 ½ –12 ounces = 1 ¼ cups
12 ounces vacuum = 12 ounces = 1 ½ cups
#1 = 11 ounces = 1 1/3 cup
#1 tall = 16 ounces = 2 cups
#1 square = 16 ounces = 2 cups
#2 = 1 pound 4 ounces = 1 pint 2 fluid ounces = 2 ½ cups
#2 ½ = 1 pound 13 ounces = 3 ½ cups
#2 ½ square = 31 ounces = scant 4 cups
#3 = 4 cups
#3 squat = 2 ¾ cups
#5 = 7 1/3 cups

#10 = 13 cups
#300 = 14–16 ounces = 1 ¾ cups
#303 = 16–17 ounces = 2 cups
Baby food jar = 3 ½ to 8 ounces

What are the difficulties in making money on a homestead?

Making money while working your homestead is difficult because the usual products of a small homestead are only bought buy a few who have a special interest. Fast and easy grocery stores are hard to compete with in the mainstream world. The key to maintaining an income is to be extremely flexible in what you are doing, and find a niche market. A niche market is a special customer who needs a very special item that is hard to find . . . for instance a certain gourmet restaurant wants special Japanese herbs for their dish. By selling them for cheaper than shipping the herbs from Japan, you have found a niche market. Unlike a large department store that sells everything, you would specialize in a specific category of products. The greatest tool a small business has is the Internet. For a small monthly fee a website can be made to advertise and sell products all over the world.

Stay-at-home jobs:

Writer
Illustrator, photographer, designer
Musician
Mail-order products
Internet-sold products

▾ Farmstand business

▲ **Plant nursery**

Newsletter publisher
Artist
Bed and breakfast
Carpenter
Freelance consultant
Crafter
Family tree researcher
Financial advisor
Day care
Pet sitter

Homestead products:

Aquatic plants
Berries and fruits
Gourmet food products
Herbs
Seeds
Useful animals
Organic fresh food
Organic processed foods
Craft supplies
Furniture and handmade items
Organic insecticide
Herbal preparations
Fishing supplies
Flowers
Firewood
Plant nursery

Most homesteaders have to work at least part time off their homestead, but you can significantly increase your chances of quitting your day job if you can decrease your expenses. No matter how you live, you have to pay taxes for your property and your income if you make enough. You will also eventually have medical expenses at some point. Make a plan of how much you need to make to pay your taxes and either pay for health insurance or have a savings account (unless you live in Canada). Find an ingenious way to make that amount each year. You should always have a savings set aside no matter what.

TRACTORS AND TRUCKS

What can a tractor be used for?

A tractor can be used for anything that a truck or horse can't pull, and they can save you much backbreaking effort. They can plow fields, pull heavy loads and any other work you need done around the homestead.

What do I look for in a tractor?

You will probably want a used tractor because new ones are incredibly expensive. And you will not need a huge industrial-type tractor; you will only need a small tractor. Of course you will want the best tractor for your money—be sure it's a size and horsepower you're comfortable with. A tractor in the 30-horsepower range is sufficient for most homesteads; Ford's "N" series tractors are a good example. Also, never underestimate the capabilities of an old tractor. They're extremely rugged machines. While some makes and models may be difficult to get spare parts for, old tractors have become collectible—on the good side, this means more after-market parts are being made; the bad side is that prices are going up.

Buying a used tractor is more practical for small homesteaders

What do I need to watch out for?

1. Listen to the engine for any knocking or spluttering noises when it is warmed up and idling. A diesel tractor will be somewhat louder than a gasoline engine but should not have any serious knocking sounds.
2. Look for oil and fluid leaks from the engine, gear train and hydraulic system.
3. The tires should be free from cuts, cracks, and too much wear; there should be tread left on them.
4. Test the brakes—each side should brake at the same time and they should be equal. But, be somewhat familiar with the tractor before testing the brakes. Some makes and models have separate brake pedals for each wheel—a steering brake of sorts. If you apply unequal pressure to both pedals, you'll get uneven braking and the tractor will seem to pull to one side. Also, most old tractors do not have power brakes . . . which is quite a bit different from modern vehicles.
5. Even if the tractor is made to shift while driving or when idling, it should still have a problem-free clutch and you should have no serious difficulty shifting gears.
6. Watch for missing safety equipment.

What about add-ons and accessories?

Once you have a good solid tractor to work with, you'll find yourself wanting it to do more jobs around the homestead than it's set up for. A lot of old tractors were scrapped over the years, but that doesn't necessarily mean their accessories were.

Front-end loader: Many homesteaders rightly consider a front-end loader indispensable . . . something that can be expensive, but doesn't have to be—if you shop around. Local farm auctions can be a good source of accessories, but the competition can be fierce. One great way to find parts is on the Internet, especially auctions online. Be patient and pick up a part when you find it and it will be much cheaper.

Tractor forks

Forks: One thing you'll probably want to do is make (or buy) a set of forks that will either interchange with the bucket on your front-end loader or a set that attaches to the bucket. There are many companies that give away free shipping pallets—setting heavy items on pallets and being able to move them easily with the tractor is a huge labor-saver.

What can a truck be used for?

A truck can be used for hauling, usually for large loads. It must be able to haul a 16-foot stock trailer, pull other cars out of the mud, and carry heavy loads in the truck bed. It has to be able to plow through snow.

What do I look for in a truck?

A 4x4 three-quarter-ton pickup (any smaller doesn't have power or weight enough) with very good tires should be able to do most homestead jobs. For bad roads you will need 10-ply (not 10-ply rated) aggressive tread mud/snow tires.

What equipment do I need for my truck?

Get a spare tire, a four-way tire iron, a good high-lift jack, a winch, a cell phone that has a lighter re-charger (if possible), a shovel, and a 20-foot tow chain. During fire season carry a tank of water and an ax. If you live off road get heavy studded tire chains with two black rubber bungee cords per tire to use on slippery roads.

AUTOMOTIVE MAINTENANCE

Warning: ***All of the fluids (except water) used in a vehicle's engine are toxic. Fumes can cause brain damage, and antifreeze can burn your skin and make you blind. Antifreeze tastes good***

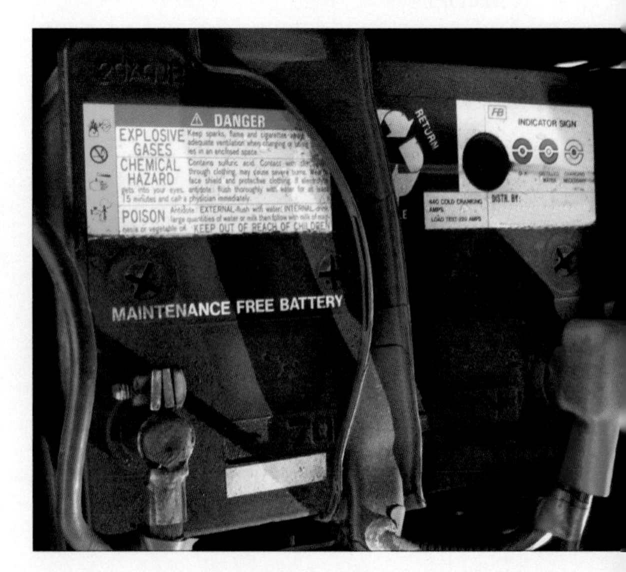

to animals but is fatal to them. Don't spill these substances and make sure you dispose of them properly. Keep them away from kids and pets and wear gloves.

How do I replace a battery?

1. Check the vehicle owner's manual to find out what battery size you need. On the top of the battery, a CCA number and an RC number will be displayed. A higher CCA number means it is better starting in cold weather, and a higher RC number means that your battery has a bigger reserve in case the engine fails. Get a new battery using these facts.
2. Put on safety gloves and glasses.
3. Look at the vehicle's manual to find out the grounding system it has. Some will have a negative system and others a positive system.
4. For a negative grounding system, remove the battery cable connected to the negative post first. If it is a positive system, remove the cable on the positive post first. Use a wrench to loosen the nut and bolt.
5. Remove the other cable, and write down which ones were on the positive and negative.
6. Use a ratchet wrench to loosen the clamps that hold the battery in place.
7. Lift the heavy battery out carefully.
8. Clean the connectors on the cables with baking soda, water, and a wire brush if they have any corrosion on them. Then dry and clean them off with a clean cloth. Make sure they are completely dry.
9. If there is any corrosion or dirt in the battery tray, use the baking soda, water, and wire brush to clean it also. Wipe it off and dry it.
10. Put the new battery into place and clamp it down.
11. Put the battery cables back on, in the ***reverse order*** from which they were removed. If you removed the negative first, put the positive back on first. If you removed the positive first, put the negative back on first.
12. Use a wrench to tighten the nut and bolt on the cables.

How do I check the fluids?

1. Check the oil frequently or you risk permanently damaging the engine. Wait until the engine has cooled for at least 15 minutes to get an accurate reading. Find the dipstick, which will have a loop for a handle, and will be inside a long metal tube. Pull it straight out.
2. Wipe the dipstick clean with a cloth rag, and put it all the way back in. Pull it out again and look at the oil at the end. The oil should be between the two lines (the add and full lines). If you are unsure, look at your owner's manual.
3. If it is low, add some more oil by removing the oil cap and pouring it in. Don't put in too much. If your oil looks dark or dirty, you need an oil change.
4. Check the coolant or antifreeze when the engine is cold. On some vehicles it can be checked in a coolant-recovery tank, while others you will have to look in the radiator. If it is low, add a 50/50 mixture of water and antifreeze. You can use a hydrometer to test the mixture. If the coolant is brown, you may need a radiator flush.
5. If you have an automatic transmission, you can check the fluid when the engine is running and warmed up, and parked on level ground. Let the engine

▲ Checking the oil level

idle, put on the parking brake and put your foot on the brake. Change through all the gears to make sure the fluid is in all the pumps.

6. Put it in park, leave the brake on and the engine running. The transmission dipstick is near the back of the engine. Pull it out, wipe it clean with a rag and reinsert it all the way.
7. Pull the dipstick out again and look at the levels at the end of the stick. It should be between the two lines. If it is low, add more fluid.

How do I change a tire?

1. Turn off the engine, put it in park and apply the parking brake. Put a large rock, block, or log in front of the wheel that is diagonal from the tire you're changing.
2. Loosen each lug nut (called breaking loose), by turning it counterclockwise with a lug wrench without removing it.

▼ Loosening lug nuts

▲ Jacking up the vehicle

3. Use your owner's manual to find out where to place the jack. Use the jack to raise the vehicle enough that the tire is not touching the ground.
4. If there is a hubcap, remove it, and take off the lug nuts completely. Remove the highest nut last, and keep the nuts organized so you know which ones went where.
5. Take the new tire and align the holes with the bolts, starting with the top bolts first. Make sure it is flush against the hub. Put the lug nuts on loosely.
6. Tighten the lug nuts as much as you can by hand, placing them on in a crisscross pattern so that the wheel will tighten evenly (for example, top left, bottom right, bottom left, top right, etc.). Then tighten them further with a lug wrench.
7. Lower the vehicle, and remove the jack. Tighten lug nuts one more time with the wrench.

Rules for driving with a 4x4 in the snow:

1. Drive slowly and under control even if you have a 4x4 truck.
2. Shift down when going up or down a steep hill—avoid using brakes.
3. Use momentum rather than speed to get past holes and snow drifts.
4. Keep your eyes ahead.
5. Don't spin your tires when you get stuck. Shovel out and rock back and forth.

How do I put chains on a 4x4?

1. Find a dry, flat area if you can. In bad mud or ice you can chain up the rear

wheels, but in very slippery conditions of mud and deep snow you should do all four.
2. Place the chains in front of each tire with the hook ends next to the tires. When you drive over it the loop will be pulled over the tire. Make sure the chains are wide and straight and centered so that you will only have to move the truck forward 18 inches to be in the middle of the chain.
3. Move the truck forward (have someone guide you), then shut it off and put on the parking brake. Make sure it is not going to roll or slip.
4. Reach under the truck and hook the inside first, making sure to hook it short enough, then hoist the chain up from the front of the tire to the top to hook the outside as tightly as possible. Make sure both are hooked very tight.
5. Fasten all hooks—even bungee hooks away from the tire or they can puncture tires. If your chains do not have tightener plates use two short black bungee cords to tighten it by making an X across the wheel.

GETTING WATER IN THE WILDERNESS OR AT HOME

We never know the worth of water till the well is dry.

~*Thomas Fuller,* Gnomologia, *1732*

Is all water contaminated?

Mountain stream water is often contaminated with ***Giardia***, a parasite. Your underground water may have agricultural chemicals in it. You can purify both of these with a home distiller or purifier. Rainwater is most likely to be pure, although it may have acid in it. Get all of these sources tested.

How much water do I need?

Even a single person needs a tank that can hold at least 1,000 gallons of water. A large family will need at least twice that. Even if you have a well or a year-round spring, having a store of water is essential to your survival.

Where do I find water?

Note: Beware of water from streams, ponds, and bogs without thorough purification. These are open to the air and susceptible to contamination. Never drink salt water.

1. In limestone caves.
2. In a dry canyon that cuts through sandstone.

▲ A mountain stream may look pure . . .

3. In areas with lots of granite.
4. Dig a hole on the greenest, grassiest hillside.
5. Dig a hole for groundwater in valley floors with loose soil.
6. Dig a hole in low forests, seashores, and river plains.
7. On clear nights collect dew with a cloth or sponge.
8. Dig in dry streambeds in the mountains.
9. Melt snow in the sun.
10. In the desert, watch where animals, ants, and bugs go to drink.
11. In the desert, dig where cattails, greasewood, willow, elderberry, or salt grass grow, or where it looks damp.
12. Collect rainwater.
13. Dig behind windblown sand hills at the back of an ocean beach.
14. Even ocean fish contain fresh water. Dice the fish and lay it on a cloth, then wring out the moisture.
15. Condense ocean water.
16. Use condensation to collect desert moisture.

Which plants contain water?

1. Plants with fleshy leaves or stems have water inside (don't drink if the juice is milky or colored).
2. Cut off the top of a barrel cactus and mash the pulp inside, then drink.
3. Desert oak and bloodwood roots can be pried out, chopped into two-foot lengths, stripped of bark, and the water sucked out.
4. Some vines have edible sap. Cut a deep notch as high as you can reach, cut it off at the base, and let the water drip into your mouth or cup.

What is a spring?

A spring is a place where water comes out of the ground. If you are looking for one on your property you might not even

▲ . . . but almost all surface water is contaminated.

notice it—it could just look like a marshy place. If your spring produces 100–150 gallons per day is more likely to be reliable through any season, and won't move or dry up. You should always wait a year to watch it and see if it lasts. You can build a dam, put in a hydroelectric generator, or you can

▼ Collect dew from plants

▼ Salt grass

▲ Barrel cactus

▲ Spring fitted with a pipe

tap the spring. If it is uphill from the house, it will provide one pound of pressure per two feet of elevation. You simply dig out the hillside and install a pipe for the water to come out. You can also tile the area to prevent erosion.

What kinds of dams can I build?

You can build a barrier dam or a contour dam. A barrier dam is built across a flowing stream, blocking it completely where there is room. This is good for water storage and ponds. A contour dam is built on a very small slope, and is used for irrigation. To build a barrier dam, you will need to be very careful to provide good spillways.

What is dowsing?

Dowsing is a way to find underground water. Some people can dowse and some can't, and no one knows why. Several

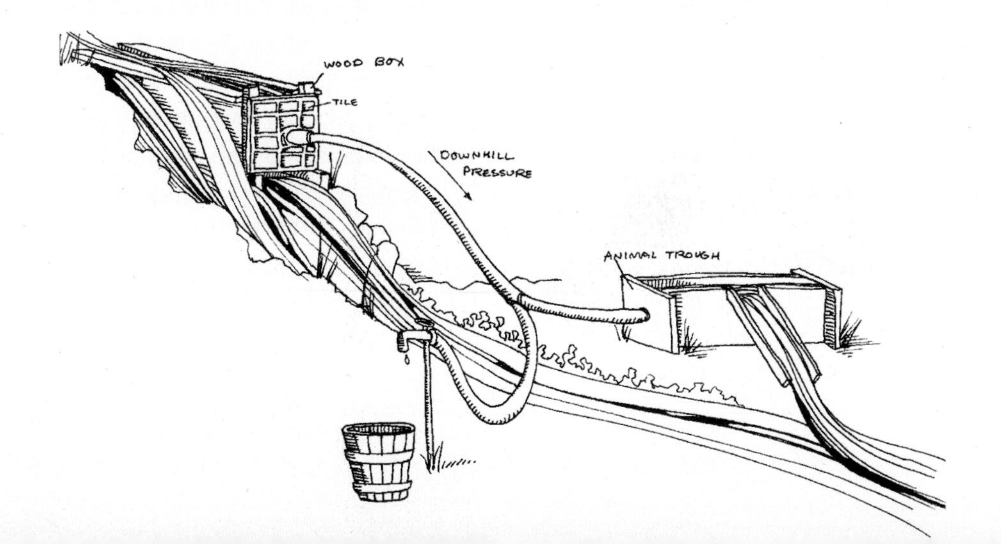

Barrier dam

methods can be used, with several tools. A coat hanger, two sticks, or one stick broken in a V shape work well. The dowser holds the ends of his tools loosely in each hand, so that the stick(s) can move. Then he walks around where water might be and waits for a sign from his indicator. With a hanger or V-stick, the point will dip down towards the ground. For two sticks, they might dip or they might come closer together to form a V or cross in the middle. There are professional dowsers and this is a proven method of finding water underground.

What is an Archimedes screw?

Invented by Archimedes, it is basically a screw inside a tube that is turned so that the water can be pulled up gradually. Unlike a regular pump, it has to be put at an angle, so it cannot carry the water for any great height.

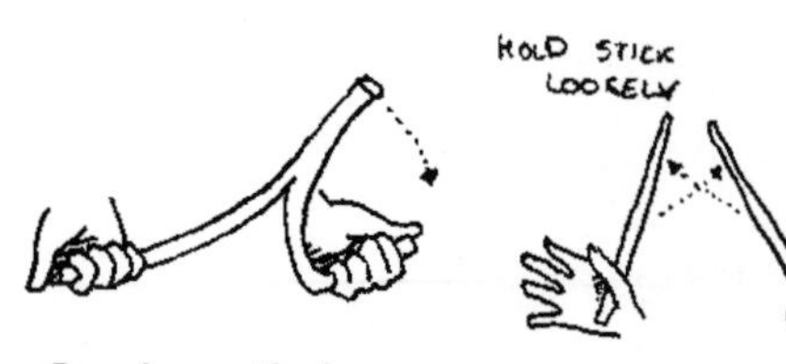

Dowsing methods

How do I make an Archimedes screw?

1. Make a perfectly round beam, and then divide it off into equal sections all the way around, based on how wide your screw will be. The math formula to calculate this is that you want the total diameter to be ⅛ the length of the beam. So figure out how wide your spirals are going to be, and how wide your beam is, and then divide the beam into reasonable widths.
2. Draw vertical lines all the way down the beam, dividing it off into equal sections the same size as the other lines. This will make perfect squares all over the beam.
3. Starting at one end of the beam, draw a diagonal line from the corner of one of the squares to its opposite corner. If

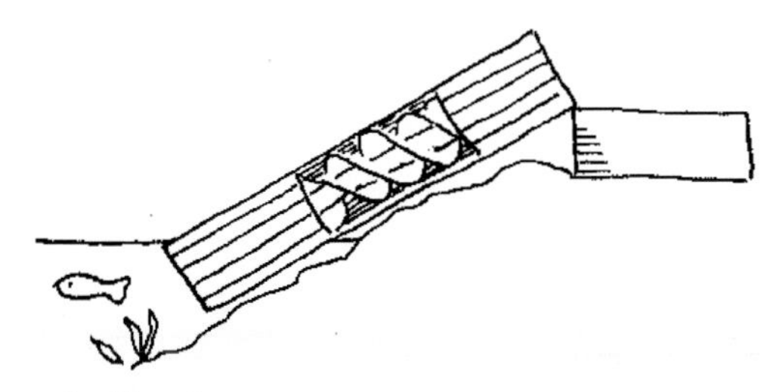

▲ Archimedes screw

you lay the beam down in front of you and start off on the left, every single diagonal line would run from the top left corner to the bottom right corner in every single square.

4. After drawing all these diagonal lines, you will see that they will connect spirally all the way down the beam.
5. Find a flexible wood such as willow and make a very thin strip with a uniform width. Waterproof it with pitch, and then fasten it to the beam along the spiral line.
6. Continue fastening waterproof strips to the beam along the spiral lines until all of them have been covered and you have a screw.
7. Enclose the screw in a tube made of wood boards, somewhat like making a barrel. Soak them with pitch on the outside and bind together with very tight bucket hoops or strips of iron. It has to withstand the pressure of water.
8. To figure out the angle it should be set up at, divide its length into five parts. The length of three of those parts is how high the head of the screw should be from the bottom.

▼ Construction of an Archimedes screw

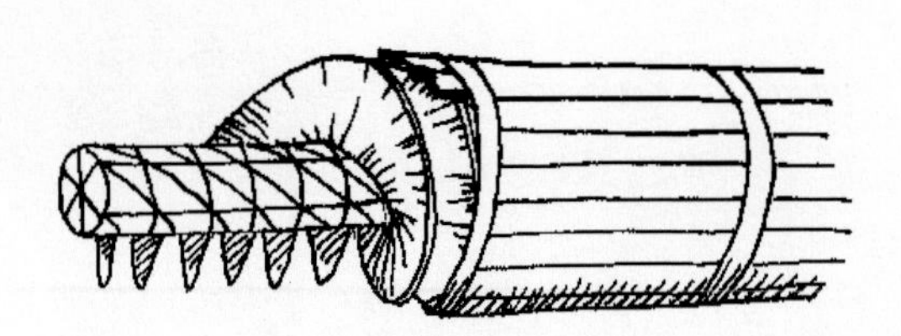

Why is hand digging a well so difficult?

Hand-digging a well is extremely difficult. First, you have to locate the most likely place to have water, usually by dowsing. Then two people must dig a very deep hole, while a third person hauls up the buckets of dirt. Once they reach gravel and rock, a pickax must be used to break up the hard stuff to get to the water. If you are lucky, it will be an artesian well that would fill the hole and run to the ground surface. If you're not lucky, the well water will be contaminated or it could just be too deep to get to. As the hole is dug, it is lined with brick and mortar, concrete, or fieldstone to hold the walls up, so that it is watertight. The top should be covered to keep out contaminants from the air and a hand pump installed.

What is a percussion drill?

A percussion drill is a heavy bit that is attached to a rope that is lifted by hand or machine and then dropped. Water is added to the hole so that the bit makes mud as it cuts the earth. When several feet of mud is collected, the bit is taken out and a bailer is attached. A bailer is a hollow tube with a door in the bottom called a flap valve. When the flap valve hits the mud it opens and mud flows in, then it closes so mud can be lifted to the surface. The drilling and bailing is repeated until water is reached. If the hole might cave in, a metal pipe or rock wall is inserted into the hole, and when the hole is finished then a smaller permanent steel pipe (or casing) is inserted. A drill can be made out of logs, powered by a group of people or animals pulling and dropping, or a truck can be used to pull out, and then the rope released to let the bit drop.

How do I use a pump?

A safe and energy-efficient well system includes a cistern to hold water. For a depth of 50–300 feet, you will need a small pump to draw the water up into the cistern, only about 100 watts. This could be a low voltage DC submersible pump, which can supply all the water needed by the average household. If you don't use a cistern then you will need a ½ horsepower 110-volt AC pump, which will use 1,500 watts. The other alternative is a hand pump or a windmill. To decide your pump size, decide how many gallons you actually use by timing it with a stopwatch and a gallon container.

Priming a hand pump:

Keep a jug of water handy because in order to pump water a hand pump needs water to create suction. Add the water to the upper cup at the base of the crank, and refill it when necessary.

If a well stops giving:

Wells stop for several reasons, including the pump malfunctioning, the water level dropping, the screen getting plugged, or too much sand in the hole. Every time you open a well it should be disinfected, unless you have a really good water purification system.

How do I make a very simple rainwater collection system?

The simplest system is one in which a roof or other sloping surface is equipped with a gutter which carries the water to a covered storage container with an outlet tap. Or you can use a hillside that has a V-shaped barrier near the bottom, which allows the water to pour into a tank through an outlet in the corner of the V. This water can then be purified before drinking, or can be used directly for washing clothes,

▾ Dug well

▲ **Pioneer-era sweep well uses fulcrum and weight to lift water**

watering plants, etc. The old term for this system is the "rain barrel."

What equipment do I need for a really good rainwater collection system?

1,500-gallon (a family should have 2,000 gallons) cistern of non-toxic metal or plastic (or homemade cistern)

½-horsepower shallow-well pump

Plastic PVC and CPVC piping

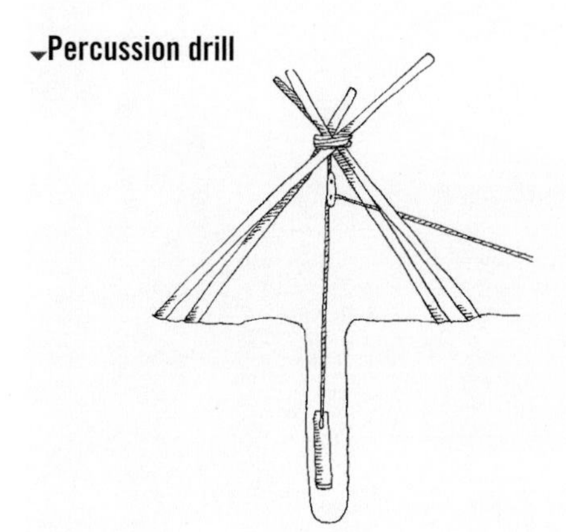

▼ **Percussion drill**

Water purification method

Roof washer

Reduced pressure backflow prevention device (if you are in the city)

20-gallon diaphragm pressure storage tank

How does a really good rainwater collection system work?

1. The roof of the house is where the water is collected. The bigger the roof the more water you will get. The rain washes off the roof into the gutters. Metal roofs are best because they easily wash and collect the most water.
2. The gutter is covered with a leaf screen and directs the water into a roof washer.
3. The roof washer is a device which, at the start of each rainfall, directs 1 gallon of water per 100 square feet of roof into a separate tank. For a 1,200-square-foot roof, that would mean that the first

▲ Windmill pump and cistern

twelve gallons are sent to the garden. This is so you don't drink roofdirt.

4. The drinkable water is then sent to the cistern. A mosquito screen covers the entrance hole of the cistern, and an overflow pipe at the very top directs extra to the garden.
5. From the cistern the water goes through a valve, and a second check valve. From the valves the water enters the house and goes to the pump.
6. The water leaves the pump, goes through a valve and diverts through another valve and into a pressure tank.
7. The water continues to a water purification system (for example: through a 20-micron filter and then into a 5-micron filter, then through an ultraviolet sterilizer).
8. After purification, the water goes to a valve that leads into plumbing fixtures. If you are in the city, the city water

▼ Hand pump

▼ Rain barrel

will also enter the house at one end, and go into a city-approved reduced pressure backflow prevention device, then through a valve and into plumbing fixtures.

Rules of rainwater collection system maintenance:

1. Periodically replace the water purification system.
2. Clean the cistern and tank every year.
3. Inspect the backflow device every year.
4. Periodically test water for purity.

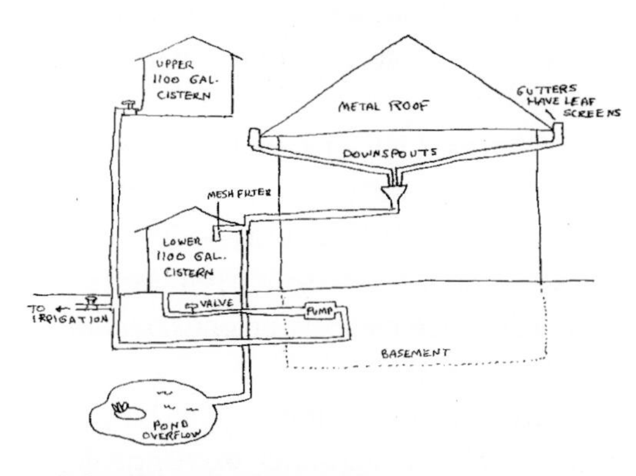

▲ Rainwater collection system

How do I condense water in the desert?

1. This process takes about 24 hours per quart of water. Dig a hole in a sunny place three feet across and 15–18 inches deep or deeper. The ground should be moist if possible, preferably a depression in a creek bed.
2. If there is green material nearby, use it to line the hole to make it watertight, weighing it down with flat rocks. In the very center of the hole put a container to catch moisture.
3. Lay a clean garbage bag or plastic sheet covering the hole and seal the edges lightly with dirt. Put a stone in the center of the sheet right above the container to weigh it down.

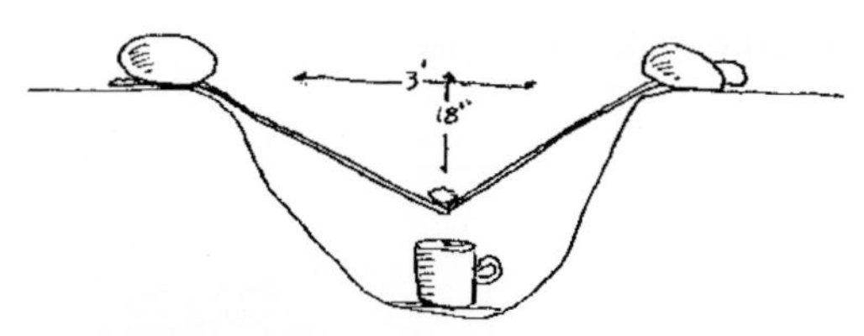

▲ Collecting condensation

4. Moisture from the ground and green material will condense on the underside of the plastic, run down to the lowest point and drip into the container.
5. Urine, food waste, tea leaves, and other moist things can also be put into the hole. One to four pints can be collected in a day. Animals such as snakes and rodents will go into the hole but will not be able to get out, so watch out. Change the site every 2 days.

What should I know about dehydration?

Dehydration is the greatest threat to a human. You can live without food for some time but not without water. If your water situation is critical, avoid sweating and evaporation. Even if you are hot, do not remove your clothes because your water will evaporate. A human generally needs two cups a day to live, but in the desert you need one gallon a day because you will lose moisture so quickly. Keep your heart rate low and don't let your temperature rise any more than the heat of the day. When you have water, drink a lot—and when you find water again, drink it slowly or you will get sick.

▾ Purifying water by boiling

What types of water contaminants are there?

Iron and manganese: Chlorination, a greensand filter, or water aeration can remove it.

Nitrates: If the nitrate levels are between 10–20 milligrams per liter, then it will only be bad for babies—adults can drink it. Higher levels can be removed through reverse osmosis, anionexchange, and distillation.

Chloride and sulfates: Can be removed with an acid-base-exchange unit.

Fluoride: In small amounts it's not harmful, but excess can be removed through reverse osmosis, distillation, or ion exchange using bone char or activated alumina.

Metals: pH levels lower than 7.0 should be treated through reverse osmosis, distillation, or by running water though soda ash or limestone chips.

Radium and radon: Ion exchange and reverse osmosis can remove radium, and granular activated carbon and aeration can remove radon gas.

How do I purify water?

Rainwater and water stored from the tap do not need to be purified. Any other water needs to be purified because it

▲ **Purifying water with a portable pump filter** ▸

can contain giardia or cryptosporidium parasites or E. coli and salmonella bacteria. Even stored water that has been contaminated by flooding or city water can have this danger. Boil (a high boil) it for ten minutes at least, then pour it back and forth between two containers to air it out. Or if you have an emergency pump filter then you can use that. Iodine or chlorine does effectively get out all bacteria if the instructions are closely followed. For muddy water, strain through thick cloth.

How do I choose a water purification system?

A slow sand filter can be part of any purification system, but should not be the only method. If you can't get a reverseosmosis, chlorination, or other kind of purifier then use the slow sand filter with distillation.

How do I make a slow sand filter?

A slow sand filter is a large concrete box that water is pumped into, and then soaks through 6–12 inches of fine sand, 27–36 inches of more course sand, 6–8 inches of gravel and then into a perforated pipe. Sand can remove iron and manganese, and some bacteria.

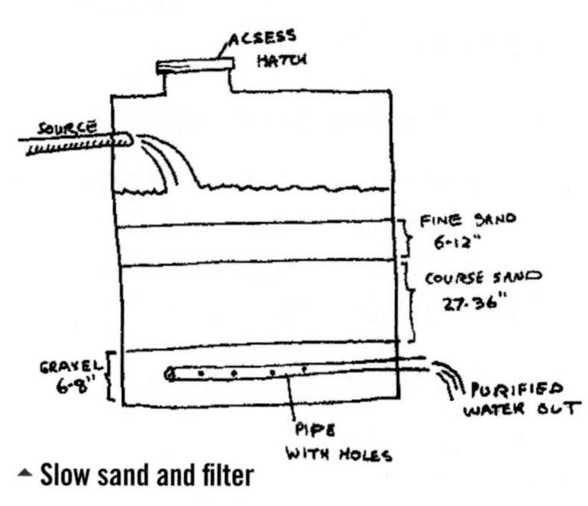

▲ **Slow sand and filter**

How do I distill water?

If water is heated above 212°F (100°C) it becomes steam, and this temperature is called its boiling point. To purify using distillation, dirty water is heated to 212°, when it turns into steam and the dirt stays in the container. The steam then goes into a cooler where it returns to its liquid state. The cooler can be anything that can catch the condensation, such as a spiral copper pipe or even a pot lid. Distillation can remove heavy metals, poisons, bacteria, viruses, nitrates, and fluoride. It can't remove oil, petroleum, alcohol, and things that don't mix with water. Clean the boiler now and then as it collects impurities. You can use a regular still, or you can make a solar still using a sheet of glass to condense the water.

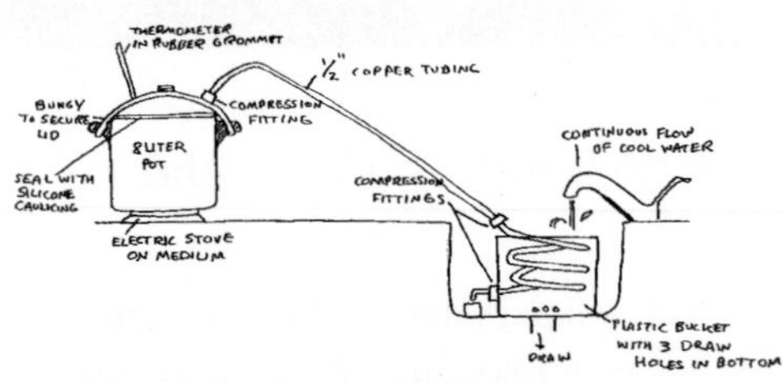

▲ Simple pot still used for distillation

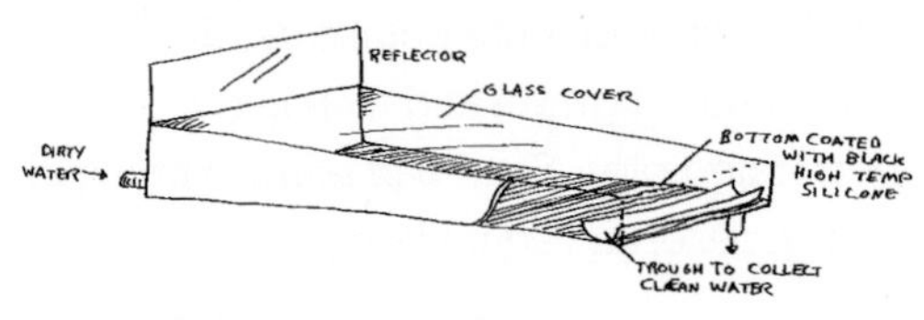

▲ Solar distiller

How do I soften hard water?

Drop several bits of charred hardwood in to the pot of boiling water while it is purifying. Skim away the particles and strain with a clean cloth.

How do I desalinize salt water?

Dig a hole and line it to make it waterproof. Build a fire and put in rocks to heat. Pour salt water into the hole and when the rocks are hot put them in the water. Put a cloth or thick mat over the hole and as the water boils the steam is soaked up, and the cloth or mat can then be wrung out into a container.

How do I make an underground cistern?

1. Dig a hole ten feet deep and eight feet in diameter, and then pour a concrete floor. This will hold about 4,500 gallons of water.
2. When the floor is dry, plaster the entire thing, floor and walls. Plastering is a difficult process because the plaster dries very fast, and is best done over a layer of wire mesh. The type of plastering needed is cement-based bonding plaster with vermiculite (you can buy it, but to make it see below). The walls are dirt and so you will not need to sand down the plaster, but you will need to put on at least three coats or more. The entire thing must be sealed and watertight.
3. Cover with a cement lid or top, but if that isn't possible, make a wood lid and seal it with plaster. In the center of the lid make a 2-½-foot square hole, with a lid to cover it.
4. If you build the cistern uphill, you will get ½ pound of pressure per 1 foot of downhill drop.

How do I plaster?

In traditional plastering for a watertight area (or even fire-proof), the plaster needs to be a mixture of Portland cement and vermiculite (a mineral), and sometimes a little sand. If you add too much water it will be sloppy and impossible to apply, and too much sand or vermiculite will make it crumble. In a flat, shallow mixing box, first put in the dry ingredients and mix well. The proportion is 1 part Portland cement plaster to 3–5 parts vermiculite and a little sand. Add just enough water to make a thick paste, and stir it up until it is a good consistency. Don't mix too long or it will start to get dry—you need to get it on the wall quickly. Traditionally it is placed on a board for easy carrying as it is being applied with the putty knife or plastering tool.

DISPOSING OF YOUR WASTE

A human being: an ingenious assembly of portable plumbing.

~*Christopher Morley,* Human Being

What are my human waste disposal options?

Outhouse: Suitable when there is no existing plumbing or waste disposal and you are not within city limits.

Septic tank absorption field: These can only be used in places with a lower groundwater table, bedrock far from the soil surface, and in dirt that has a reasonable percolation speed (not too fast and not too slow), as well as other regional factors.

Mound system: A septic tank that is placed in a mound in areas where the groundwater table is too high for a conventional septic tank.

Lagoon: Suitable for areas where the percolation rate is too slow for a regular septic tank. Uses aerobic treatment instead of anaerobic treatment, and requires energy to run a compressor or stirrer.

Composting toilet: A toilet that puts the waste into a tank where it composts material naturally.

What health codes do I need to consider?

Most cities or towns will not permit you to have a privy if you have access to the city sewer. This is for good reason—the bacteria in human waste can

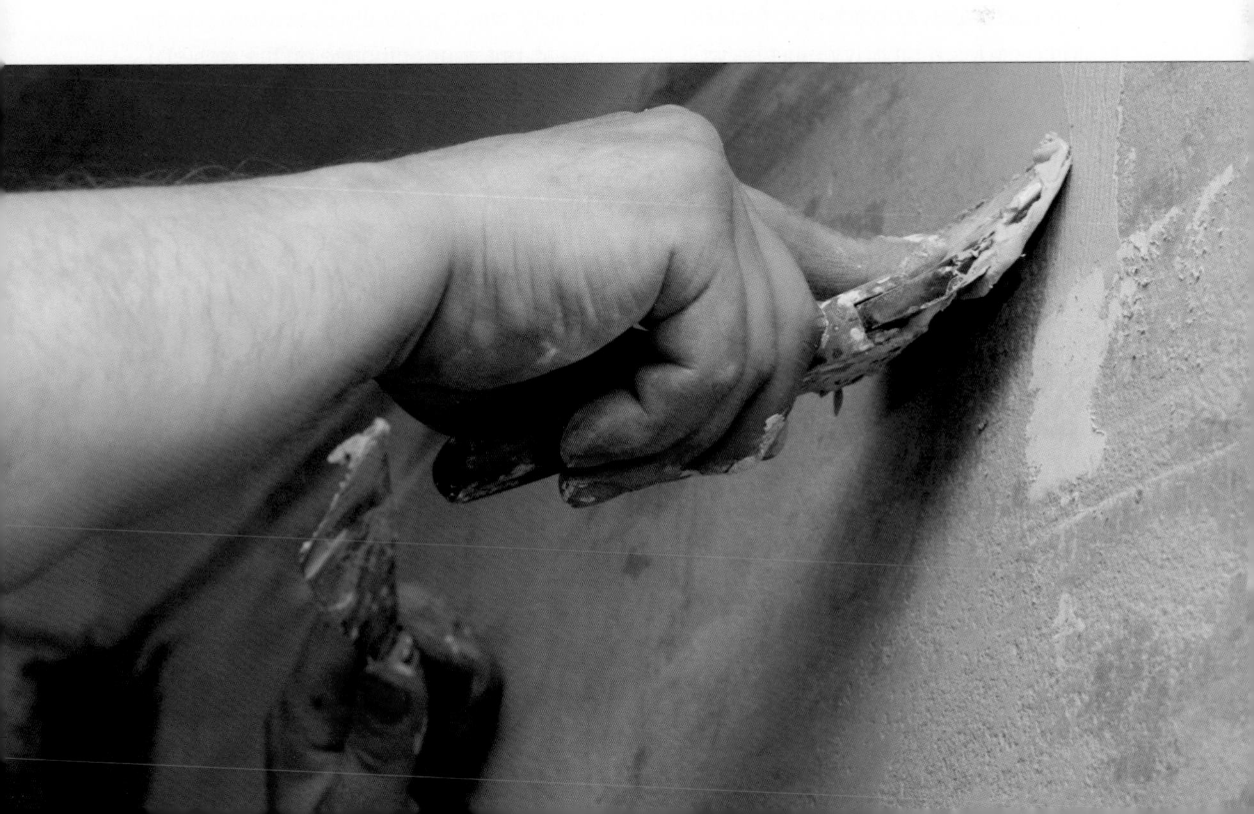

kill if it gets into your drinking water or the ground water. According to Ontario building codes, a pit privy must be well ventilated, adequately weather-proofed, and the bottom of the pit must be at least 36 inches above the high-ground water table. Another 6 inches of dirt must also be added on ground level, extending out 2 feet from the base of the privy. It also has to be 50 feet from a drilled well, lake or river, 100 feet from a dug well, and 10 feet from the property line. Check your local health department for your own specifications.

Where do I put the outhouse?

Besides taking into consideration the health codes, it is best to keep the outhouse 100 feet from your house, and downwind from the prevailing wind. So if your wind generally blows towards the west, put your outhouse on the western side of your house. It should also be away from poorly ventilated areas like in the trees, or in a small valley.

How do I design my outhouse?

An average outhouse is 4 feet square and 7 feet high. The seat is 2 feet high

Locating your outhouse

with a hole 12 inches by 10 inches. The pit under the outhouse averages 5 feet into the ground and 4 feet square. You will also need a pit ventilation pipe going from the pit out through the roof of the outhouse. six-inch PVC sewer drainpipe works well for this.

How do I maintain my outhouse?

Every time you use the outhouse you need to cover your waste with some kind of dry material such as bark chips, sawdust, lime, etc. Keep the lid closed so gases will escape through the pipe and not vent into the outhouse. Once a month pour a solution of 1 tablespoon of yeast dissolved in a quart of warm water into the pit. This will break down deposits. Twice a year remember to clean out the vent pipe or it will stink.

How do I make my outhouse comfortable?

Switching from bathroom to outhouse can be traumatic. Keep it stocked with toilet paper material, and for women keep a ceramic container with a lid for feminine products. Also have a container with a handy scoop for the dry materials for covering your waste. If you don't have running water in your house, the old way is to keep a basin of water and soap on the porch with a cloth for washing hands. If you live in a cold climate use a non-metal seat and insulate the outhouse.

What is humanure?

Humanure is human waste used as manure. Although vegetarians are perfect candidates for using their own waste for the garden (since meat in the diet is a perfect breeding ground for bacteria), meat eaters may also use their waste with a well-designed system. With proper design and hygiene, humanure can be simply and easily harvested with little of the disgusting odor you might expect.

How can I build a simple waterless composting toilet?

Use a five-gallon bucket with a fitted toilet seat that can be removed. Next to the toilet, keep a supply of wood chips or sawdust, and every time the toilet is used, add a thin layer on top of the waste you just added. When the bucket is full, dump it in

the compost area and cover with a layer of woody material. Get a backup bucket for when this bucket is too full and you have removed it. The compost pile it is dumped into needs to have a concrete base covered in a large layer of soil with a layer of leaves and grass and other absorbent material on top. The concrete is essential in order to avoid leeching. Some people have put the pile in their greenhouse with a concrete floor and held it together with a wire fence. The heat from the pile warms the greenhouse all winter. Others simply put it in a bin just like regular compost—it should not be airtight because oxygen is necessary for decomposition.

How do I take care of my toilet?

Someone in the household must be responsible for the compost. This includes making sure the lid stays closed, that there is enough cover material (such as sawdust) available, making sure it doesn't need to be emptied, and that it is not waterlogged or breeding flies. Every so often the tank that is being used must be raked flat and more straw, weeds, or sawdust must be added on top to reduce the smell, and because they help speed decomposition.

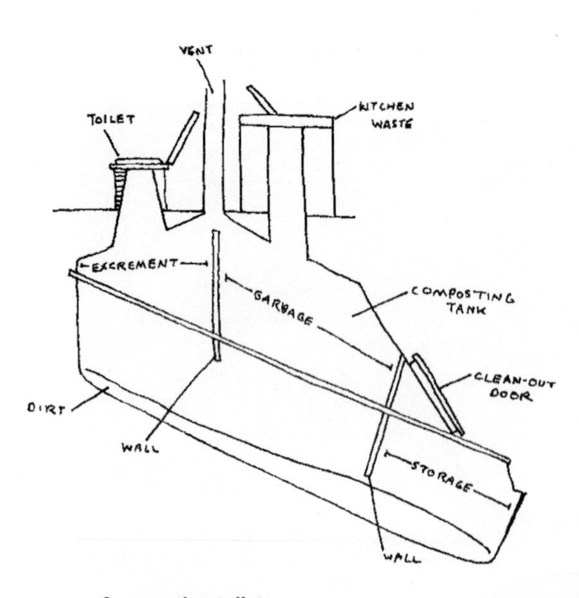

▲ Composting toilet

How is a more complex composting toilet designed?

Two chambers are needed, one where fresh materials are dropped, and one to age the compost. At the bottom of the chambers a layer of hay bales can be put in to absorb liquids. Every time someone uses the toilet they must dump clean, organic material on top of their waste. The chambers are on a slant so as the material dropped from above composts it can slide down towards a door and you can rake it out. Obviously this would have to be in a basement with a pipe going all the way down from the toilet to the tank.

▼ A warm compost pile will kill odor-causing bacteria

How do I compost the humanure?

The optimum mix for the least stinky and quickest-composting pile is one where you add all your compostable material to it—all your food and plant waste, or else there's not enough carbon and too much nitrogen. Get a compost thermometer and measure the heat. It should rise above the temperature of the human body (98.6°) for an extended period of time, and hopefully above 120° for at least one day. If it is too cold, add different materials to help the pile heat up.

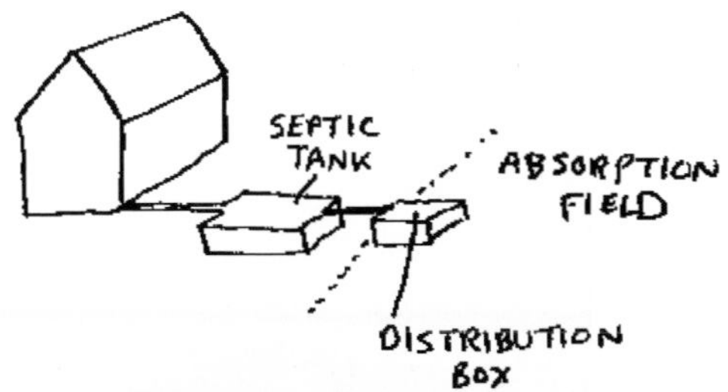

▲ In-ground septic system

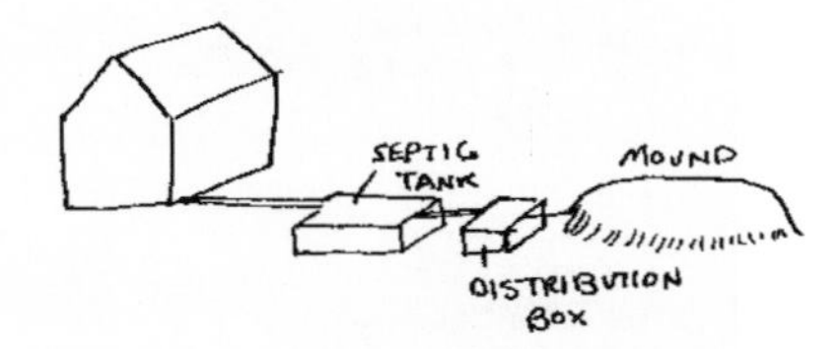

▲ Mound septic system

How does a septic tank work?

A minimum septic tank capacity is 1,000 gallons. For more than three bedrooms, 250 gallons are added per bedroom. The tank must be watertight and resistant to corrosion, settling, frost, etc. They are usually made of concrete or fiberglass and coated in bitumastic (a chemical sealer). In the tank the solids are separated from the liquid and go through a natural anaerobic digestion. The solids then turn into sludge at the bottom of the tank, while the liquid flows out to the absorption field where it percolates into the soil. The soil acts as a final filter, removing bacteria, particles, pathogens (disease-causing material like bacteria), and chemicals. The sludge must be pumped on a regular basis or it will back up and clog the percolation field. No kind of kitchen waste or foreign objects besides toilet waste should ever go in the tank, and heavy vehicles should never drive over the tank.

▼ Septic tanks

How does a lagoon work?

The aerobic treatment process is not as stinky as the anaerobic process because the high level of oxygen (since it is done in the open air). This process utilizes a lagoon, or a large pool at least 900 square feet (with five people or more add 175 square feet per person) square and 3 feet deep. Air is mixed with the waste with a stirring agitator or air compressor, which feeds the good aerobic bacteria. The liquid must still be discharged to an absorption field for final filtration. The problem with the lagoon system is that although it is less smelly, it does still smell and it is a hazard since it is an open pool of sewage.

Graywater system:

Graywater is the water that goes down the drain when you take a shower or use your sink—it is not toilet waste. With good planning this water is valuable for watering plants, which means you can use it twice, as long as you use a good graywater treatment system. Take an inventory of the gallons of water you use from laundry, dishwasher, bath, sinks, etc. Then decide if you are going to use aerobic or anaerobic treatment. A graywater tank is designed somewhat like a small septic tank, which drains into raised planters. Some people just water the garden with it straight from the source, without pretreatment, but that isn't very sanitary.

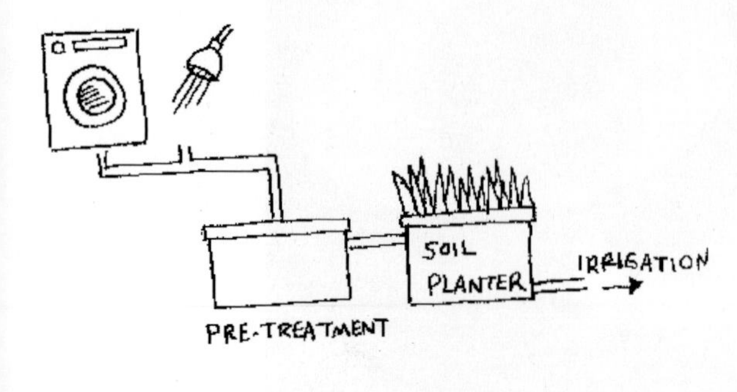

▲ Graywater treatment system

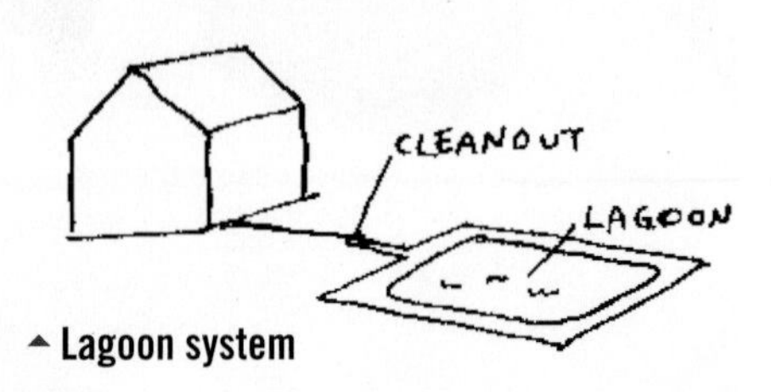

▲ Lagoon system

PRIMITIVE HEAT AND LIGHT

Firelight will not let you read fine stories but it's warm and you won't see the dust on the floor.

~Irish Proverb

How do I prepare to make a fire?

If you are outside, make a ring with rocks, and clear all the burnable material around the ring for at least ten feet. Find some water and bring it nearby, or build the fire near water. Make the fire in the center of the circle. Before you start, gather tinder (small pieces of dry stuff) and place it nearby. If you are using a fireplace, it can be helpful to clean out some of the old ashes.

How do I start a fire using flint?

Use a piece of flint and strike it with a stone over a spongy piece of dry wood until the spark lights the wood.

How do I make a fire if all I have is a pocketknife?

1. Take a rectangular piece of bark, big enough to put your knee on with plenty of room left over, and cut a triangular notch in the edge of it.
2. Find a 1-to 2-inch stick, about a foot long, and sharpen one end to a point.
3. Take a sturdy stick, and tie a cord of twisted grass, or rope or yarn or whatever is handy, to each end, making a bow with a loose string
4. Take a flat stone, preferably with a hollow in the middle of it.

▲ **Long ago, the hearth was the center of the home.**

5. Find a piece of smaller bark, square.
6. Place the big piece of bark on top of the smaller one with the notch in the middle of the small piece.
7. Kneel down on one knee, placing that knee on the end of the big bark.
8. Stand the pointy stick straight up on the big bark right next to the tip of the notch.
9. Twist the cord of the bow halfway around, making a loop, put the loop around the pointy stick, and put the stone on top, making a primitive drill.
10. Place your left hand on the stone, pressing down firmly. Hold the bow in your right.
11. The drill should work by doing a sawing motion horizontally with the bow, while holding the stick steady with the stone, so that the pointy stick twists back and forth, drilling into the bark.
12. Keep working the drill quickly and smoothly until a black powder forms in your notch, and it starts to smoke.

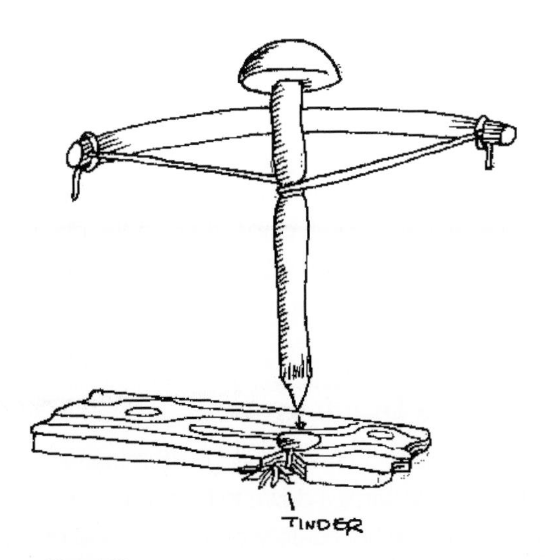

▲ **Fire drill**

13. When you see a red coal in the black powder, quickly add the tinder, and stop drilling.
14. Keep adding bigger and bigger sticks as your fire gets as big as you need. You don't need a very big fire to cook on or keep warm.

How do I find out how hot my fire is?

Hold your hand three inches above the spot you want to cook over. Count the seconds until you feel like you have to take your hand away because it is too hot. If you have to move your hand before one second (one-thousand-one), it is between 450–500°. Two to three seconds and it is 400–450°. Six or more is too cool to cook over.

How do I use a pan over a fire?

You can support a pan or pot with three large stones around the fire and set the pan on them. Or hang a pot from a stick poked into the ground and supported by a rock. Or suspend a stick crosswise

▲ **A luxurious campfire kitchen**

between two forked sticks stuck straight up in the ground, making a rack for a pot. Or use a Dutch oven.

Dutch oven size and capacity:

5 inches = 1 pint
8 inches = 2 quarts
10 inches = 4 quarts
12 inches = 6 quarts
12 inches deep = 8 quarts
14 inches = 8 quarts
14 inches deep = 10 quarts
16 inches = 12 quarts

Seasoning:

When you buy a new Dutch oven, you have to season it. Wash it in hot water and dry it to get off the factory coating. Spread olive or vegetable oil or vegetable shortening (like Crisco) over the inside and outside of it, legs included (don't use a spray). Put the lid on and turn it upside down in the oven, with aluminum foil on the rack below to catch drips. Turn on the oven to 350°F and bake 1 hour. It will smoke and stink so open the windows. Let it cool in the oven.

Using a Dutch oven:

Get a good fire going 30–45 minutes before and burn it down to form hot coals. Arrange the hot coals evenly under and on top of the oven. When you're done with it, scrape out the uneaten food and turn on the hot water (never use cold water on a hot oven!). Scrub with a plastic scrubby or brush (don't use soap!). Dry, then coat with light coating of olive or vegetable oil. Never keep the lid on tight when storing the Dutch oven. Stick something between the lid and so the air can circulate.

Soup:

For soup for twenty or more: Dig a hole twice as big as the kettle, make a fire in it, and let it burn down to red-hot coals. Make a hole in the coals, set the Dutch oven in it, and cover it all with dirt. In 4–8 hours the soup should be done.

Bread:

Grease and preheat the oven, then set a loaf in it that is already risen and ready, or pour in batter for cornbread (don't need to preheat) or cake. Set the oven in a hole, as for soup, and make sure it's level. Cover with coals and dirt and cook for three hours.

Biscuits and pie:

Grease and preheat the oven. For biscuits, put chunks of firm dough in the bottom while the pan is sitting on the coals. Turn them over when they're brown and then put the lid on. Put coals on the lid, and cook ten minutes. Cook pies by putting a pie pan right inside the oven.

How do I put out a fire?

1. Get water and sprinkle it on the top. Then stir it around with a stick, sprinkle again, and stir again.
2. Keep putting water in the fire until you don't feel any more heat, even underneath, and all the sticks are completely out. If you have no water, get sandy dirt and stir it into the fire. Every last spark should be killed. Scatter it all so that no one would know you were there.

Features of woodstoves:

Catalytic combustion: Burns up the gas given off by burning wood. This gets rid of creosote and uses less fuel. These kinds of woodstoves are for heating. To work well, the chimney must remain hot, the bypass to the honeycomb has to be engaged when the stove has been 550° for at least fifteen minutes, and it has to be cleaned twice a year.

Non-catalytic stove: Preheats the combustion air and puts it in the firebox where it mixes with flammable gases. This is clean and efficient, but is a newer design and thus more expensive.

How do I install a woodstove?

1. The stove should be put in a central location, but not where it can block traffic or an exit—a basement is a good choice. It also must have enough room to store some wood and load up the stove. In very cold climates some people put the stove on the north side of the house so that the coldest side of the house stays warmer.
2. Some houses need the floor braced for the extra weight, so test the strength of your floor (especially if it is wood). Again, a basement with a concrete floor is a good choice.

▲ A properly installed woodstove

3. If the legs are less than six inches tall, and it is on a wood floor, you should lay stone, brick, or concrete at least 8 inches thick under it. On a concrete floor, if the legs are shorter than 6 inches, place a layer four inches tall. Even if the legs are longer than 6 inches, and the floor is concrete, put down a layer 2–4 inches thick.
4. If you need to place it near the wall, put it at least 36 inches away, unless you put heat-resistant material on the wall, in which case it only needs to be 18 inches away.
5. The indoor stovepipe should be at least 24-gauge blued steel (not galvanized) with no more than one elbow and as short as possible. Join

the pipe sections with the crimped end closer to the stove.

6. The outdoor pipe must be three feet above the roof and two feet above any part of the cabin within ten feet. If it is too short there won't be enough draft for the fire and too much smoke—too long and it pulls the fire up the chimney.

What other things do I need for my woodstove?

You will need a fire extinguisher, a container for kindling, an ash bucket, a poker, and small broom and stove attachments. If you can't get an extinguisher, keep a full container of water nearby.

Cleaning a chimney (or stack):

1. Creosote is the gunk that gradually builds up in your chimney. If you can scrape off ¼ inch or more of it out of the stack, or if it's hard and difficult to scrape out, it's time to clean it. You should also clean it if the stove is burning less efficiently, and in spring and fall.
2. Cover the floor to protect it. You will need to cover your face with a dust mask or cloth, and wear gloves. You will also need a wire brush specially designed for cleaning chimneys.
3. Climb up on the roof to the top of the chimney and use the brush to clean out the pipe as far as you can. If it doesn't reach all the way down, tie a rope to it and lower it further. Use a flashlight to check your progress.
4. Some debris may get stuck halfway down or the buildup may be particularly hard near the bottom. Go back inside and use a chisel or hammer to chip deposits loose.
5. Clean out your stove and pick up all the debris. If you got more than 2 gallons of gunk then you need to clean more often than once a year.

How a wood cookstove works:

On one end is the firebox with vents and dials. The dials regulate the amount of air that goes through the vents and into the firebox, and the more open the vents are

the hotter your fire will be. The ash door can be used for a surge of air but that is for more experienced people. At the back of the firebox is a sliding mechanism (called a slide unit) that directs the smoke around the oven box before going up the chimney so that the oven is heated more consistently.

To heat the cookstove:

It takes about 15 minutes to heat the oven to 350°. First the slide mechanism needs to be directed to the chimney (each stove is different on which direction this is), or else there will be no draft to pull the smoke out and it will go into the room. After starting the fire it must be watched for temperature drops. At first, use quick-burning wood with a lot of heat. After it heats to 350°, use a slower-burning piece of wood.

To bake with the cookstove:

If the stove gets too hot, open the door briefly, but not very long or you will lose too much heat. Some stoves have hot spots that need to be watched. If bread bakes badly, there is soot buildup around the sides, so just scrape it off. Woodstoves need to be watched in order to maintain a steady temperature and good results.

Cooking on the wood range:

The circles on the top are not burners. The whole stove surface is hot, and is divided into temperature zones. The hottest is directly over the firebox (either on the right or the left), the medium temperature is in the middle, and the coolest is on the front opposite the firebox. Remember to watch your stove for temperature changes.

Grilling:

Remove one of the circles over the firebox and place a heat-resistant grate directly over the flame. The fire needs to be hotter than during normal grilling because usually a barbecue is a bit closer to the fire.

Cleaning and care of a cookstove:

Clean only when the stove is cool.

Use wet soapy water to clean decorative trim, metal backsplash, and warming oven.

Use a non-abrasive cleaner or a razor blade to remove baked-on food residue.

Avoid spilling or splashing since you won't be able to clean it right away.

Don't put a wet pot on the stove—cast iron rusts and left too long pits will form.

Use a metal sanding pad to remove bad spots, and only work in one direction or it will look scratched.

After buffing with metal use a ***very*** thin layer of cooking oil over the whole thing to season it.

Periodically clean out the small opening under the decorative nameplate right under the oven.

Clean out the sides of the ash compartment periodically.

Scrape off ash that collects on the top of the oven box (see below) monthly in summer and weekly in winter.

Inspect the gasket around the top of the stove every year and replace if worn.

Cleaning the oven box:

Remove the panels that make up the cooking surface (set on newspaper), then scrape the ash that is collected on the top of the oven box into the ash pan slowly (too fast and it will go in your face). Use a long-handled scraper and scrape the sides of the box. The soot will fall to the bottom; use the little door (disguised as a nameplate) to remove it.

Fire safety:

Don't store the bucket or extinguisher behind the stove.

Don't burn corrugated boxes, plastic, paper, or treated wood from power poles or railroad ties. Don't use kindling as a major fuel—put in heavy wood.

Never use kerosene, charcoal lighter fluid, or lantern fuel to start the fire.

Don't throw an icy log into the stove or it can explode.

Know the signs of a chimney fire: air sucking noises, a loud roar, and the stovepipe shaking.

Have a fire drill and figure out where you would escape from areas of the house during a fire.

Don't use anything in the oven that you wouldn't put in a regular oven.

Don't position handles over the stove that you couldn't put over a gas burner.

Don't ever put a plastic bowl on the stove even if you think it's safe.

How do I start a fire in a woodstove?

1. Make 4–5 balls of paper and put them in the firebox near the air supply.
2. Put sticks about the width of a finger on top of the paper, and then put 3–4 pieces of wood the width of your wrist on top of that. Curvy sticks work best because they won't roll.
3. Put slightly bigger sticks across the previous sticks, so that they form a tic-tac-toe pattern.
4. Light the fire and open the air intake holes, or leave the door open.
5. After 3–4 minutes rearrange the wood and add two bigger sticks and close the door.
6. In 15 minutes when the two big sticks are in the fire, add your seasoned firewood.

What do I do if there is a chimney fire?

Get everyone out of the house and call a fire department (if possible). If you want to try to put it out yourself, close the dampers then open the stove and spray it with the extinguisher so that the chemical will be sucked up the chimney.

How do I know if I have enough wood to warm my house?

A family goes through 7 cords of wood a year for heating. A cord is a pile of logs 8 feet long, stacked 4 feet high and 4 feet wide. A stack, or face cord is ⅓ of a cord. A typical family uses 5–8 cords per winter for heating exclusively with wood. When buying a cord of wood, make sure that it really equals a cord, and is not just a pile of wood in a pickup truck. You need to

▲ Split firewood stacked properly for drying

have access to that much wood, as well as the skills to cut it and chop it and the equipment to haul it in. If your house is not energy-efficient, you could be using double the amount of wood you would otherwise. If you are heating with your cookstove, then you will use more wood. Softwood burns faster and hotter, and dry wood is slower and has a steadier temperature. You may want to use softwood in the heat stove during the day, and hardwood at night and in your cookstove.

How do I make firewood?

Cut ¾ through the top of the log in a series of firewood lengths—typically about 1 to 1 ½ feet long. Then roll the log over and cut through the other side of each cut. Don't saw into the ground because that dulls the blade. Use an ax to split each round chunk of wood into smaller pieces.

How do I prepare firewood?

If the wood is 8 inches or more in diameter, it needs to be split. Even smaller

▼ Comfortable outdoor kitchen

wood can be split if you want it to dry faster. A good woodshed is simply a ventilated shelter with a raised floor and some kind of walls to hold the wood in a stack. When the wood is dry enough for burning it will start to crack on the ends.

MODERN CONVENIENCE AND ELECTRICITY

Concentrate all your thoughts upon the work at hand. The sun's rays do not burn until brought to a focus.

~Alexander Graham Bell

What is the easiest method to conserve energy?

When building, designing, or improving your house, one of the best things to do is to design the house to conserve energy. This makes it easier to power and heat your house without much work and expense on your part. Dirt is one of the best insulations, but simply sealing windows and doors to minimize heat loss (or cool air loss) can also improve your energy consumption.

Keeping cool in summer:

Make an outdoor kitchen, cook in the evening, or eat mostly raw or grilled foods.

Open all your windows at night to bring in cool air, and close them in the morning.

▾ Laundry equipment

Use heavy insulating curtains in the sunniest windows.

Take a cool shower, go swimming, or take a nap in the afternoon. Work at night.

Drink lots of water and fruit juice.

Keeping your home warm in winter:

Seal up your windows, either with plastic or heavy blankets.

Seal around doors with plastic or cloth.

Only use one door, hopefully in a separate entryway.

How can I conserve energy used by my electric well pump?

Water pumping can be a major drain on your power consumption during winter. Designing an efficient way to use your well by installing a cistern can make the pump you need much smaller.

What are the cheapest energy-saving appliances?

Buying a very expensive propane appliance does not save you money in the end because you have to keep buying propane. The best course of action for electric appliances is to purchase regular, energy-efficient models of any regular brand. Even a smaller photovoltaic system can power a TV, and with a bit more investment it will be easy to power your house reliably.

Appliance replacements:

Electric baseboard heat: passive solar design

Space heater: woodstove

Electric stove: wood cookstove

Hot water heater: solar or compost hot water

Clothes dryer: clothesline

Dishwasher/washing machine: hand washing

Shop equipment: replace AC with DC, or use hand tools

Refrigerator: If you have a spring with a temperature lower than 40°F, put the food in a waterproof container in the water. A large-scale version is to have an insulated stone or block springhouse with a cement trough with spring water flowing through it, called a springhouse.

Block springhouse

Appliance	Average Wattage	Average Hours per Year	kWh per Year
Dehumidifier	257	1467	377
Attic Fan	370	786	291
Infrared Heat Lamp	250	52	13
Humidifier	177	921	163
Portable Space Heater	1322	133	176
Blender	386	39	15
Coffee Maker	894	119	106
Dishwasher	1201	302	363
Garbage Disposal	445	67	30
Microwave	1450	131	190
Mixer	127	102	13
Range and Oven	12200	96	1171
Range/Self-Clean Oven	12200	99	1208
Toaster	1146	34	39
Waffle Iron	1116	20	22
15 cf. Upright Freezer	341	3504	1195
15 cf. Upright/FL Freezer	440	4002	1761
12 cf. Refrigerator	241	3021	728
12. cf. FL Refrigerator	321	3791	1217
14. cf. Fridge/Freezer	326	3488	1137
14. cf. FL Fridge/Freezer	615	2774	1829
Curling Iron	40	50	2
Hair Dryer	750	51	38
Shaver	14	129	2
Radio	71	1211	86
Color Television	300	2200	660
Sewing Machine	75	147	11
Vacuum Cleaner	630	73	46
Clothes Dryer	4856	205	995
Hand Iron	1008	143	144
Automatic Washing Machine (Including energy for hot water: 2500 kilowatts)	512	208	107
Water Heater	4474	1075	4811
Desktop Computer/Monitor	150		

What is a compost water heater?

A compost water heater uses biomass (a mass of decaying organic matter such as plant material) in a large enough pile that it produces enough heat to make hot water, and to kill fungus and bacteria living in the organic matter.

What is a heat exchanger?

With proper design, a composting toilet can be linked to your regular food waste receptacle and a heat exchanger installed. A heat exchanger is a container or material which absorbs the heat from the hot compost and transfers it to what you want—such as water or air, thus making hot water or warming the room. A simple heat exchanger could be a flexible plastic pipe that is attached to your cold water source. The water will sit in the pipe, warming until you turn the tap, where it will flow out. Then you will have to wait for the water in the pipe to heat up again. A 1.5-inch pipe 100 feet long will hold 9 gallons, enough for a few hot showers or a load of laundry.

How much biomass do I need?

You will need a large amount of biomass to create enough heat to make hot water. If you keep this pile in your greenhouse it will heat the greenhouse all winter. One of the best designs is 35 feet of woven wire fencing 5 feet high that is made into a circle. A great deal of manure (either animal or human) is needed in order to have enough decomposition to create the heat.

What do I need in order to produce methane power?

How much manure do your animals produce? You will need at least 240 pounds of fresh manure per day, although more is better. Also, these animals must be confined in order to gather the manure while it is still fresh. You will also have to make sure that your system is safe by having a warning system in case of a leak—when you mix methane and oxygen they can explode. A methane generator does not work well in wet, chilly weather, so it is best done where the winter is fairly short and doesn't reach extreme temperatures.

What is a digester?

A methane system (also called biogas) consists of a digester, which is basically a semi-underground tank or holding area that is covered so that the gas emitted by the manure cannot escape. This digester can be both a place to put your human waste (thus combining it with your composting toilet) and it can be used for biomass heat, and the waste can then be used as fertilizer, so that it is super efficient. You will also need a system of pipes to take the gas to be burned either as a heat source or in a generator.

How can I use a plastic bag digester in a cold climate?

Because the simple digester needs to be made with a plastic bag and needs to be kept very warm, it is difficult to use in cold climates. It would be a waste to produce heat with biogas only to use it to heat a room in order to make more biogas. It could be possible to make this digester

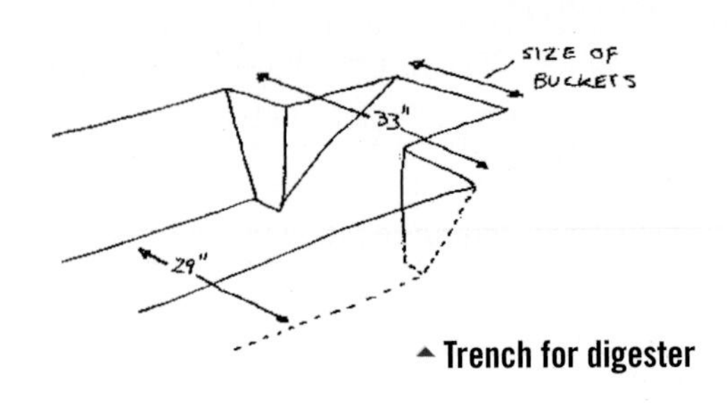

▲ Trench for digester

inside of a greenhouse heated with a large manure or compost pile

How do I build a plastic bag digester?

1. Dig a trench in the ground 32 feet long and 33 inches deep. The bottom of the trench will be skinnier than the top so that the walls angle outwards. The bottom should be 30 inches wide while the top should be 3 inches wider.
2. Get eight old buckets all the same size or two cement pipes 3 feet long and 1 foot in diameter. At each end of the trench dig a ramp coming up from the bottom and make it as wide as your buckets or cement pipe. Make sure there are no stones in the trench and that all the walls are flat.
3. Get a polyethylene plastic tube 92 feet long and 5 feet wide. It should be 1.000 gauge (that's how strong it is). Lay it out flat on a surface with no sharp or bumpy objects. Fold it in half and cut it into two 46-foot pieces. One of the easiest ways to buy this is to purchase a roll of heavyweight poly tubing from a plastics supplier. For about $86 you can buy 1,000 feet of 4-foot-wide polyethylene tubing, which is used to make plastic bags for packaging.
4. Take off any sharp or pokey clothing (such as rings or belts), and grab hold of one end of one plastic tube. Crawl through the second tube with it, making it a double-layered tube. Fold it and put it away safely.
5. Make a ¾ inch cut through both layers of plastic, 13 feet from the end of the tub so you can insert a 1-inch PVC screw (a male adapter).
6. Inside the tube, first place a 6-inch plastic disk with a 1-inch center hole, then a rubber disk (used tire works well) and put in the PVC screw to hold them in. On the outside of the tube place another rubber disk, then another plastic disk, then a 1-inch PVC screw (a female adapter) and screw it together.
7. Put a 4-inch PVC tube into the female adaptor, attach a 90° PVC elbow to the end and then attach another 4-inch PVC tube to the end of the elbow.
8. Put the plastic buckets or cement tubes in the ramps in the trench. To use buckets, remove the bottoms of the four buckets and connect them together.
9. Take the whole plastic bag and lay it in the trench. Put the two ends through

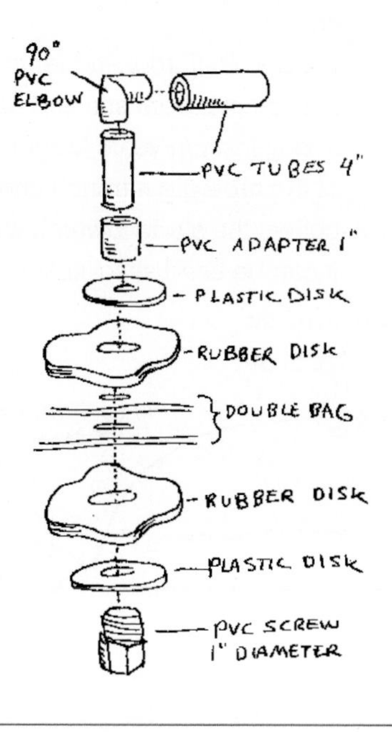

▸ Pipe for outgoing gas

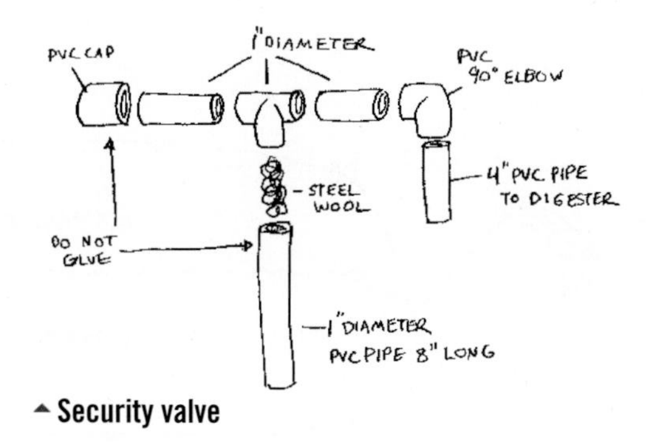

▴ Security valve

the two plastic or cement pipes at each end of the trench.

10. Make a security valve by making a T of 1-inch PVC pipe. The long leg of the T should be 8 inches tall, and each arm of the T should be 3 inches wide. Inside the top of the long leg put a ball of steel wool—don't glue it because you have to change the wool every six months.
11. Plug one arm of the T with a PVC cap, and put a 90° elbow at the other end. At the end of the elbow put a 4-inch PVC pipe.
12. Attach the pipe coming out of the elbow to the outgoing biogas valve, which is attached to the big plastic tube, with a transparent plastic hose.
13. Drive a wooden pole into the ground right next to the safety valve. Fill a transparent plastic bottle with water and put the bottom of the T into it. Bind the bottle to the pole with rubber belts.
14. Tie one end of the plastic double bag with rubber belts so that it is ***airtight.*** Do the same for the other end but put a plastic hose into it. Connect the hose to the exhaust pipe of a vehicle, making sure it is tightly attached.
15. Start the vehicle and allow the exhaust to fill the plastic bag. The water in the plastic bottle will start to bubble when the biodigester is at its maximum capacity.

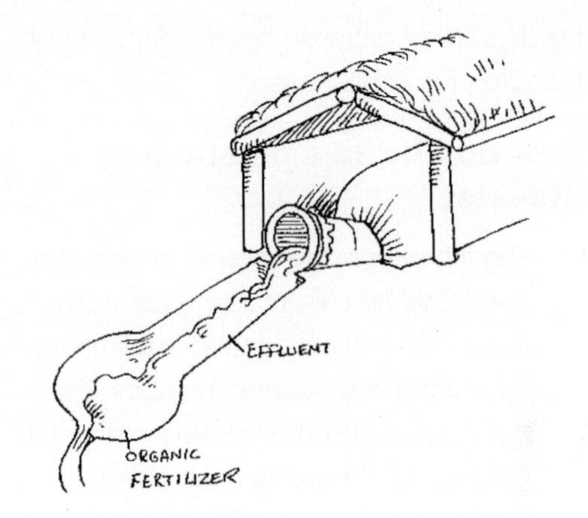

16. Remove the hose from the vehicle and use it to fill the bag with water. Fill it up until the water level reaches just above the height of the buckets or pipes and the gas can no longer escape out the sides. The bag will be about 75percent full of water and 25percent full of exhaust. This creates the bell-shape where biogas will form.
17. Open both ends of the plastic bag. If you filled the bag with water properly, the exhaust inflating the bag will stay in because it rises to the top, and the buckets or pipes tilted upward won't let the water escape. Thus it will hold its shape.
18. Build a small roof and fence for the bag to protect it from sun and animals and it may last ten years longer. At the end of the tube dig a trench and a hole to collect the sludge, which will run off so it can be used as fertilizer.

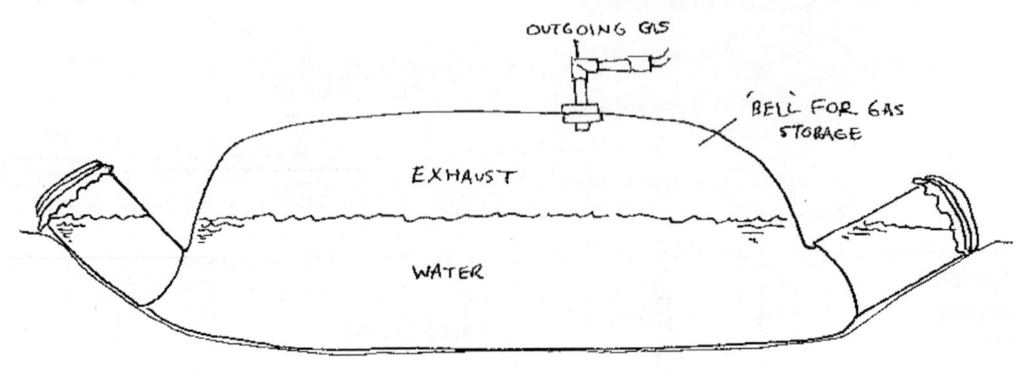

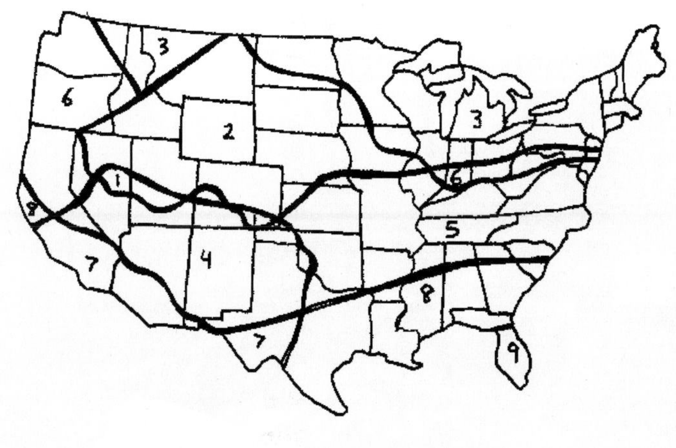

▲ **Geographic viability of solar power**

19. Thirty days after you installed the plant and you have been maintaining it properly with manure, remove the PVC cap from the other side of the T-shaped security valve and connect it to a burner with plastic or PVC pipe (the gas is corrosive). Only the pipe used to burn the gas should be galvanized iron.

How do I maintain my mini-biogas plant?

Every day you must add 40 pounds of very fresh manure from bovine animals,

▼ **Passive solar wall and window design**

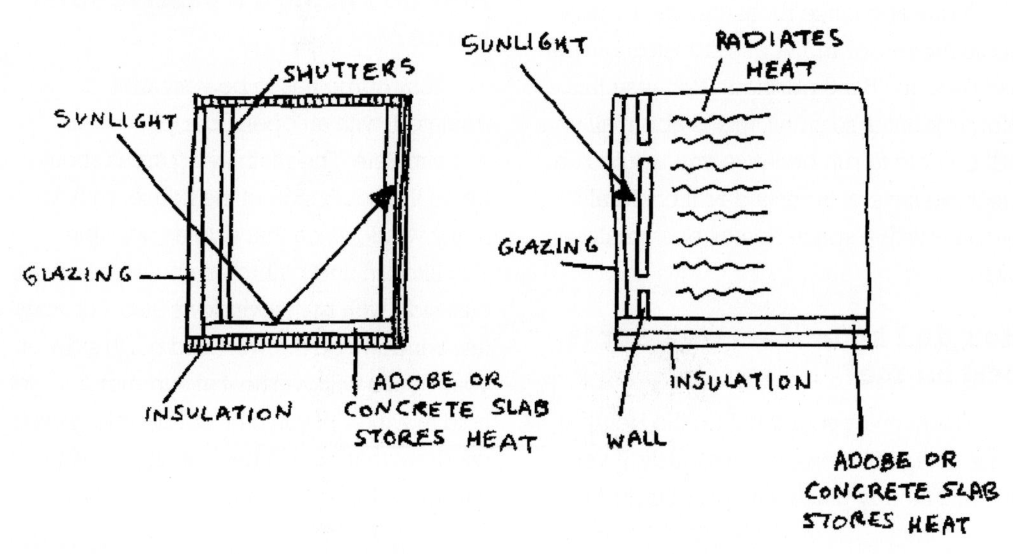

▲ Sunroom

goats, pigs, or your own manure, mixed with 198 pounds of water. This is a ratio of 1 bucket of manure per 5 buckets of water. Never use chicken manure because it won't make gas properly. Pour the mixture into one end of the plant through the buckets or cement tubes. Change the ball of steel wool in the safety valve every six months.

How do I burn biogas?

A gas appliance measures its usage in cubic meters, or m3. A gas stovetop burner uses about 0.5 m3 per hour. Because this biogas plant is so small, this is about all you will be able to run because you would run the stove several times a day. It could also be used with a space heater or a small gas oven.

How do I know if solar power is best for me?

The sunnier your location the better it is for using solar energy, obviously. If you live in a cloudy place with only summer sun, solar may not work well for you. Solar power is also simple—pretty much anything that utilizes sunrays or heat is solar-powered, which makes it cheap. You can use either a passive solar design, in which you just design your living space to utilize the heat and light of the sun, or a collector system that gathers the solar rays and converts them electricity.

How do I design a passive solar home?

Your home has to be very well insulated, with an open floor plan so air can circulate. The main living areas should be on the south side of the house, with as many windows on that side as possible. The kitchen and bathroom should be on the east side with many windows also. Put very few windows on the west and north side, or your house will overheat in summer and get cold in winter. If you can, get double-glazed windows that hold in heat, and put in drapes, window shades, or shutters. Cover the windows at night during cold weather or in hot weather to regulate the temperature.

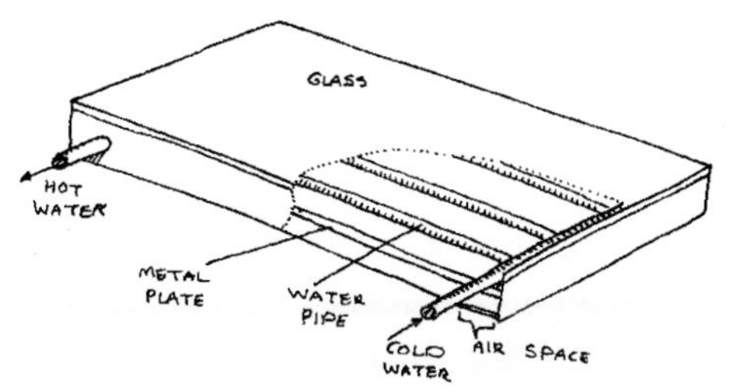

▲ Homemade passive solar hot water heater

What is passive thermal mass?

A thermal mass is a large area of heat-absorbing material, usually stone or brick that absorbs solar energy during the day and radiates it back at night. Wood, carpet, and furniture don't do this efficiently. Most people do this by creating a sunroom on the south side of the house with either the floor or the far wall made of stone, brick, or tile. Another way is to make the south side all windows, and then put a wall inside this window with smaller openings to let light in. The wall absorbs heat from the sun and radiates it into the rest of the house.

▸ Commercial passive solar hot water heater

Parts of electricity generating solar power system:

Solar array: A single solar module is a glass sheet enclosing either single-crystal or poly-crystal solar cells on top of a waterproof backing material and edged with an aluminum mounting frame. An array is several modules connected together.

▾ Array

Several solar arrays connected together make a photovoltaic array.

Charge controller: A device that controls battery charging. It prevents overcharging during the day and discharging during the night. A cheap charge controller is simply a relay device that opens and closes the circuit to the battery.

Battery bank: One or more batteries wired together matching the voltage rating of the inverter and the solar array. Car batteries do not work because they can't last under the charge/discharge cycle. A lead acid 6-volt golf cart battery works very well, and with more money an L-16 electromotive battery is ideal.

Inverter: If you are using 120 volt AC lighting and appliances you will need an AC inverter. Buy a larger unit and scrimp on other parts because you will need it (a marine or RV inverter is not big enough). A small inverter to run power tools, computers, lights, and appliances has a modified square (or sine) wave design. Laser printers, light dimmers, microwaves, well pumps, washers, and refrigerators need a large capacity sine wave inverter. Those cannot operate from a 12-volt battery bank and so you would need a solar array and battery bank designed to operate at 24–28 volt DC.

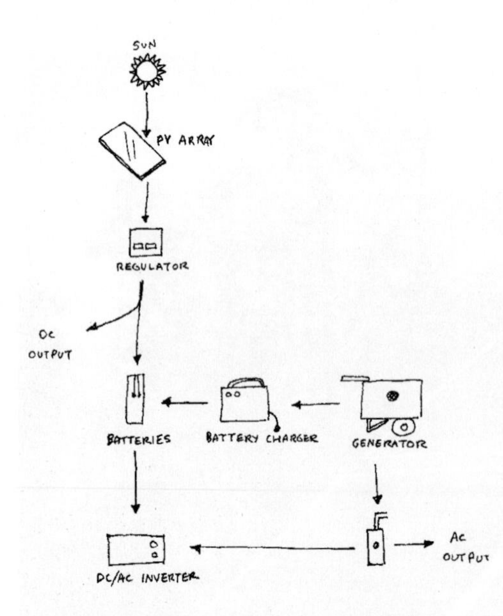

Safety fuses and circuit breakers: Fuses and circuit breakers are necessary safety devices that must be DC rated. Don't go cheap because using an AC fuse in a DC battery circuit is a fire hazard.

How do I set up an array?

The array can be put on a tilted rack, mounted on a pole, or put on the roof of your house or garage. They should face within 15° south in the Northern Hemisphere and the tilt angle can be 30° in the south to above 60° in the north. A steep tilt in winter and a shallow tilt in summer will improve performance. A pole-mounted array can be hooked to a tracking device for longer solar exposure but not if you are in a residential area.

How do I set up a battery bank?

You need a battery bank big enough to store energy for at least 3–4 days, which means you might need many batteries. If you get a diesel generator then you can get by with less, because after two days you can run the generator, recharge the batteries, and turn off the generator. The liquid electrolyte batteries need to be put in a battery box and should not be exposed to freezing or very hot temperatures—optimum is 77°F.

System arrangements:

Simple system: The sun shines into the array, which sends the power to a regulator. The regulator sends it to an inverter, which is also hooked to the battery, or to the DC electrical equipment.

The inverter converts DC to AC, and sends it to the AC electrical equipment.

System with a generator: The regulator sends the power to the DC electrical equipment, and to the batteries, which sends it to the DC/AC inverter, then to a transfer switch, and to an AC appliance. The generator goes to a battery charger, which also goes to the batteries, and to the transfer switch.

How do I buy a photovoltaic system?

Pick a specific module and standardize everything. Make sure you like the brand so you can upgrade to the same kind in the future. To size a system, a single-family house usually uses 1,000 kilowatt hours per month, or 34 kilowatts per day, if they are conserving their energy very frugally. In most places you can collect 5 ½ hours of sun in summer and 4 hours in winter. If you don't improve your efficiency you will need a 6–7 kW array to power this, which can cost $25,000 USD. A 4,000-watt inverter will take care of a house if heating, dryers, hot-water heaters, and cookstoves are not hooked to it. To lower the wattage of a well, use a cistern. For a backup system, you can get a 2–3 kW array and hook it to the main and backup circuit breaker panel. Match the solar system to the loads it can handle and during a power outage do without the other stuff. A photovoltaic system will probably only harvest 65–75 percent of the advertised rate, so plan for that also.

How do I install one myself?

If you are not an electrician, then you can really get burned, literally. The wiring of a solar electric system is specified in the National Electric Code and they are supposed to be inspected by a local building inspector. The codes are important because lots of people have already made the mistake that you

probably will and that saves you the trouble. But, many good solar power kits come with good instructions, so proceed at your own risk if you want.

How do I check the battery?

Always use caution when tampering with the battery. Regularly check the water level and fill it with distilled water only, clean the battery tops and terminals, monitor it closely with a voltmeter so that it never runs down all the way. Always keep the battery charged, even when you're not using it, and protect it from freezing.

How do I know if hydropower is best for me?

What is your water source? Is it an ocean, lake, river, underground, waterfall, or stream? For water power you need a stream, which you must measure to find out how much it drops and how fast it is going in feet per second. With those numbers you can build a water wheel, which can both pump water and produce electricity. Or you can buy a small hydroelectric generator, which is about the same price as a small diesel generator. Or you can use a non-electric approach and use a water wheel to turn a large beam that uses gears to turn whatever it is you need to turn.

Types of wheels:

Overshot wheel: The water is directed over the top of the wheel, turns it as it runs over it, and then is dumped down to a lower stream. It has an efficiency of 60–80 percent.

Undershot wheel: The water flows under the wheel, turning it as it goes. It has an efficiency of 60–75 percent.

Pelton wheel: A smaller, high-speed wheel that the water is pushed against at a higher pressure.

How do I generate electricity?

Because it's difficult to regulate the voltage produced by a water wheel, using

Overshot water wheel ▸

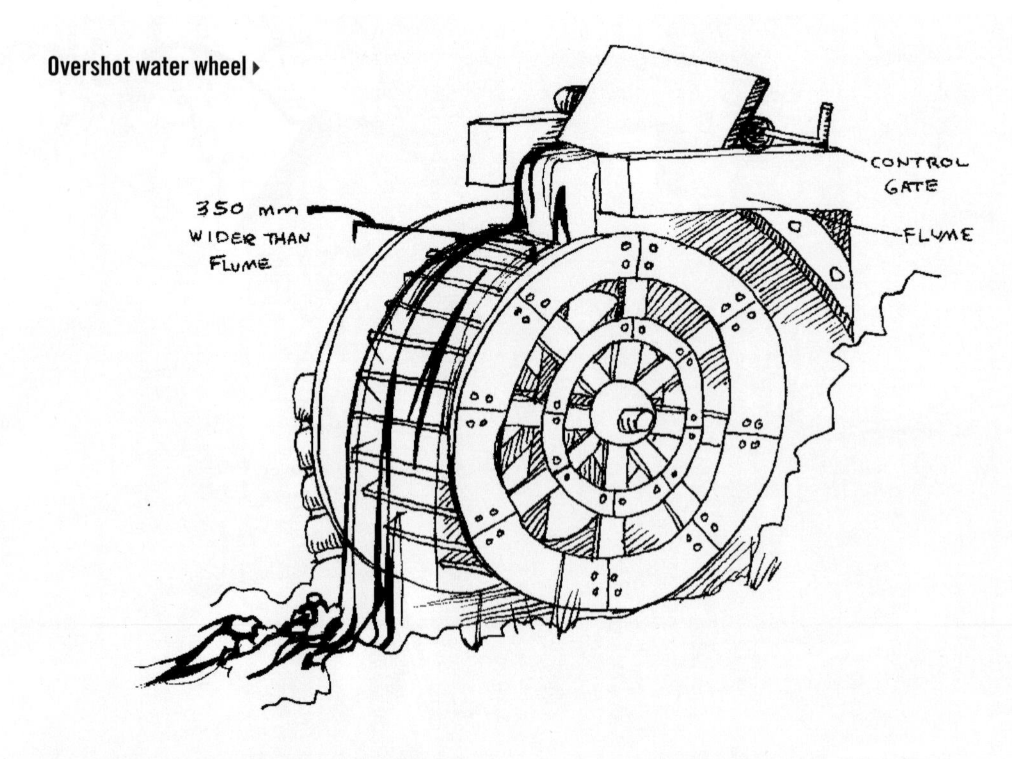

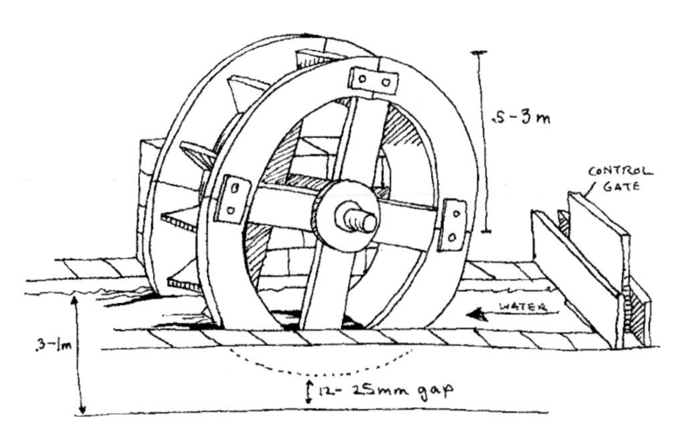

▲ **Undershot wheel**

▲ **Pelton wheel**

AC can be difficult. DC is better because you can store the electricity in batteries and even use car parts to build it. If you have a solar, wind, or diesel system you can link them together. Generally a stream is considered to have "low head" (head is how much pressure the water has) if it has a change in elevation of less than 10 feet, while a stream with "high head" is more than that. If it has a drop of less than two feet it just won't work. A mathematical formula is used to find out the head of a stream.

To find out your Net Head:

1. First find out your Total Gross Head, which is the vertical distance that the water is moving, or its drop in elevation. If you have an altimeter (some watches even have them) you can use that, or hire a surveyor. Measure the drop from the water source, to where it will come out of the turbine. One way to do this is get a 20–30 foot hose, a funnel, an assistant, and a measuring tape. Have the person stick the funnel in the end of the hose and put it right under the surface of the water at the source. You hold the other end downstream and lift it until water stops coming out of it. Measure the distance between the water and how high you had to lift the hose, and that is your Gross Head.
2. Normally an expert would then use the Gross Head minus Head Losses to find out the Net Head. Head Losses are any obstructions or bends in the stream that cause the water to slow down, and there are many formulas for these. It is not necessary to do this for the small homesteader.

To find out Minimum Water Flow:

1. The easiest way is to first dam the stream with logs or boards to divert it to a place that you can put a five-gallon bucket.
2. When the stream is flowing into the bucket, use a watch to calculate how many minutes it takes to fill. If it fills in 2 minutes, then your flow rate is 2 ½ gallons per minute.
3. Repeat this several times a year because your stream's flow rate will fluctuate depending on the season. The

possible, with a pot still you will have to run it through several times. With a pot still you will also need to use wheat or sugar, because fruit is not recommended. Please don't drink the ethanol you produce because there are many other processes that prevent poisoning, though they are not included here.

Steps for making ethanol with grain:

Warning: It is illegal in the U.S. to do this for making beverages, and you will need a permit to do it to make fuel. Please check with your local government.

1. Soak the grain for 24 hours in water, change the water, and soak for another 24 hours. Keep the temperature of the water between 62–86° F (16–30° C). Use a screen to take off any floating debris.
2. Spread the grain out on a wet surface and keep it between 62–86° F (16–30° C) until it sprouts, which will take 7–10 days. Turn the grain over daily and keep it moist.
3. Allow the sprouts to grow about 5 millimeters long, then spread them out very thin and let them dry out in a dark place, such as a basement. Using an oven to do this is not recommended because the enzymes needed for fermentation can die.
4. Put the grains in a bag and bang it against a hard surface to knock off the sprouts. Put the grain through a sieve to shake out any debris. This is called the "malt".
5. To grist the malt, use any kind of grinder or rolling pin to crush the grain exposing its center, the starch. It should only be broken in 3–5 pieces.
6. Mashing converts the starches into sugars. Heat the malt in a big pot to 143–145° F (61–63° C) for 45–60 minutes in 4.5 liters (19 cups) water per kilogram of grain. Stir it occasionally.
7. Raise the temperature to 167–170° F (75–77° C), being very careful not to let it boil over. Use a sieve to strain out the grains, leaving the liquid in the pot (it is called wort or mash). Rinse any remaining materials from the grains by putting them in a cloth straining bag over a bowl that has a little of the wort in it. Rinse the bag in the wort to clean the grains, squeeze it out, and put the wort back with the rest.
8. Cool the liquid to below 86° F (30° C) and add yeast, either store-bought or your own wild yeast. You can buy yeast especially made for fuel producers, such as Red Star Ethanol Red. For wet yeast, use 3 cups per 25 liters (6.6 gallons) of wort. For dry yeast, let it soak beforehand in warm water for 1 hour, 4–8 grams (.14–.28 ounces) per gallon of water, and then add it to the mash.
9. Yeast also needs food to grow. Per 20 liters (5.3 gallons) of liquid, add 4 teaspoons salt, 4 teaspoons lemon juice, and 5 kilograms (11 pounds) sugar.
10. The less oxygen the yeast has during fermentation the more it will make ethanol. Therefore once you have added the yeast and the yeast food seal the container so that air can't get in. A fermenter would have an airlock that you can use to let out a little bit of carbon dioxide. Keep the temperature at 77° F (25° C). If you added enough yeast, within 4 hours it will have a tall layer of foam on the top.

11. It should stop bubbling within 5 days. Basically about half the sugar will convert to carbon dioxide and bubble away. So if you put in 5 kilograms of sugar to feed the yeast, the weight should drop by 2 kilograms. Let it ferment until it does.
12. Turn off the heat and let it settle for a couple of days. Siphon the clear liquid (called the wash) into the still.
13. Turn on the heat of the still. When it is at 140°F (60°C), turn on the cooling water to the condenser. The first 100 milliliters liters (almost half a cup) of wash that comes out of the still needs to be thrown away because it contains methanol. The temperature needs to get exactly to 173°F (78.4°C), the boiling point of alcohol. This will separate the water from the pure alcohol.

▾ Wind turbine

▲ **Old-fashioned moonshine still**

14. Keep the process steady so that the fermented liquid (the wash) coming out of the holding tank equals the amount of steam coming out, and that any water coming out of the still is slightly warm (adjust the cold water valve to do this).
15. When it is done, hopefully you will have 198 proof ethanol, which only has a little water. 60 percent water is 100 proof alcohol, which is drinkable—any higher can kill you. Use a hydrometer to find this out, and follow the instructions carefully for an accurate measurement.
16. The best distillation you will be able to get with a pot still is around 96 percent, so you will need to dry it using lime. This dries the alcohol further, but don't do that step until you are ready to make biodiesel because the alcohol will absorb water from the air and return to 95 percent fairly quickly.

Why make biodiesel when you could just use ethanol?

Ethanol is not as efficient as biodiesel so you need much more of it. You also have to convert your gasoline vehicle to run ethanol, while biodiesel fuel only needs a regular diesel engine. Principally people have been using ethanol to replace gas and biodiesel to replace diesel. For the non-mechanical homesteader however, biodiesel is much easier and applies to any diesel-run engine, such as a truck, tractor, or generator.

How do I prepare the ethanol for making biodiesel?

1. Before it can be used in transesterification, the ethanol needs to be made more dry using quicklime. Mix 35 pounds or more of lime per gallon of water to be removed. This can be measured using a hydrometer.
2. Let it sit 12–24 hours, stirring occasionally to allow it to "slake". This is a chemical reaction that will form calcium hydroxide, which will sink to the bottom leaving purer alcohol on the top.
3. This is then distilled (for a second time) exactly at 173°F (78°C), the boiling point for pure alcohol, in your home still. When it is decanted the residue will be wet calcium hydroxide and lime with some traces of alcohol.
4. To get the alcohol out of the residue, continue distilling. The temperature will rise above 173°F, and when it reaches 208–212°F (98–100°C), all the water has been removed.
5. The lime (or calcium oxide) can't really be used for this purpose again . . . at least, I haven't found a way of recycling it yet. If you are interested in a

reusable alternative, you can invest in a molecular sieve, or zeolites. It is an artificial material that will absorb all the water in your ethanol and can be reused.

What is biodiesel?

Biodiesel is made of vegetable oil in a process called transesterification. Basically ethanol (moonshine) or methanol (wood alcohol—extremely caustic) is mixed with lye, which makes sodium methoxide. The sodium methoxide is mixed with vegetable oil, which then settles leaving glycerin on top, which is skimmed off, leaving biodiesel (methyl esters) on the bottom. The biodiesel is washed and filtered and then used in a regular diesel engine. It is cleaner than petroleum and you can do it yourself. But, it is also dangerous to your health to make it—make small test batches before trying a larger amount.

Copper still

How do I stay safe when making biodiesel?

Please don't make biodiesel in your house! Do it outside in a shed or other building because there is significant danger of fire and other problems. Wear protective gloves, aprons, clothes, eyewear, and a mask. Don't inhale any vapors—the area needs to be well ventilated. Potassium and sodium hydroxide are caustic substances. Keep a hose running nearby just in case of accidents, and children and animals need to be kept far away. If you spill anything on you, wash the area immediately in running water for a long time.

Tiny test recipe of biodiesel:

72 grams, 2.5 ounces, or .07 liters of 199 proof ethanol

250 grams, 8.8 ounces, or .26 liters of vegetable oil

Hydrometer

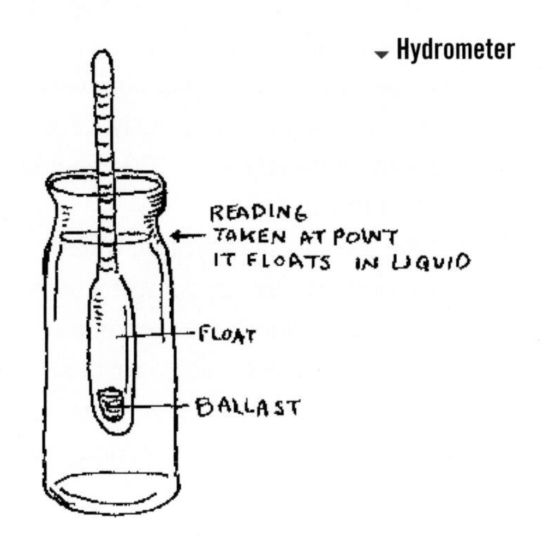

2.5 grams potassium hydroxide OR 1.82 grams dry sodium hydroxide (regular lye).

The lye should be only 1 percent of the weight of the oil, and the ethanol about 20 percent of the oil. When using sodium hydroxide, make sure it is dry and don't use wet utensils in it.

How do I make biodiesel?

1. Calculate how much of each ingredient you will need using the recipe. Making vegetable oil and lye is written about elsewhere. Potassium hydroxide is also known as caustic potash, and is simply lye that has been processed further—it can't be made at home like regular lye.
2. Heat the oil to 148°F and hold it there until you can turn off the heat and it stops bubbling immediately. This is how you know that the oil is extremely dry. Any more than 0.5 percent water can kill the chemical reaction you will be trying to achieve.
3. Warm the ethanol slightly, and then dissolve the potassium hydroxide in the ethanol while stirring slowly. You can use regular lye, but it will dissolve much more slowly. It is possible that you may need more ethanol to get a reaction, which is why doing a small test run is important.
4. Add the ethanol/potassium hydroxide to the oil and stir it vigorously. Keep it at room temperature for 120 minutes while it reacts.
5. Let the mixture sit overnight while separation occurs. Do not use it unless you see the glycerin has separated from the biodiesel. This might take place immediately, in a few hours, overnight, or never. If never, don't use it.
7. Siphon off the biodiesel, which will leave glycerin at the bottom. You can store the glycerin for use in herbal mixtures and soap. Let the biodiesel sit for a week so that any soap residue will settle.
8. Next it is necessary to wash the biodiesel so it will be easier on your engine. One simple way is to first add a tiny amount of vinegar in it to neutralize the pH more.
9. Find some kind of translucent container made of PETE plastic, such as a 5-gallon bucket. Install a valve 3–4 inches from the bottom of the container.
10. Fill the container with water halfway up to the valve. Fill the rest of the bucket with biodiesel.
11. Stir it gently so that you don't create soap suds, and then let it settle 12–24 hours. The oil and water will separate. Use the valve to get the cleaned oil out.
12. Repeat this process 2–3 times, without adding any vinegar. After the third washing, reheat the oil very slowly, and any remaining water and impurities will sink to the bottom.
13. Test the pH level of the water with litmus paper, and it should be pH 7. The transesterified and washed biodiesel will become clearer over time as remaining soap drops out of the mixture. It should look like clear vegetable oil with a light brown color, similar to filtered apple cider. There should not be any film, particles, or cloudiness.
14. If it is still not clear, let it settle another week.
15. Dispose of the wastewater from the process in your graywater system.

How do I use biodiesel?

Before using biodiesel in your diesel engine, slow down the injection time by 2–3°. Biodiesel has a higher cetane number (the measurement of how well it ignites when compressed), and the fuel will burn cooler. Biodiesel works as a solvent, cleaning out the dirt left behind by petroleum diesel. Get a new fuel filter and check it frequently at first because as the diesel cleans the engine it may clog the filter. Also, check any engine made before 1996 and replace any natural rubber parts in the fuel system or they will rot. Viton parts work best.

What about motor oil?

Of course an engine does not simply need fuel, it also needs lubrication from oil. Usually this has been made from synthetic and mineral oils, but it can also be made from vegetable oils. Unfortunately this development is fairly recent and the specific mixtures are patented. However you may be able to experiment with your own concoction. Heat canola seeds and crush them to release the oil. Then you must mix in small amounts of sunflower, soybean, and castor oils to get the right consistency. The trick with the oil is that unlike mineral oils, vegetable oils tends to become solid at freezing and don't work when it gets too hot. The small amounts of other kinds of oils help to give the canola oil the characteristics to combat its tendency to lose its effectiveness when in an engine (where heat and cold and pressure happen all the time). This kind of oil runs cleaner, is nontoxic, and in some ways more efficient than regular motor oil.

What can I use a generator for?

A regular fuel-powered generator is excellent and sometimes necessary to

▾ **Small generator**

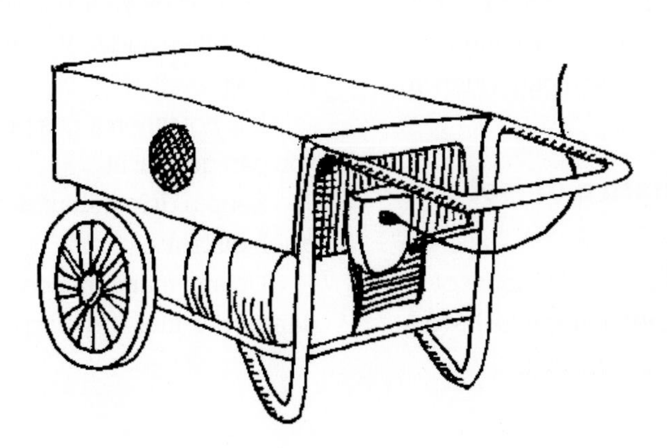

provide back-up power. For a homestead a generator would not be practical for supplying all the power, but for those using solar power or hooked to the regular grid, a generator is a simple and relatively inexpensive way to create emergency electricity.

What is the benefit of diesel instead of gas?

Diesel engines have no spark plugs or carburetors to replace, they burn less than half the fuel, and they outlast gasoline engines. Diesel fuel is safer to handle and cheaper to buy than gasoline—or free if you make biodiesel. The only maintenance required is changing the oil (in a scheduled and timely fashion), and changing the fuel, air and oil filters. However, you won't need tune-ups.

How do I choose a generator?

The bigger the machine the better it will be built. The slower the RPM's the longer the generator will last and the quieter it will be. A large generator will cost more but will outlast a smaller generator because it will run at half the RPM's. However don't get one that is too big for you because you need to be running it as close to full capacity as you can, when you do use it.

How do I maintain a diesel generator?

Change the oil every 100 hours of operation. Most diesel generators have a large fuel filter in the fuel line between the fuel tank and the engine, and another filter (usually a canister type) on the engine itself. These must be cleaned and changed regularly. Don't start your generator for a short period of time, because they don't operate well during rapid temperature changes. Start the generator before letting it carry a heavy energy load, and let it run a bit after the energy load is removed. Run it at a high load so that there will be enough combustion pressure the keep the piston rings firmly against the cylinder walls. Change the air cleaner often.

Rules of running the generator:

Only run it when you will use a large portion of its power, and coordinate your activities to run all at the same time.

Get an automatic shutdown kit for safety.

The enclosure or shed the generator is in should be properly ventilated with fresh cooling air. It may need to have a blower plugged directly into the generator so it will run at the same time.

Don't run it in very hot temperature because heat kills a generator.

Don't stick your fingers into things; keep clothes and tools away while it is running.

It is hot when it has been running and you can get burnt.

Keep a drip pan under the engine to catch oil leaks.

Do any maintenance outside to avoid dangerous fuel and oil spills.

3 Building Shelter for Man and Animal

TEMPORARY AND NOMADIC SHELTER

The log at the wood pile, the ax supported by it;
The sylvan hut, the vine over the doorway, the space cleared for a garden,
The irregular tapping of rain down on the leaves, after the storm is lulled,
. . .The sentiment of the huge timbers of old fashion'd houses and barns.

—*Walt Whitman*

How do I make a pit shelter?

Dig a pit that you could lay in, one arm deep. Put dry leaves or grass in the bottom to make a comfy bed. Lay branches over the top, covering the hole but leave a space to crawl in. Crawl in. If your feet get cold, cover them with dirt.

How do I make a tree shelter?

Find a forked tree, and clear the ground around it. Use a rock to dig a ditch one hand wide and one hand deep all around where you will sleep, to drain rainwater if it rains. Cut or find a long pole as thick as your arm and twice as tall as you. Put the pole leaning in the fork of the tree. Pile up leaves or pine needles to make a soft bed in the circle made by your ditch. Place the big pole so that it forms the main support of your roof. Lay thick sticks leaning on the pole to form a wall on either side. Weave thin sticks into the big sticks. Go in and close up the door with more sticks.

Why would I want to make a tipi?

A tipi is the most well-designed portable tent. It is big, self-ventilating, won't blow down, and you can have an indoor fire. If you are keeping the tipi in one place then it is good because the poles are difficult to transport, having been designed for prairie travel. The smallest tipi is 10 feet in diameter, but for families it is common to have a 15–20 foot tipi.

How do I build a 10-foot tipi?

1. Use 22 square yards of heavy canvas or any other tightly woven heavy fabric, 100 feet of 3/16-inch clothesline, and some string. The material may not be in one big piece, so sew it together to make a rectangle 20 feet long and 10 feet wide.
2. Lay the rectangle flat on the ground. In the very center of one long side, peg a string that is 10 feet long and attach a pencil to one end. Pull the string tight to the left corner and draw a half-circle to the top right corner. The widest part of the circle should reach the other long-side edge.
3. At the top center where you pegged the string, mark two triangles coming from that point with their tips pointing into the middle of the fabric. The top edge of the triangle is 6 inches, and the height into the fabric is 12 inches. 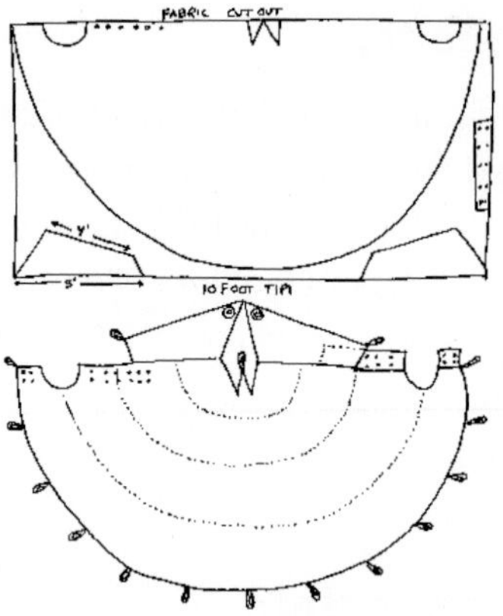
4. Cut out the half circle and the two triangles using a sharp knife or strong scissors. On the scrap fabric cut out two smoke-flaps, which are irregular-shaped rectangles, 5 feet on one side, 4 feet on the other, and 2 feet wide on one end and 1 foot on the other end.
5. Sew the smoke flaps on either side of the two triangle cutouts. Mark a 2-foot space 10 inches from what will be the bottom of the tipi to be the door.
6. Two inches away from the edge, on either side of the smoke flaps, make a double row of holes. The holes should be 1½ inches apart from each other, and each pair should be 5 inches from the other pairs. Don't put holes where the door will be. You can strengthen the holes by making grommets.
7. Cut a 4-foot rope and attach its center to the point between the two triangle cutouts. Cut another 10-foot rope and fasten it at two points at ⅓-foot intervals around the bottom, then sew it to the bottom all around.
8. Cut 12 pieces of rope, each one 15 inches long and fasten about 8 inches apart all around the bottom of the tipi to tie tent pegs or stakes.
9. Cut a hole for the door big enough for an adult to crawl through. To make a door, cut a green sapling ¾ inches thick and 5 ½ feet long, and another one 22 inches long. Bend the long one into a U shape, and fasten the short one across the ends to form a D. Lay a piece of canvas across it and draw a D shape to go over it, leaving a flap at the top.
10. Cut out the D-shaped canvas with the flap, and sew it around the edge of the wood, encasing the wood except where the flap is. Hem the flap and cut two holes in it to attach a lacing pin.
11. Cut 9 lacing pins, which should be smooth, round, straight hardwood, each 1 foot long and ¼ inch thick. Use them to push through the holes in the tipi to overlap and hold it together.
12. You will need 12 poles, smooth and straight, 13–14 feet long, and 1 inch thick at the top. Two poles are for the

smoke vent and may be skinnier than the others. You will also need 12 pegs 15 inches long and 1½ inches thick.

How do I put up a tipi?

1. Tie three poles together 1 foot higher than the canvas (11 feet up) and then spread them out into a tripod smaller than the circle made by the tipi.
2. Lay the other poles in a circle around the tripod, leaving the two smoke flap poles and one extra pole out. Bind them all together with a rope, letting the end hang down inside for an anchor.
3. Fasten the two ropes, at the center of the tipi between the smoke flaps, to the thicker pole that you left out. Tie it 10 feet up, then raise the pole to its place on the tripod, bringing the tipi cover with it, opposite where the door will be.
4. Pull the bottom corners of the tipi cover around until they meet and fasten the sides together with the lacing pins through the double holes.
5. Pull all the bottoms of the poles out to spread them out, making the tipi cover stretch tight over them. Put the end of each smoke flap pole into the holes in the smoke flaps. Brace the ends of the poles into the ground on the outside of the tipi.
6. Hang the door up and put a stake inside the tipi for the anchor rope, then put in all the stakes for the outside of the tipi.
7. Make a fire pit inside the tipi by digging a hole 18 inches wide and 6 inches deep. By using the door and the smoke flaps you can control the flow of smoke. Keep the smoke vent turned downwind. You can even raise the bottom of the tipi in warm weather to bring in air.

What is a yurt?

Ancient Mongolian nomadic herdsmen designed the traditional yurt or gher. The basic shape of a yurt is a circular lattice frame, several roof support poles, and a

◂ **Interior of a small yurt**

▲ Yurt on top of a wooden platform

protective outer wall cover, traditionally made of felt. Modern yurts have made great strides in longevity and warmth by substituting vinyl, canvas, and insulating layers instead of chilly wool.

How do I get a yurt?

There are now many different companies that make pre-built yurts and ship the pieces to you, and which offer many different options. You can spend as little as $2,000 to upwards of $20,000 on a good yurt that is suited to your climate. There are yurts for snowy weather and yurts for tropical weather. Setting these pre-made yurts is fairly simple, especially if you have a helper handy. For permanent shelter, most people build some kind of foundation, either using a concrete slab or a simple elevated wooden platform. Most people also heat their yurt with wood, and some pipe in running water and solar electricity as well. You can build your own yurt but, like a tipi, it requires some complicated sewing and it may be worth it to just buy it.

How do I care for my yurt?

A really good yurt that is properly cared for may last 10–15 years. Most yurts deteriorate because of wet conditions and mold. Keep the yurt well sealed and use a good waterproofing layer all over it.

How do I build my own yurt?

1. First you have to make the lattice wall. For a 10-foot diameter yurt you will need 65 rods 1 inch in diameter and 5 feet long. These are traditionally willow but hazel is better. If you cut it yourself, do it between October and March. Green oak is the best and may be purchased from sawmills.
2. Remove any bark. Don't let the wood dry before building because they will pull straight as they dry.
3. Drill seven 3.5-millimeter holes 9 inches apart down each rod, leaving

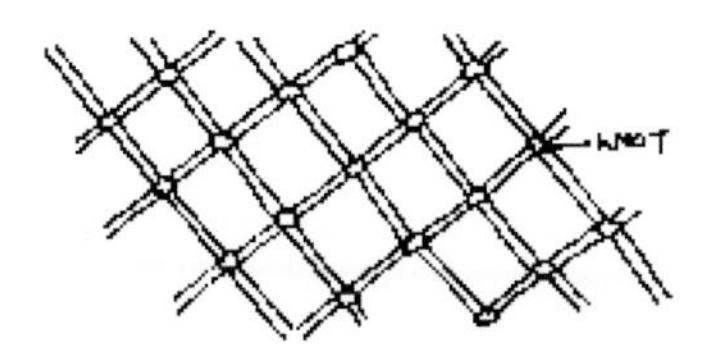

▲ **Yurt on top of a wooden platform**

2 inches at one end and 4 inches at the other without holes. Make sure that these holes are accurate and exact.

4. Take 24 rods and tie them together in a crisscross fashion with strong nylon string to form a lattice. To do this you tie a knot at one end, and slightly melt the other end to make it pointy (for easy threading). Push it through the holes you drilled in two rods and pull it very, very tight. Then knot it.
5. As you reach the end of the lattice, some of the rods will have to be shorter in order to form a straight edge. You can cut them.
6. Take 24 more rods and make another section of wall.
7. The door can be a simple frame of wood, and is attached between the two lattice frames.
8. The hardest part is the roof wheel. This is the center tip of the yurt roof. To make the template, first draw a circle 2 feet 6 inches in diameter. Inside the circle draw another circle 2 feet in diameter. Divide these two circles into quarters. Trace one of these quarters onto cardboard and cut it out.

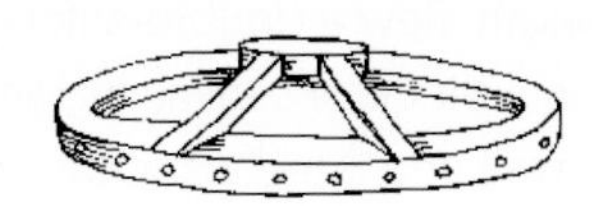

▲ **Roof wheel**

9. You will be making a wheel with 8 sections. Place the template on the timber and cut out 8 pieces.
10. Glue and clamp together the sections. Once they are dry you can drill a screw into each section from the one below for extra strength.
11. Sand the outer and inner rims of the wheel.
12. Mark the outer rim of the crown with 32 equally spaced points just below the middle of its width. Drill into each mark to a depth of 1 inch, angled upward 25°.
13. Cut out eight spokes of 1 ½-inch-square timber 13 inches long with one end cut at a 25° angle and the other end at a 115° angle.
14. Make an octagon of wood 2 inches thick and 5 inches across. Screw the octagon, rim, and spokes together to form a wheel with a raised center. It is a good idea to varnish this.
15. To make the roof supports that attach to the roof wheel, you will need 32 roof poles each 5 feet long. These will rest on the V-sections formed along top of your lattice wall. One end should be shaped into a round peg 1 inch wide and 3 inches long, to fit into the holes around the rim of the roof wheel. The other end should be shaped into a 45° V-shape to fit into the top of the lattice. At the V-shaped end a loop of string should be attached to slip over the tops of the poles.
16. Assemble the lattice and roof frame on your foundation with the door. Get some waterproof canvas (and flame retardant as well if you want) and

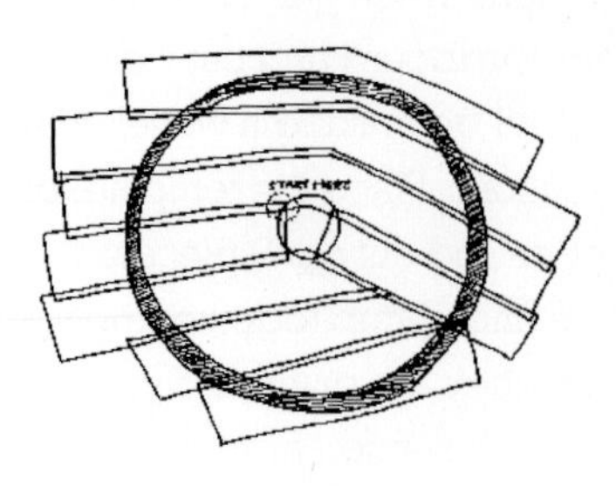

▲ **Roof sewing pattern**

polyester thread. Make sure to always sew a double seam.

17. Measure the height and circumference of the walls, excluding the door. Cut out a wall cover the same height and width but make it 3 feet longer. This will probably take two widths of canvas. The bottom of the yurt should overlap on the outside about 1 inch to encourage rain runoff, and there should be a 1 ½-inch hem on both ends.
18. You can also sew 6–8 inches of plastic all around the bottom edge of the yurt to repel water.
19. At the ends of the wall panel fit brass grommets and put a 3-foot string loop through each, to fit over the top of the crossovers of the wall lattice. This is to hold it in place.
20. The roof cover is sewn with eight pieces of canvas. You want to start from the center and work your way out, because each seam must overlap from the top going down, in order to keep out water.
21. First lay the piece of canvas over the top of your frame, and pin it to the next one, making sure to overlap the top seam over the bottom piece. Sew these two together, lay them on the frame, and add another piece. Keep doing this until the pieces cover the whole roof.
22. To cover the crown, make a three- or four-pointed canvas star with a center that will more than cover the crown. Make a couple of tucks to create a more concave shape in the center. Each point of the star should have a grommet with a string for tying it down.
23. The tension band is a piece that goes around the edge of your lattice wall and the ends of the roof poles to hold them in place and add a bit of strength. Sew a double-stitched hem on a piece of canvas as long as the wall and 8 inches wide. Sew a 1 ½-inch hem at each end. Put in two brass grommets at each end, one in each corner. At each end tie

▼ **Yurt outer waterproofing fabric**

▲ Traditional gher

a 10-inch piece of rope between the two grommets to form a loop. Tie a 3-foot rope to each loop.

Setting up the yurt:

1. Set up the two wall lattices in an approximate circle.
2. Tie the ends of the lattices together, and the other ends to each side of the doorframe.
3. Adjust the walls until the house is perfectly circular. If you use two poles end to end you can make sure that the house is the same diameter all the way around.
4. Tie a strong rope around the top of the walls. Don't leave this step out! Skipping this step can wreck your yurt.
5. Get your buddy to hold the crown (roof wheel) over their head in the center of the yurt.

▲ Permanent yurt

6. Fit the 32 roof poles into the holes of the crown rim, and slip the loops onto the tops of the lattice wall.
7. Tie the tension band tightly around the entire yurt at the point where the wall and roof poles meet.
8. Have your buddy pull on the crown hard to make sure it is secure.
9. Fit on the wall cover.
10. Fit on the roof cover and tie it down by passing a rope through the grommets.
11. Tuck the plastic skirt under the bottom of the wall poles
12. Fit on the door.

CONSTRUCTION SKILLS AND MAKING LUMBER

Give me six hours to chop down a tree and I will spend the first four sharpening the ax.

~Abraham Lincoln

What are common road problems?

The main problem with roads is water, obstacles, and ruts. Snow, ice, sleet, and rain degrade your road by removing dirt through run-off, creating puddles, and turning into mud. These can be avoided through adequate drainage.

How do I clear the ground before making the road?

There is a special plow that you can use to clear the ground and get it ready for you to create your road. Plow the area, remove the stumps and other obstacles, and then make the road.

How do I make a dirt road?

1. Drain water from the road by making one side of the road taller than the other. The slope should fall ¼ inch per foot of road (for a 20-foot road that would make the high end 5 inches higher than the low end).

2. If water cannot just run off on its own, you will need a ditch for the water to drain into. The bank of the ditch should slope down at least 3 inches per foot from the edge of the road, and the ditch itself should be at least 1 foot wide and 1 foot deep. If you have a lot of water, make it bigger.
3. If you need to direct the water across the road to send it downhill, you will need to make a grade dip. This is a shallow ditch running across the road that is lined with rock. On the downgrade side (the lower side), the dirt from the ditch is piled into a speed bump which blocks water from running down the road and directs the flow through the ditch. The grade dip should be 70 feet wide including the ditch and the mound. The ditch should be 50 feet wide and the mound 20 feet and have a rise of 45°. If the uphill grade of the road (or the steepness) is more than 5 percent, the grade dip should be widened 5 feet per 1 percent of grade rise (a grade of 8 percent should have a grade dip 65 feet wide).
4. For a low area where puddles or snow piles form, dig a turnout (or shallow notch) into the outside slope so that water can flow out. Line it with rock so the road won't erode.

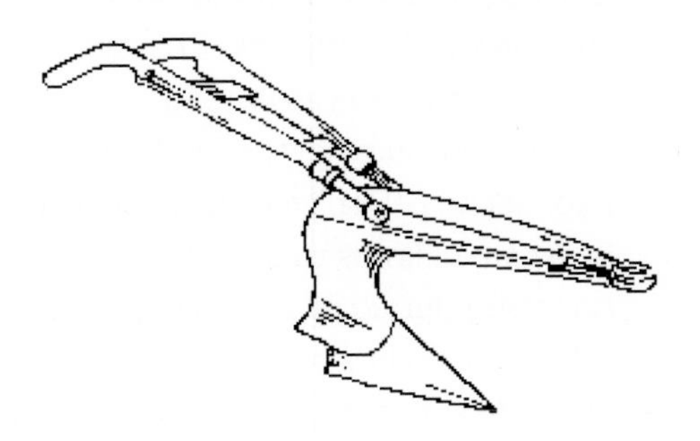

▲ **Road-making plow**

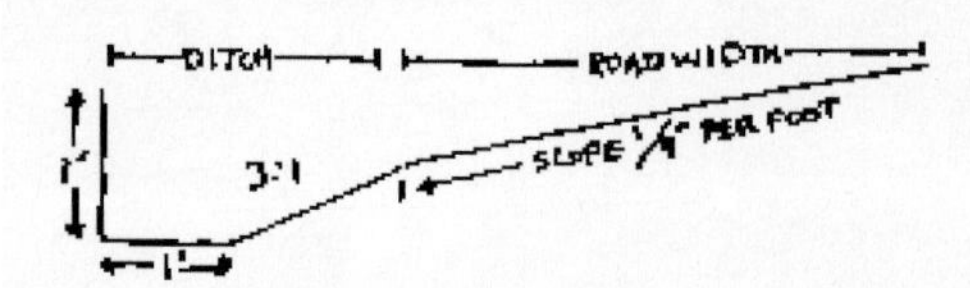

▲ Ditch and angle of slope

How do I fill in ruts?

You can fill in ruts when the dirt is just moist enough to make a crumbly ball (too dry or too wet and it won't work right). First you will need a big log or beam. Attach a chain to a vehicle or horses making one side longer than the other so that the log or beam will drag at a slight angle. The long end should be on the side of the road that slopes downward. Bolt the chain to the log or beam so that the chain does not slip and drag it down the road, which will fill in holes. Use rocks to fill in stubborn holes.

Chainsaw safety rules:

1. Don't use a broken saw or one with problems.
2. Keep the chain sharpened and maintain proper chain tension.
3. Refuel the saw outside after it has cooled, at least ten feet away from anything.
4. Keep fuel away from anything with a flame.
5. Use funnels and spouts so you don't spill any fuel.
6. Don't saw on a tilted or unstable surface.
7. Don't ***EVER*** start the saw on your knee.

▼ Grade

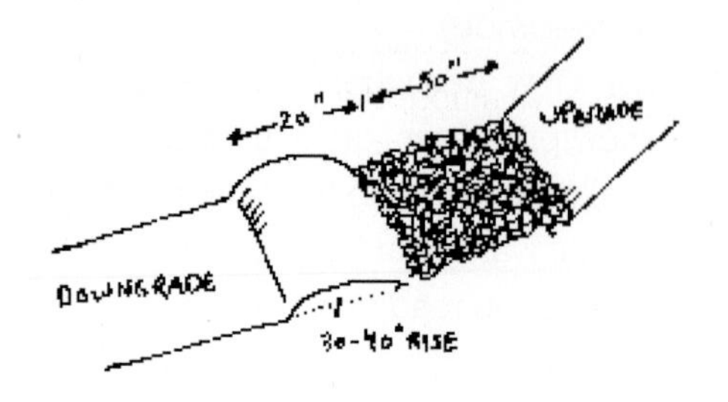

How do I adjust the chain tension?

In the first thirty minutes of cutting, make sure that the lower chain span just touches the bottom bar rails. If it isn't, raise the bar tip while tightening the bar fasteners.

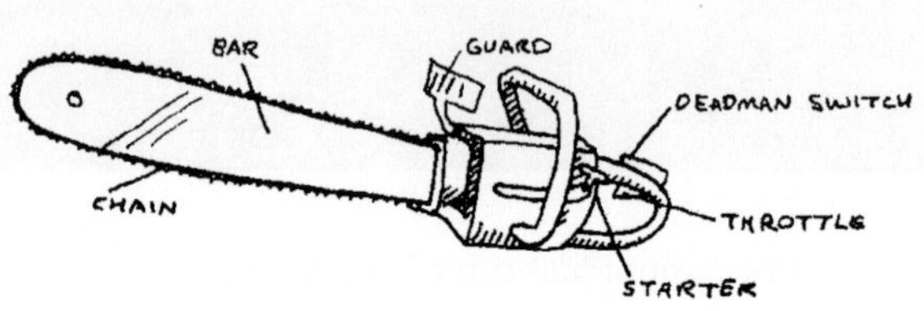

Chainsaw power guide for log sizes:

Log up to 18 inches: 55cc–67cc
18–36 inches: 68cc – 85cc
36 inches or larger: 86cc–120cc

Safety rules for cutting trees:

1. Always follow ax and chainsaw safety rules.
2. Don't be overconfident with your cut—always provide lots of escape routes in case you are wrong.
3. If you're not sure how to cut down the tree, get help. Don't try too much alone.
4. Be aware of limbs whipping and bouncing during all stages of cutting.
5. When cutting limbs, stand on the opposite side of the trunk from the limb being cut.

6. Don't hold a running saw with one hand. Shut off the saw if you need to do something.
7. Have good footing at all times and keep well balanced so you can react quicker.
8. Logs roll downhill so if possible work on the uphill side.

What should I do before I cut down a tree?

1. Note how tall the tree is, which direction the wind is blowing, and which way the tree is leaning. If the tree is leaning, cut it when the wind is not blowing against the lean and plan to chop it using the lean to direct its fall.
2. Remove loose dead limbs or cut it in such a way that the there is no way the limb could hit you if it flew off. Remove low limbs, underbrush, and anything that could stop you from running away in any direction.
3. Be sure that the tree cannot become lodged in anything as it falls. It ***must*** hit the ground, because hitting other trees can cause limbs and branches to hit you and kill you.
4. Plan several escape routes so that if the tree falls in an unexpected direction you will have another way to go.

What problems are there to consider?

A tree with imbedded material (such as barbed-wire) needs to be cut above where the material is. Don't try to cut through it because it can injure you and hurt your saw. A dead tree is very dangerous to cut down because it will break and will not fall as predicted.

How do I cut down a tree with a chainsaw?

1. For a tree 8 inches in diameter or less, just make one cut. To make a cut, make a notch on the side facing the direction you want the tree to fall (the fall side) by making a horizontal cut and then making another cut above it angling into the tree toward the bottom cut.
2. For a tree 8 inches or more, make a notch (called an undercut) at least ⅓ of the trunk's diameter on the fall side of the tree. Make the lower cut of the notch first or the chain will get pinched or bent.
3. Sometimes if you want to control the tree's fall it is possible to use two heavy logchains with hooks and a come-along (cable hoist puller). In the direction you want the tree to fall pick a large tree as an anchor, with a distance that is less than the length of the logchains. Wrap one logchain around the anchor tree a couple of times and use the hook to secure it. Wrap the second chain around the tree you are cutting, at least 1 foot above where you will make the back cut (step 4), and secure it with the other hook. Make each chain end into a loop with the other two hooks, and attach the loops to the come along. Ratchet the

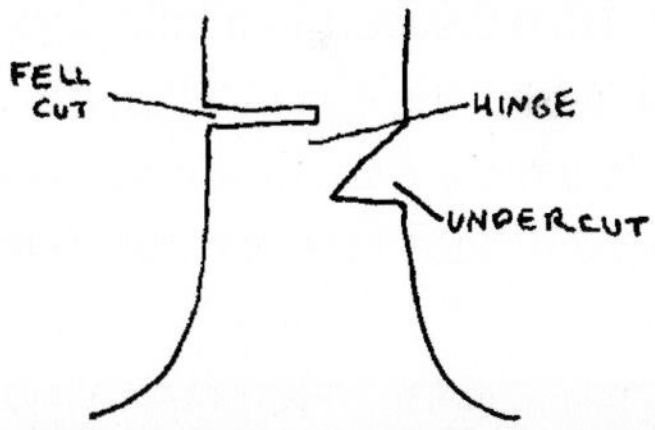

▲ **Undercutting a tree**

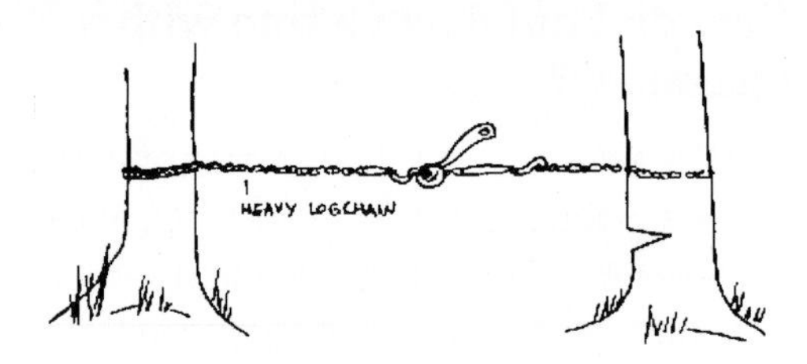

▲ **Controlling a fall**

come-along until it is so tight that the tree bends slightly.

4. Make another cut on the other side of the tree 2 inches above and parallel to the horizontal cut in the notch (this is called a felling or back cut). The tree will start to fall when you are several inches from the inner face of the notch. Leave a narrow part connected to act as a hinge for the tree.
5. If the saw starts to get stuck (or binds), you've misjudged the cut. Remove the saw quickly if you can, and if you can't, shut off the engine. Remove the bar and chain, and drive a flat wedge into the lateral cut. Wiggle the bar out, drive the wedge a little deeper, and get out the chain. If that doesn't work, a more dangerous way is to use logchains the same way as in step 3, which may relieve enough pressure to get them out. Another way is to get out of the way and wait for the tree to fall on its own (usually in 24 hours it will).
6. The leading cause of death when felling timber is if the butt of the tree hits you, if a limb bounces back, or if the top breaks off. As soon as the tree starts to fall, run away fast. If it doesn't fall, push it with a pole.

How do I limb a tree?

Start at the base of the tree. Limbs go all around the tree, so start by cutting off the ones on the top all the way down the tree to its tip. Then cut the ones resting on the ground. As you remove the limbs on the ground the tree will start to sag and roll. You do ***not*** want a tree to roll on you so be ready to jump out of the way quickly. Chainsaws will quickly become dull if you cut leaves, dirt, or water, so be cautious.

How do I buck a log?

Bucking is when you cut the log into the lengths you need to build with. If the log is flat on the ground, cut it from the top, then roll it over and cut it through from the opposite side. If the log is resting on

▾ Broadax

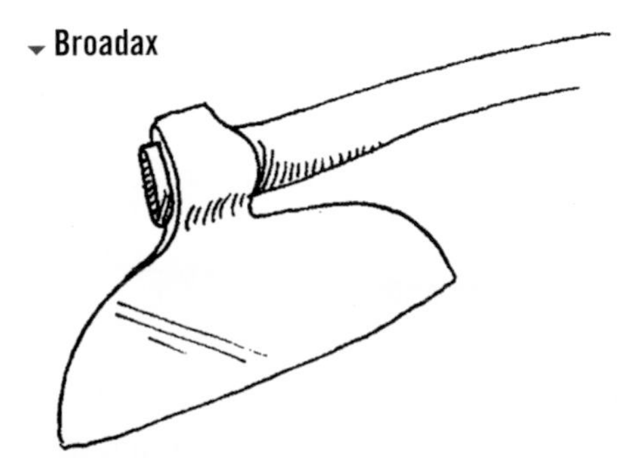

one end, cut ⅓ through the log from the underside to avoid splintering, then cut the rest from the top. If it is supported at both ends, cut ⅓ through the top, and then cut the rest from the bottom.

How do I use a broadax?

Make the blade go straight down, with as much of the blade as possible hitting the wood at once. If you don't, when you hew timber it will be curved in, and not straight.

▾ Poll ax

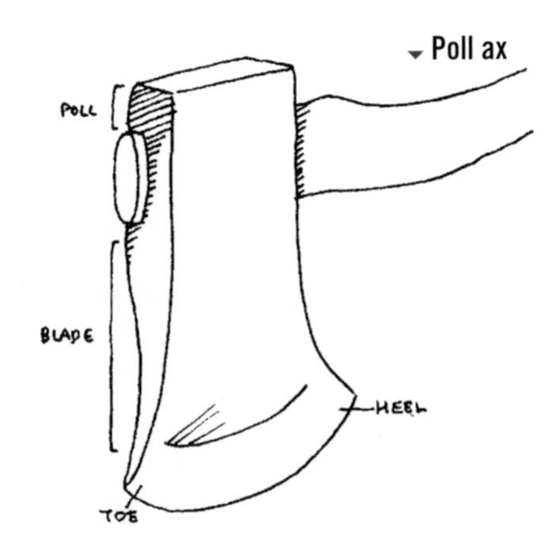

How do I use a crosscut saw?

Put your thumb on the grip, and your index finger pointing ahead to steady the

▾ Crosscut saw

saw. Let the saw do the work, and pretend you are just the operator. This will help you keep up a steady momentum.

How do I make timber with a broadax?

1. Making timber is hard work, but not too difficult to figure out. Use three axes to make timber: a cutting-down ax, a scoring ax, and a broadax.
2. Cut down the tree with the cutting-down ax. Cut off the limbs on one side first. Leave the rest, because they'll hold the log steady.
3. Take the scoring ax and chop notches down the log, about 10 inches apart.
4. Take the broadax and chop each section of round log off, so all that is left is a flat wall. You have to cut straight down, with as much of the blade hitting at a time, to make your cut straight. This is called hewing; it makes a round log into a square beam.
5. Hew the other side, and then cut off the limbs so you can turn the tree over.
6. Hew the rest, and you have a piece of timber.

How do I make a plank with a beetle?

(A beetle is a wood tool used to hammer a wedge or glut.)

1. Cut a log in half and prop it upright with stakes hammered into the ground.
2. Put a glut where you want your plank thickness, and hammer it with a beetle.
3. Keep adding the wedges (the gluts), until the board splits right off the log.

How do I air-dry lumber?

1. Make a foundation that will elevate the lumber 1–2 feet off the ground. Use square timber already dried, railroad ties, or some other sturdy material. The foundation should be 3–5 feet wide and as long as your boards.
2. Make stickers or spacer boards that are ¾-inch think and 1–2 inches wide, or up to 4 inches for softwood. Stickers can be narrow boards you already

◂ A home sawmill

▾ **Air-drying wood**

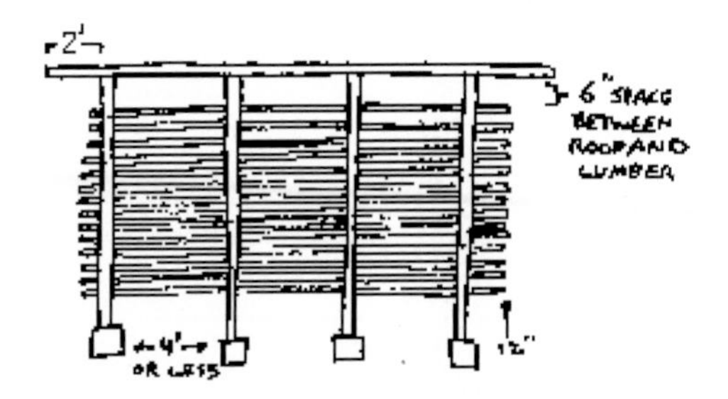

have, as long as all the stickers are the same thickness. If they aren't the same the boards can warp.

3. Put stickers over foundation timbers and stack boards alternately. Boards should be 1 inch apart, perpendicular to the stickers. The stickers should be completely vertically stacked in a straight line all the way down the pile. This is important for straight wood.
4. Cover the top with old plywood, boards, slabs, corrugated metal, or anything that will protect from both water and direct sunlight. Boards used for rough applications can have 15–20 percent moisture. Furniture needs 7–9 percent moisture, which can only really be gotten with a solar kiln.
5. Once it is dry, store the lumber where it will be used. Outdoor lumber can be put under a tarp outside and indoor lumber needs to be brought indoors.

What is shrinkage and warping?

Radial shrinkage is when the board shrinks on its height (a skinny board gets even skinnier). Tangential shrinkage is when the board shrinks along its width, and longitudinal shrinkage is when it shrinks along its length. Shrinkage is all right as long as it's even all over the board. Uneven shrinkage causes warping.

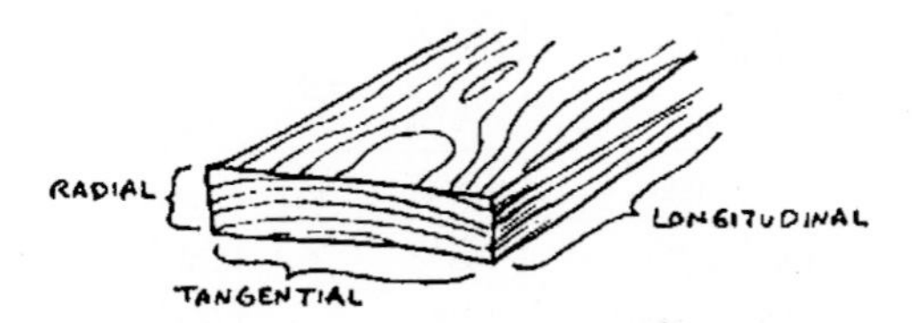

▴ **Board shrinkage**

▾ **Warping**

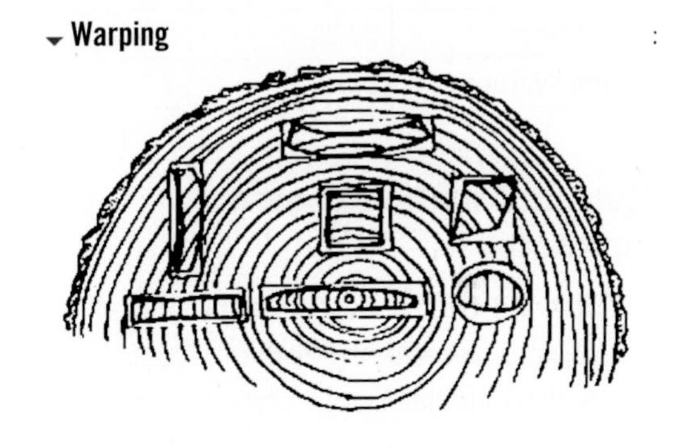

Drying characteristics of softwood:

Name	Radial Shrinkage	Tangential Shrinkage	Tendency to Warp	Tendency to Split	Time to Dry
E. White Pine	2.1%	6.1%	Low	Medium	60–165 days
Jack Pine	3.7%	6.6%	Medium	Medium	40–165 days
Ponderosa Pine	3.9%	6.2%	Low	Medium	15–120 days
Blue Spruce	3.8%	7.1%	Low	Low	20—120 days
Eastern Red Cedar	3.1%	4.7%	Low	Low	20-120 days
Douglas Fir	4.8%	7.5%	Medium	Low	20–120 days

Drying characteristics of hardwood:

Name	Radial Shrinkage	Tangential Shrinkage	Tendency to Warp	Tendency to Split	Time to Dry
Green Ash	4.6%	7.1%	Medium	Medium	60–165 days
Linden Basswood	6.6%	9.3%	Medium	Low	40–120 days
Paper Birch	6.3%	8.6%	Medium	Low	40–120 days
Butternut	3.4%	6.4%	Low	Medium	60–165 days
Black Cherry	3.7%	7.1%	Low	Low	60–165 days
Eastern Cottonwood	3.9%	9.2%	High	Low	50–120 days
American Elm	4.2%	9.5%	High	Low	50–120 days
Hackberry	4.8%	8.9%	Medium	Medium	30–120 days
Hickory	7.4%	11.4%	Medium	Medium	60–165 days
Sliver Maple	4.8%	7.2%	Medium	Low	30–120 days
Sugar Maple	4.8%	9.0%	Medium	Medium	50–165 days
Northern Red Oak	4.0%	8.6%	Medium	Low	60–165 days
White Oak	5.6%	10.5%	Medium	High	70–200 days
Pecan	4.9%	8.9%	Medium	Medium	30–120 days
Sycamore	5.0%	8.4%	High	High	30–120 days
Black Walnut	5.5%	7.8%	Low	Medium	70–165 days

How do I make my own concrete?

1. You will need to calculate the ratio of sand to cement. For the projects outlined in this book, a good ratio is 4 parts sand to 1 part cement. Another mixture is 3 parts sand, 2 parts gravel, 1 part cement, and 2 parts hydrated lime.
2. Mix the cement on a hard surface or in a wheelbarrow, ideally covered with a tarp. Measure out the sand and cement and mix it together well. Then form it into a mound with a crater, like a volcano. Don't breathe in the Portland cement.
3. Fill a bucket with water and add some dish soap. Pour a small amount of

water into the crater, and then starting from the outside, turn the mix into the middle so the water is soaked up.

4. Mix and repeat until the mix feels like cement. Drier is better so you can add more water later. The mix should last 30 minutes to 1 hour depending on the temperature. After three days the cement should set, and the slower the better. The longer it takes the stronger it will be, so keep the temperature around it down by covering it with plastic or canvas.

How is cement made?

Limestone or clay is heated in a kiln at 2,500° F. This forms large, glassy cinders called *clinkers*. The clinkers are ground into a powder, and when the powder is mixed with water a chemical reaction takes place, which turns it into a super-strong artificial stone that gets harder in contact with water.

BUILDING A HOUSE FROM AVAILABLE MATERIALS

When one has finished building one's house, one suddenly realizes that in the process one has learned something that one really needed to know in the worst way—before one began.

~Friedrich Nietzsche

What kinds of insulation are there?

Log homes and underground houses have a lot of natural insulation, but using a further insulating lining inside your house will double the efficiency of your heating source. Fiberglass is super cheap, and animals won't eat it, but it is not something you can make yourself. Straw, cardboard, and sawdust are cheap, well-insulating and easy to get, and although rumors warn you will get bugs and your house will burn down, that is not necessarily true—it is about the same as any other kind of insulation. The alternative is compacted clay, which is one of the best insulators, clean, and cheap.

Insulation types:

Sawdust: must use a vapor barrier and it must be bagged and sealed in plastic.

Feathers: enclose in mesh bags.

Seagrass: good if there is low fire risk, must be dried and partially compacted.

Straw: good if low fire risk, use dried and partially compacted or in straw bales.

Paper: soak shredded paper in 1 part borax to 1 part water.

Trees: trees planted around the house provide shade to keep house cool.

▲ **Cabin chinked with clay**

Clay: compacted clay is the best insulation possible, with none of the problems.

How much attic insulation do I need?

Attic insulation should be 10 inches thick. Check for moisture leakage, find the source, and seal it before installing insulation or it will just help everything rot. For every 150 square feet of attic space without a vapor barrier, or every 300 square feet with a vapor barrier, you will need 1 square foot vent to help keep moisture down.

How do I seal windows and doors?

All window frames should be caulked or weather-stripped to keep out moisture and cold air, and to improve the efficiency of your home. During the winter use either plastic taped around the windows, or put up wooden storm windows inside or outside for further insulation. Door frames can also be caulked or sealed, and if the doors are drafty, either build an outside door, storm room, or put a sweep on the door (a vinyl piece that sticks to the bottom of the door).

How is the foundation weatherproofed?

The place between the foundation and the sill (the bottom beam of the house) will get gaps over time, especially if you have a cabin. This should be sealed every year with mortar. You can also skirt the house with plastic if you want.

How do I make a sod roof?

After you have laid wood on the entire roof, covering it, put a wooden log or metal pole all along the edge of the roof where the water will be running off. Staple or nail down a plastic moisture barrier, all the way to the edge (even covering the pole). Then lay sod down on top, planting grass or some other ground cover.

Thatching materials:

Water reed: the most durable material with the longest life.

▲ **Alaskan trapper cabin with a sod roof**

Water reed thatch

Longstraw: straw from old-fashioned long-stemmed wheat varieties that are cut when slightly green. Then it is dried and threshed, and made into *yealms* (a tight layer of straw which is level at both ends), a process called *yealming*. It takes a long time, and is done on the ground. Because birds attack it on the roof, netting is put over it.

How do I thatch a roof?

Contrary to popular belief, thatching does not absorb much water—it runs clear down the stems. The eaves of the building should be wide in order for the run-off to fall clear of the walls. There are two types of ridges (the peak of the roof), a wrap-over ridge and a butt-up ridge. A wrap-over has a thick layer of straw that is folded over the edge of the roof peak and is fastened on both sides. A butt-up ridge forces the straw together from both sides of the roof to form a peak.

1. Bind the straw to the rafters.
2. When binding, keep the straw straight. One way to do this is to put wooden stakes up as you go.

Straw thatch

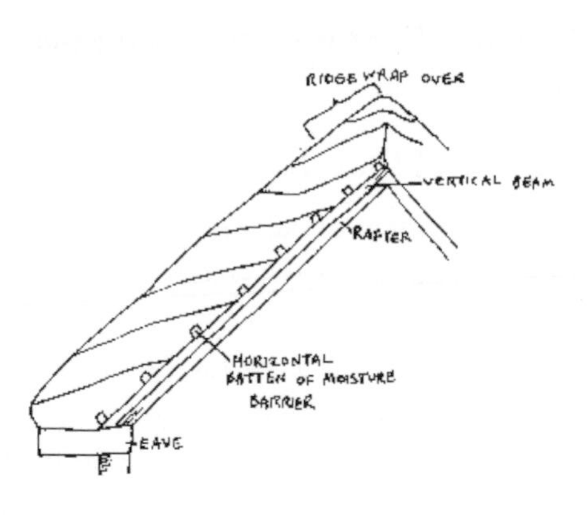

▲ **Cross-section of a thatched roof**

3. Bind and trim the roofline. At the top of the roofline the tops of the grasses will be sticking up and need to be trimmed off.
4. Trim the eaves and edges.

How do I make a shingle?

Set the blade of a shingle frow on a block of cedar, and drive it into a rectangle

▲ **Binding the straw to the roof**

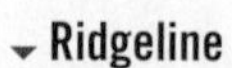

▼ **Ridgeline**

▼ **Eaves**

piece of wood with a maul. This makes one shingle.

How do I find cheap windows?

If you are building a log house and had many long logs so that you didn't have to make windows a certain size, then one cheap way is to ask your local building supply where reject windows go. Reject windows are ones that people bought and then didn't use—not necessarily warped or bad windows.

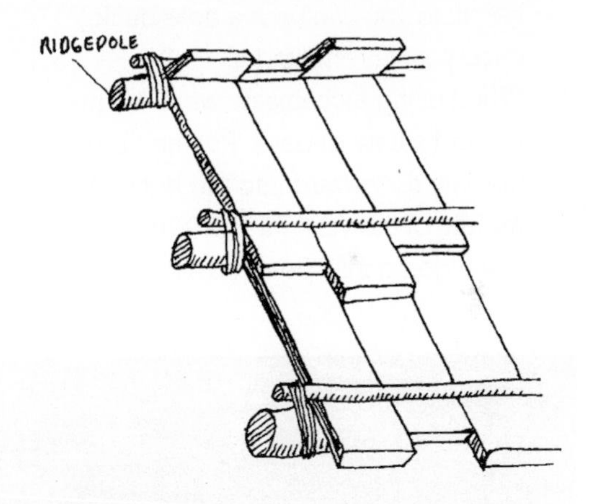

▲ Easy plank roof

▼ Shingle frow

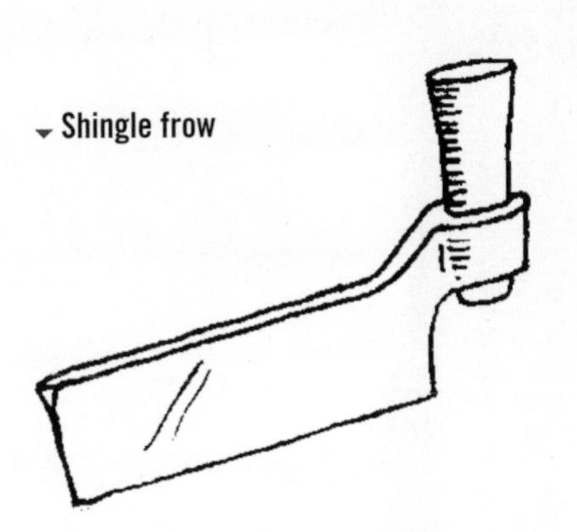

How do I cut an opening for the window in a log home?

1. Measure the first window very carefully, including space for the frame. With chalk make a mark at the first top corner on the wall that you want your window to be. With a level mark the second top corner, then make the vertical line and measure the other two bottom corners. This is important because you can't put a hole back once you've removed it.
2. When using a chainsaw wear a hard hat and safety glasses. Put the tip of the saw downward into the log in the middle of the vertical line. With the chain going full speed cut downward to the corner, then upward to the top corner.
3. Do the same to the other vertical line. If this is a log house, have someone there to catch the logs as they fall. In a log home, along the top of the window frame cut several vertical cuts and use an ax to chip out the chunks. Otherwise just cut a horizontal line.
4. On the bottom edge of the window frame there should be a slight slant down and outward to allow water to drain off. In a log home the log must be shaped, and the cuts done in the same way as the top of the frame.
5. Measure the window again and trim the hole if necessary. Then put in the window.

What kinds of woods are used for carpentry projects?

Ash, black: baskets, cart axles

Ash, white: rake, pitchfork, ax handles, levers

Basswood: paintbrushes, thread

Beech: planes, levers

Black birch: lumber

Chestnut: fences

Elm: harrow

Hickory: bends easy, rakes

Ironwood: beetle, glut

Maple: floors

Pine: floors

White oak: awls, buckets, doors, door frames, anything strong

White birch: sugaring basins, brooms

White pine: dugout canoe, won't rot when wet

Yellow birch: lumber

Yellow cedar: baskets, cloaks

What is bamboo?

Bamboo is a type of grass that is very big. It looks more like a tree and grows very fast, making it especially useful for the homesteader.

▲ Bamboo house

How do I grow bamboo?

Plant bamboo in late spring, in warm soil, well after the last frost. The soil must be well fertilized and kept very moist the first year (but not so much later on). You can either transplant whole baby plants or propagate them by planting bamboo stems in the ground. The giant variety, Moso, must be planted 10 feet apart, and smaller ones in clumps 6 feet apart. It can't

▲ **Dovetail joint**

survive below-zero temperatures. Because bamboo is a grass it can really take over and the more you cut it the more it will grow. Either keep cutting it or get some goats.

What can I use bamboo for?

Water pipes, fences, sheds, handles, utensils, sailboats, weapons, fishing poles, furniture, screens, windows, wall coverings, baskets, mats, screens, birdcages, etc. Many houses around the world are made of bamboo.

What is a dovetail joint?

A dovetail joint is the strongest joint to make, time-consuming, and used primarily for furniture and cabinets. The finished joint should look like a there are two notches cut in each side of the end of the wood, making the end "flare".

How do I make a dovetail joint?

1. On the part of the wood to be connected to another, measure a half triangle cut on each side of the wood, 1 inch long and $\frac{1}{8}$ inch wide on the biggest end. The point of the triangle should start on the end of the wood. This should leave you with a flared projection on the end of the wood.
2. Cut the triangles out, and then trace the angles on the wood you are connecting it to. This should make a flared shape on the wood.
3. Cut out the flared shape, and fit the other piece of wood into it. Drill a hole through the middle of the shape, and fit the other to it.

What kinds of doors can I make?

The simplest door is half-log siding, or tongue and groove, 2x6 lumber strengthened on top and bottom with a horizontal 1x4 and a brace with an X of

two 1x4s. In a cabin the frame needs to be done the same as the window frames, so that it can settle with the house. You can make hinges by using a strip of strong leather and fastening it to the door and door frame with long strips of wood.

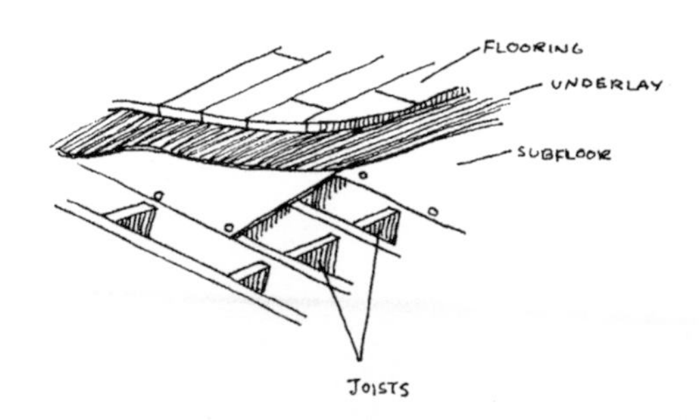

How do I make a smooth wood finish?

Finish wood by rubbing with a piece of glass, and if you want it very fine, polish with pumice. Then oil it with fresh or softened linseed oil (oil from flax seeds). Pumice is a volcanic rock that has many air pockets in it, and is often used like sandpaper. If you don't have pumice, find some other rock such as sandstone.

How do I build the support for the floor?

1. Bridging is a framework of boards that connect the floor joists together. A replacement for bridging is concrete. If you use bridging, lay down sheets of plywood. This makes up your subfloor.
2. On top of the subfloor is underlay, or a layer of plywood or hardboard sheets.

Hand plane

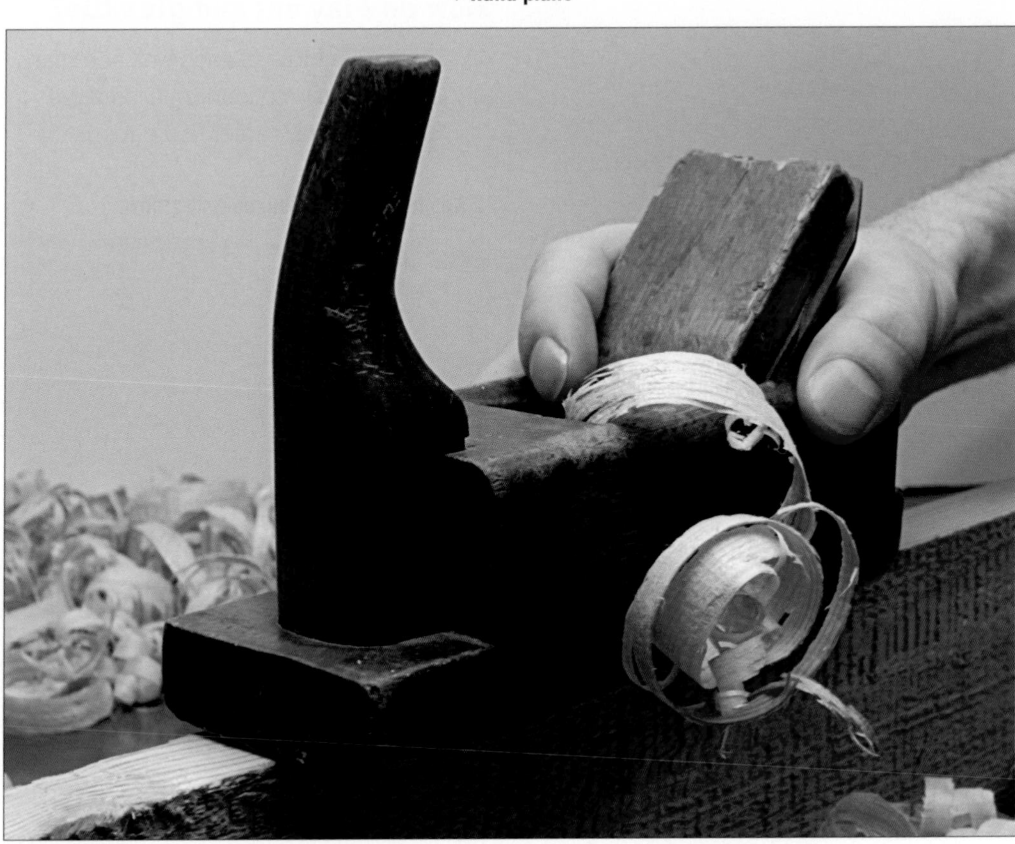

▲ Tongue and groove

The underlay should be staggered so that they edges do not fit perfectly together, and there should be a ⅛-inch crack between each sheet for expansion. The nails should be spaced 6 inches apart.

3. On top of the underlay are the tongue and groove floorboards. Hardwood is best because it wears longer.

How do I make a tongue and groove joint?

1. Cut a tongue (a thin, rectangular piece of wood) the length of your board.
2. Using a plowplane the same width as the tongue, cut a groove down the center of each thin side of the board, so that the tongue will fit halfway in on its side.
3. Fit the tongue into the groove on the right side of the board.
4. When assembling your wall or floor, pound the boards gently together with a hammer.

How do I lay a wood floor?

Plank flooring is one of the easiest types of flooring to make yourself. Basically you make lumber between 3–9 inches wide, and they don't all have to be the same. Then you make tongue and groove joints and fit each board together. One side of the board has the tongue, and the other the groove so each piece interlocks.

How do I stain a floor?

Your floor will last much longer if you finish the floor with varnish or shellac. If you want to stain the wood, it has to be done on raw wood or over other stain, so do that before finishing the wood.

How do I lay out and glue tile?

1. Find the middle of each wall and mark off the room into four equal squares. Starting at the center of the room, lay

▼ Wooden floor with tongue-and-groove

▲ **Applying tile glue to the subfloor**

the tile down in one of the four squares. Use an L-shaped spacer to make sure each tile is the same distance away from each other.

2. Repeat in each of the other three squares until all the tiles are down and aligned. If you reach the wall and you have less than 1 inch of space to put a tile, move the center line of the room 3 inches closer to the wall.
3. To cut tile for fitting, mark where it needs to be cut, and use tile-cutting pliers carefully and slowly so that the tile doesn't crack. Fit those tiles in.
4. Pick up the tiles from one of the four squares and set them carefully aside. Apply glue to the subfloor in an area of

only 3 square feet at a time. Lay the tile starting at the center of the room, using your spacer, pressing them into the glue while keeping them straight and at the same height.

5. After completing 3 square feet, use a board to check the height level to make sure it is even. Push down any that are too high. Repeat this process for all the tiles.
6. When the glue is dry, pour a large amount of grout on the tiles and spread it with a grout spreader. Work it into all the seams. Remove the excess as you go, and clean the tiles with a damp sponge.
7. Keep the tiles damp for 2–3 days, and sponge them down. Cover them with plastic to keep the moisture in.

Does my location help determine my building materials?

Your building materials will be dictated by your location and climate. It will not work to build an adobe house in a rainy area, but it will work to build an underground house. It will not work to build a cabin if you have no wood. Don't forget that now it is almost impossible to build a house any way you want without being up to code, but it is possible to build all of these structures in a professional manner. Check your local regulations.

Types of structural materials:

Bamboo: quick-growing wood, can be used to build any structure (is actually a grass).

Wood: if you have heavy forest, use old growth, not young trees.

Adobe: adobe is a special mud mixture that is packed down into a house shape.

Dirt: either underground or as sod stacking.

Canvas: yurts and tipis are warm in winter and cool in summer.

Stone: with lots of stone available, a well-insulated house can be made with rocks and concrete.

What are the considerations in building a house?

If you are building your house, you must consider what you will be doing in it. A homestead house is used for many different activities: making things, integrating with your garden (through attached greenhouse or containers), food and seed storage, cooking, bathing, sleeping, eating, etc. It is important to utilize the sun and the shade properly, in order to minimize the energy needed to keep warm and cool. For humans, the best rule of thumb is to go with what is natural—when it is hot, live outside. When it is cold, burrow up and stay close.

What should I consider when modifying an existing house?

Houses that are already built are often not properly insulated and are exposed to the sun. To make your house more energy efficient, seal all cracks in doors and windows (with caulking or clay mud), insulate the walls and ceilings, attach a greenhouse to the sun side (even if it is a small-window greenhouse), install a solar hot-water heater on the roof, attach a shade room to keep cool, and plant trees and shrubs to provide more shade.

What is a greenhouse room?

A greenhouse room is a room that is partly or completely made of windows, and which utilizes the sun to heat the room,

grow plants inside, heat water, and give natural light with no electricity. The windows can either be clear or translucent. The warmer your summer, the fewer windows you will want because it will get very hot, and in winter window rooms usually have to be heated. In a cold climate, the greenhouse will have to be sealed from the rest of the house. Bathrooms and outdoor kitchens work well for this kind of room, especially in a place where the temperatures are not extreme.

What is a shade room?

A room can be made that is basically a slatted roof covered in vines and ferns which provides cooling shade and air circulation. Plants can be grown in this room, especially if you live in the desert.

What is a cellar?

A cellar is basically a dark, cool room, usually under the ground, to store food. The old-style cellar had an outside door and was also used as protection from tornadoes. The best cellar has a temperature of 32° F, which is optimum for most food storage. A finished basement can be utilized as food storage if it is not heated in the part you will store the food. Make sure that your water table will be lower than the floor throughout the year or you will have flooding, and don't use pressure-treated wood. An underground room must be supported by more than wood, with either stone or reinforced concrete. However, it is better that the floor itself is dirt, especially if being used for food storage. Cement will crack and absorb moisture.

What is an underground house?

An underground house is the most efficient house of all. Completely enclosed by dirt, it keeps warm in winter and cool in summer. The roof can be used to grow plants and harvest solar heat and light, and water can be easily harvested from rainwater. The simplest form of underground house has all the rooms facing into a center courtyard, which is open to the fresh air and may have a pond, and it has an outdoor kitchen. The structures can be made of domes for maximum strength.

How can I save energy depending on the season?

In the winter, close off the greenhouse and shade room and utilize any other solar heat that you might have. Close off rooms you are not using, and localize your activities near your heat source. In summer, use the outdoor kitchen, greenhouse, and shade room to live, letting cool air drift through the house in the cool part of the day, and closing it off in the hot part.

How do I know where to build my house?

Light: Face the house to the north and it will be brighter inside.

Heat: In a cold climate, it is advantageous to face the house to the south, putting big windows on that side and small or no windows on the north and eastern sides.

Landscape: You can utilize hills to increase the insulation. Build the north wall against a hill if it is a cold climate, but if you have lots of moisture do not put it in a valley.

Water: Figure out where water drains, and put the house on higher ground.

People: Situate the house away from other houses, and streets and highways.

Privacy: Plan your privacy; make sure that you have room to block the view of your house with fences and plants.

Solar: If you are using solar power, make sure that the house is planned for it.

Steps to building a house (this is for a conventional house, but useful for homesteaders):

Note: This planning scheme is for a regular house with non-natural materials such as Styrofoam and polyurethane. Substitute your own materials, but if you are within city limits you may run into difficulties if you use alternative methods.

1. Figure out where you want to build the house. If the lot slopes more than 3 feet, you may need to make a topographical plan so it can fit into the slope. Stake out the shape of your house.
2. Clear the lot of trees, brush, rocks, roots, debris, and anything else in the way.
3. Get your electricity hooked up to the property, or if you are generating your own power, bring in the equipment and plan how to set it up.
4. Install a well and a septic system if you want, or dig your outhouse. If you are building a basement, dig it now.
5. Put in the footings. Pour a mass of concrete to support the foundation of the house (a building inspector must check footings before they are poured). It does not cover the whole floor of the house, but simply outlines the shape of the house.
6. A foundation is made of brick, concrete block, rocks, or poured concrete, and makes up the base of the wall. Or it can be the basement wall. It must be high enough to allow water to go away from the house. The crawl space should be at least 18 inches high, and the wood framing should be 8 inches above the soil. With a basement, the foundation wall should be at least 8 feet high. Waterproof the foundation, and install

▲ **Concrete basement wall and foundation**

▲ Footing

▲ Rock foundation

▲ Block foundation

a footing drain connecting to a dry well, storm sewer, or other method.

7. Install the sewer line and water pipes under the concrete (called rough-ins, or roughing in), and any wiring that needs to be installed should be put in a conduit and roughed in.
8. Some areas require slab perimeter insulation, which is usually a piece of Styrofoam that runs around the perimeter of the house. The slab is the floor of the basement, or the concrete slab under the floor of the house, right above the footings. On top of the insulation put in a 4–6 inch layer of

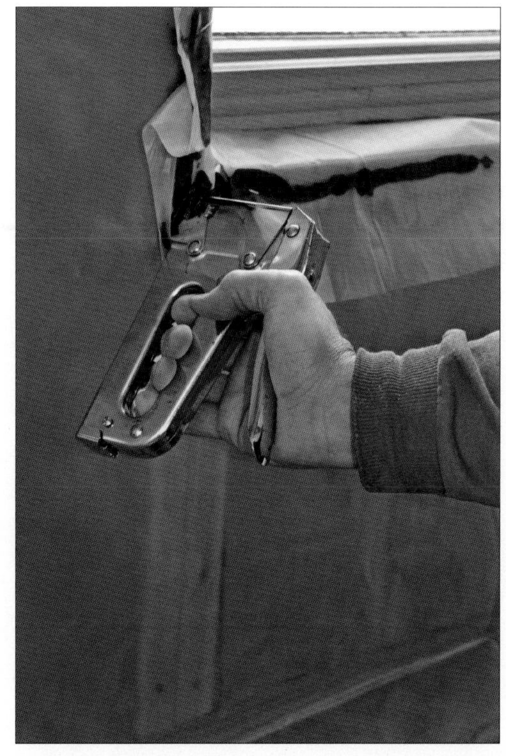

▲ Vapor barrier

gravel, a thin vapor barrier of some kind, and a #10 wire mesh on top of that. Some places require an inspection of this layer before pouring the slab.

9. For a regular house, build the framing, the exterior siding, trim, and veneer. For other houses, complete the building with whatever materials you are using.
10. Build the chimney, fireplace, and then install the roofing.
11. Install all electrical, plumbing, heating, phones, alarm systems, etc.
12. Install the insulation. Some areas require inspection by both utility and building inspectors before putting in drywall.
13. Put in the hardwood flooring and underlay, and then install the drywall. Drywall is ½ inch x 4 feet x 8 feet gypsum wallboard sheets. In wet areas

▲ Traditional house

like bathrooms, use waterproof board or paint drywall with enamel paint.

14. Prime all walls with a flat white latex paint. If you discover drywall problems, fixing them now is called "pointing up."
15. Install doors, moldings, cabinets, and shelves.
16. Paint the house inside and out. It is important to paint the outside by this time to prevent warping and mold.
17. Install counters, vinyl floors, ceramic tiles, wallpaper, etc.
18. Finishing the plumbing is called "trimming out." Install the final plumbing fixtures, and then install heating, air-conditioning, switches, lights, appliances, etc.
19. Hardwood floors should be sanded and sealed with polyurethane or other sealant, and for three days no one should go inside the house.
20. Build the driveway and make landscaping. Get final inspections made and get homeowner's insurance.

Cabin styles:

Adirondack: a half-house, in the old days the front was left open to let the heat of a campfire inside. The roof does not have a peak and it slopes downward in the back. Quick and easy to build, a 12 feet wide by 8 feet deep Adirondack needs 60 logs each 8 inches in diameter.

Alaskan trapper: has a steep roof with a ridgepole running from front to back. The roof sticks out in the front and covers a small front porch. With such a tall roof the one-room cabin feels bigger. An addition can be added to the back end of the house easily.

Appalachian: a one-or two-story cabin with the ridgepole running along the width of the house. While Alaskans are long and skinny, Appalachian cabins are wide and short. They are traditionally made with squared-off logs.

What is a cabin kit?

Cabin kit manufacturers have special mills which plane out imperfections and precision-fit the logs together. They also dry and treat these logs for longevity, and they often have good plans and come with all the lumber and windows. They save time too. However, they cost much more (not only do you pay for the kit, you have to ship it too), and you don't get the experience of doing it from scratch.

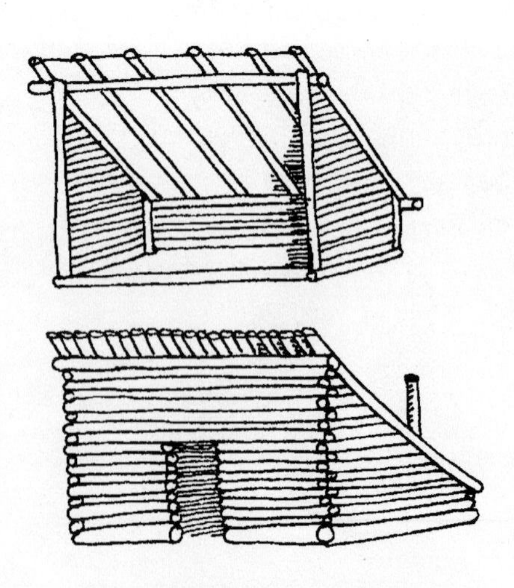

▲ Adirondack lean-to and cabin

▲ **Appalachian cabin**

To make a log cabin foundation and floor without a basement:

Note: This is to build a cabin in which the logs are flattened on top and the notches are squared. This makes a very tight cabin, but if you are pressed for time or do not have the time or tools, ignore hewing the logs and simply follow the traditional cabin log stacking which leaves the logs round with the notches farther in and curved. Don't make a log cabin out of hemlock with the bark on, or it will rot.

1. Make a foundation out of stone by digging a ditch the shape of your house. Make a stone wall inside it, and cement it with concrete.
2. To make the floor, place two sill logs (logs that support the floor) on the

Perfectly consistent kit cabin logs ▸

foundation wall on opposite sides of the house and secure them with large anchor bolts. Cut square notches in the sills on both ends of these logs (the notch is like an L shape, and the other logs will fit into this L).

3. Take 8-inch logs and hew them flat on top. Cut the ends square to fit into the notches, and make sure they are secured.
4. If your floor is more than 10 feet long, you will need a center floor girder to rest the ends of the logs on.
5. Lay 1x8 rough lumber on the floor to make a subfloor and secure it. Then lay wide pine boards, or oak or maple. Fasten with trunnels (wooden pegs hammered with a mallet) or nails.

To make a log cabin foundation and floor with a basement:

1. Dig a hole the size of your house and make a stone floor and walls all the way up. Make your foundation extend two feet above the ground.
2. Dig a trench all around your basement to the same level as the basement floor, and place 4-inch, flexible, perforated plastic tubing all the way around. Connect a pipe to it leading away and downhill from the house. Fill the trench with gravel. There should be 3–5 inches of gravel under the tubing, and at least 4 inches on top of it. This is to keep moisture out.
3. To make the floor, place two sills logs (logs that support the floor) on the foundation wall on opposite sides of the house and secure them. Cut square notches in the sills on both ends of these logs (the notch is like an L shape, and the other logs will fit into this L).
4. Take 8-inch logs and hew them flat on top. Cut the ends square to fit into the notches, and make sure they are secured.
5. If your floor is more than 10 feet long, you will need a center floor girder to rest the ends of the logs on. You will need to make a frame for an entry hole into the basement, which will be fastened to the sills. Don't put boards over the hole.
6. Lay 1x8 rough lumber on the floor to make a subfloor and secure it. Then lay wide pine boards, or oak or maple. Fasten with trunnels (wooden pegs hammered with a mallet) or nails.

How do I make the cabin walls?

1. Cut enough logs to make your house. You will have to estimate how long and how tall you want your house, and how

▾ **Dovetail-notched logs**

▾ Saddle-notched logs

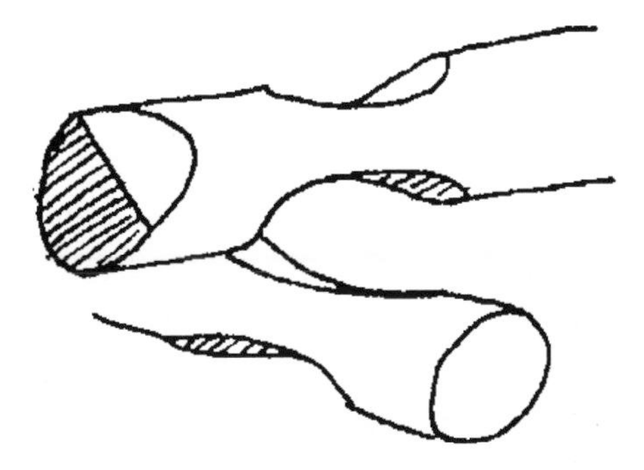

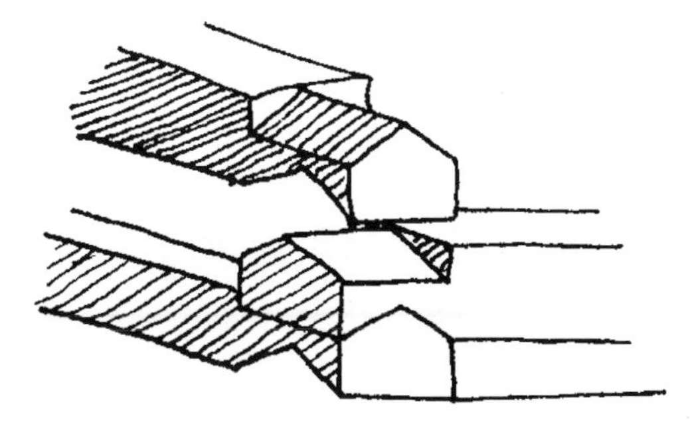

wide each log is, in order to figure this out. You can hew your logs square, hew only the tops and bottoms (as described above), or leave them round and chink it (or fill in the gaps).

2. On every log, you will need to cut a notch that will fit in with another log. When you lay the logs, you go around in a circle, and the logs fit together at 90° angles to form a corner of the cabin. The notch is shaped like an L, and must fit the L notch in the next log as tightly as possible.

3. As the cabin gets taller, put two poles on either end of one side of the house, leaning against the top. Place the log at the bottom of the poles. Fasten two ropes to the inside of the opposite side of the house, and pull them over to the side you are working on. Slip them under the log, tie them together, and put the rope back over the house to the other side. Pull the rope so that the log rolls up the poles to the notch. Repeat for all logs.
4. Alternate the wide and narrow ends of the logs, and make sure everything is level and fits right before continuing to the next log. Cut the holes for the windows and doors as you go (you can do it later, but it's easier this way).
5. If you want a loft, build to the top of the walls, and then make notches 2 feet apart in the top log where you want your loft to be supported. Lay tie beams (notched beams resting their ends on either wall near the roof), and then build up the wall more to the height you want. You can lay flooring on the tie beams later.

How do I build the roof and finish the cabin?

1. Lay two longer logs, which have been hewed flat on one side, on top of two

Rafter-style roof with gabled ends

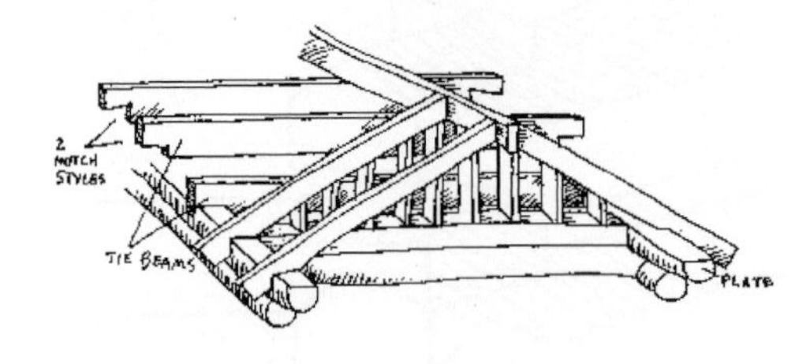

Rafter installation

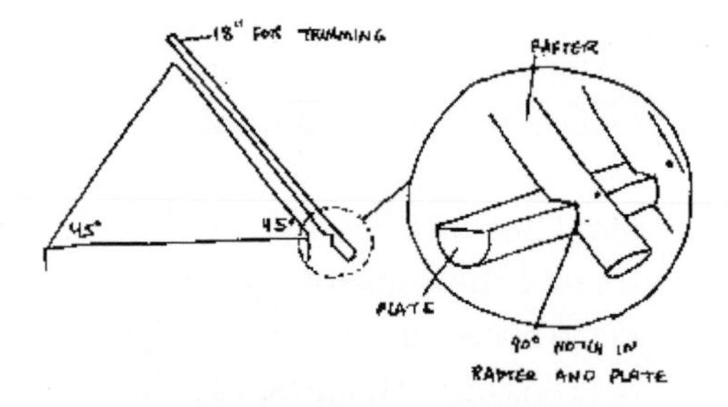

sides of the house extending past the walls (making eaves).

2. Think about where you want the triangular frame for the roof to be, and put the triangles on the walls where you ***didn't*** put the flat logs.
3. Lay a long ridgepole from the tip of one triangular frame to the other, extending out as far as the long logs on each side.
4. Make notches on the top log of the walls (the ones you just laid that were hewed flat), and lay tie beams in these notches all the way down the house.
5. Lay rafters down both sides of the roof, resting one end on the ridgepole and the other end on the top log of the wall. Notch the wall end to rest on the corner of the log.
6. Put a support between the rafters on each side, extending across from one side to the other, about halfway up the rafter. These supports cross under the ridgepole.
7. Lay boards on top of the rafters to support the shingles.
8. On the bottom of the roof, lay down a layer of tar paper, and then lay down shingles, planks, or some other kind of roofing.

9. Cut a groove around the window and door holes. Make door and window frames, and put a long strip of wood on each side to fit into the groove. Make the doorstep of a hardwood like oak. Fasten with trunnels or nails.

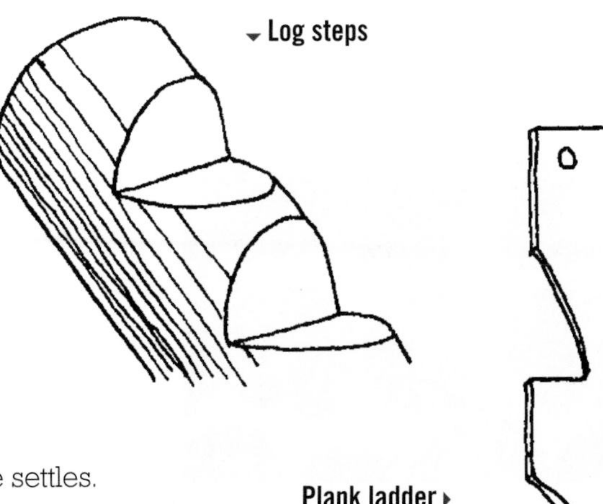

Log steps

Plank ladder

10. Chink (seal the openings) with clay or concrete every so often until the house settles. Then you will not have to do it so frequently. Insulate and waterproof your roof from the inside, either with insulation and plastic, or straw and other materials.
11. When installing windows and doors, your cabin will settle over time and windows can buckle. Leave a slot above the window and fill it with soft insulation, and made by, fastening the window to a buck (a window support), cut a slot under each nail so that it can travel down the slot.
12. Wait a few days for the logs to thoroughly dry then buy real log sealant, not stain (or make your own). Do two coats because the first coat will soak in.

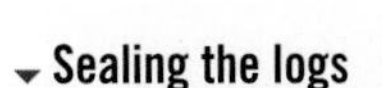

Sealing the logs

What is a cordwood house?

A cordwood house is when, instead of stacking whole logs, round discs of wood are stacked and sealed together with concrete. It is very energy efficient, even better than a regular house.

Rules of cordwood:

1. Remove the bark from the wood or it will shrink and bugs will grow in the gaps.
2. Make sure the wood is dried long enough or it will shrink and you will have holes.
3. If it is really too dry (which is hard to achieve anyway), moisture will be absorbed and the wall can break.

How do I build a cordwood house?

1. Building a cordwood house is very similar to building a stone house. Cut the logs into circles (called face cords) the width that you want your wall, usually at least 6 inches wide, kind of like slicing pepperoni.
2. Remove the bark and allow the wood to dry and shrink, but not dry enough that it splits.
3. Mix up the mortar the same as for building a stone wall. Start stacking the cords on their curved sides, rather than flat, and fill the spaces with mortar.
4. Build the windows and doors in as you go. This is actually a relatively quick process, and just takes time for the mortar to dry. Don't seal the wood when it is done, it lasts longer if it is left alone.
5. Build the roof just like you would a cabin roof.

How do I build a mud house?

This method of building is actually called *wattle and daub*. First a frame is made of timber, and then a woven lattice of oak, poplar, or hazel or other material is woven between the beams, just like a basket. The pipes and electrical must be added before the mud, and for easy access a chase (or box) in the walls can be made. The mud is then spread into the lattice in

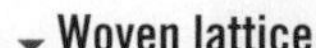
Woven lattice

▲ Mud spread over a lattice

▲ Timber-framed wattle and daub

▲ Interior of a mud home

▼ Wattle and daub English home

◀ Adobe desert home

▲ Stone house with thatched roof

layers until it fills all the spaces. The floors can be made of mud plastering.

How do I make the mud?

A more modern mud mixture is 20–30 percent clay, 5 percent cement, and the rest regular mud. It is mixed with water to create a smooth paste. Another mud mixture is mud, a little sand, and chopped rice straw. The straw is chopped into 2-centimeter pieces. Some add cow dung to increase the stickiness. The mixture is splashed upon the structural frame layer upon layer, not exceeding 9 inches. Too much clay will cause too much shrinkage.

Stone house (similar to building a stone wall):

1. Dig a hole for the foundation and throw in some rocks for support. Don't put too many rocks or it won't be strong enough. The hole should be 1–2 feet deep depending on the frost level, and twice as wide as the wall of the house will be.
2. Mix cement just wet enough that it will fill in the holes between the rocks, and pour cement into the foundation hole. It will take about two days for it to set up.
3. To make the walls, stack the rocks with cement in between. Get plenty of rocks handy before mixing the cement. Make sure the cement is thick enough to hold its shape, but not too thick that it won't fill spaces between the rocks.
4. You can use a hammer to shape the rocks better as you stack them. Clean the rocks before stacking them, and then dip them in water so the cement will stick better. Stack a rock on, then put ½ inch of cement between it and the other rocks.
5. Smooth and remove any cement that protrudes from the rock face or it will channel water into the wall.

6. Make sure you work evenly all the way around the wall—use a string if you have to. Save the flat rocks for rock floors, and use the largest rocks for the base of the wall. Make the wall between 8–12 inches thick.
7. When you get to the height that you want to have door and window frames, put the frames on and build the rocks around them. Drive large nails through the frames to secure them to the rockwork.
8. When you get nearer to the top of the wall, lay the stones around large threaded bolts, which will protrude from the top. They will be used to attach the "header," the board the roof will be attached to.
9. Drill holes in the header that will line up with the bolts exactly. Pour a little mortar on top of the wall and lay the board down with the bolts through the holes. Tighten the bolts so that the mortar is squeezed out a little, to make a tight seal.

▲ **House built with stones shaped with a hammer**

What is a timber frame house?

The difference between a timber frame house and a typical house is that the regular house has many slender boards that are cut and nailed together. A timber frame house is made of a few posts and beams that are carefully cut to lock together. While a typical house's frame is concealed with walls and ceilings, a timber frame is so beautiful that it is not covered up.

What is a mortise-and-tenon joint?

The most common joint in a timber frame house is a strong mortise-and-tenon joint. This joint is also used in chairs and tables because of its strength. The mortise is the slot or hole, and the tenon is the tongue at the end of a board that fits into the mortise. A

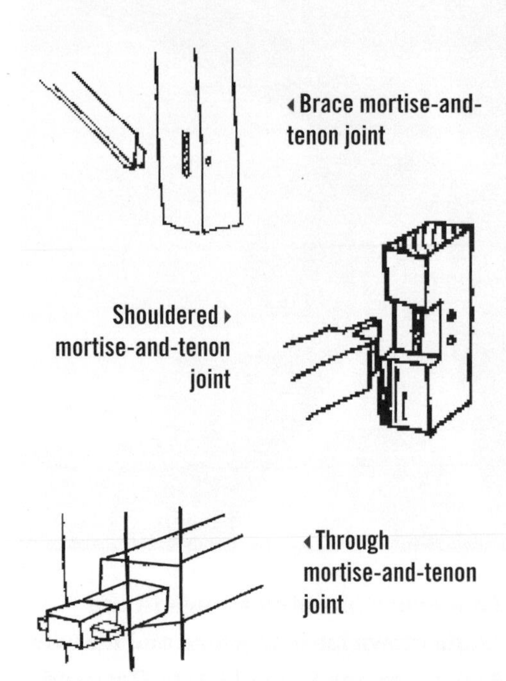

blind mortise-and-tenon joint is when the tenon is completely hidden in the mortise. A through mortise-and-tenon is the opposite—the tenon goes all the way through the wood and out the other side.

How do I build a timber frame joint?

1. After cutting the logs, use an ax or adze to hew the logs flat. A cabin's logs are sometimes done the same way but instead of leaving two rounded sides, hew all the sides flat to make the log into a square beam.
2. Fit the timbers using mortise-and-tenon joints into a series of frames called "bents," or a square with a peak that will support the roof. Shape them with a small saw and chisel, and hammer them together using a beetle. These frames can be highly artistic.
3. When there are enough frames, raise them using ropes and pulleys to an upright position on top of the foundation (the same as for a cabin), brace them, and then connect together with purlins and other connecting pieces. A purlin is a long beam going all the way down the roof, sitting in little square slots on the bents.

Is it difficult to build a timber frame house myself?

Timber frame houses are really very simple but take some skilled cutting work for the joints. They also require some heavy pulling to raise the bents, but you could try oxen and horses. Barns were

traditionally built this way, which is why a barn is "raised" rather than built. The beams are beautiful and sturdy.

BUILDING A SAFE BARN AND FENCE

A farmer travelling with his load
Picked up a horseshoe on the road,
And nailed if fast to his barn door,
That luck might down upon him pour;
That every blessing known in life
Might crown his homestead and his wife,
And never any kind of harm Descend
upon his growing farm.

~James Thomas Fields

What flooring is best for horses?

Stabled horses require lots of ventilation to prevent health problems. The flooring is probably the biggest consideration. The floor must be absorbent, easily cleaned, and resist degradation due to pawing. It must not be slippery. Before building the floor the topsoil should be removed to prevent settling.

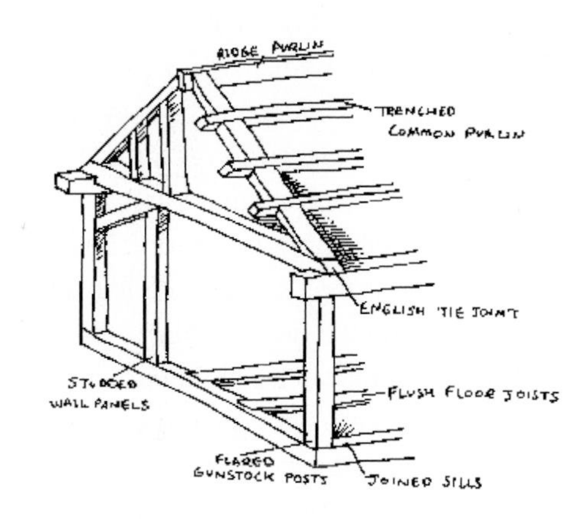

▲ **Construction of a timber frame joint**

Floor materials:

Clay or clay/sand: High maintenance because it can become slippery. It stays relatively warm but holes develop and it is difficult to clean. Should be placed over a well-drained subfloor of crushed rock or gravel. For clay/sand, combine ⅔ clay and ⅓ sand, which allows good drainage and is easier to maintain. Make sure it is well mixed, leveled, and packed.

Sand: Sand is inexpensive because it requires no additional bedding such as straw. Must be cleaned and the sand changed regularly. It can cause colic in horses because horses will eat some of it. It is better as an underlay for some other kind of material for drainage.

Limestone dust: Place over a good base with lots of drainage; make it level and hard by watering and packing well. Can be almost as hard as concrete, so lots of bedding is necessary. Should be 4–5 inches thick and put over 6–8 inches of sand or other drainage material.

Wood: Rough cut hardwood at least 2 inches thick and treated is low maintenance but can be slippery. Boards should be placed over a base of 6–8 inches of sand or gravel and spacers made between large planks for drainage. Pack the cracks with gravel or clay.

Concrete/asphalt: Easy to clean and maintain but there is no drainage. Lots of bedding is needed for softness, traction and absorption. If the floor is unsealed asphalt you will need 20 percent less bedding than if it is sealed or concrete. Horses kept on concrete or asphalt need to be outside for at least four hours per day. Wood planks or rubber mats can be put on top of it for softness, which can help.

Rubber floor mats: Expensive, but easy to clean and adds additional softness. Should be level and well packed at least ⅝ inch thick. They should fit tightly to each other and the stall walls and durable enough to resist pawing. Bedding should also be used.

Sawdust: You can use any kind of sawdust except walnut. Don't even use sawdust that was milled right after walnut was milled—always ask. It is deadly to horses. You should also make sure the sawdust is not too fine or the horse can inhale it and have lung problems. Otherwise sawdust is great bedding, especially if you put wood chips or rubber mats under it.

What should I consider when designing my barn?

1. Orient the barn at a 45° angle to the prevailing wind so that the barn won't act as a wind tunnel. If this is not possible have entrances from two directions so you can close it off.
2. Put the barn close to the manure pile and opening into a pasture. The barn should be laid out to minimize work—water and feed should be nearby and manure removal should be easy from all stalls.
3. If you have a tractor or other equipment to go in the barn, make sure all doors are much wider than they are. Twelve feet is recommended for a center aisle, but 14 feet is much more comfortable.
4. Use vent windows 2x2 or 2x4 feet high on the outside walls. Place them high enough that the animals can't reach them. A simple way for them to open is to use a hinge or chain so they open at a 45° angle.

5. Use natural lighting as much as possible, and install a light in each stall with its own switch.
6. For horses a stall should be at least 12x12 feet with a 10-foot clearance to the rafters. The bigger the stall the easier it is to clean. Walls into the center aisle can have bars for ventilation but between stalls they should be solid to provide privacy.
7. Stall doors should have a wood lower section and bars above. Animals like to see out. Hinges should swing freely, and the bottom should not be made of metal because it is too noisy.

Wide center aisle and open layout

8. Make one stall a wash stall with a drain in the center of the floor (a slope of 4° going into the drain) and have a hose for bathing. Storage, a sink, and counters are also nice to have.
9. A tack room separate from the feed room will help keep the tack in good condition. Have storage space and counters, and a small refrigerator can store medications (these can be kept in the house but it is safer and handier in the barn), and a sink is nice.
10. The feed room should be rodent-proofed just like your grain food storage with sheet metal. If possible a storage space should also be added for mucking tools such as pitchforks and shovels, away from the feed.

How do I make single-pole fence?

Drive two stakes in the ground, making an X. Then a little farther down, make

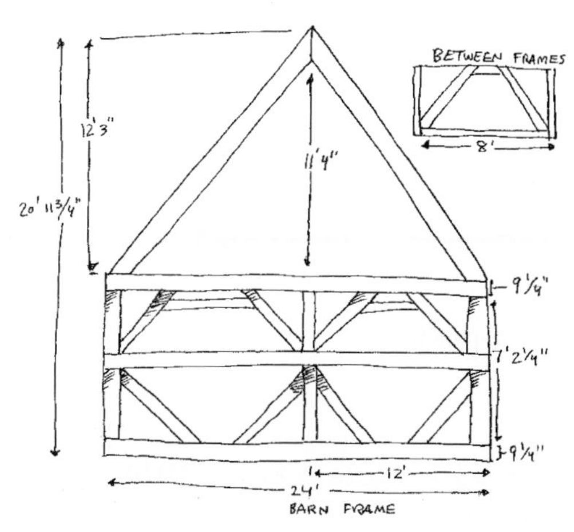

▲ Barn plan

another X. Keep making Xs the entire length of the fence, then lay poles from one X to another horizontally. Chestnut is a good wood for this.

How do I make a stump fence?

When you have cleared out a field, lay the stumps, roots and all, along the fence line.

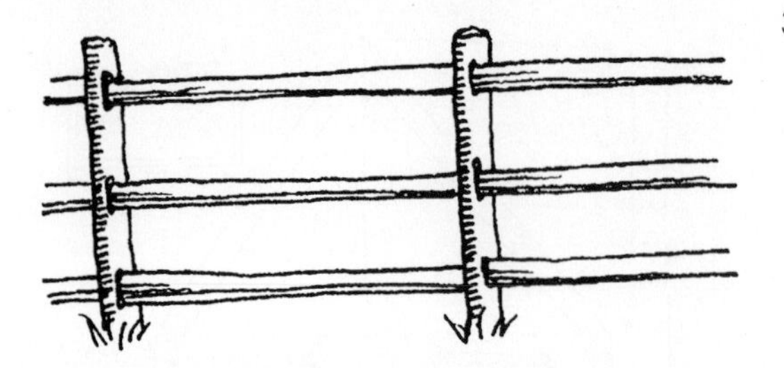

▲ Single-pole fence

How do I make a stone pile fence?

1. Plow the dirt where you will have your fence, then shovel it down past the soil to the subsoil (layer under soil).
2. Pull your stone boat along the ditch, rolling the biggest stones into it the hole, end to end.
3. Go around again, piling up the rest of the stones to make a wall. This is very hard to do, and takes practice. The trick is to have each stone touch as many other stones as possible. It should be about two stones deep, made so each leans against each other. Little stones fill in the center. If you have enough, use the flattest ones as capstones, laying them end-to-end down the wall on top.

What materials do I need to build a stone wall?

A wet wall is made with mortar, while a dry wall is made without. For a wall in a field you can make it dry, but for a house you will need mortar or the wall will have to be very thick. For a wall 8 feet high it will

▼ Variation of the single-pole fence

▲ Stone pile fence

need to be at least 24 inches wide on the bottom.

How do I build a freestanding stone wall?

1. Dig a footing twice as wide as the wall will be, and deeper than the frost line.
2. For a footing 6 inches deep, pour 3 inches of concrete, and then put ½ inch reinforcement rods in quickly while it is still wet. Then pour the rest of the concrete.
3. When the concrete is dry, wet it down and spread an inch of mortar on it. Put down whatever rocks fit in best, trying to keep the top fairly flat. Don't use the prettiest rocks—save them for the top of the wall. Press the rocks in firmly.

▼ Stone wall

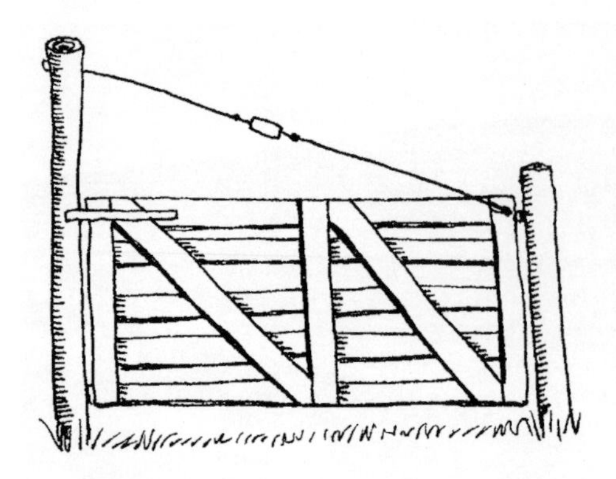

4. Trim off excess mortar with a trowel, and spread it into the spaces between the rocks.
5. If you want, use a small stone hammer to knock off sharp points sticking out. Watch out for flying pieces.
6. Lay only one or two layers every two days, and then cover it with wet burlap or a tarp. The day in between, brush the mortar with water and cover it to hold the moisture in. So one day lay rocks, the next day brush, the next day lay, etc.
7. Tie a string along the line where you want to go so that you can keep it straight. If the stones touch the string then it will change the direction of the wall. Use a plumb bob (a weight tied on the end of a string) to make sure the wall is vertical.
8. Use the flattest stones for the top of the wall so it will be as straight as possible.
9. When it is all dry, use a wire brush to get the excess crumbly mortar out.

4 Horses and Other Animals

ANIMAL BASICS AND HEALTH

I have been studying the traits and dispositions of the "lower animals" (so called) and contrasting them with the traits and dispositions of man. I find the result humiliating to me.

—*Mark Twain,* Letters from the Earth, *1907*

Animal Behavior

Understanding how an animal thinks is important for the well-being of you and the animal, even if the animal is a wild creature you encounter in the wilderness. Humans have the tendency to think of everything as a human—to "anthropomorphize" things. The only creatures that we can really do this with are house pets like cats and dogs. Animals learn by cause and effect . . . over time and through repetition animals can learn behavior based on results. Much of what an animal does is based on instinct as well. I believe that an animal does feel attachment to a human—for example, a dog loves his master, but it is also important to remember that the love the dog feels is doggy love. It is an instinctual love that is descended from the wolf pack and that is the way the dog is relating to you. No matter how long an animal lives with you, it will still react in the ways that it was designed to react. A horse may be your closest friend but still requires very specific training and handling methods so that you don't create bad behavior that comes out of the natural reactions that horses have. Anything that an animal does in relation to you will be your fault.

Biosecurity

While most small farmers don't have to worry too much about biosecurity, you still need to protect yourself against foreign health risks to your animals. It is just common sense to protect their health by preventing exposure to diseases from other animals, especially lately with avian flu, mad cow, hoof and mouth, etc. Small farms and homesteads tend to have hardier livestock—raise your animals to be hardy and build their immune system just as you would your children.

1. **Limit access.** Unless you are making money from agritourism, you should limit the amount of people who come in and have access to your animals. If you must have lots of visitors, they should clean their shoes before walking around. It is especially dangerous to have guests who have been overseas less than a week before—they should wait longer before being exposed to the animals.
2. **Change your clothes.** When you go off the farm, and especially if you have visited an animal show, auction, or another farm with animals, change your clothes and shoes before handling your animals. Pick one pair of shoes that

you only wear on your own farm, and another pair for going places.

3. **Careful purchases.** When you are going to buy a new animal, be very careful whom you are getting it from. Do your research and buy only from breeders with a good reputation. Make sure you know how healthy their herd is. Sometimes it's better to buy from someone who has had animal health problems but who knows about it and fixed it, rather than someone who has no idea how healthy their herd is.
4. **Quarantine.** When you buy an animal, or if one of your animals leaves the farm for any reason, quarantine it for at least twenty-one days. When it is by itself you can observe it for disease, and there is no risk of spreading anything to the rest of your livestock.
5. **Keep clean.** You don't want mice and rats and other pests to eat your animals' feed. Not only do they make a big mess, they also poop in the food and spread disease. Use prevention to keep these critters out by keeping food in rodent-proof containers and keeping the area clean. You should also clean anything that comes in contact with manure, and it is recommended that you disinfect it.
6. **Observation.** Keep a close eye on all your animals every day so that you will notice any unusual behavior or symptoms, such as blisters around the nose and lips, blisters around the hooves, or staggering. That way you can catch it right away and hopefully stop it from spreading. If any of your animals dies for an unknown reason, take it to a vet and have him or her find out the cause.

First Aid Kit

Note: Everything in the kit should be kept cleaned and sterilized.

Contents of regular first aid kit:

Toolbox

Restraint equipment (for individual animals)

Digital rectal thermometer with string and alligator clip

Stainless steel bucket

16-, 18-, & 20- gauge hypodermic needles, 1–1.5"long

5cc, 12cc, & 60cc syringes

IV complex hose

Obstetric chains and handles

Neonatal resuscitator

Neonatal esophageal feeder

Stomach tube (sized for individual animals)

Small refrigerator for medicine

Flashlight

Funnel and soft rubber tubing

Cotton balls

Q-tips

Cotton rags

Contents of wound kit:

Rubber gloves

Antiseptic cream

Betadine cleansing solution

Hydrogen peroxide

Nonstick gauze pads

4"x4" standard gauze pads

Roll cotton

Vet Wrap

Adhesive tape

Worming:

Giving a shot is the easiest way to worm. Oral paste is cheaper but harder to administer. There is a square spot on the rear of the horse, from the base of the tail to the highest point of the rump on the near side of the horse, running alongside the spine. This square is where you give the shot. Talk to a vet to help set up a regimented schedule for worming.

Horse vaccination and worming schedule:

February: 5-day Fenbendazole interval or daily dewormer

March: Tetanus, EEE/WEE, Intranasal flu, West Nile Virus, VEE (optional)

April: Ivermectin interval dewormer

May: Ivermectin or Moxidectin daily dewormer

June: Potomac & rabies (optional), Moxidectin interval dewormer

August: Pyrantel pamoate (double dose) interval or daily dewormer

September: Intranasal flu

October: Ivermectin interval dewormer

November: Ivermectin or Moxidectin daily dewormer

December Moxidectin interval dewormer

To prevent behavioral and medical problems:

Keep the stable and the horse stall and the feed and water containers clean. Don't introduce bad smells into the water or feed. Use fresh feed and non-moldy hay that smells good. Make changes in the feed gradually every now and then. Feed the right amount to your horse. Watch the horse's behavior carefully.

▾ Proper and improper filing

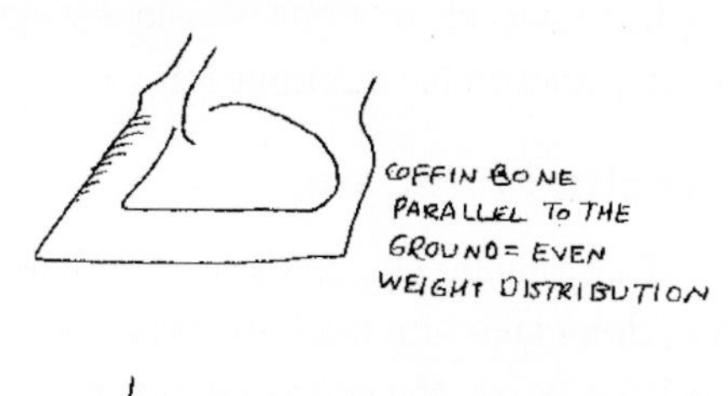

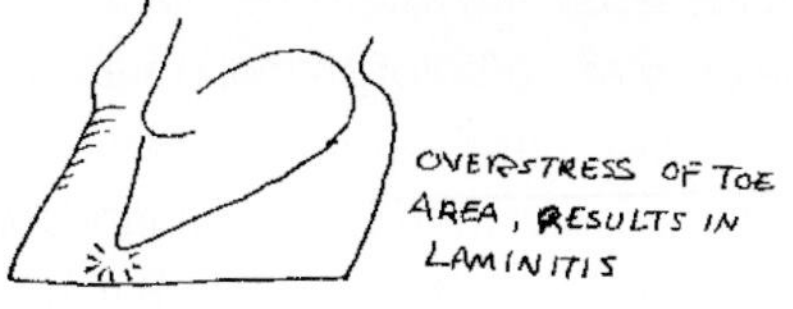

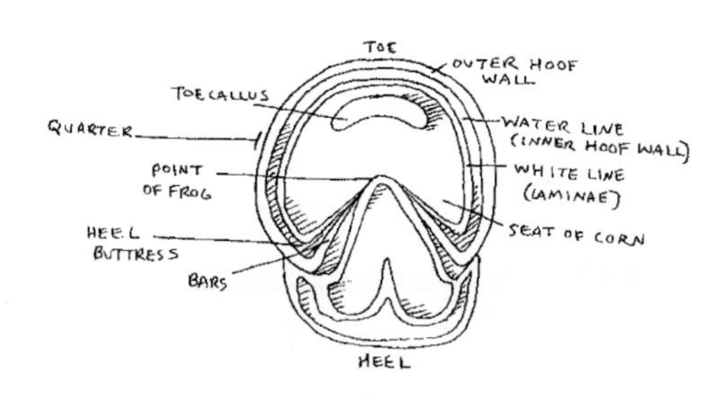

▴ Parts of a horse hoof

Preventing laminitis:

Laminitis is a serious disease in the foot of the horse, more commonly the front feet. Without treatment it can lead to even more serious problems like founder and sinking. To prevent laminitis, don't ever give the horse a large quantity of grain suddenly, suddenly change to lush green pastures from a sparser pasture, work on hard surfaces, let him drink large quantities of very cold water, eat black walnut shavings or beet tops, or have high or prolonged doses of anti-inflammatory medication. Proper hoof care is obviously extremely important, including correct filing, cleaning, and shoeing.

Recognizing laminitis:

Mild case: A sawhorse stance, shifting feet, and reluctant to move.

Moderate case: You can feel a pounding pulse above feet, the sole of foot is painful, and irregular hoof growth.

Severe: Unwillingness to stand, loss of appetite, and separation of hoof wall from sole.

Taking a horse's pulse:

1. Squat down next to the horse's left front leg. Place your index finger around the left side of the fetlock joint at its lower edge. Apply pressure with the finger and run your finger from side to side around the joint until you feel a cord-like bundle snap underneath your touch.
2. Apply pressure to this bundle (a vein, artery, and nerve) for 5–10 seconds until you feel a pulse. If you press too hard you may stop the blood flow, and if you press too softly you won't feel the pulse at all. Practice until you feel a pulse.
3. Usually a healthy horse's pulse is difficult to find. If it is easy for you to find the pulse then there is a problem and you should see a vet. Repeat steps 1–2 on all the legs.

Cleaning a horse's sheath:

A male horse needs to be cleaned down there every 6–12 months. A vet can do it when the horse is sedated, or you can do it at home. Use disposable glove that reaches up to your elbow, an old towel, a bucket of warm water, sheath cleaner, and lots of paper towels or clean old towels. Put a dollop of sheath cleaner in your gloved hand and reach right up in with the wet towel. The grime inside is called smegma and it will take several towels to get it out. Then remove the whitish gunk that forms in a small pocket at the tip of the penis, called the bean. Rinse with clean water and towels. When the water and towels are clear you are done.

What is choke?

Choke in horses is when something gets caught in the throat, not the windpipe. It is usually not fatal, and it happens when the horse eats too fast or doesn't chew his food enough. Horses that eat pelleted grain, dry hay cubes, and unsoaked beet pulp are particularly susceptible. The horse will stand still in his stall, won't eat or drink, large amounts of green discharge will come out of his nose, and some horses may look like they are in pain or panic.

Treating choke:

Take all the food out of his stall and put a muzzle on him because if he continues eating he is at risk of pneumonia. If you feel a bump on the left side of the neck below the throatlatch, it is possible to massage it out. Or you can rinse his mouth with a garden hose and if he swallows it may dislodge the bump. If he has symptoms for more than an hour call the vet. There is a high chance that choke may recur within 72 hours, so ask the vet for anti-inflammatory to reduce swelling and feed gruel to your horse instead of hay.

Tying up:

Tying up is exercise-related muscle degeneration with an unknown cause. The horse will suddenly be unable to move (or won't want to), will sweat a lot, at rest will have a heart rate over 60 BPM, tense hindquarter muscles, a swishing tail, and the horse will show signs of pain, especially if you press on his hindquarters.

Treating tying up:

Discontinue all exercise immediately, and ***don't*** take him back to his stall—don't walk the horse anywhere. Give him a dose of zylazine/butorphanol (relaxes and relieves muscle cramps) and phenylbutazone to reduce inflammation and pain. Give the horse water. If you are

▲ Percheron

▼ Haflinger

▲ Clydesdale

▲ Fjord

◀ Fell

Cut a finger hole in the skin and hold it by the hole.

11. Use the knife to gently remove the skin by holding the knife at an angle and pulling with the other hand. Do not cut or push with the knife.

Fish fillet rules:

The teeth in the fish can be sharp so be careful not to stick your hand in there or you can get cut. Don't poke yourself with the spines in the fins. Freeze or can any fish that you are not going to use right away.

Rendering tallow:

1. Set up a big pot over a heat source. It is best to do this outside because of fire risk and the greasy coating this process creates. Build a fire without high flames or flying embers under the pot, or the tallow may burst into flame.
2. Have a lid handy that fits completely over the pot so if the tallow does catch fire, simply cover it to seal out oxygen. Do not try to put it out with water.
3. Fill a big pot ¾ full of chopped fat (goat, deer, elk, or whatever animal). The chunks should be no bigger than ½–1 inch in size. Cover the fat with water and bring to a rolling boil.
4. As the water boils away the tallow will begin to be extracted from the fat. Keep up the boil the whole time, and when most of the water is gone you will see the mixture change.
5. Watch the pot closely. You will see the bubbles becoming smaller and less violent, and the color will change from light, muddy brown to a dark clear liquid.
6. The fat is rendered when it turns into brown crisps which look like bacon, and it will smell like bacon. This will take about 3 hours, and when the water is completely evaporated a light, white smoke will come off it. Take the pot off the flame immediately or it can catch fire, and be very careful not to spill or it can cause an explosion.
7. Let cool 10 minutes then strain out the impurities. When it is still warm you can use it to make tallow candles in a candle mold.

Neat's-foot oil:

Boil the hooves, horns, and scrap pieces of hide for a few hours. Let it cool and the oil will rise to the top. Use as it is or mix with fat.

Gelatin:

1. Use goat or sheep leg bones, or chicken heads, feet, necks, or backs (3 pairs of feet make 1 pint of gelatin). Put them in a pot and cover with water.
2. Simmer the bones for at least 4 hours. Goat and sheep should be boiled for 6–7 hours, while chickens need a bit less.
3. Let the bones cool, then skim off the fat from the top of the water (see neat's-foot oil, above), and the sediment from the bottom, and remove the bones. The liquid left over is the gelatin.
4. To clarify the gelatin, heat until the gelatin melts. For every quart add ½ cup sugar or honey, and the shells and slightly beaten egg whites of 5 eggs. Stir it in, then stop stirring when it gets any hotter.
5. Let it boil 10 minutes, add ½ cup cold water, and boil 5 minutes more. Remove from heat, cover with a lid and let it sit in a warm place for 30 minutes.
6. Get a large clean dishcloth and fold it to make several layers. Soak it in hot water and wring it out, then strain the gelatin through the cloth. You will have to squeeze it through the cloth (hold the cloth like a bag).
7. If it doesn't get clear, strain it through again. To set up, or harden, the gelatin after using it in a recipe, put it in the fridge. If it doesn't set up then you added too much liquid. Pineapple, papaya, mango, and figs all need to be cooked before using or they will also prevent the gelatin from setting up.

DRAFT HORSE CARE

Men are generally more careful of the Breed of their Horses and Dogs than of their Children.

~*William Penn,* Fruits of Solitude

Buying a Horse

What you can expect to pay:

Weanling (under 1 year old): $500–$2,500

Yearling (1 year old to 2 years): $1,000–$6,000

Two-year-old (2 years to 3 years): $1,500–$10,000

Riding horse (3 years and up): $2,000–$20,000

Draft team (matched w/ harness): $6,000 - $20,000

Thoroughbred foal for racing: $5,000–$50,000

Thoroughbred 2-year-old: $50,000–$3 million

Thoroughbred stud: $100,000–$3 million

Considerations when looking at a horse to buy:

1. First you need to know how to handle and care for a horse. If you don't have any experience with horses you need to take a class or ask a horse person to give you at least some beginning knowledge.
2. Consider what you can pay, and what you need the horse for. Buy within your range, but get the best you can. Also, make sure you do your research on the breed of horse and make sure it is suited to the work you want to do, whether it is riding or plowing.

3. Is the horse calm and sensible, but alert? Is he submissive but not timid? Pay careful attention to the horse's behavior around his owner and around you. The horse should be friendly. Think about spending lots of time with this horse and whether you will work well together.
4. Is the horse in the right condition to do the work you have for it? You should always get the horse checked out by a vet for health and for recommendations on what the horse is able to do, just as you would if you were buying a car.
5. Does it move smoothly and balanced (not stiff or crooked)? This is another thing the vet can look for, but you can learn to watch the way horses walk and recognize conformation, or good walking gaits.
6. Does it have any bad habits? If so, ***don't*** get it. First of all, it doesn't matter if the owner is honest about the habits. It's not worth the problems.
7. Is the horse good with your level of experience? With such a powerful and unpredictable animal, it's not a good idea to learn as you go. For your first horse, sometimes getting an older, patient, and boring horse is the best.
8. If you are going to ride it, is it your size (heels shouldn't go past underline)? Not only will a mis-sized horse be difficult to ride, but also it will be more difficult for you to handle.

Choosing the type of horse:

There are stallions, colts, geldings, mares, and fillies. The stallion is obviously the untouched and virile stud that gets other people's mares pregnant when he gets loose. Beginners should not even consider having one, and really the only purpose for having one is for breeding. A colt is a young male horse, which means that he will eventually be a stallion or you will have to pay for gelding, which can be expensive. A gelding is an ex-stallion who has had his testicles removed. Today there are the options to "proud cut" the stallion leaving testosterone-producing tissue, or "rig," in which they retain one testicle. Both of those options makes your stallion more manly, but also more difficult to handle. The gelding is generally the gentlest, even more so than mares because if a mare is fertile she has a cycle every three weeks which makes her have a sort of PMS, but at least she can give you foals.

Ways to buy a horse:

Private treaty: Seller sells directly to buyer. It may take longer to find the horse you want and the price may be higher.

Breeding farm: A farm that breeds horses for sales may have an auction or sell all the time. Make an appointment if they sell all the time.

Auction: A way to sell off stock quickly. Good and bad horses are sold, so you have to be ***really*** careful. Read the terms and conditions.

Draft horsepower:

1,200-pound horse: 180 pounds for 10 hours, and brief moments of 1,800 pounds.

1,600-pound horse: 200 pounds for 10 hours, and brief moments of 2,400 pounds.

2,000-pound hours: 220 pounds for 10 hours, and brief moments of 3,000 pounds.

A 1,600-pound horse is more efficient than any other size, and has enough overload ability to break logs loose (the 2,400 pounds).

Breeds:

Belgian: Mature early, and relatively low maintenance. Dominant color is chestnut/bay, and white socks and blaze or stripe is desirable. Average 17.3–18.2 hands high. Can weigh 2,800 pounds, but averages 2,100 pounds.

Clydesdale: Gentle and enjoy people, common color is bay; preferred markings are white socks and blaze. Averages 16.2–18 hands high, and weigh 1,600–1,800 pounds (stallions can weigh up to 2,200 pounds).

Fell: Black, brown, bay, or gray, without much white. Averages 13.2 hands.

Fjord: Very gentle, long-lived, agile, and intelligent. They are usually a brown dun color with a zebra-like stripe on the back. Averages 14–14.2 hands and between 900–1,200 pounds.

Haflinger: Cooperative, with chestnut coloring. Ranges from 52–59 inches tall, weighing an average of 800–1,300 pounds.

Percheron: Calm disposition and intelligent. Usually black or gray, no markings. Averages 16.2–17.3 hands high. Weighs 2,700 pounds or more, averages 2,000 pounds.

Shire: High endurance and patient. Black, brown, bay, or gray with blaze and white socks. Have feathers on legs. Averages 17.2 hands, and between 1800–1,200 pounds.

Suffolk: Gentle, high endurance. Chestnut color with white on face and feet. Averages 16.1 hands or more.

Welsh: Reliable and small, any color. Section A ponies do not get taller than

Belgian

▲ Shire

▲ Welsh

▼ Suffolk

out somewhere, wait until ALL symptoms have disappeared, then slowly walk home (don't ride) and stop every 30 minutes for a 10-minute rest. Watch the urine for discoloration 1–2 days after the tying up—if it is dark then there may be kidney damage.

Foundering

First measures:

Foundering is when your horse or goat gets too much rich food too soon. They need to be eased onto a new diet and any sudden change can cause the animal to bloat. Its belly will swell and it will burp a lot. If it stops burping then it can die because the gas is building up. Keep the animal moving, and if it is still swelling, raise the front feet 6 inches higher than the back ones by having it stand on an elevation. Call a vet right away, and if you can't get a vet, then you will have to take drastic measures.

Drastic measures:

Get a rubber garden hose 4 feet long or less, and a piece of black plastic pipe big enough to go in the animal's mouth. Put the plastic pipe in, and then stick the garden hose through the pipe and down the throat. The pipe is so the animal can't bite down. Stick the hose down until it reaches the gas, and stay out of the way because the gas will come out of the hose. Or, loop a rope around the bottom jaw of the animal and push it back as far as you can until the horse gags. Stay out of the way or the gas will get you.

Really drastic measures:

If that still doesn't work, and the animal is on the ground and can't get up, and it's gasping for breath, then the animal will die if you don't do something. Stick a pocketknife into the left side of the animal, right behind the ribcage below the spine, in the middle of the dipping spot that caves in naturally. This is where the gut is. The knife should be pointing towards the right front knee of the animal, and you will have to watch out because you can get sprayed and hit with gas. Turn the knife to make space for the gas to come out.

Poisons found in water:

Algae: Beware of blue-green algae when it's blooming. It causes convulsions and sudden death to all animals.

Salt water: When an animal drinks salt water or is deprived of water the results can be blindness, deafness, or paralysis to mostly poultry and ruminants.

Fluoride: Animals exposed to feed or water or airborne factory waste with fluoride over a period of time may begin having abnormalities of skeleton and teeth.

◂ Muzzle to prevent eating too much rich feed

Poisons in soil:

High alkaline soil: High alkaline soil contains selenium that can cause blindness, staggering, cracked hooves and lameness.

Imbalanced soil: When animals forage on imbalanced copper/ molybdenum ratio soil, the results can be severe scours and emaciation to ruminants.

Poisons in feed:

Cottonseed feed food coloring: Too much food coloring in cottonseed feed contains gossypol, which can cause weight loss, weakness, and loss of appetite to young ruminants.

Ammonia: If there is a sudden addition of urea or ammonium salts in the feed it causes muscle tremors, weakness, difficulty breathing, and death to mature ruminants.

Moldy feed**:** Moldy feed contains mycotoxins that can cause lameness, paralysis, listlessness, jaundice, and internal bleeding to all livestock, especially horses and poultry.

Fluoride**:** Animals exposed to feed or water or airborne factory waste with fluoride over a period of time may begin having abnormalities of skeleton and teeth.

Copper**:** When an animal eats too much copper from feed or plants it can cause liver damage, diarrhea, pain, dehydration, jaundice, and blood in the urine to sheep.

Eating blister beetles**:** Blister beetles (found mostly in the Southwest) accidentally eaten contain the poison cantharidin, which can cause oral ulcers, abdominal pain, shock and blood in the urine to horses and sheep. Blister beetles in feed can be prevented by not crimping when cutting hay.

▲ **Blister beetle**

Poisons from farm supplies:

Lead: Eating paint, batteries, grease, oil, etc. can cause lead poisoning, leading to dullness, lack of coordination, blindness, and convulsions to all animals, especially dogs.

Poisons from household supplies:

Chocolate**:** Unsweetened baker's chocolate (which has theobromin) ingested by an animal causes nervousness, vomiting, diarrhea, seizures, and sometimes a coma to all animals, but especially to dogs.

Houseplants: When an animal eats a poisonous houseplant it can cause vomiting and neurological symptoms to, mostly, dogs and cats.

Cleaning supplies and medication: Ingesting cleaners causes severe vomiting and diarrhea. Tylenol, Advil, Aleve, etc., cause death to cats and dogs.

BUTCHERING AND ANIMAL BURIAL

You have just dined, and however scrupulously the slaughterhouse is concealed in the graceful distance of miles, there is complicity.

~Ralph Waldo Emerson

Putting an animal down:

There are times when you will have to put an animal down because of injury, age, or illness. In those situations you may also want to use the resources available in order to not waste anything. While you couldn't eat the meat, you could make leather or gelatin. Butchering can be sad and stressful. If you are butchering a wild animal then you may not feel so much remorse for the creature than if you raised it yourself, but it is still sad. If you raised the animal from a baby and it had a name, then it will be much more difficult to eat. It doesn't matter if the animal is for milk or eggs or meat, they all deserve love and kindness, and they need to be killed humanely, without terror, with as little pain as possible. Don't kill the animal in its home in front of its family.

When to kill an animal:

It is time to butcher farm animals for meat when the temperature is 40° during the day, when the pasture doesn't make enough food for the animals, or when a sheep or goat is nine months old or younger. Chickens can be killed any time, and deer are usually hunted in early winter so that's when you would butcher them. Because this is a chapter about working animals like horses or oxen, if you had to put the ox down it is probably because of injury or illness, which means you won't be eating it and you will have to do it whenever necessary.

Preparation for butchering:

Decide what you are going to keep of the animal. Hooves can make gelatin, and intestines can be used for catgut (used in stitching up wounds). You can also save the hide for making leather. You will need a gun, butchering knives, a gambrel, a big container to hold guts (like a bucket), a hose connected to running water, and a large clean bowl. It is also a good idea to have a rope or chain to hoist the body, hooked to the gambrel, up high enough for you to work comfortably. Another very good idea is to get an experienced friend help you and teach you how.

Slaughtering methods:

According to the University of Iowa, there are only a few acceptable methods of killing an animal, and most require a veterinarian:

Small animals (rabbits/rodents): carbon dioxide, barbiturate overdose, and anesthetic overdose.

Dogs and cats: barbiturate overdose, anesthetic overdose.

Birds: barbiturate overdose, anesthetic overdose.

Farm animals: barbiturate overdose, anesthetic overdose.

Notice that it is never recommended to shoot the animal or chop off your bird's head as most homesteaders practice. First of all, injecting an animal with drugs before butchering makes it unusable for meat. Second, it is incredibly expensive. However, before you go out and start shooting animals with your gun, you must remember that if you get it wrong you will cause a great deal of pain in the last moments of your animal's life. A more selfish reason is that if you stress out a cow or goat before you eat it, the meat won't taste as good. Learn to do it right or don't do it at all. Practice your aim on an inanimate target first so that you will only have to shoot once.

How to shoot livestock:

Shoot a goat with a .22 rifle in the back of the head. A larger animal needs to be shot in the front, hopefully with a hunting rifle. To find the right spot, draw a line from the tip of each ear to the opposite eye, making an X. Shoot at the center of the X. Try very hard to shoot a big animal on the first shot or it will charge towards you. When butchering a cow, you usually have to wait to remove the skin until after you have gutted it, but in a situation where you won't be eating it this is not necessary.

Why would you have to bury an animal?

Sometimes livestock die from an infectious disease, or a disease that prevents you from utilizing it for meat, or if you don't eat your livestock animals will just get old and die. Eating meat contaminated this way can cause you to get sick or even die. In most places, the requirement is that you dispose of the body on your own property within thirty-six hours after death. The only way to have someone else dispose of it somewhere else is if you hire a licensed rendering company.

Before meat processing got more efficient (or regulated), you could pay a licensed rendering company to get your animal, liquefy it, and turn it into meat and bone meal for feeding to animals. This is not a good choice because of transmission of diseases, and the ethical implications of this option are obvious. Another option is to incinerate the animal in a special high temperature incinerator. These require permits because they stink and use a lot of fuel. Some large farms compost their animals above ground in special facilities, then spread the compost on the fields. For the small homesteader burial is the best method, but if the ground is frozen you might have to cremate the animal. If you do decide to burn the body, then make sure you burn it down to ashes. One good thing about cremation is if the animal was sick then there is no chance of spreading disease.

You must dig a hole that will put the animal at least 4 feet below the surface. The land can't be dug up again—it must be kept as an animal graveyard. The burial pit must be at least 100 feet away from anything. The soil must be deep and fine textured. There should be no danger of groundwater being contaminated.

What to do right away after killing an animal for meat:

Note: Pigs need special procedures for cleanliness. Don't follow these instructions for hogs.

1. If the animal is not where you want to butcher it, put a noose around its neck and drag it to where you want. This needs to be done right away because the next step is best done with the heart still beating.
2. Hang the animal upside down from the hind feet and slit the throat by sticking a big knife in and pulling it outward. Make sure to sever the arteries and veins. Any time you make a cut, avoid cutting into the hair—instead keep your knife between the flesh and skin and cut out. It is so important to hold the hide away from the meat as you cut (for example, hold it away with your left as you cut with your right). Your hand, the hide, and hair should never touch the meat, in order to avoid contamination.
3. If this is an uncastrated male, remove the head and testicles or the meat will

be tainted and taste bad. To remove the head, use the slit in the throat to cut all the way around. For a goat, twist the head until the bone snaps. With a larger animal use a meat saw to cut the spine.

4. Make slits between the Achilles tendon and the ankles and insert the gambrel. Remove the front feet, and hoist the animal to a height convenient for work on the animal's rear end.
5. To remove the skin, starting at the slits at the tendons, cut around the foot (cutting out), and be careful not to cut the tendon. Slice a line down each leg from that point in the center of the leg. Then where the two lines meet, slice down the center of the body to the neck.
6. Starting at the junction between the two leg cuts and the body cut, use the skinning knife to separate the skin from the flesh. You will have to pull the skin with one hand as you go so they will come apart.
7. If you are going to save the hide, be careful as you go. Skin the whole belly, then work around the legs from front to back.
8. Start the top of the Y (the junction of the cuts you made earlier) and skin up over the crotch. This is the tightest spot so be very careful if you are saving the hide. If you leave the fat on the body it will be easier to skin.
9. Skin over the anus to the tailbone. Pull the tail sharply and it will separate from the spine. The rest of the animal will be easier to skin. Raise or lower it to be a comfortable height.
10. To skin the forelegs, start on the outside of the leg and work around to the front. Then skin the neck and inner forelegs and the skin should come off.
11. Cut around the anus with the sharp pointed knife, being careful not to poke any holes in the intestine. When the anus is free, pull it out slightly and tie it off, unless it is a goat (not necessary for a goat).
12. Cut down the belly (from the inside out), holding the guts away from the point of the knife with your other hand. Cut through the belly fat down to the sternum, and then cut the meat between the legs.
13. Cut out the penis if it has one, then place a very large container underneath to catch the guts, which will be bulging out of the hole you just made. If this is a ruminant animal, green liquid may flow from the neck, but this is just cud. Don't let any cud get in your container.
14. Cut through the fat surrounding the guts and sever any tissue connecting them to the rear wall of the body cavity. Pull the anus through from the inside,

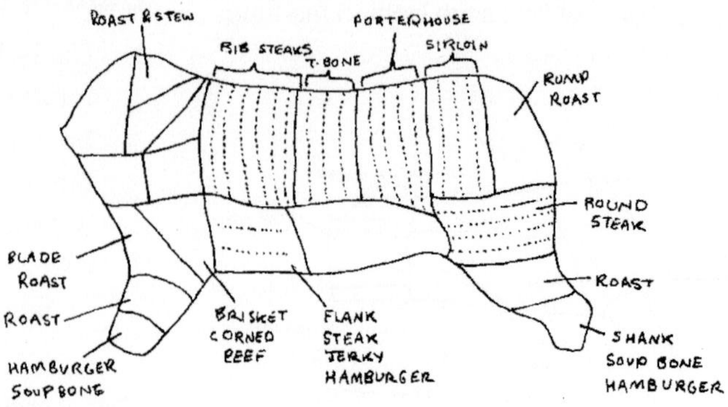

Cuts of beef ▸

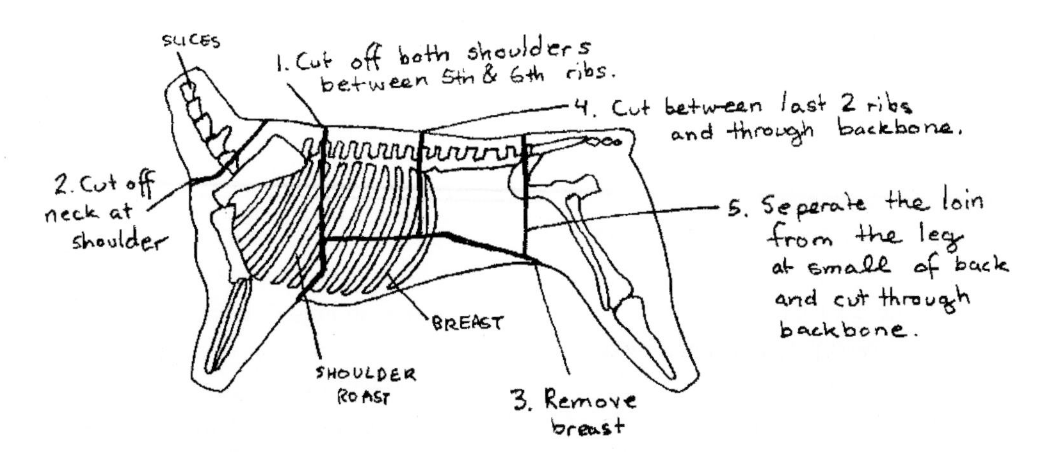

▲ **Anatomy of goat butchering**

and then take it out of the body cavity through the big hole. Take your time with this step and go slow—if you poke a hole in any of the intestines or the bladder then it will contaminate the meat because liquid is still in there.

15. Pull the intestines and bladder out of the body. Just reach in and lift them over the sternum and out (they go into the big container), and most of everything will be hanging out of the body. Strip off as much belly fat as you can, which can be fed to chickens. Cut out the organs you want to keep and put them into another bowl that you have handy (the liver, kidneys, etc.).
16. Cut the flesh connecting the stomach to the body and let it fall into the big bucket. Cut out the diaphragm and the connecting tissue behind the lungs and heart to remove them. Separate the heart from the lungs, and squeeze it to get out the blood. Put the heart in the bowl to keep. Put the lungs in the bucket.
17. At the neck, cut out the windpipe and make sure the hole is clear all the way through the body cavity. Hose the whole thing down with cold water.
18. To get the tongue, cut under the jaw in the soft space in the middle. Then reach in and cut the tongue loose at the base. To get the brain, saw the skull in half with your meat saw. If you want to save the head for headcheese, skin it, remove ears, eyes, nose, and anything else that is not meat or bone, and brush the teeth to clean them.

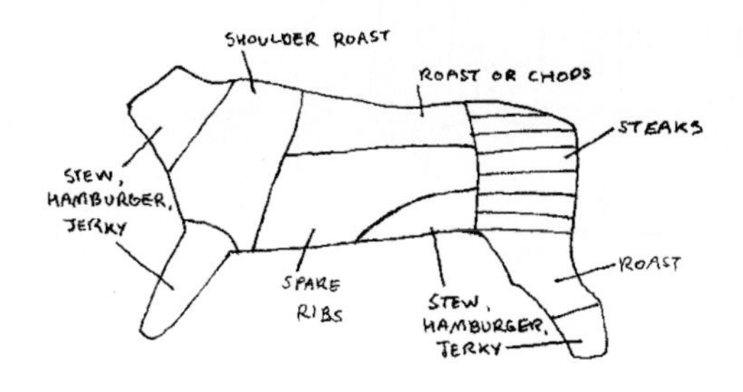

◂ **Cuts of mutton**

Cutting up a goat:

1. Cut behind the shoulder blade to remove the front legs, then cut off the bottom half of the leg at the elbow. The bottom half is not good for anything but soup bones. The shoulder can be packaged as it is, or the meat can be taken off the bone and rolled and

tied for roast, or it can be chopped for stew meat or stir fry.

2. Take as much meat from the neck as you can and use it for soup. Saw through the backbone between every rib to get chops, or take the bone out completely, or just cut out the whole muscle bundle along the backbone. When the whole muscle is taken it is called "backscrap" and is the best-tasting meat.
3. Use the meat saw to cut the ribs from the backbone, then cut the ribs in half with your knife and package. Cut out the meat under the backbone—this is the tenderloin.
4. Cut off the rear feet, then cut off the legs at the knee. The bottom half of the leg are the shanks. Remove the leg at the pelvis and use it for roast. Take out the bone now or package as it is.
5. Go over all the bones and get any last bits to be put into sausage and jerky.

How to kill a chicken:

1. Catch the chicken, either by sneaking up on it at night, or by catching it some other way. Have a large bucket ready for catching blood, and if you are going to remove feathers then have a large pot or bucket starting to boil.
2. There are many ways to kill a chicken. A common way is to do it with an ax on a chopping block. Hold the chicken with one hand and use a heavy-headed ax to chop the head off in one blow. The chicken may move its head at the wrong time, but the goal is to cut as close to the head as possible. Another way is to hold the chicken just below the head and swing it so the body twirls around, and on the third swing the head will separate from the neck. Or to prevent flopping, tie the chicken by the legs upside-down to a tree branch. Cut off the head with a sharp knife. The least messy way is to cut off the top of a milk jug, two inches below the handle. Nail it to a wall, with the small end down. Pick up the bird by the ankles. Put it head downward into the jug so that its head pokes out the small hole (that used to be a pouring spout), hold the head and stretch out the neck a bit and cut at the base of the head with a sharp knife. A shot from a .22 will kill a chicken also.
3. If you used a method that prevents the bird from moving, then the bucket should be ready under the chicken to catch the blood. If you used the chopping method then you will need to quickly stick the bird into the bucket or the chicken will flop around all over for a long time, scattering blood everywhere. Let the blood completely drain out.
4. If you just skin a chicken you do not have to go through all the work needed to remove feathers. To remove feathers, while the bird is draining, go make sure the boiling water temperature is between 130–180° (hotter is better). Lay out newspapers or other cover to pluck on.
5. Grasp the bird by the ankles and immerse the body in the hot water. If you are using a roasting-style pan, then soak the breast first and flip it. Every feathered part should have an equal time in the water, including the feathered knees. It should take about 30 seconds.
6. Put the bird on the newspaper (but don't ever let the skin touch the lead-filled ink), and start picking out feathers

as soon as you pull it out. Start with the wing and tail, then the body and then the pinfeathers. The big ones need to be pulled in the direction they grow.

7. You will be left with a bird covered in down and fine hairs. Use a gas stove flame, candle, propane torch, or alcohol burner to singe these off. Make sure you can't set anything else on fire.
8. Now the chicken is cleaned or drawn. Put it in very cold water to cool it, and either get ready near a faucet or get a bowl of cold water. Also get a sharp knife, a cutting board, a box lined with wax paper or a bucket, and a container to put chicken pieces in.
9. When looking at the bird, if it feels very light and the muscles are thin, and the innards look strange and have abscesses, then there is a strong possibility the bird had tuberculosis (or TB), so don't eat it!
10. Cut off the head and throw it out. If there is no head, cut off the top of the neck if it got dirty, and discard.
11. Feel for the knee joint where the scaly leg joins to the feathered knee, bend the knee and cut across the joint until the foot is off.
12. Put the chicken on its back with the rear end facing you. Cut across the abdomen from one thigh to the other, being very careful to not cut into any intestines. Reach inside between the intestines and the breastbone until you reach the heart, and gently loosen the membranes that hold the innards to the body wall. The gizzard, heart, and liver will come out in one big mass and put them in the box or bucket. Watch out for the green sac embedded in the liver—if it breaks it will make the chicken taste bad.
13. Scoop the lungs out, which are stuck to the inside of the ribcage, with your fingers and put them in the box or bucket. Cut around the vent (the anus) with lots of room so you don't cut into the intestine. Throw that into the box or bucket.
14. The kidneys will be stuck inside the cavity against the back. With your fingers, scoop out as much soft kidney tissue as you can; you won't be able to get all of it, and that's ok.
15. Cut the skin down the whole length of the back of the bird's neck. The crop (the thin-skinned pouch on the bird's esophagus) tears really easily, so you will have to pull the skin away. Pull the crop, esophagus, and windpipe out of the bird and put it in the box or bucket.
16. Dispose of the innards by burning, composting, or if you they had no chicken diseases, feed them back to the chickens. Don't feed to dogs or cats or they will start killing chickens. If you want to save the giblets (heart, liver, and gizzard), cut the gallbladder off of the liver and rinse the liver, and put it in a giblet bowl. Cut the gizzard open along its narrow edge ¾ of the way around, and dump out food and grit. Peel off the yellow inner lining in the gizzard by separating it at the edge you cut at and pull it out. Rinse the gizzard and put it in the bowl, then cut the heart away from the arteries, rinse, and save it. Cut the neck off at the base and put it in the bowl also. Wash the feathers and save.
17. You can clean the outside of the chicken with soap and water if you rinse well, which removes any remaining dirt.

How to kill any other bird:

Turkeys can be killed the same as chickens, except the bird is five times bigger so it will take longer. To scald a turkey, it is easy to use a metal garbage can and heat the water to 140–180° for 30–60 seconds. To remove turkey pinfeathers, put under a faucet and use pressure and rubbing to remove them.

Ducks are difficult to catch because they are easily injured, so grasp it by the neck and pull its body into your armpit, holding the wings down with your arm. Keep the duck away from your face, and kill just like a chicken.

Hold geese the same way as a duck, but make sure that you are holding the neck behind the head to keep the beak away from you. To scald a goose, heat the water to 155°, and soak for as long as 4 minutes. To clean water fowl, you will also need to remove the oil sack on the tip of its rear end, and all the yellow area around it.

Killing a rabbit:

1. Hold the rabbit's hind legs as high as your chest with your left hand, and hold it around the neck with the right. Pull the head down and bend the head backwards as hard as you can until you feel the neck snap. The other method is to whack it in the back of the head as hard as you can.
2. Hang it by the hind feet with the feet spread as wide apart as the body.
3. Cut the skin off from the hock joint of each hind leg and peel back so you can see the Achilles tendons, and stick hooks, ropes, or chains, or a gambrel, to hook in the tendon.
4. Cut off the front feet and tail. Use a small sharp knife to cut the skin between the legs, starting with the hock joint on the inside of the leg and working up the inside of the other leg to the other hock joint.

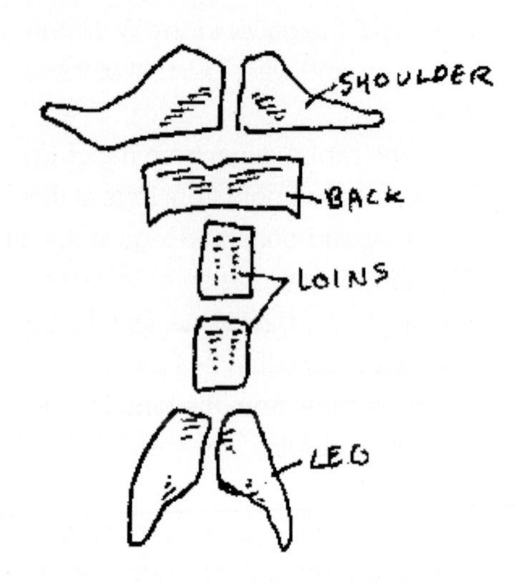

▲ Cuts of rabbit

5. Cut the skin from each front leg to the neck, and peel the skin off the body starting at the hind hocks. It should peel like a banana, or as if you were pulling off its shoot, so that it will be inside out. Cut around the anus and any place that the skin does not come off easily.
6. Put the skin aside for tanning. Cut around the anal vent and down to the breastbone down the middle of the belly. Don't cut into any intestines!
7. Pull out the intestines, then reach into the chest and pull out the heart and lungs. Look at the liver, and if it has white spots or any discoloration then it has an infection and you can't eat it and neither can any animal. If it has white cysts on its stomach or intestines then it had tapeworm and it must be cooked very well to eat it.
8. If you want to save the liver, cut away the gallbladder very carefully, and

don't spill the gallbladder. Wash the liver well, and wash the rest of the body.

9. Take the rabbit down from hanging up, then cut off both front legs at the shoulder and both hind legs at the hip.
10. Cut through the ribs on both sides parallel to the backbone. Divide the rest of the back into 2–3 pieces. Cut away the meat from the hind legs and haunches or loins.

Aging:

Goat meat should age 1 week at 40°, and longer if it is colder. A larger animal needs to be hung at least 2 weeks at 40°. A goat can be hung whole, but a big animal needs to be cut in half. To halve it, have someone help you. Face the belly while your helper holds the body and helps guide the saw from the back.

Meat grinder plate sizes:

1/8 inch: hamburger, bologna, franks

¼ inch: hamburger, salami, pepperoni

3/16 inch: course ground hamburger, breakfast sausage

3/8 inch: chili meat, first sausage grind

½ inch: chili meat, vegetables

How to fillet a fish:

Note: This technique can be used for a medium-sized fish such as coral trout, barramundi, and salmon, and can be used on any fish with scales.

1. Put the fish on very clean, flat surface, belly down. If the head is not removed already, hold on to its head with the hand not holding the knife (if you are right-handed, then the head would be towards your left). You will need a thin flexible knife and a broad flat knife, both sharp.
2. Use the thin knife to cut downward just behind the head in a diagonal line until you reach the backbone, and run the knife along the top of the bone until you reach the top fin nearest the tail. Use a sawing motion and do not try to cut off too much fillet at this point.
3. When you're near the tail, hold the knife flat against the backbone and push the point through the side of the fillet. With the knife sticking out the other side, cut through the remaining fillet towards the tail. Peel the fillet back with one hand while cutting with the other using small slicing motions. Turn the fish on its side with the backbone facing you.
4. Starting just below the Y bones coming off the backbone, cut a ¼ inch deep the length of the fish until you reached the point you stopped with the first fillet.
5. Cut deeper this time, skimming along the edge of the ribs from front to back again, stopping at the same point as before.
7. Make a cut along the top of the belly by following the white line in the skin and lift off the fillet as you cut.
8. Turn the fish over and repeat steps 3–7 on the other side. After you're finished, there will be sections of meat behind the rear dorsal fins. This portion does not have Y bones in it so run your knife along the backbone all the way to the tail to remove a boneless fillet.
9. Repeat on the other side and you should have a total of five boneless steaks.
10. To remove the skin (this is optional): Hold the skin in one hand and use the flat knife to slice a small portion of the flesh away from the skin.

12.2 hands, Section B ponies do not exceed 14.2 hands.

Reasons to use a draft horse instead of a tractor:

Horses do not destroy your topsoil and do minimal damage to the land.

Horses make their own fuel because you can grow their feed, and they make manure for the fields.

Horses cost less to purchase and maintain, can be sold for profit and don't break down so often.

They reproduce baby copies of themselves.

Buying a horse supports a farmer while buying a tractor supports a corporation.

Horses work well on hills and bumpy terrain, and can go through mud that tractors would get stuck in.

Horses can work hands-free if well trained, while you follow behind doing your work, with no helpers needed.

Horses repair themselves, you just keep them rested and a while later they feel better. There are no parts to buy.

Horses prevent accidents related to machinery because they are slower and less mechanical.

A horse is your friend: intelligent, quiet, and sensitive.

Working with rare breed draft horses ensures the survival of an animal that has served humans for centuries.

Things to remember when using draft horses:

Horses need continual training and education.

You need continual training and education because you are the cause of all horse problems.

You must have absolute patience and understanding. A horse is not a machine, it has feelings.

Horses need frequent exercise, health checks, vaccinations, etc.

Unless you're a farrier, the horse will need shoes from a professional.

Tractors don't eat when they aren't working—horses do, and you have to be around for them.

If you only have a couple of acres don't get draft horses, just use draft ponies or mules.

What a horse can do:

A 1,500-pound team of draft horses can plow 1 ½ acres, single row cultivate 7 acres, harrow 8 acres, mow 7 acres, drill 8 acres, rake 14 acres, plant 8 acres, and haul a wagon with 1 ½ tons for 20 miles—all in one day. However, this is when they are in peak condition and worked consistently. It takes time to get a horse back into condition when they are only worked half a year.

Nutritional needs of horses:

Calories: Calories come from hay and fresh pasture.

Calcium: Found in alfalfa hay and nutritional supplements.

Phosphorus: Must be given at a 1:1 calcium/phosphorus ratio or it will cause bone disease. Timothy, grass hay, grain, and supplements contain phosphorus.

Copper: Found in supplements.

Zinc: Found in supplements.

Vitamin D: For horses, only found in sunlight.

Vitamin C: Found in supplements, only needed by pregnant mares.

Protein: Hay (or forage feed) contains protein.

Forage nutritional percentages:

Type	Calories per KG	Protein	Fiber	Calcium	Phosphorus	Copper ppm	Zinc ppm	Vitamin D
Alfalfa	2	17%	23.5%	1.24%	.24%	16.1	28	1,810UI per kg
Bahia grass	1.75	8.5%	28.1%	.45%	.2%	2.1	6	None
Bluegrass	1.58	8.5%	28.1%	.45%	.2%	None	None	None
Clover	1.9	20%	18.5%	1.35%	.3%	8	15	None
Orchard grass	1.95	7.6%	33.9%	.24%	.3%	16.9	34	None
Timothy	1.77	8.6%	30.9	.43%	.2%	14.2	None	1,763UI per Kg

Kinds of forage:

1. You can turn your horse out to pasture. To use this as your ***only*** food source, you need 2 acres per mature horse. It should be weed-free, dry, and clear of debris. If you manage the pasture well, you may be able have less acreage and enough forage for the horse, especially if you rotate. The best bet is to offer a pretty good pasture and throw grass hay out for them. They'll eat the hay if they need it.
2. There are many kinds of hay. It's not as good as pasture, but will keep your horse healthy. Alfalfa hay is the best kind, with protein, energy, and calcium, but may need a grain supplement. Grass hay can be used as the horse's only food source. Grain hay can be used if the grain hasn't been harvested and the plant is still green (yellow grain hay with no grain is straw). If you buy it you will need about 1 bale of hay per horse per day at most.
3. Straw is the stem of any type of grain after the seeds are taken off. It is usually used for bedding, but can make good feed. The best for feed are oat straw and chopped-up alfalfa (called chaff—in the U.S. it is sold covered with molasses and called A & M or O & M).

Kinds of grain:

Corn: The most common grain fed to horses. Be careful and feed less of it so you don't overfeed. Needs to be cracked, flaked, or rolled to make it easier to digest. Don't grind it.

Oats: You'll be less likely to mess up when feeding oats. Feed by the weight of the feed, and not by the volume.

Barley: Barley is in-between corn and oats in the health category. Needs to be processed like corn before feeding.

Estimating the weight of the horse:

Note: The girth here indicates the measure around the horse just behind the forelegs. This is an estimate only.

Girth (inches)	Weight (pounds)
80	1540
75	132
71	1100
Girth (inches)	**Weight (pounds)**
65	880
53	660
41	440

How to feed your horse:

Water: Give your horse an unlimited amount of water. Try to keep the water at 45–75°, make sure it's clean and fresh, and keep the water container clean. Limit the water given just after exercise—wait until the horse is cooled down.

Forage: This is the bulk of a horse's meal. It can be almost any plant material, including grass, hay, straw, etc. A horse needs about a ton of good hay to feed it through the winter. Never go directly from a sparse pasture to a heavy pasture—use the same gradual principle as grain.

Grain: A working horse may need 15 pounds of grain a day, but never give more than 5 pounds per meal. Grain adds some extra nutrition and vitamins to the diet. Put a large, smooth stone in the center of the feedbox and your horse will slow down and not throw his food around. Never change feed suddenly! Do it gradually, over at least two weeks.

Salt: Horses need salt, so put a salt block or loose salt in the pasture or the feed bin.

How much to feed:

First consider the weight of the horse. He will need 1.5 percent of his body weight in feed (this includes weanlings). At least $^{2}/_{3}$ of this must be forage. The goal is to maintain the same body weight the horse already has, unless the horse is too fat or thin. Look at the horse from the rear. The back should be round and smooth, with the backbone not apparent. There will be a bump if the horse is thin, and a dent if the horse is fat. If your hay or grass does not meet the nutritional requirements then you will need to add grain. When adding grain make sure to know how much phosphorus it has and add more calcium to the diet in order to keep the 1:1 ratio. An adult horse can maintain an ideal body condition on only forage grass and a salt block, even when living outside with a simple shelter. Winter temperatures can sometimes make it difficult for the horse to maintain body weight, so keep an eye on him. Breeding stallions and mares, pregnant and lactating mares (see below), and foals all have special food requirements.

How much to feed a working draft horse:

Draft horses can be fed just like other horses until they start working, when you will need to add grain to the diet. To estimate calories, hay adds 1,000 calories per pound, oats adds 1,500 calories per pound, and vegetable oil adds 4,000 calories per pound. The following calorie guidelines are for a 2,000-pound horse:

Maintenance: 22,500 calories

Farming and light logging: 33,750 calories

Heavy logging and plowing: 45,000 calories

Broodmare: 26,000 calories (last three months of pregnancy)

Lactating mare: 39,000 calories

Lactating mare at heavy work: 62,000 calories

Dangers in roughage:

Some kinds of roughage in grazing can contain dangerous elements to the horse. Some hybrid sorghum grasses can cause Cystitis syndrome or Prussic acid poisoning, leading to death. Alfalfa hay can have blister beetle poisoning, but it can be prevented by using hay from early cuttings rather than during midsummer, and not using crimped or conditioned hay. Fescue can contain endophyte fungus which can cause a mare's milk to dry up (agalactia) and early foal death, so all pregnant mares should avoid fescue at least 90 days before foaling.

Pregnant mares:

Pregnant and lactating mares require more food than normal. Remember that the horse should be kept looking healthy, the back should be level, ribs cannot be seen but can be easily felt, fat can be felt around the tailhead, and bony areas do not look obviously thin. Mares kept in stalls or dry lots need 1 pound of hay per 100 pounds of body weight.

Feed percentages in lactating mares:

With alfalfa hay: 50% oats, 45% cracked corn, 3% molasses, 1% bone meal or dicalcium phosphate, .5% ground limestone, .5% trace mineralized salt.

With good quality hay or grazing: 40% oats, 40% corn, 15% soybean meal, 3% molasses, .75% ground limestone, .75% bone meal or dicalcium phosphate, .75% trace mineralized salt.

Pounds for lactating mares:

Lactating mares need much more feed than pregnant mares: Eleven to twelve pounds roughage and 13–14 pounds grain and other ingredients. This will be

about 2–3 percent of her body weight. Make sure that as you increase her diet; allow 7–10 days for her to adjust gradually. By the fourth month her needs will decrease and she can begin to return to a normal diet while the foals are creep feeding.

Orphaned foals:

A newborn foal should receive 2–3 liters of colostrum every hour for the first 6 hours after birth. Divide into doses and bottle feed, or if that is not possible have a vet use a stomach tube. It is incredibly important that the horse receives the antibodies to protect against disease or it may die. You can milk 200–300 ml of colostrum without robbing a newborn foal, or it can be milked from the mare just before she dies. If kept frozen it should keep for one year, and thawed to room temperature before use. Foals nurse every 1–2 hours during the first week, and 4–6 hours the second week, and in a few weeks will start solid food. Use a nurse mare (if the mare will accept the foal) or goat (put the goat up on hay bales) or manually feed with a bottle. Use mares, goats, or powdered mare's milk or foal formula (see below). Don't use straight cow's milk, and don't use a big calf nipple.

Foal formula:

Cow's milk: 24 ounces cow's milk, 12 ounces saturated limewater, 4 teaspoons dextrose.

Evaporated milk: 4 ounces evaporated milk, 4 ounces warm water, 1 teaspoon white corn syrup.

2% cow's milk: 8 ounces 2% cow's milk, 1 teaspoon white corn syrup.

Creep feeding foals:

Creep feeding is when you feed the foals separate from the adult horses so they can eat solid food before they have been fully weaned. This should be done within the first two months, and will make foals gain 2 ½ to 3 pounds per day. A creep feeder is basically a fence that small foals can get under but adults cannot. With more than four foals the trough should be made so that they can stand on both sides with their tails toward an exit. The height of the creep should be about 4 feet when the mare's weight is between 1,000–1,300 pounds. In a stall, a foal feeder trough low to the ground can be made that has bars that are too small for the mare to get through.

Creep feeder size:

1 foal: 8x8 feet
2 foals: 10x10 feet
3 foals: 12x12 feet
4 foals: 14x14 feet
5 foals: 16x16 feet
6 foals: 18x18 feet
7 foals: 20x20 feet
8 foals: 24x24 feet

Weaning foals:

Foals are weaned at about four months old. Before you separate them, they should be used to being out with horses in a herd situation, should no longer be dependent on the dam nutritionally and should be accustomed to the pasture that they will be weaned in. The foal should have been creep feeding for at least two months. The pasture should be right next door to the pasture that the mother is in so that even though the foal cannot nurse they can see each other through the fence. If either the mare or the foal starts to get thin, go slower with the separation.

Estimating foal weight:

At birth, the foal weighs about 10 percent of the dam's weight. Around 7–28 days, measure the foal's heartgirth (use a cloth tape measure around his barrel just behind the withers), subtract 25, and divide by .07 to estimate his weight. Around 28–90 days old, use the same formula as before but at 10 percent of the result.

Grooming

Equipment needed:

Dandy brush
Body brush
Small, soft face brush
Rubber currycomb: raises dust and loose hair from the coat.
Metal currycomb: for cleaning brushes.
Mane and tail comb

▾ **Body brush, metal currycomb, and hoof pick**

Hoof pick with brush
Hoof dressing
Kitchen towel
Show Sheen or detangler
Fly repellant
Bot knife

About grooming:

Keep the grooming kit in a basket or container in the same place you keep your tack. The kits should be kept clean and each horse should have their own. Have a routine that you follow each time you groom, either starting from the head or starting from the feet. Starting from the feet is a good idea because the feet are very important for a horse's health.

Grooming steps:

1. Pick up each hoof and use the hoof pick to remove caked mud and dirt. Clean the crevices on each side of the frog to prevent thrush. Use the brush to scrub the sole and remove small particles. Check that the frog is firm, with no discharge, and make sure the shoes are on tight with no missing nails. Check that the shoe clenches have not

▴ **Cleaning the hoof with a hoof pick**

risen (the nails aren't sticking out the top side of the hoof).

2. Check the body and legs for lumps, bumps and heat using your hand. If you do this every time you will notice new ones immediately.
3. If the horse is caked in mud, use the rubber currycomb in a circular motion against the direction of the coat. Sometimes you may have to hose the horse down with water. Then use a dandy brush to remove the worst dirt, following the direction of the hair and using the metal currycomb to clean it after each stroke.
4. If the horse is fairly clean, or if the mud is mostly gone, use the medium soft body brush in the direction of the hair. Use sweeping movements, and after every other stroke use the metal currycomb to remove dust and hair from the brush. Bang the metal currycomb against the heel of your boot to clean it.
5. Use the body brush on the legs. Check the front legs for bot eggs (from botflies), small yellow dots attached to the hair. Use a bot knife to remove them. If the horse eats them he can get sick.

6. Use the dandy brush around the fetlocks if they are muddy. Inspect the hollow at the back of the fetlocks for fungus or scratches and treat them.
7. Spray the tail with Show Sheen or detangler and let it dry for a few minutes. Separate the hairs of the tail with a comb, starting at the bottom and making sure comb goes easily through each section of hair before moving on to the next.
8. Spray the mane with detangler and let it dry for a few minutes. Use a stiff brush to brush the hair off the neck to remove the scurf that collects at the base of the mane. Then comb down with the mane comb.
9. Use a retired kitchen or hand towel and wipe down the neck, body, and quarters for a final rub.
10. Once a week apply hoof dressing by massaging it into the coronary band on each hoof. Then brush it over the hoof wall.
11. During the summer apply a generous amount of fly repellant whether the horse will be outside or not.

Fly spray formulas:

Myrrh: Mix ½ teaspoon oil of myrrh, 2 cups water, ½ cup cider vinegar, and ¼ teaspoon citronella oil.

Eucalyptus: Mix 2 cups apple cider vinegar, 2 cups cold tea (sage or chamomile), 20 drops eucalyptus oil, 20 drops citronella oil, 10 drops lavender oil, 10 drops tea tree oil, and 10 drops cedar oil.

Clipping:

During the winter the horse grows a thick winter coat that will make them

◂ **Body brush**

sweat when working. If the horse will be wearing any kind of tack or doing any kind of work it is important to clip the horse according to the work he will be doing.

Trace clip: Hair is removed from the underside of the neck and belly, and left on head, back, and legs. This is used for horses in harness.

Blanket clip: Hair is removed from the neck and flanks, leaving the hair on back and hindquarters and legs. Used for extreme weather and outdoor work.

Hunter clip: Leaves hair only on legs and saddle area. Used for hunting horses or for working in wet and muddy conditions.

Full clip: The whole coat is clipped off. This is used for show horses.

To remove a loose shoe:

1. You will need a low-slung hoof boot or a handmade hoof slipper (see below). This prevents slipping. You will also need clinch cutters, a hammer, and show pullers.
2. Place the narrow blade of the clinch cutter against the bend in the clinches (or bent nails coming out of the hoof wall). Bend them straight by tapping on the cutter with the hammer, or simply cut them off.
3. Pick up the horse's hoof and try to use the shoe pullers to pull the nail heads out. You probably won't be able to get them all. Put them somewhere they won't get stepped on.
4. Slip the shoe puller's jaws between the shoe and the buttress of the hoof's heel. Push the tool's handle inward toward the center of the sole to loosen the shoe. ***Don't*** try to pry the shoe off outwards.
5. After loosening both heels, pry the shoes loose in the same way, towards the center of sole. Repeat this motion wherever the shoe is still nailed until it comes off.
6. If any nails are still in the hoof wall, pull them out with shoe pullers, then remove the shoe. Put on the hoof boot or the hoof slipper. Keep the horse confined until the farrier can come.

To make a horse slipper:

Use 1 square foot of padding (a layer of roll cotton, four layers of sheet cotton, trimmed bath towel, sweatshirt, etc.), put it over the sole of the foot, bringing the edges up around the hoof wall. Secure it with elastic bandage and cover the bandage with duct tape.

Hoof trimming tools:

Apron/chaps: A leather cover that protects your legs from the horse's hoof and legs, and help you grip.

Hoof knife: Used to trim away the loose dried-out sole of the foot and loose and ragged frog (***don't*** cut live flesh).

Nippers: Trims the growth of the hoof wall. A full-grown horse will have about 3–3 ¾ inches of hoof wall.

Rasp: Like a file, used to level the bottom of the hoof wall.

Hoof gauge: Matches the pairs of hoofs in their angle to the ground so that the two front hoofs are the same, and the two back are the same (fronts are usually 50–58°, backs are 2° steeper).

Trimming the hoof:

1. Hold the hoof between your knees for a front foot, or across your lap with the hock under your arm for a rear foot. Hold the nipper cutting blades parallel

to the bottom of the horse's foot, with one handle in each hand.

2. Trim off the hoof wall to the toe starting at one heel, then start at the other heel and trim to the toe. Use only half of the nipper's cutting blades to cut to make an even trim. Don't use the line of the sole as a guide—trim in a straight line from heel to toe.
3. It is popular to trim the heels to the widest part of the frog, but leave a little bit more, and don't cut into the sole.
4. Use the rasp on the bottom of the hoof wall. Make the hoof level and smooth. Pull with one hand and push with the other hand starting from heel to toe, and then toe to heel, with equal pressure from both hands, keeping it flat to the hoof and in contact with both sides of the hoof wall at the same time.

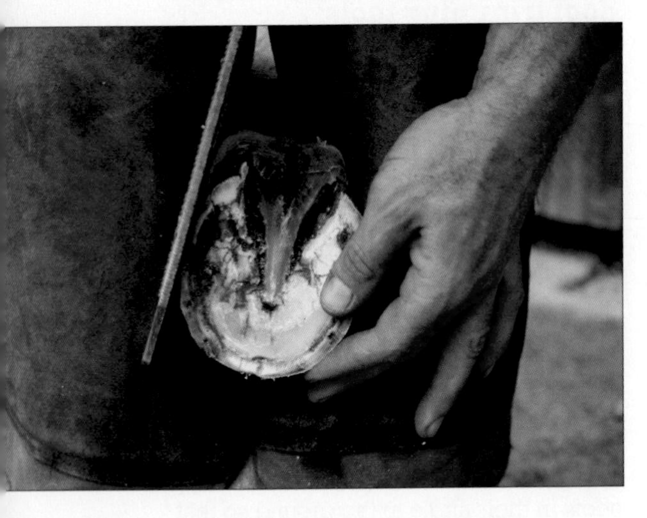

Stop every few strokes to check your work.

5. Check the bottom of the hoof by holding the horse's leg by the cannon bone. Let the hoof relax and look across it from the heel. The bottom of the hoof wall should be flat, with no dips or high spots. The bottom of the

hoof and the back of the leg up to the knee should form a T.

6. Move the horse's leg and hoof to the front and put it on a stand or on your knee. Use the rasp to round off the bottom edge of the hoof wall and remove any flares or dishes in the hoof wall. Do not rasp the wall thinner than half the thickness of the good wall, and do not rasp up to the hairline (which removes the protective covering, the periople).
7. Trim hooves regularly, every 4–10 weeks. The hoofs should never be 1 inch longer than normal.

The parts of a horseshoe:

Nails: Made of soft steel, with four sides and a tapered shaft.

Frost nails: Nails made with a head that provides temporary traction on hard surfaces.

Borium: A coating for horseshoes that provides traction on pavement or ice and increases the life of the shoe.

Calks: Projections attached to the ground side of the shoe to increase traction, alter movement, or adjust the horse's stance.

Clinches: The parts of the nails visible outside the hoofs, which are folded down against the hoof like a clamp.

Clips: Flat projections extending upward from the outer edge of the horseshoe to prevent the shoe from shifting, stabilize the hoof wall, and reduce the number of nails.

Hot fitting: Holding a hot shoe against the bottom of a hoof until it burns the high spots of the horn so that the farrier knows what needs to be removed to make it level.

Scotched: A horseshoe with an outer edge that is sloped down and outward from the hoof at the same angle as the hoof, also called Scotch-bottom shoes, they are often used with draft horses.

Rockered toe: A horseshoe that has been bent upward at the toe to ease and direct breakover, for which the hoof must be specially prepared.

Rolled toe: A horseshoe that has been rounded or beveled on the outer edge of the ground surface at the toe to ease breakover, for which no hoof preparation is needed.

Caring for stallions:

Stallions hardly ever need their sheath cleaned, unlike a gelding. In fact, they may never need it—the best way to prevent infection is to not breed with a questionable mare. A breeding stallion may need additional carbohydrates in his diet but does not need extra protein. Check his body condition and weight once a week during breeding season and adjust his diet accordingly. In a stall, a stallion will relieve himself in the same spot continually if there is a bit already there, so leave a small bit for easy cleanup. A stallion needs lots of exercise and should not get stressed during breeding season. Watch your vaccinations carefully and consult a vet—fever will decrease his sperm count.

How to breed:

The best way is to pasture a very healthy mare in with the stallion and let nature take its course. To do this, the mare and stallion must both be healthy with strong bodies and normal hormone levels. If they are not, then this may not be the best choice. Mares stay in heat for 2–10 days, and the signs are aggressiveness, frequent urination, opening and closing her vulva

(called "winking"), and raising her tail. Some mares don't show any signs. A mare that is too thin has a poor chance of getting pregnant and will not provide enough nutrition to a foal.

Pregnancy care:

Besides good feeding (see below) for the next 340 days of gestation, you can also vaccinate your mare 2–4 weeks before her due date in order for the foal to carry the same vaccination (talk to a vet). Also make sure that she is receiving all her scheduled dewormers. Mares do not look obviously pregnant until their final months of pregnancy (the 10–11th month).

Vaccinations advised for pregnant mares:

Tetanus: Fatal to horses, so all horses should receive the vaccine. A new foal can also get the vaccine with two doses 30 days apart.

Encephalitis: There are three types of it, Eastern, Western and Venezuelan. It is spread by mosquito and infects the nervous system including the brain.

Potomac horse fever: If you live near the coast or rivers in a moderate climate then consult your vet.

Rabies: All animals should be vaccinated for this; it is fatal and can be spread to humans.

Rotavirus diarrhea: Consult your vet, it is a virus carried by the mother that can kill a newborn foal.

Influenza: Not 100 percent effective but will help prevent or lessen equine flu in young horses.

Rhino: Can cause abortions, it is like a common cold.

Strangles: If you will be visiting a breeding farm the mare should receive it, but consult the vet.

Feeding brood mares in early pregnancy:

In the first stages of pregnancy a horse will consume the necessary nutrients if they have quality pasture. If it isn't, a substitution of high quality hay will also work, from 1.5 to 1.75 percent of their body weight. They will also need good water and a mineralized salt block. If the mare is in good shape this should be fine without grain, but a mare in bad shape will need supplements.

Feeding brood mares in late pregnancy per pound:

With alfalfa hay: 50% oats, 45% cracked corn, 3% molasses, 1% Bonemeal or dicalcium phosphate, .5% ground limestone, .5% trace mineralized salt.

With good quality hay or grazing: 40% oats, 40% corn, 15% soybean meal, 3% molasses, .75% ground limestone, .75% Bonemeal or dicalcium phosphate, .75% trace mineralized salt.

Daily feed for pregnant mare:

With alfalfa hay for 1,100-pound pregnant mare: 13–14 pounds hay, 4–5 pounds grain and others.

With good quality hay or grazing for 1,100-pound pregnant mare: 11–12 pounds grass hay, 5–6½ pound grain and others.

Pregnancy duedate signs:

2 months before: The fetus moves to a delivery position. The mare's belly will look much bigger and she'll take bigger steps. Her udder and milk veins will enlarge.

1 month before: The mare will have a loss of appetite and may prefer to be by herself.

2 weeks–2 hours: Sticky, clear yellow drops will drip from the teats and harden, called waxed teats. Her tail head, croup, and perineal areas will become relaxed as hormones release in preparation for delivery.

Hours away: The udder will fill with milk and become warm, and the waxy buildup will drip off—delivery will be any time.

Equipment for foaling:

A bright flashlight

At least 8 feet of cotton string

Scissors

A cup with Betadine or iodine solution

Clean towels

Vet's phone number or someone knowledgeable

Enema tube, mild soap, and lubricant or foal laxative

Foaling and afterbirth:

1. When you have noticed that the udder has filled with milk and the waxy buildup has dripped off, the mare may also become restless, urinate frequently, sweat, get colder, breathe faster, etc. Some of these signs are similar to colic but keep watch. Her water will break and she will go into labor.
2. You should already have your foaling equipment ready, so now wrap the mare's tail to keep it out of the way and clean manure out of the stall. Take out any objects that can get in the way.
3. Some vets suture the vulva during pregnancy to prevent infection. These should be removed now.
4. The mare will lay down and may get up a couple of times, then finally lie down and push with the contractions. Her body will go stiff and all four legs will strain and she will grunt. Watch from a distance to make sure everything is going all right
5. The foal's feet will appear within 15 minutes after the mare lies down. You should see both together, one before the other. If you do not see two feet, gently reach in and straighten the other foot so that the bottoms of the feet are facing downward towards the mare's legs.
6. The foal will slowly emerge, and the most difficult part is the shoulders. The foal's front feet should be uneven, one farther forward than the other so that its shoulders are slanted to make it easier.
7. You can help with the shoulders by wrapping the foal's legs in a towel and gently pull them down towards the mare's hocks ***only*** during the contractions. DON'T pull when she is not contracting or you can tear something or hurt the foal.

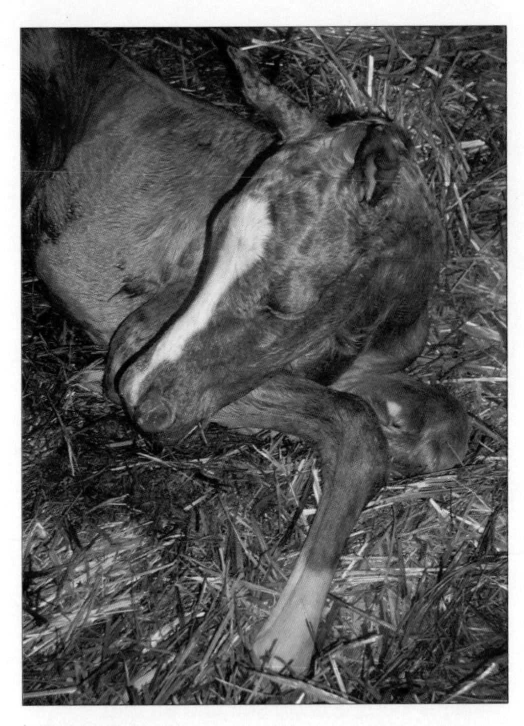

▲ Newborn foal

8. Once the shoulders are out the rest of the foal will slide out quickly. Clear the membrane from the foal's head. If he does not sputter and cough then you need to help him breathe.
9. Let the mare and foal rest, once the foal is breathing, for at least 20–40 minutes. Don't cut the cord and don't startle the mare or she will stand up and break it. This is a good time to check the mare's health and you can imprint the foal now (see below).
10. After leaving them for this long the placenta will drop out of the mare when she stands up. If it doesn't within 4 hours, then call the vet.
11. If the cord does not break on its own when the mare stands up, cut the cord three inches from the stump and tie it with a piece of string. Then dunk the whole end of the cord in iodine solution.
12. Give the mare fresh hay, water, and grain and make sure she is comfortable. Watch the foal during this critical time (see below).

Problems during delivery:

Mare pushes 45 minutes and no feet come out: Call your vet right away or else the foal may not get enough oxygen.

Foal's feet are upside down: Get the mare to stand up. If the foal is upside down this may turn him around. If not, then the foal is breech and you need to call your vet right away. Walking the mare may slow labor until the vet arrives.

Labor stops completely: Call the vet. If her placenta is too thick the water may not break and no labor will happen. Or if she felt threatened for some reason then labor can stop.

Water broke but nothing happens after 20 minutes: If the foal isn't moving then the hormones that stimulate labor weren't released. Tug on his leg and he should tug back, stimulating labor. If he doesn't, call the vet right away!

The foal should stand up within a couple of hours ▸

The shoulders are out but the mare can't get the hips out: Call the vet right away because their pelvises have become locked.

The mare stands up quickly after birth and blood goes everywhere: Blood comes out of the cord no matter what, but if it is coming from the foal's navel, pinch and tie it with cotton string.

Mare is still lying down but the placenta came out: Break the cord and tie it off.

New foals:

If the mare was ill before birth the foal may be malnourished. The foal should be able to walk, stand, suck, and nurse within 2–3 hours. He should also pass his first meconium (dark, hard manure) within 6 hours. Some foals are constipated, so a mild soap and water enema or a foal laxative can help. Some small, weak foals may be fine at first but weaken after 24 hours and a vet will be necessary. Keep the foal in a dry, warm place and when it is stronger let it run in a small pasture during good weather. If it is an orphan keep it with other orphans, a pony, goat, or horse. Don't treat the foal as a pet. Get the foal all routine vaccinations and deworming at 60 days, with a booster after 4–6 weeks, and control parasites every 8 weeks.

Foal problems:

Hasn't nursed within 6 hours: Show him where the nipple is, and halter the mare so you can hold her while he drinks. If he still can't do it call the vet because he may need special help or have a problem.

Mare won't let the foal near: There is too much going on around her. Rub fluid from the foal's body (amniotic fluid) on the nostrils of the mare and go away. However, if she might harm the foal, restrain her and call the vet for advice.

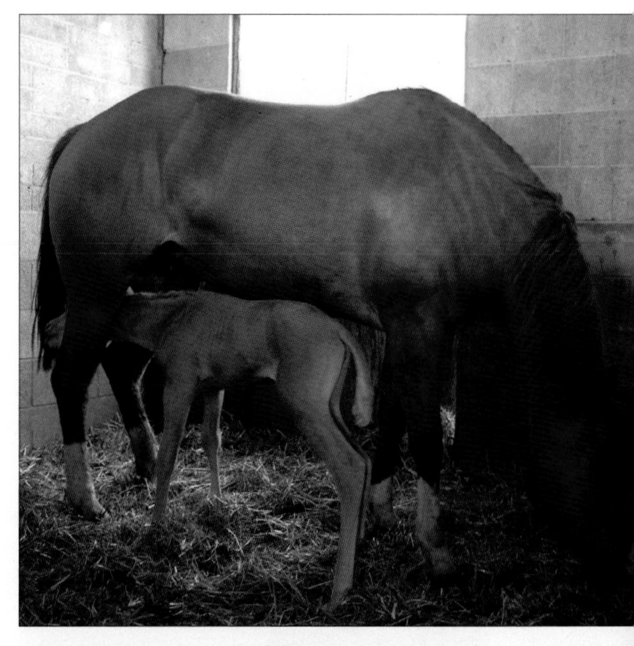

▲ **The foal should nurse within a couple of hours**

Foal is weak, makes a barking sound, wanders, has convulsions, and walks like a drunk: Call your vet if these last an hour after birth, because they may be signs of brain damage.

Has frequent, watery bowel movement: He may have a virus so call the vet!

The whites of the eyes are yellow: Call the vet right away; it could be jaundice, a ruptured bladder, or a milk problem.

Legs are deformed: Call your vet if the legs do not straighten out when you let them out to pasture the day after. If the joints are puffy call the vet right away because it could be an infection.

Navel is swollen or drips urine, testicles are huge, or the filly looks like she has testicles: Call your vet; it could be a hernia.

Foal is solid white and has blue eyes: Called lethal white syndrome, it causes abnormalities. Call your vet; the foal may have to be euthanized to prevent suffering.

Imprinting and haltering a foal:

Use your hands and a clean towel to rub her neck, body, and legs until she is completely relaxed. Then run your hands over the ticklish spots, the gums, ears, tail, hooves, girth, and teats. Then take the foal's halter and run it over her body before slipping it over her head. Then remove it and replace it a few times. This should take about 20–25 minutes. Make sure the mare can see her foal and she is comfortable.

HORSE TRAINING AND HANDLING

We shall take great care not to annoy the horse and spoil his friendly charm, for it is like the scent of a blossom—once lost it will never return.

~Pluvinel

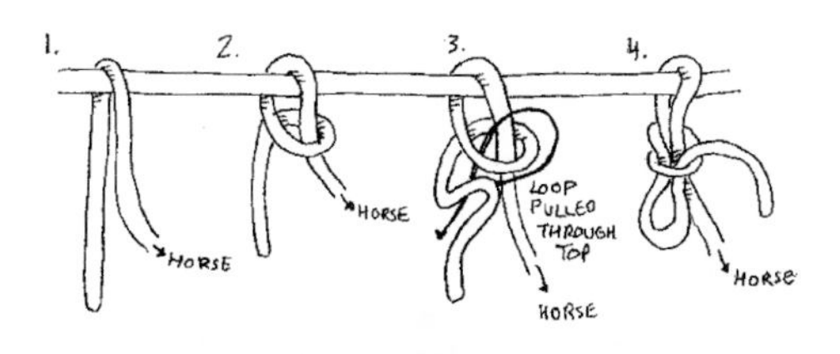

▲ **Quick release knot**

Handling:

Note: Inexperienced people must get some experienced advice and help when buying a horse and equipment, handling a horse, etc.

Respect and learn from the horse.

If the horse doesn't respond to something, try something else.

You are the leader—be positive and confident.

Recognize warnings that something is wrong.

In a battle of strength the horse will win.

Use intelligence to work around problems.

Have everlasting patience.

About horses and handling:

Horses are ***not*** color-blind and they can see almost all the way around themselves, except right in front of their feet and a space behind them. They hate to step on unstable ground or anything unreliable. A mare will be more moody and have difficult times during her heat cycles, unlike a gelding, but will be more sensitive to your feelings. Be aware of what the horse is doing even when your back is turned, and keep alert at all times. Horses should *only* be tied with a quick release knot. It is dangerous to you and the horse to use any other knot because you might need to be able to untie the horse quickly.

Lifting the hooves:

Run your hand down the leg to the bottom of the cannon bone, just above the fetlock. Put pressure on either side of the leg with your thumb and forefinger. The horse will lift his foot for you if you find the right spot. When you left a rear leg, keep the hoof turned upwards instead of down and support it with your legs. If the horse doesn't pick up his foot, lean against him slightly to push his weight to the other leg, which will encourage him to lift it.

Putting on a halter:

1. Have the halter unbuckled and the lead rope attached before you approach the horse.
2. Stand on the left side of the horse and pass the end of the lead rope under his neck with your left hand. Put your right hand over the horse's neck and grab the end. Pull the rope down and catch it in your left hand.
3. Pass the halter buckle under the neck of the horse and grab it with your free right hand.
4. At this point you should have the rope and halter strap in your left hand with your right arm over the horse's neck and the halter buckle in your right hand.
5. With your hands on either side of the horse's head, position the noseband so the horse's nose will slide into it easily. Raise the halter into position.
6. Bring the halter strap over the horse's head, placing it right behind the ears and fasten the buckle. Keep on holding the rope in case he tries to walk away.
7. Remove the loop of rope from around the horse's neck and lead away. The halter should fit properly, not too loose or too tight. The noseband should lie

A submissive horse and an angry horse

two fingers from below the cheekbone, and have two fingers of space between the nose and the noseband.

Vocal communication of horses:

Neighing: greeting, fear, or anger—depends on the tone.

Nickering: greeting, calling foals, or courtship.

Squealing: excitement, warning, or resentment.

Snorting: alarm or playing at being frightened.

Grunting/groaning: great effort or pain.

Body language of horses:

High-held tail: excitement.

Clamped-down tail: fear and submission.

Ears pricked and forward: startled or saw something in the distance.

One ear back, one forward: attention is split.

Droopy ears: lack of attention.

Ears back: submission, sleepiness, fear, or anger—depends on body language.

▲ Startled or fearful horse

Nose high and top teeth exposed: smells something unexpected.

Tense, tight mouth: upset or confused.

Head lowered, moving the tongue around the teeth: submission.

Nudging: friendliness, seeking attention.

Thrusting with head: aggressiveness.

Jerk head back: startled.

Shaking head vigorously: get rid of an annoyance.

Pawing with hoof: warning, investigating, frustration, or anger—depends on body language.

An interested horse ▲

▼ This horse is possibly fearful and angry

▲ A bored horse chewing on a fence

How to ride bareback:

1. Use a fence, trunk, person, or other handy elevated place to get on the horse (since you have no stirrups).
2. Sit on the widest, most comfortable part of your behind. Your legs will be further forward than in a saddle and you should not be leaning into your crotch. Your weight should be evenly balanced and your lower legs should be relaxed.
3. Grip the horse with your thighs and knees, and hold onto a handful of mane. Start off with a walk, and then move up to a trot. It will be bouncy but maintain good posture and stay balanced.
4. Move up to a canter, and as you both learn to ride bareback you can let go of the mane and use the reins only.

Types of harness:

Buggy harness: for pleasure driving or training.

Farm harness: for practical work.

Logging or plow harness: a lightweight harness for skidding, dragging or plowing. It is like the farm harness but instead of a britchen and holdback assembly, has a crupper and long lazy straps.

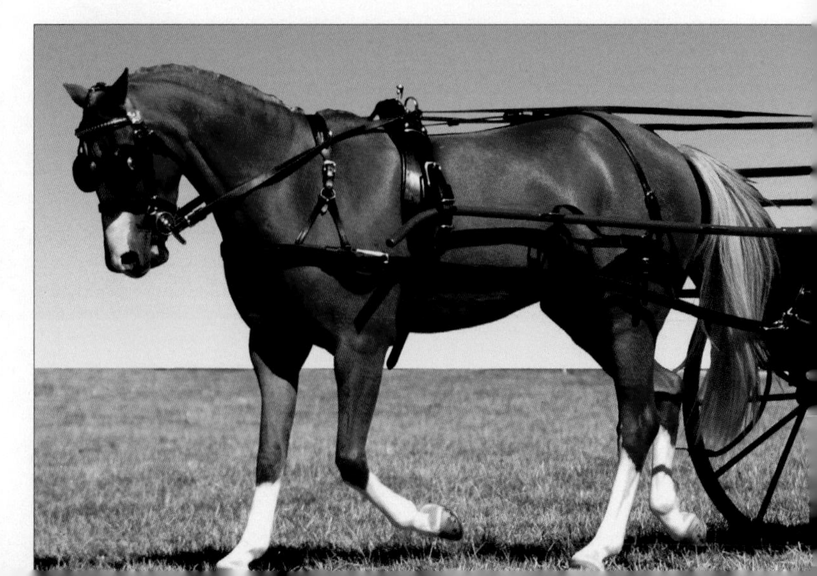

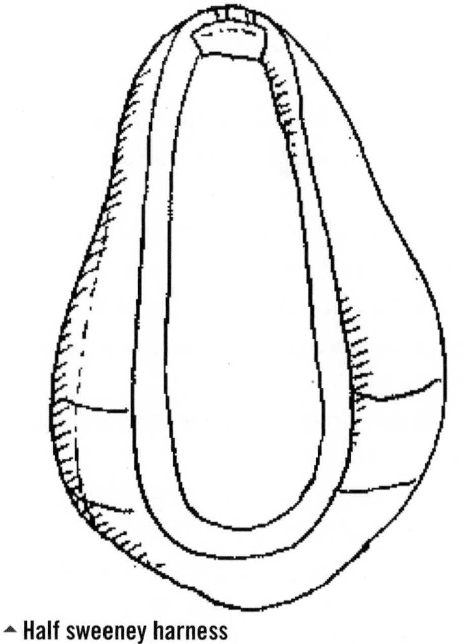

▲ Half sweeney harness

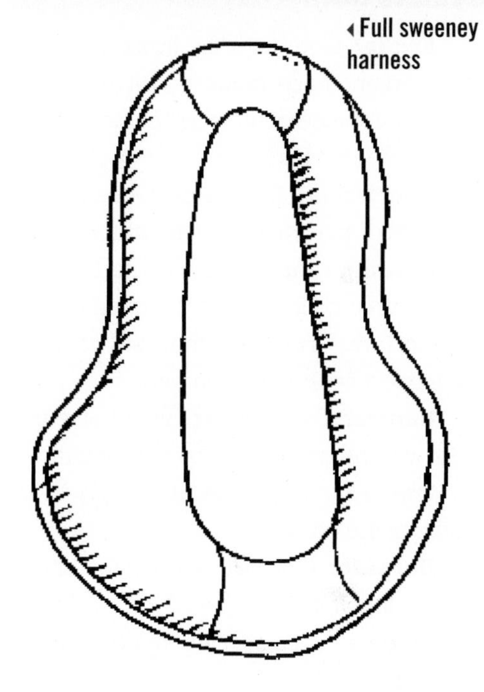

◂ Full sweeney harness

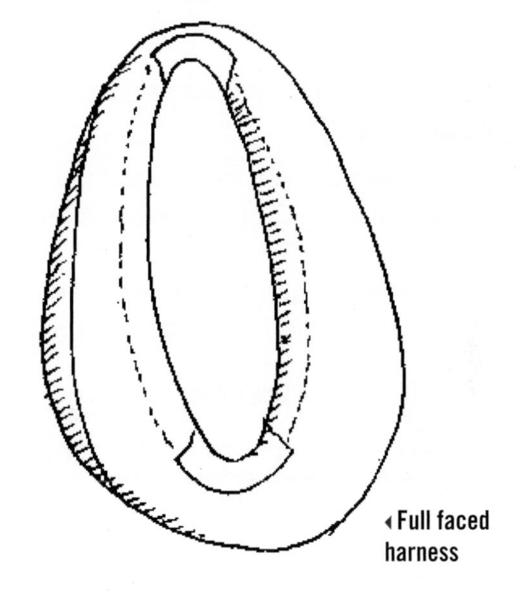

◂ Full faced harness

Harnesses come in pony size (800 pounds and less), regular horse size (800–1,400 pounds), and draft horse size (1,400 pounds and up).

Collars:

There are three shapes of collars: full face for thin-necked horses such as buggy horses, half sweeney with less stuffing by the withers for muscular necks which are used for draft horses or thick-necked horses, and full sweeney for very thick necked heavy draft pullers. The collar should have enough space at the bottom

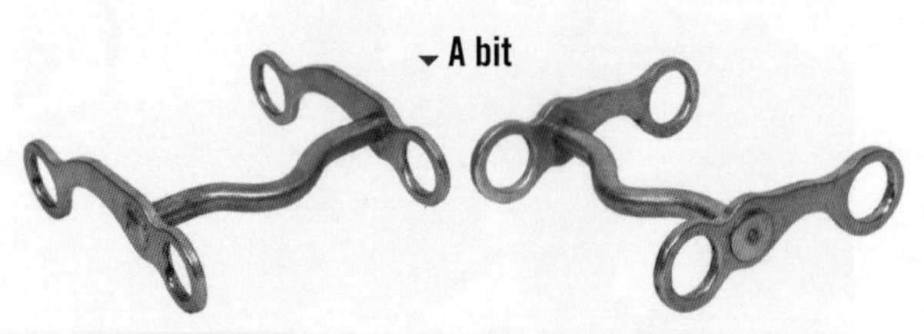

A bit

of the neck to let you slide four fingers in between easily.

Parts of a plow harness (from head to tail):

Bridle: Goes over the horse's forehead and nose, holds the bit in place.

Bit: A piece of metal in the horse's mouth to direct his head.

Check rein: Also hame. Goes from the bit over the housing to the other side.

Leading rein: A chain going from the bit to a strap connected to the trace chain.

Collar: A ring of leather that goes on the horse's neck so he can pull.

Hames: Two pieces of wood on either side over the collar, held by the topstrap.

Topstrap: Holds the hames together on top of the collar.

Housing: Covers the top of the hames and the topstrap.

Meeter: Connects the hames to the crupper.

Tug hook: Holds the chain that goes from the hames to the trace chain.

Parts of a plow harness

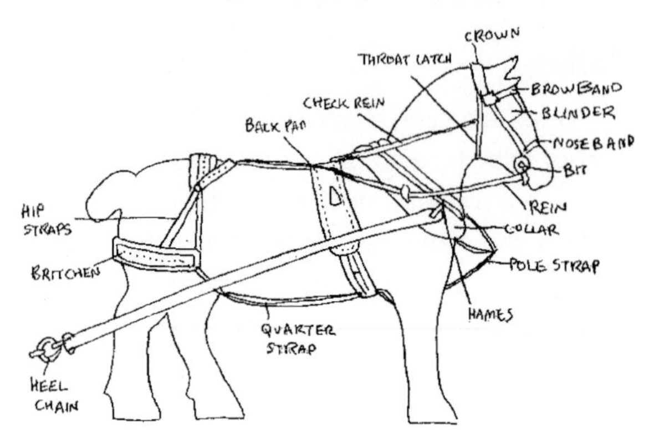

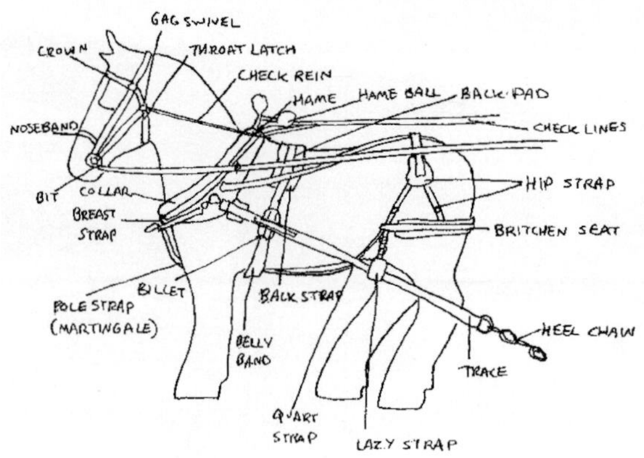

Parts of a shaft harness

Martingale: Strap connecting the collar to the girth.

Crupper: Runs along the back of the horse and under its tail; backbone of harness.

Lead guide: Strap from the crupper to the leading rein behind the housing.

Backstrap: Goes over the horse behind the lead guide to connect to the trace chain.

Girth: Goes from ends of backstrap under the horse's belly.

Loin strap: Strap just in front of the tail connecting to the traces.

Trace chain: Also traces. Goes from backstrap to spreader; pulls the plow.

Spreader: Keeps the traces apart behind the horse's legs.

Parts of a shaft harness (from head to tail):

Blinker: Also blinder. Sits on the bridle over the eyes of the horse so it can only see ahead.

Saddle: Sits just behind the housing; holds the chain that carries the shafts.

Single tree

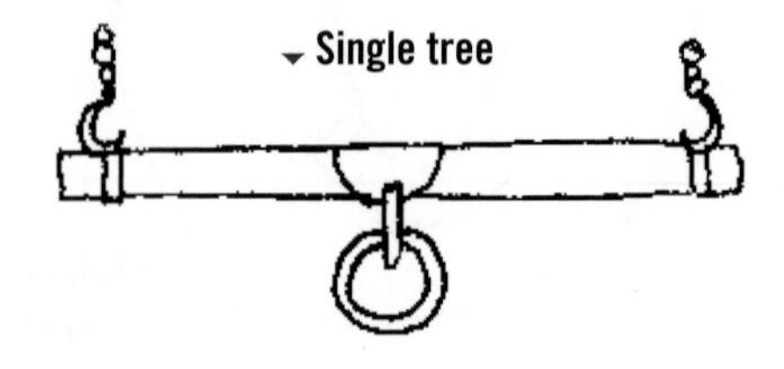

Saddle cover: Underneath the saddle; adds support.

Ridge tie: Underneath the saddle cover; connects to the ridger.

Ridger: Strap underneath the ridge tie; connects over back of horse to the girth.

Staple: Connected to the chain going to the harness; holds the shaft.

Shaft: Long pole for pulling the cart or wagon.

Loin straps: Three straps over back of horse behind the saddle, to help carry shafts.

Quilter: Piece of wood on shaft connected to ends of loin straps and chain from staple.

Measuring cinch size:

1. To find out what size cinch your horse needs, put the horse on a level surface. Run a rope around the horse just behind the withers where the front cinch would be.
2. Pull the rope snug but not tight enough to press into the skin. Hold the point where the end meets the rope to mark it. Take it off the horse (don't let go of your mark) and pull it straight and measure with a tape measure.
3. Use the inches and divide by 2. Subtract 3 from the result, and if it is not a whole number round up. This is the horse's cinch size in inches.

Inspecting a saddle:

1. Squeeze the saddle close to the pommel at the very front. There should not be any movement, and you should not hear any grinding noises. If you do, there is a weakness in the tree.
2. Make sure each bar where the stirrup attaches is secure and no rivets are loose. Wiggle and pull hard to do this.

3. Look under the flaps and check for weak stitching on the girth billets, the straps the girth attaches to. This is the most important check. Flex and twist the straps for dryness and wear, and if there are any cracks don't use it.
4. Check that the girth and any elastic ends are in very good shape, and check the buckles for wear or rust. This is the second most important check.
5. Check stirrup leathers for dryness and wear, especially where the leather folds. Flex and twist and pull to check for strength. Replace leather with cracks or unraveling stitching.
6. Check all the leather on the bridle by flexing and twisting, particularly where there are folded areas. Any cracks or separation mean you have to get new ones.
7. Check the bridle's stitching. If you have an awl and some tools this can be fixed yourself.
8. Check the bit for rough edges and replace if it shows roughness or wear.
9. Check the buckles on the bridle for rust and being bent.

Inspecting any tack and harness:

Check the leather for dryness, cracks, and wear. Once leather has a crack it needs to be replaced. Look for stitches unraveling and if possible use an awl to repair them. Check buckles for rust and bent tongues. Keep the leather in good condition by using leather dressing on a regular basis.

Harness care:

1. Take the harness completely apart and inspect the whole thing. Do any necessary repair work.
2. Fill a washtub ¾ full of warm water (not hot) with a handful of laundry soap. Put the harness in the tub. Get a board 10–12 inches wide and 6 feet long and put one end in the tub. Rest the other end against a bench or stool.
3. Put each piece of harness on the board so the water will run down into the tub, and scrub with a scrub brush. Put each piece on the floor on newspaper to dry if inside.
4. Warm the harness oil and when the harness is still slightly damp, apply with a rag or paintbrush.
5. When the oil has dried (it can take all night), wipe off excess oil with a rag. If the harness is in bad condition add another coat. Too much oil will wear out your harness.
6. Apply leather dressing (see below) and reassemble the harness.

Leather dressing:

5–6 ounces beeswax
8 ounces lanolin
8 ounces cedarwood or other oil

Heat all ingredients together to 160° in a double boiler and mix thoroughly. Pour into containers and use as a conditioner and waterproofing for leather.

What is a trailer?

Horses have to be transported behind a truck in a trailer. The best trailers are spacious, have room for at least 2, and have been inspected for safety. The trailer should be stocked with extra halters and lead ropes, emergency flares and reflector triangles, flashlights, jumper cables and fuses, spare tire and a jack, duct tape, first aid kits, etc.

During the trip:

Leave the horse untied or use a long-line. This prevents respiratory stress.

Put him with his buddy in the trailer to prevent stress.

Keep the trailer spotlessly clean. Each time you stop clean out the manure.

Have the horse practice loading and unloading before ever taking a long trip.

The risk of no ventilation is worse than too much cold and rain. Blanket the horse and leave the windows open that are not going to cause a strong blast.

Drive slow and steady—put a ½ full glass of water on the dash and if it sloshes to ¾ full then you are driving too rough.

Provide lots of water at least every four hours, even if they are not drinking.

Put hay in the trailer and wet it down a little bit to prevent dust.

After the trip:

Hose out the trailer and make it ultra-clean.

Let the horse rest for at least a week after a long trip.

Keep the trailer in good shape and check floorboards, ramp, brakes, and hitch.

Training rules:

Do ***not*** be afraid of your horse.

Throw out any preconceived ideas you have about training.

The horse can do no wrong: you influence everything the horse does.

Recognize the communication of the horse.

Do ***not*** use any pain—including hitting, jerking, pulling, tying, or restraint.

Base the training on: you would ***like*** him to do it (not: he ***must*** do it)

Training the foal before weaning:

1. The foal should already be comfortable having a halter on its head. When the foal is three days old, attach a lead line to the ring on the halter and stand behind her.
2. Pull the line until her head is pulled at a somewhat awkward angle, her head being pulled back towards you. After a while the foal will take a step towards you to relieve the pressure. Tell her what a good foal she is.
3. Stop for the day, and repeat every day for longer and longer, until the foal follows you around. After this it should be simple for her to learn to stand tied. Praise her frequently and she will train even faster.
4. Think of everything that could be scary to a young horse (or even an older horse). Trailers, the pasture, cars, getting new shoes, blankets, clippers, scary objects, etc. are all good opportunities. Expose the foal to the different situations, and rub scary

objects over her as an imprinting device, massaging until she is relaxed, then praise her. Every time she stands still and is calm, reward her for it. Doing this while they are young makes later training easier, and even during later training use the imprinting method to reduce fear.

5. When imprinting with objects or lifting and tapping hooves, do not let go when the foal struggles, or else you will imprint on the foal that if she fights she gets her way. With the initial imprinting you simply rubbed her down. Now you are introducing objectionable things and she must get used to this.

Training the yearling:

1. Backing up: Pull the lead rope towards her chest so that the halter is applying steady pressure to her nose. Put your hand on her chest until she takes the tiniest step back, and then praise her very well. Do this progressively until she backs up comfortably.
2. Squaring up: When backing up is easy, lead her forward at a brisk pace, stop quickly and face the horse. Back up 1-2 steps until the back feet are square, or the hooves are in a straight line to each other. Then ask her to step forward so the front feet are square also (squaring up is when the four feet make a box). If she is confused, pick up the front feet and set them down where you want them.
3. Pivoting: Face the horse and push the lead rope under her chin towards her rear, and use your other hand at the same time to apply pressure to the outside of her left shoulder. Take a step forward and she will get uncomfortable and turn away by taking a step. Do this

and increase the steps until you can go all the way around. Make sure she is not crossing one leg in front, instead of behind. Then stop using your hand to apply pressure and use it only if she arches her neck or gets crooked.
4. Lunge line: Use a lead rope and lunge whip (use imprinting if you need to), and stand at her hip. Tap the whip on her rump gently in a rhythm and cluck your tongue as the signal to move forward. Do this continuously until she steps forward. Then praise her profusely. Work up to the point that you can go in a full circle without stopping. If she stops or pauses, tap and cluck.
5. Go and stop commands: When she can go around without stopping, turn your body sideways, lower the whip and say "whoa" while applying pressure to the lead rope. Do this until she stops, and then praise her. Put her on a longer lunge line and do it again.
6. Trotting: The signal for trot is the same as for forward, except faster and louder clucks until the desired speed is reached. Use lots of verbal praise, but it can take as long as six months to learn this.
7. Lope: Go from a walk then to a trot in both directions, then step close to the hindquarters and give the verbal cue (a kissing noise) and raise the whip. Do this until she goes into a lope

and verbally praise her. At this point she may rebel, so just stop gently and start over when she is calm. To prevent running or bucking, slow to another gait when you notice frustration.

Preparing a young horse for starting with a saddle:

This is to prepare a horse that is too young for a saddle, but big enough to have a saddle on him, to get him used to people handling him and riding. Basically you use your hands to simulate different riding equipment and care procedures for 5–10 minutes a day until he accepts it. To find out if the young horse is ready to begin real training, feel his knees. There should be no space in between the bones.

Starting a grown horse:

Note: If your horse is completely green, spend a few days getting used to the bit and long lines. Practice on an experienced horse before trying any of this.

1. Get two light 30-foot long lines, a snaffle bit, a saddle, saddle pad, one stirrup leather, and a halter.
2. Put the halter on the horse and bring her into a pen with a clear area at least 50 feet in diameter. Carry the long line with you.
3. Stand in the center of the pen and rub the flat of your hand on the horse's forehead. Do not pat.
4. Move toward the rear of the horse, while staying out of reach of any kicking. When you are behind her or she flees, throw the end of the line toward her rear. She will start going around the pen.
5. Because the horse is retreating, you must advance. Continue throwing the line from the center of the pen two times per circle, or as many times as needed to keep her going. Be aggressive—keep your eyes focused on her eyes, and your shoulders square with her head. Stay away from the kick zone.
6. Get the horse to canter five times around one way, and then go the other direction. Watch the body language of the horse carefully.
7. When the horse is ready to stop, the ear closest to you will slow down or stop moving, while the outside ear will still be moving. The head will start to lower and turn a little towards you, until it is near the ground. She will come closer and be running her tongue around outside her mouth.
8. Coil the line and put your eyes down submissively. Don't look at her eyes, and turn slightly away, showing your back. If she comes to you, that's great. If she stands and looks at you, start moving closer to her in curves, don't make a beeline for her.
9. If she leaves you, do a few more laps, and then repeat the whole thing. She should come up to you and reach out to your shoulder with her nose.
10. Give her another rub on the forehead when you can approach her head. Then walk away, moving in semi-circles 10 feet wide. She should follow you or at least keep her head in your direction. If she doesn't, start all over with laps.
11. The horse should follow you to the center. Start on the near (the left) side, and use both hands to massage her neck, withers, back, hips, fore and rear flanks. Repeat on the other

side, and then pick up the feet like you would normally.

12. Bring all your equipment into the pen and put it in the center on the ground. Let the horse look it over, then move it and the horse in opposite directions until she follows you and not the tack.
13. Snap the line on the halter, with the line over your left arm 3 feet from the snap. Place the saddle pad very gently on her back just before the withers. Slide them back into place. Pick the saddle up with the irons and girth over the seat and move slowly along the near side from the neck to the shoulder with the saddle resting on your right hip.
14. Place the saddle on her back and move past her head to the off side (rub her forehead while you're there). Fasten the girth at mid-fetlock point without hesitation, but slowly and smoothly. Go smoothly back to the near side, rubbing her forehead again
15. Stand near the foreleg. Place the front buckle on the front billet, then draw it tight enough so that it won't turn if she bucks, but not too tight. Place the back buckle on the back billet, tightening it a little more than the first. Go back to the first one and level them up.
16. Unsnap the line and step backward cautiously, with the line in hand. Stay out of the kicking range and use the line to move her around the pen. Don't encourage bucking or submitting.
17. When the horse shows signs she wants to stop, make sure she is traveling comfortably with the saddle, then let her come back to you. Put the bridle on and put the reins under the rear of the saddle with plenty of slack. Put the extra stirrup leather through the off-stirrup iron so it hangs halfway through. Move to the near side and pick of both ends of the leather, carefully. Buckle it through the near iron.
18. Take both lines at the snap end. Put one over the seat of the saddle, letting the snap reach the ground on the off side. Put the second snap through the near iron from back to front, snapping it on the near-side bit ring. Do it also on the off side.
19. Go to the near side. Pick up the two lines and move at an angle backwards, keeping out of the kicking range, toward the rear of the horse. Move her forward and swing the right rein over her hips to the long lines. Go slowly if you are inexperienced. The goal is to make a little communication with the mouth, but cautiously. If you are experienced, ask her to canter in a circle then trot both ways. Practice a few turns and stops. Face her away from the center and rein back one step.
20. The horse is ready to be ridden. Make sure the saddle is adjusted and the girth is tight. Snap a line on the near-side bit ring. Then pull yourself up and rest your belt buckle (or stomach) on the pommel but don't get on. Have the horse led in circles to the right and left, and if she's happy, put your feet in the stirrups and get all the way on. Circle again.
21. If the horse is still happy, make bigger circles. Unsnap the line carefully and ride at a walk or trot around the pen in both directions, stopping to rein back a step between them.

22. Do not rush—this does not have to happen in one day. If the horse doesn't want a rider, wait until another day.

Fixing bad habits:

If you went out a bought a horse with a bad habit, or your horse developed a bad habit, the first thing to remember is that the habit was caused by humans. Behavior problems are usually caused by pain, from saddles and bits not fitting properly, injuries, improper health care, etc.

OXEN, CATTLE, AND WATER BUFFALO

And so we plough along, as the fly said to the ox.

~Henry Wadsworth Longfellow

Buying cattle:

To have cattle you need a fairly large barn, at least two acres of good pasture and a place to store several tons of hay. There are hundreds of breeds of cows, and they are divided into two groups: beef and dairy. Dairy cows make more milk and are gentler for easier milking. There are a few breeds that have been bred to do both, such as Shorthorn and Brown Swiss. For a small homesteader dairy breeds are probably best because they can still be used for beef and the steers could be used as oxen. If you are inexperienced with cows buy a four- to five-year-old cow instead of a heifer because she will be more experienced in giving birth. You could even get a water buffalo, which are very gentle draft animals and make lots of meat. Don't get a cow with mastitis, tuberculosis, brucellosis, birthing problems, gentleness problems, or an extra long udder.

What is an ox?

They say that oxen have been used for over 8,000 years (give or take a few years). Any breed of the bovine family, either male or female (bull, cow, or steer) can be used as an ox, although steers are preferred. They are trained to be guided entirely by voice commands, body language, and whip pointing. It usually takes four years to train the oxen fully and for them to reach full strength. It is simpler and cheaper to use an ox than it is to use horses. They don't need a harness or shoes, nor as much medical care and they are gentler. And you can eat it when it gets too old to work (if you don't mind tough meat).

Training oxen:

1. Start off with a pair of calves that have been picked for intelligence. Halter-break them the same way you would a foal, and teach them how to lead. In Asian countries most often people use

▲ Yoke of oxen

▲ Water buffalo

just one ox or one water buffalo, but it is nice to have a matched pair.

2. Measure the oxen for a yoke and make one. As they grow you will have to make increasingly bigger ones. Get them used to wearing it. Always put them on the side they started off with—never switch the left ox over to the right or they will get very confused. If one ox dies you will have to replace it with an ox trained for that side.
3. The oxen need to learn four voice commands: "gee gee" for "go right," "haw haw" for "go left," "whoa" for "stop" and "geddup" for "go," and "back" for "back up." Start off with "geddup" and "whoa" by saying the commands while using the lead rope to tell them what to do, until they do it without the rope.
4. Use a "gad" or whip to indicate directions when teaching "gee gee" or "haw haw." You can make a gad out of a 4-foot piece of hardwood sapling for a handle, and attach a thin 2-foot leather strap on the end. Never use it to punish—it is simply a physical aid to touch the ox. Work at least 30 minutes every day, right before dinner.
5. Once they have directions down, attach a pole with a noisy sled or cart, or even just a log, so they get used to pulling things. After they are 18 months old they should be ready to start pulling more heavy loads.

Handling cattle:

Cattle are color-blind with 360° vision and so they can startle quite easily. Their lives are based on habit and so they sometimes seem stupid but they aren't. Be very patient, quiet, and gentle, and utilize their herd instinct—if you get one cow to do what you want they all will follow. If you use food as a reward you can call them and they will come without any work on your part. Cows with calves are very protective, and a cow in labor can be violent so you might want to tie her for safety. Be careful around cows in heat, and never ***ever*** turn your back on a bull. Never let your kids around a bull, and if a bull threatens you, eat it. Don't breed a bull younger than eighteen months old.

Managing water buffalo:

Buffalo are more disease resistant and can live on less than other bovine animals. The mother cows will easily adopt orphaned calves. Their herd instinct is very strong, and they are smart enough to pull up fences with horns if they feel the need, so fences need to be very strong. Although intelligent and placid, they can be startled easily so be extra quiet and calm. The best identification mark is an ear tattoo, although in North America they are rare and so losing your buffalo is not a problem.

▲ Water buffalo are used in warm, wet climates for raising rice

Because they are so gentle the horns do not need to be removed, although sometimes it is nice because they can accidentally cause injury. Buffalo milk is thicker than cow's milk and has a butterfat of 6–8 percent, compared to a cow's 3–5 percent. But unlike goat's milk it can be separated and made into many things. They provide just as much meat as a cow, and can work as draft animals. Heifers are also used for draft work.

Food and water:

Like all other animals cattle need grass or hay, water, and salt. Cows have four stomachs and are made to eat with their heads down low. They need fresh (and unfrozen) water all the time, and a salt block that is sheltered from the weather. The best dairy cow pasture is lush wirh clover and alfalfa, and grasses and herbs like lavender, mustard, rosemary, and sage. In the winter they need tender green hay that was cut just after it bloomed, 2–3 pounds per 100 pounds of body weight. Every cow needs about 2 tons of hay over the winter. Don't give her more than she can eat until the next milking or she won't eat the old stuff. If you have only good quality hay or grass she will still give milk, but if the quality drops so will her milk supply. If you have no fenced pasture you can stake a milk cow with a large bucket of water and move her three times a day, which helps utilize more pasture.

Supplement grain:

After a cow gives birth you might want to supplement with grains in order to increase her milk production. Give oats, barley, corn, wheat, peanuts, soybeans, sorghum, peas, or legumes, and start off gradually, slowly increasing it as her milk supply grows. Near the end of her lactation start to taper it off. The most she will ever need is 1 pound of grain per 3 pounds of milk she makes. With quality grass and ½ pound of grain she will give about 90 percent of her milk-producing capacity. Cornstalks, washed and sliced root vegetables, and whole sunflower heads

Jersey ▸

▾ Holstein

make great supplements. If you feed grains make sure they are chopped or ground, and it is better to mix the grains together. Alfalfa hay is a grass that also has protein and so it may not need a supplement.

Health care of cattle:

Downed cow: Sometimes cows can't get up from the position they are in, or they are injured. If it's a position problem, get several people to pull her legs out and to a level surface. If it is an injury, make sure she is fed and watered until she can be moved.

White muscle disease: A calf gets very stiff joints and has difficulty moving to the point that it can't follow its mother and dies of starvation. There is an immunization for this.

Scours: A calf with diarrhea has scours. Don't change his diet—instead give him Kaopectate, which is a mixture of pectin and kaolin clay.

Heel fly: A fly that lays eggs in the heel of a cow which hatch and crawl through the cow's blood vessels, bore holes through the skin, and fly away. It can only be prevented with medicine.

Bang's disease: All of a sudden a calf will be born dead. Humans can catch the disease, which is called undulant fever. Prevent with a vaccination to heifers when 4–8 months old.

Udder cuts and problems: The same chapped teats, cuts, and other problems that happen to goats can happen to cows.

The life of a diary cow:

A cow can have a calf every year and constantly give you milk until she is 10–16 years old. Some cows have been known to breed until 20, but this is rare. Some of the best dairy breeds are Ayrshire, Brown Swiss, Guernsey, Holstein, Jersey, Red Poll, and Devon. Don't breed a cow until she is 13 months old—be careful, because a three-month-old can conceive. The basic yearly cow schedule is to breed her, and 2–3 months before she calves dry up her milk. Her gestation period is 285 days (9 ½ months). Usually she will go into heat for 18–24 days, but it will be 30–60 days after she calves, so keep an eye on her. Then you can breed her immediately. If you breed in the middle of May she will have her calves at the beginning of March.

Is my cow in heat?

A cow in heat will be restless, stamp her feet and twitch her tail excessively, and she may make lots of noise. Her vulva may get red and swollen, and other cows will try to mount her, even females. A normally peaceful cow might suddenly run through a fence, or bulls might struggle to get to her. She might try to mount you.

Giving birth:

When your cow is close to calving, bring her into the barn and put down a thick layer of clean bedding. She will be close when you see her udder swell with milk, and the skin around her tailbone gets loose. Check her frequently, and although she should be able to do it alone, if she doesn't give birth after 3–4 hours of labor call the vet. Make sure that the calf can get up and get milk fairly soon also, and take a big towel and rub it dry. A cow may hide her calf (called "brushing up") and go away to feed it during the day. Make sure you check on the calf, and if you want to move it just pick it up and the cow will follow. Keep your cow milked at least once a day or she will have too much milk and might get mastitis. If you are going to vaccinate, do it as soon as possible after birth. If you find the calf and it has become chilled, bring it in the house and warm it in a bath. Keep it wrapped up and dry after it has stopped shivering.

Feeding young calves:

A calf can be taken away right after birth. The best way is to separate them right away and bottle-feed the milk that you milk out of the cow. The first four days you will get colostrum, which all needs to go to the calf. Once the colostrum taste is out, you can drink it too—you will have

starts drinking the milk. Use this method only if you don't have a bottle.

Amount of milk to feed calves:

At birth: About 1 quart per day.

Birth–2 weeks: Work up to 2 quarts a day.

2–3 weeks: Work up to 3 quarts a day.

Too much milk causes scours so unless you have a really big calf he might just be greedy. Feed 5 times a day at first, and then go to 4 times a day for a few days. Then go to morning, noon, and night, and gradually work to twice a day until he is weaned.

plenty of milk for both of you. Never give the milk cold. If you don't have colostrum, use milk mixed with an egg yolk a few times or until the cow starts to thrive. If you can't do that, use fresh goat's milk mixed 50/50 with water. If using a bottle, make sure that the nipple is on very securely. If using a pan, it needs to be 6 inches deep and 1 foot across. Make the calf lie down and get someone to hold the pan. Get milk on your fingers so he sucks on them and then slowly lower them into the milk until he

Weaning:

Taking a calf away from the cow right away instead of letting it nurse on demand will make your calves very easy to wean. An on-demand calf will look better and gain weight faster than a bottle-fed calf, but will have a hard time leaving its mother for grass. Either way, wait for the calf to start being interested in other foods by offering grass and hay or creep feeding. As the calf eats more and more grass, feed less and less milk until the calf is only eating hay and grass. He should be weaned by 4–5 months old. During summer he will need at least an acre of good pasture. In winter a weaned calf will eat about 2 pounds of hay per 100 pounds of body weight per day. three pounds of cornstalks and cobs equals 1 pound of hay.

◂ **Milk cans**

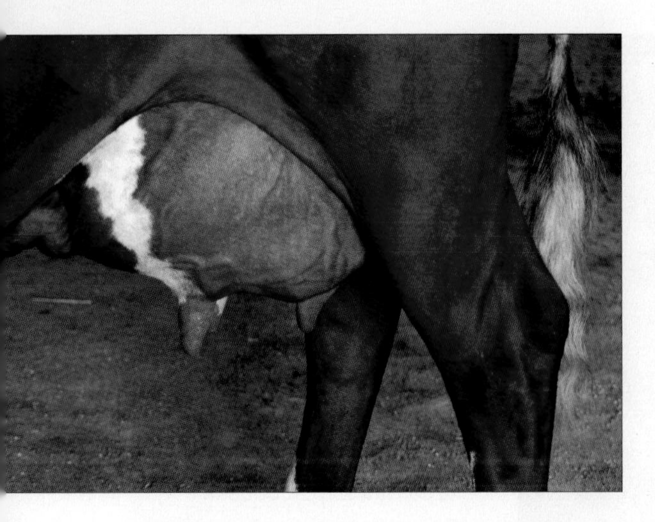

Caring for milk equipment:

Stainless steel without seams is the best and most expensive container for milk. Food-grade plastic and glass will work also if they are seamless, but don't use commercial plastic milk jugs for storing milk. Rinse buckets, containers and utensils in lukewarm water right after using them or you will get milk deposits that are difficult to remove. Then wash by scrubbing thoroughly in warm, soapy water, then rinse again in scalding water and air-dry upside down. Any cloth used for straining should be rinsed and boiled directly after. Use homemade soap and rinse the cloth in water with a little lye in it to sterilize it completely.

Steps to milking:

1. Clean the milking utensils in warm soapy water (bucket, strainer, or storage container).
2. Put the cow in the milking stall. Brush the fur to get out loose hair and dirt, put down fresh bedding and make the cow comfortable. Look her over for problems.
3. Put some feed in the stanchion's trough. Milking is done every 12 hours, and the best time is in the early morning before they go eat. Always be on time or the cow will get over-full and be in lots of pain.
4. Wash your hands and dry them, fill a bucket with water 120–130°F, and get a fuzzy cloth.
5. Wash her udder and teats. This helps the milk let down, removes dirt and bacteria, prevents milk from souring quickly, prevents a weird taste in milk, and makes the teats easier to grip. Wait 1 minute after washing to start milking.
6. The cow needs to be happy and relaxed for the milk to let down. Put your thumb and forefinger around the teat near the top of the udder, pushing up slightly, and allow the teat to fill with

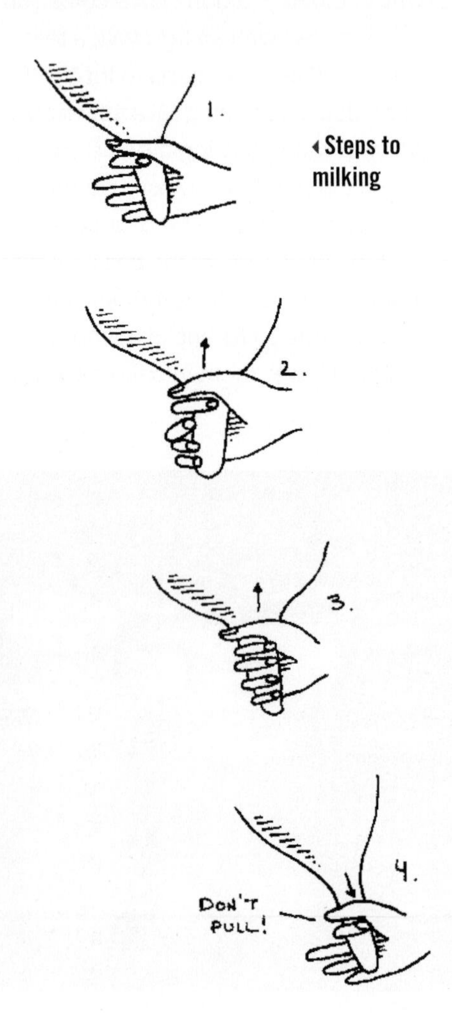

◂ Steps to milking

milk. Then close your hand around the teat and squeeze the milk out while pulling down, or stripping. Keep your hand away from the nipple hole or the milk will go all over. Squirt the first 3 squeezes into the ground because the first milk has more bacteria in it. Make sure you completely empty the udder or she will produce less and less until it is gone.

7. Sometimes the cow may need to poop while you are milking. Grab the bucket quickly and get out of the way if necessary.
8. Strain the milk. Use a regular kitchen strainer lined with several layers of clean fabric such as a dishcloth, muslin, or even a cloth diaper.
9. If you choose to pasteurize, put the milk into a double boiler on your stove. Use a thermometer to heat the milk to 161°F, and keep stirring so it is heated evenly. Maintain that temperature for 20 seconds, then quickly remove from heat and put the pot in very cold water, stirring constantly until it gets down to 60°.
6. If you don't pasteurize, put the storage container inside another container full of cold water. It should be chilled to 40° within 1 hour. Store any milk in the coldest part of the fridge.

Milking problems:

The milk won't let down: If you have washed her udder with the warm water and waited a minute and the milk doesn't let down, try massaging either with the cloth or with bag balm, or gently pat the udder like a calf butting her. If she still won't let down, suck on the teat a little and the milk will come down.

Drying up milk: If you want to have a calf every year, you have to dry her up after ten months of milking. However, it's not necessary and you can wait until she dries up by herself. To dry her up purposely, gradually milk less and less, leaving the udder with a little milk in it and milk less often (see below). Watch carefully for mastitis.

Mastitis: The first sign is usually flaky, lumpy, or stringy milk. Check the milk at least once a week by squirting it into the cloth. Feel the udder for tumors, or large hard areas, or an abscess—a red, tender swelling of the whole side of the udder. The udder will feel warmer than usual, and it will be more difficult to milk. If it gets worse, the milk may turn yellow or even brown or pink from pus and blood. ***Don't drink the milk!*** Don't throw it somewhere that an animal can lick it, and wash your hands very well after touching the animal because it is infectious. Get antibiotics in her quickly. A cow that has had it once is susceptible to it. A bruised udder or trying to dry up too quickly so that she gets bruised from being too full can all cause her to be likely to get it.

Drying up:

Forcing her to dry up improperly can risk mastitis. Stop giving her grain and supplements, reduce her water a little bit, and watch her carefully. As you milk her, don't strip her as well, and take less milk until it starts to slow down. Then milk her less often by milking her only once a day, then going to every other day, then every three days, then once a week until she is not making any more. If you plan to dry her before calving then you should start three months before her due date. And remember, if there are any signs of mastitis, don't drink the milk and don't let any animals drink it!

BEES

The keeping of bees is like the direction of sunbeams.

~Henry David Thoreau

Bee clothing:

It is sometimes better to buy a bee outfit because you know that it will be safe. But to make your own you will need:

Very thick leather gloves that are easy to move so you can handle frames easily.

Gauntlets that cover the top of the work gloves and come up above your elbow. They need to be tight around your arm, made of canvas, and baggy so the bees can't reach you.

Work boots with your pants tied tightly to the outside of the boots. If bees get in your pants they will crawl up your legs quickly so it is even better to have an additional cover over the top of your boots and pant legs that is tightly sealed.

White, one piece, baggy coveralls. If they are made of rip-stop nylon then the bees can't walk on you.

A construction hard hat, jungle-style hat, or bee-proof straw hat with a firm, wide brim. The bee veil is difficult to make, but it goes over the top of the hat, sits on the brim, and is tied firmly with a drawstring. Make sure the drawstring is secure because if your hat comes off then you will get stung on your face. If you do have to use a hard hat or other hat without a back neck protector, you will need to make a cardboard protector which you can fasten on with duct tape.

Tools:

You will need a hive, a smoker (although a stick with burning rags might do), a hive tool (for a standard hive), a bee brush, and a feeder. The smoker is the most important, it has a chamber in which you build a small fire and put it out, which creates billowing smoke. A bellows-like pump pushes smoke out.

Bee skep:

A domed house made of coiled straw. The straw is gathered into long bundles and wrapped around with a thin strip of wood. Then it is coiled and sown, and a hole the height of one coil is cut out for a door. These are easily made, but are not recommended anymore because of the higher risk of disease. They are not easy to clean out or examine, while the new hives are much more accessible and hygenic.

Bee hive:

A standard hive has several layers of parts. At the bottom is a hive stand, a platform that keeps it level. On top of that is the bottom board, a thin piece which holds up the brood chamber. The brood chamber is where the bees live; the queen sits there and lays eggs and the cells hold baby bees. Supers, or honey supers, are shallow boxes that sit on the brood chamber and usually hold honey, although sometimes the bottom one holds more baby bee cells. It is better to have a shallow super than a deep one because they can get too full of honey and are difficult to carry. Each super has ten vertical frames which can be removed, and the bees build wax cells on the frames. On each frame is "foundation"or a flat sheet of beeswax that has six-sided hexagons imprinted on it as a template for the bees to build cells. The foundation can be bought or you can make your own.

Making foundation:

You will need a roller, either by finding a rare antique one or making one yourself out of hardwood maple and carving perfect hexagons all over it. Then use an 8–10 inch board, dip it in hot beeswax twice, let it cool, and then cut down the edge. Peel off the sheet of beeswax and use the roller to imprint the hexagon cells onto it.

The best bee situation:

Bees do better in a milder climate. You probably won't see much honey until the third year. Feel good if you get 1–2 gallons per summer per hive, but with practice you should get 4–5 gallons. It is difficult to do and better if you can learn it in person, because a professional knows how to stay calm and continue extracting honey even

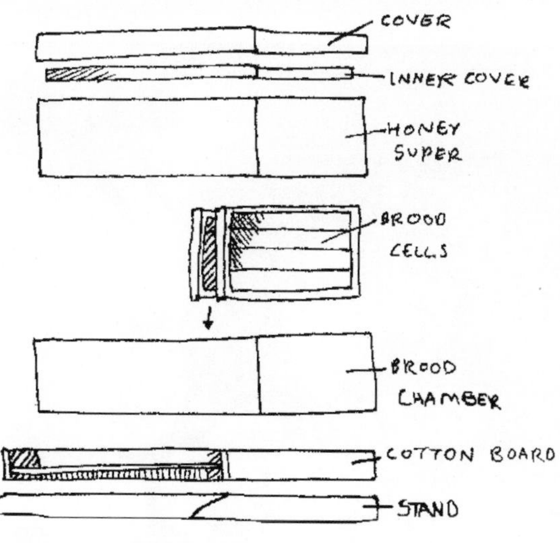

▲ **Parts of a beehive**

▲ **Frame with wax built on foundation**

when stung multiple times. Point the door of the hive towards wherever you want them to go and the morning sun, and away from places that people and animals tend to be so that their flight path won't cross you. Also don't put the hive near motors, tractors, or other vibrating noises. In some areas used bee equipment is illegal because of the spread of disease, so contact your area's department of agriculture. After setting up your bees you shouldn't have to see them too much until you actually harvest honey.

Bee stings:

The treatment for bee stings is under First Aid, but as a person working around bees you are at special risk. You are certain to get stung many times over the years and you can become somewhat immune to the sting. However, you can suddenly get an allergic reaction for no reason. Before getting bees, get tested for allergies to bees and bee stings. Get protective bee gear and always work with someone so that if you do develop an allergy they can get help. When you do get stung, scrape the stinger out quickly with your fingernail so that less venom will enter your skin.

Types of bees:

Honeybees: They can only sting once because they leave their stinger in your skin and then die. This makes them more careful with their stinger and less likely to get you. There are three breeds: Italian, Caucasian, and Carniolan. Italians work hard, Caucasians sting less, and Carniolans are the gentlest.

Wasps/yellow jackets/hornets: They don't eat flower nectar, they eat bugs, fruit, and other foods. They are great to have in a garden because they eat thousands of pests in one season, but they will also sting for any reason, and over and over again.

African bees: There is a lot of hype about these bees since they escaped from a lab in Sao Paulo, Brazil, because they tend to attack as a swarm and have killed a few people. You can outrun them if you can run 15 mph for 5–7 minutes, or 8–10 city blocks. Don't ever use bug spray because it just makes them angry—the only way to deter them is with huge quantities of soapy water from a fire truck.

Buying bees:

In the old days a swarm would be found by accident and then encouraged to move into a hive that the homesteader made. You can either buy bees from a supplier or buy a whole hive from a local person. The last option is the easiest since the bees will already be producing a fair amount of honey. Once you have one or two hives you can make as many as you want be moving some bees (see below).

How to move bees:

To move an entire hive, plug the door of the hive very, very tightly with a folded window screen, grass, or cloth. Don't use wood or other nonporous material or the bees can suffocate. Make sure it's very secure because if the plug comes out while you are walking you will be in the middle of an angry swarm. Once you have one hive of bees you can create more by moving only a few bees. Around May 1st, use your most well-established hive and take out four frames that have some brood cells (cells from which worker bees will hatch), some honey, and some bee bread (or

Calming the bees to remove them from a frame

▾ **Brood cells and a newly hatched bee**

pollen), which looks yellow and grainy. If there aren't many worker bees crawling the frames them then scrape some more onto it. There should probably be one queen cell per frame also—if not, don't worry they will make one. Put the frames into the new hives and stuff the door loosely with grass and leave it alone. There are some issues with splitting a hive to make a new one that need to be considered. When you remove brood, they can produce no new brood until they have produced a queen, which can take two weeks. Then they need time for maturing and mating, which is another four weeks. So during the most important time of year for making honey your two colonies will be so busy building up the colony that you will only have enough honey to support them over winter. You can also lose a new colony if they fail to feed royal jelly to the queen on the first day of hatching. This means that a queen can only be raised from brood (eggs) that are less than forty hours old.

Handling bees:

In your smoker, start the fire with crumbled paper and then add tinder such as pine needles and dry grass. You want fuel that will create smoke. Once the fire is burning, close the lid so the fire will stop but it will smolder, and pump it once in a while to keep it going. At the hive stand to one side of the entrance and blow some smoke in the door. Wait a minute and take off the cover and blow more smoke in. Anytime that the bees start to get agitated with you, use more smoke. Don't hurt a bee or it will release a panic odor which will make all the bees sting. In the winter the bees will cluster in a ball to keep warm so you won't need a suit.

Recognizing things in the hive:

Brood cells: have dark-colored caps (unlike honey cells) and contain baby bees.

Queen cells: are an inch long, hang away from the rest of the comb, and look like a peanut shell. They contain baby queens.

Drone cells: stick out like the queen cell, but not as far, and have bullet-shaped tops. They contain baby drones.

Worker cells: the smallest cell, they are level with the rest of the comb, and contain baby workers.

Queen bee: the queen looks different from all the others. She will be an inch long and has a tapered body. The other bees won't crowd around her. Her only job is to lay eggs and she can live 7–8 years. It takes 16 days for her to hatch.

Drone bee: they don't have stingers, are very fat and big, and have big eyes. Their only job is to compete to mate with the queen. It takes 24 days for a drone to hatch.

Worker bees: they are the ones that sting and they keep the hive going. It takes 21 days for them to hatch.

Maintenance:

Going into the hive: Open the hive and carefully remove every single frame. Find the queen; if it's spring look for queen cells, find out how many bees there are, how many brood cells there are and what type, how much honey is coming, and whether they need more supers. Supers prevent overcrowding, which prevents swarming.

Water: Water should be clean, fresh, and not stagnant. If you don't have a creek, then let an outside faucet drip onto a slanted board.

▲ **Queen bee**

Heat: If the hive has lots of bees and it is hot and you see bees standing around outside the door of the hive, they can't cool the hive. Move them into the shade, make the entrance larger, or stagger the supers for ventilation. In the winter making the door ¼ wide will help them keep heat in and also prevent mice.

Food: the bees need to be near herb gardens, orchards, honeysuckle plants, clover, and alfalfa. Even cabbage and mustard flowers can make excellent-tasting honey. Don't ever use pesticides, and don't let your neighbor use pesticides. If it must happen, the best time is in late afternoon. Bees eat their own honey during the winter, and for one hive they will need 50–100 pounds. If they get low on honey, feed

▲ A swarm of bees

them 2 parts granulated sugar per 1 part water, and in the spring you can give them artificial pollen (see below). Or just dump white sugar in between the frames. Never use brown sugar.

Artificial pollen to be fed to bees in the spring:

1 part brewer's yeast
3 parts soy flour
1 part nonfat dry milk

Use as much natural pollen (such as goldenrod saved in the fridge) as you have. Put directly into the hive.

Beekeeping calendar:

Early spring: Check that they have enough food and get them started with artificial pollen. In some cold climates you

Cleaning old wax off the frames ▼

may see dead bees at the bottom of the hive, but in most places dead bees would mean the queen had failed. Check the cleanliness of the hive (worker bees should keep it very clean).

Late spring/early summer: When you think the bee population is big enough, add another section to the hive to hold the increase in comb production. This is also the time to split the hive into two hives if you want. Sometimes bees will swarm—that is, they will leave the hive as a group. They won't sting, and you may have to track them down and coax them back into the hive.

Fall: On a sunny day in the afternoon, take out the honey. Leave at least 50–100 pounds for them to eat during the winter, depending on how long your winter is.

Winter: Keep the hive very well ventilated, and protect it from wind. Check their food supply and add sugar or sugar water to keep them from starving.

How to prevent disease:

Don't buy honey from a place you don't know.

Only buy bees from a dealer with a good reputation.

Don't buy used equipment unless you've talked to the apiary person of your state.

Put in new foundations every 2–4 years.

Watch for disease every time you open the hive.

Types of diseases and pests:

Acarine: Bores holes in the air passages of bees 1–8 days old and sucks their blood. Keep the bees producing babies so your population goes down, and you may need to use a Terramycin treatment.

American foulbrood: This is the worst brood disease, a bacteria that causes larvae to rot. The cell caps will be an off color, sunken in, and punctured. The larvae will be dark brown and slimy and smell like rotten eggs. By law you must report it to the state. To prevent it, you can get Terramycin powder and give it once in fall and once in spring. It can happen when your bees can't remove dead cells fast enough. A healthy, balanced colony may be able to pull out out of it. You can tell by watching a few minutes and see if any dead stuff is being taken out of the hive.

Ants: If you have ants in your hive they are a symptom of other problems. If the colony has become weakened because of disease or a failing queen (which may cause the workers to pull back from the brood), ants will plunder the colony. Sometimes this can make the bees leave.

Chalkbrood: A fungus which causes the larvae to die, they turn white, to gray, to black and then get hard and chalky. Take out the infected combs and burn them.

Chilled brood: Not a disease, but instead the brood is too big for the outer edge to get warm in cold temperatures. You will find bees dead from all stages, instead of just one.

European foulbrood: Looks like American foulbrood but there is no treatment. Only an experienced bee person can tell the difference by pulling out the larvae with a stick and it won't be stringy.

Nosema: Bees naturally carry a parasite in their intestines called *Nosema apis*. If they don't get enough pollen then the nosema multiplies and kills them. Bees will be swollen and crawl around outside the hive with their wings shaking. Make

Clean honeycomb and honey

sure bees have enough pollen, and there is a preventative medicine called Fumidil-D.

Sacbrood: Rare virus that makes cell caps look dark and sunken. Larvae will look gray and black.

Stonebrood: Very rare fungus that makes cells green and mildewy. The bees fight this disease themselves.

Varroatosis: A mite from India that will kill European bees. They are light brown, oval, have eight legs (if you see six it is a harmless louse), and kill drone cells. Use plastic pesticide strips or the colony will die.

Waxmoths: They lay eggs on the combs and then the caterpillars eat through it when they hatch. You will see fine gray webs around paths and tunnels through the comb. If your colony is strong they will kill them.

Removing honey and making wax:

1. Some suggest heating wax in a double boiler in order to purify the honey and remove the wax. This not only ruins the color it also drives out the oils and fragrances, destroying the flavor. The best way is to hang the honey in a strainer bag and allow gravity to do a perfect job. See Food Preservation for preserving honey.
2. Take the wax and put it in a box and set it on top of the hive. Leave it one day and then remove it—the bees will have cleaned all the honey off it.
3. Wrap the wax in a thick cloth such as sweatshirt material. Put the wax in a double boiler on low heat and melt it. As it drips through the cloth it will purify.
4. If the wax looks dirty, and you have a woodstove, add a little cider vinegar and a little water and keep it at 135–140° for 2–3 days. The dirt will settle to the bottom and honey will sit just below the wax floating on top.

CHICKENS, DUCKS, AND GEESE

People who count their chickens before they are hatched, act very wisely, because chickens run about so absurdly that it is impossible to count them accurately.

~Oscar Wilde

The chicken coop:

Litter: At the bottom of the coop it is good to provide a moisture-absorbing cover such as wood shavings. It should be at least 4 inches deep, loose and dry. The coop should have proper ventilation and few water spills. Instead of cleaning it out once a week, you can pile it up until it is 2 feet deep.

Cleanliness: All houses and equipment should be disinfected before any new chickens arrive. Remove wet litter, moldy or wet feed, dirty water, or clean out nests when droppings get in. Once a year the house should also be cleaned and painted with lime whitewash.

Water: Chickens should have fresh water every day, and it should be always available.

Young birds: Keep the young birds away from the old birds because they can catch diseases the old birds are immune to.

Size: You will need 6x8 feet for 12 chickens, 10x12 for 30–40 chickens, and it should be high enough for you to walk in. You should have a 1-foot slope on the roof so snow will slide off.

Windows: Put the window on the south side in northern climates, on the side of the roof that is lower so that the roof overhang shelters the window. The window should be able to be opened to provide adequate ventilation. In warm areas the chickens can be sheltered by a tarp and be fine.

Roosts: The coop should have roosts, or perches for them to sit on. These can be a long poles or boards at least 18 inches from the wall and low enough for them to fly up to.

Doors: You will need a human door and a foot square chicken door that can be shut from the outside. You can close the little door at night for protection but you'll have to get up early to open it.

Nests: Hens like very private nests. They should be moveable (see below).

The yard: Some people let their chickens roam the property, and others make a yard fenced with chicken wire. Smaller chickens need a taller fence—at least 5 feet. They enjoy litter such as straw, leaves, cornstalks, or cobs to scratch in.

Feeding chickens:

Twenty-five light breed hens producing eggs will eat 5–7 pounds of feed. You can buy a commercial feed (preferably with 15–16 percent protein) or you can mix your own by combining ground corn with proteins and minerals. Chickens can also

▲ **Portable chicken house**

eat kitchen waste, but only once a day, and only enough to eat in 5–10 minutes. Onions and fruit peels can make the eggs taste funny. Chickens also need ground eggshells or oyster shells and grit to help them digest their food, sprinkled on top. Remember to change feed gradually over a week, increasing new feed from ¼, to ½, to ¾. Chicks need special feeding processes (see below). What most people do is feed lots of milk, as much grain as they want, a whole pile of scraps, sprinkle the egg or oyster shells on top, and provide a continuous supply of forage and lots of water. It doesn't have to be scientific. Gravel, sand, and pebbles can be put in a container for grit.

Types of feed:

Commercial feed: Probably contains chicken parts ground up labeled as "protein." For chicks, crumbles are best. For older chickens, pellets or mash work well. They come in packages formulated by age so you just pick the package suitable for you.

Starter feed antibiotic: A commercial chick feed that contains antibiotics. It should only be used for a week because the birds can become dependant on it, and it is expensive. Any feed with antibiotics should not be used for birds that will be butchered in the next week.

Home-grown chick feed: Two parts finely ground wheat, a little corn, and oats; 1 part protein such as fish meal, meat meal, canned cat food, hardboiled eggs, yogurt, cottage cheese, worms, bugs, grubs; and 1 part greens such as alfalfa meal, alfalfa leaves, or fresh greens such as lettuce finely chopped. You can add wheat germ, sunflower seeds, linseed meal, etc. Their diet should be 20 percent protein. Besides sand for grit, sprinkle ground oyster shells or egg shells.

Home-grown adult feed: The best meal is about the same as for chicks, but you can use slightly less protein, 15–16 percent. Chickens can eat peels, sour milk, pickles, meat scraps, rancid lard, overripe and damaged fruits and vegetables, pods and vines, table scraps, and stuff to throw out of the fridge. They won't eat onions, peppers, cabbage, or citrus fruit. They shouldn't eat moldy food.

Health care:

Vaccinations: Birds should be vaccinated before twenty weeks for infectious bronchitis and Newcastle disease, if you vaccinate. If you buy started pullets they should have been vaccinated for those and Marek's disease. Consult your local veterinarian for other vaccinations for your area (some places have fowl pox, but if not don't worry about it).

Parasites: Mites and lice are the common external parasites, and roundworms, cecal worms, and capillary worms are the chicken's internal parasites. Cleanliness and management will prevent them. If you get external parasites, dust with wood ashes, or dip the birds in 2 ounces of sulfur and 1 ounce of soap per gallon of water. Another mixture is ⅔ oil with ⅓ kerosene. Worms can be helped by feeding garlic and lots of grit.

Cannibalism: Birds naturally eat each other, but it is caused by stress, overcrowding, not enough food or water space, malnutrition, the wrong temperature, or the sight of blood on a chicken. The only cure is to fix the problems and then debeak the culprit bird. Use a sharp knife or toe nail clippers to cut off the tip of the beak (but not far enough to cause bleeding).

Injuries: A broken leg can be splinted with a popsicle stick and masking tape. But if the chicken does not have a disease, then you may want to go ahead and use it for meat.

Egg-bound chicken: This is when an egg gets jammed in the chicken. She will strain to lay it but won't be able to and will look constipated. Pour warm olive oil in her vent (her rear), and then try to rotate the egg out yourself.

Prolapsed vent: Usually happens when a pullet lays too early, it is when the chicken's rear is hanging out. Wash the protruding tissue with warm water and a mild antiseptic, then lubricate with petroleum jelly. Push the mass of tissue back into the vent gently, dry her off, and then separate her from the rest. Feed her lots of greens and fresh water (no grain) to slow egg production. In seven days she will be OK, but if this happens often you may want to use her for meat.

Impacted crop: Also called crop-bound, this is when the chicken eats something wrong and it gets stuck in the throat so she can't eat. She will have a fat, soft throat and move her neck convulsively. Pour a teaspoon of olive oil down the throat and gently massage the neck, working the contents up and out of the mouth. Give the bird only water for twenty-four hours, then feed solids. If this happens often the neck muscles were injured, so you will have to turn it into meat.

Vitamin D: A bird that doesn't get enough vitamin D won't thrive, will lay thin-shelled eggs, and have leg deformities or other problems. In winter if you live too far north to let them run in the sun, you can give them cod liver oil.

Diseases and vaccinations:

Only really big commercial farms have lots of diseases. Usually a small flock has very few. You can't eat a sick bird, so prevention is the key. If you do find a sick chicken, isolate it, keep it warm, and feed it well. Some things can be vaccinated, but usually a small flock doesn't need it unless your local area is having an outbreak and a vet recommends it. If the bird dies, then you will have to bury it.

Disease that can be vaccinated:

Marek's disease: A virus that causes leg paralysis, drooping wings, and weight loss. Birds may get tumors on internal organs. A bird may carry the disease but not show symptoms, but other birds may die from it. The vaccination does not work after exposure to the disease for three or more days, and infected flocks will be contaminated forever.

Infections Lyrngotracheitis: Birds gasp for air and cough up blood, and frequently die. Vaccinate after four weeks of age, and do the whole flock and any birds added later. Administer a booster every year.

Fowl pox: People get chicken pox, chickens get fowl pox. Humans can't get fowl pox. Birds get round scabs on unfeathered skin, fever, and weight loss. Birds that get it in the mouth sometimes die of starvation or suffocation. It is spread by insect bites or through wounds. All birds in the flock should be vaccinated in early spring or fall, with a yearly booster.

Respiratory diseases: Newcastle's disease, infectious bronchitis, mycoplasmosis, turkey and chicken coryza, and avian influenza all have similar symptoms, including eye swelling, runny nose, coughing, and poor weight gain. Get a blood test, bacterial culture, and virus isolation to find out what your birds have.

Gathering eggs:

1. Gather eggs at least once a day (three times is recommended) and clean out the nests once a week. If you leave them too long the chance of breakage is higher, which can cause the bad habit of egg eating. Separate dirty eggs from clean eggs.
2. To clean dirty eggs, use water as hot as you can tolerate because this prevents any microbes entering the pores of the shell. Do not soak the eggs, and if you don't have hot water don't wash them.
3. Use nonfoaming and unscented detergent (if it has a scent the egg will absorb it), such as dishwasher, laundry detergent, or borax and wear gloves. Rinse off with clean water.
4. Dip the eggs a weak solution of borax to sanitize it, and air-dry. Store in the fridge.

Selling eggs:

In most places it is legal to sell eggs without a license directly to a wholesaler or customer. However, if you are planning to sell to a grocery store or restaurant some places require a license. The best sources of information are your state's or province's agricultural department or your local extension office.

Molting:

Chickens molt, or lose their feathers, after a long egg-laying season, usually once a year and most of the time in the fall. The feathers will start falling first from the neck, the breast, thighs and back, then the wings and tail. Usually it is triggered by the days getting shorter, but by using a light in

the coop you can keep the egg production up. Other causes include temporary food or water shortage, disease, cold temperature, or sudden lighting changes. It is possible to force a molt, which can make the hen's production life last longer, if you do it at about every 14 months. The reason their egg-laying time will last longer is that during molting they don't lay, so it gives them a rest.

Forcing a molt:

Day 1: Turn off the artificial lighting, so that they are only getting about 8 hours of natural light a day. Keep giving them water, but remove all feed for 10 days.

Day 11: Full-feed cracked grain for 2–3 weeks.

2–3 weeks: Feed the normal laying ration and turn the lights on again. The chickens will be in production in 6–8 weeks.

Non-laying chickens:

You can tell a chicken has stopped laying if the comb, vent, and wattle are shrunken and pale. Her body will be smaller and the pubic bones close together and possibly covered in fat. Yellow coloring will gradually return to the vent first, then the eyering, earlobe, beak, and shank.

Preparing to hatch your own eggs:

Don't incubate eggs from hybrid or cross breed chickens because the chicks won't be the same. You will need purebred hens who are healthy, and who are being fed lots of protein and greens and very little grain. You will also need a fertile rooster. A young rooster can handle 10–20 hens, and an older rooster 5–10 hens, but don't raise too many roosters. He will be fertile between March and April, but you can make it earlier by extending their light to 14 hours a day and keeping the temperature 60°. Gather the eggs before night and keep the small end down or you can rupture the air bubble inside. Don't wash them, and mark the breed on the small end (keeping it down).

Incubating eggs:

Electric incubators are cheap and easy to use, but you can make your own.

◂ Rooster

Use a Styrofoam cooler, cardboard, or wooden box, a glass or plastic top. A wood incubator should be 11x16 inches, 11 inches high with a hinged front door. Drill 3/8 inch holes on each side, 2 near the top on the 11 inch sides, and 2 near the bottom on the 16 inch sides, for circulation. Make a tray of wire mesh on a frame 2 inches from the floor of the box, and put a water pan under it. Put the eggs on the tray and put a thermometer in with them. Use a 40-watt bulb. Keep the pan of water full so that it will keep the air humid. It is very important to turn the eggs gently, a quarter way around three or five times a day (never an even number of times, or the chick will lie on the same side every night). Mark an X on each egg to keep track. After 10 days, make sure the large end is higher than the small end. After 18 days, stop turning the eggs, and raise the humidity in the incubator. Most eggs hatch between 19–22 days.

Candling an egg:

Some eggs may not mature, and then they can rot and even explode. To find out which ones are not fertile and are not growing, wait three days after the eggs have been fertilized, make the room dark and use a very bright light behind the egg. Look through the shell. If you see a clear egg, it is not growing. If you see a dark haze or gray clouds then it is rotting. If you see a dark red circle and no veins, the embryo died. If you see a small dark center and a network of veins then the egg is good.

Hatching:

When a chick pecks a hole from the inside, that is called "pipping." The chick may start to pip on the 18th day, but it won't actually do it until there is a hole showing. Don't open the incubator, don't even touch it. Don't help the chicks get out of the eggs, they must do it on their own or they might die.

Getting a hen to brood her own:

Many hens lay eggs and forget all about them, but now and then one will get "broody" and try to incubate them. To encourage her, make her a private nest with dim lighting. The best nest is a small house 15x15 inches and 16 inches deep with a roof, that sits right on the ground. Some people put a wood egg in the nest to encourage her to lay in it. When the weather is warm she will lay one egg a day and start "setting" or incubating the eggs. Don't disturb her or the nest or she may abandon it. After 21 days the chicks will hatch. Many people take the hatched chicks into the house until all of the eggs have hatched because the hen often will be torn between sitting on eggs and taking care of new chicks. Put those chicks into a brooder, and then put them back the first night after hatching. To do this, go in late at night and slip the chicks under her while removing eggshells and wood eggs. You can add a few orphan chicks also if you want. The next day watch for any chicks that got rejected (it will get pecked at by the hen, and hide its head), and bring them into the house to be raised in the brooder.

When to help a hen:

Sometimes hens need help with their chicks or some could die. Besides the first critical night after hatching, some hens are bad moms and the chicks will have to have extra help learning how to eat food. A normal hen breaks up big seeds for her babies, but if she doesn't, use the same methods for brooder chicks to feed them. If you have several families, sometimes hens will fight each other, killing a chick in the process. Keep families in separate coops. A lost chick will chirp very loudly, but sometimes the only way the hen can save the chick is to leave the rest of her brood. In that case, go get the chick yourself and save the hen the trouble. The family should be kept away from other chickens—other mothering hens can cause fights, and adult birds may peck and kill chicks.

Brooder building:

Brooding eggs is keeping them warm like a broody, or mothering hen. You can buy a brooder, or you can make one by constructing a box with a heat lamp (or near to a woodstove) and a thermometer.

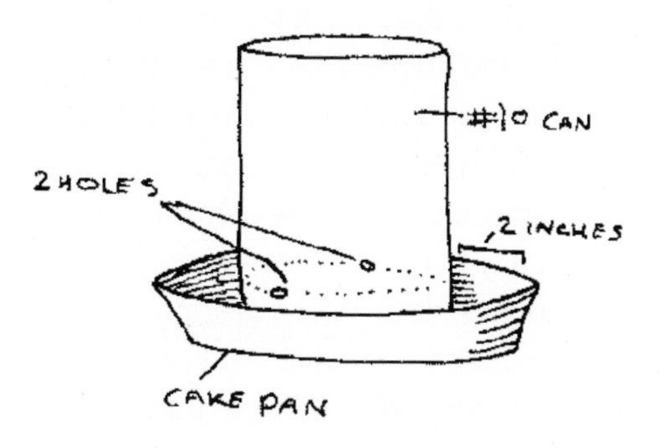

▲ **Alternative homemade chick waterer**

◂ **Newly hatched chick in a brooder box**

The bulb should be low-wattage, and the temperature should be about 95°. A box 30 inches square and a 69-watt bulb can brood 50 chicks. Put your hand down to test the heat. If it is uncomfortably hot, the lamp is too close or too many watts. As the chicks get older, tape another box next to the first and cut a door. Hang heavy cloth in the door and the other room will be cool, so the chicks can run in and out. The light should be red or green and hopefully dim, or the chicks will have to be adjusted to darkness or they might die.

Chick waterer:

Fill a quart jar (like a canning jar) with water and turn it upside down in a bowl with an edge less than 1 inch high, with a diameter only a little wider than the jar. Then stick a match or toothpick under the edge of the jar to let out a little water at a time. You will need two 1 gallon waterers per 100 chicks. Keep the water clean and full at all times, and make sure it is room temperature, not cold. Don't put the water right under the heat lamp.

Litter and heat for chicks:

Check the chicks 2–3 times per night the first week. If they are cold, they will huddle under the light. If too hot, they will scatter to the edges. If they are content, they will chirp contentedly. Decrease the heat 5° per week, so that by six weeks it is 70°. Then you can turn the heat off unless it gets chilly. Put burlap or cloth rags with no loose threads as floor covering a layer of newspaper, for the first week, and then when they know what food is, graduate to a thick layer of black-and-white shredded newspaper, hay (not straw), or wood shavings, pieces too big to fit in a chick's mouth. Stir the litter every day and remove wet spots. Doing these things prevents spraddle legs (legs turning outward) and infection.

Feeding chicks:

If you hatched your own don't give them food at first. Wait until they start pecking at the floor, then give them food. If you bought them, have food ready because they will be three days old. For the first week give food on a paper plate, egg carton, cardboard, or some other surface that will bring the food up closer to their eye level. Once they figure out what food is, put the food in a tuna can or a trough—the container should be difficult to walk in and scratch food out of, but short enough to reach in. Each chick will need 1 inch of feeding space until they are thirty days old, then they will need 3 inches. Don't put the feeder right under the heat lamp. Fill the feeder only half full to prevent throwing the food out, and so they won't waste any. Clean out old food each time you fill it, and keep it full all of the time. Chicks need small grit to help them digest food, so sprinkle sand on top of feed.

Chick dust and moving chicks:

Chick dust is a powder that comes off the droppings when they dry that can get into your lungs and potentially could even cause lung disease. When birds start to molt they can also release down into the air. Don't keep chicks in the house for very long. When they are 4 weeks they can stand on anything, so if you have a separate room in the coop (not partitioned with chicken wire, which they can walk through), transfer the heat lamp and box to there. When they are 6 weeks you can remove the box, and when they are 10–12 weeks you can put them with the other chickens (if they are big enough and weather is warm). These small chickens are called pullets and should be moved before they start laying. If the weather is warm, after 1 week you can let the chicks out on grass to run but they get tired easily and will need to come back to get warm.

Chick problems:

If your chicks get diarrhea, it is a sign of coccidiosis, a disease. Prevent by not overcrowding and by giving them adequate food. If your chicks get it add 1 tablespoon of plain vinegar to drinking water. If a chick dies remove it immediately so the disease doesn't spread. Some chicks get a pasted vent, or when droppings clog up their back end. Use lukewarm water to remove droppings, then rub petroleum jelly on the area, and make sure the chick is completely dry. Petroleum jelly can also be put on chicks that are getting pecked.

Choosing ducks over chickens:

Ducks are the gentlest of all poultry, even chickens, but they are noisier. They do better around children, but make sure that you always act and talk gently around them because will startle easily. Duck eggs taste different than chicken eggs, and ponds will add different flavors. Also, you can't eat eggs from ducks that are feeding in a polluted pond. You can cook duck eggs just like chicken eggs, except they can be a bit tougher—when frying, add a bit of water.

Watering ducks:

Ducks need lots of water. If you have a small pool 3 feet across and 12–18 inches deep that ducks can dip their heads and feet in, the water will prevent diseases and feed them with bugs. Instead of a pond, any container such as a kid pool or bathtub will also work. However, if you have a very heavy breed of duck you will need more water or they won't be able to breed since they can't do it on land. If you have fish in the pond the duck's manure will feed algae-eating organisms and help grow good pond plants. Twenty-five ducks is the maximum population per acre of water surface.

Feeding:

Ducks don't need much feed if they have water and enough forage such as a grass yard with bugs in it. Ducks need young grass to forage well, unlike geese. However, you can feed them whole ears of corn broken in two, and if you feed oats, give it to them before you give them corn. Wheat is the best grain for ducks, and goes well with oats. Only give ducks firm, round fruits and vegetables by crushing them first. Liquid milk and hard-boiled eggs are very good for laying hens, and all ducks need calcium from eggshells or seashells, and grit.

Shelter:

You will need 4 square feet of housing per duck (for ducks that are allowed outside). You can either house them in a three-sided shelter during the summer, or you can shut them in at night. You should provide good shelter during winter, at least keeping them in all night and provide a three-sided cover during the day if it is not too cold.

Breeding:

Duck male and females are sometimes different to tell apart, but basically the female will be very loud and raspy sounding, while the male will be a bit quieter. Some male ducks will also become very protective of the females. Other than that it's difficult to tell.

Mother ducks:

If you have a motherly duck, the best thing for her is to raise the chick, because the ducklings do better and learn faster when a duck raises them. Ducks will start laying between 6–7 months old, and keep laying for about three years or longer. They always lay in the spring in the morning, and are very scheduled, so let them out of the house after 10 AM. Keep a mother duck and her ducklings separate from your other ducks and keep them in at night until the chicks are 6–8 weeks old.

Brooding:

Use the same methods as for chickens, but the holders will have to be bigger to accommodate the bigger eggs. The temperature should be 99–100° for forced air, or 101–102° for still air. The eggs need to be kept very humid and moist. To do this either spray lukewarm water on them once a day or put a large sponge into the water pan and sponge the eggs gently with warm water when you turn them. Stop turning the eggs on the 25th day but keep spraying them with water until they pip. It will take 28 days (except for Muscovy ducks which need 35 days), and on that day give one last spray and leave them alone.

Caring for ducklings:

Each duckling will need 1 ½ feet of floor space until they are seven weeks, when they will need 2 ½ square feet. For every thirty-ducklings you will need a 250-watt heat lamp, a few inches higher than what you would do for chickens. Leave

▲ Duckling

the light on all night, and make sure the outer edge of the heat circle is 90°. Reduce the heat 5° every week until it is at 70°. The first two weeks the ducks should not be allowed to get wet, but they should have a drinking trough at least 2 inches deep, and only 1 inch wide. At 4–6 weeks turn off the heat unless it is super cold. By 4 weeks they should have started growing feathers so they can go outside in the daytime. Any other duckling care is just like chicks.

Feeding ducklings:

You can't use chicken feeders for ducklings; use small box tops near the heat lamp, and then graduate to rough paper, such as the bottom of grocery store paper bags. You can use commercial chicken starter feed if it doesn't have antibiotics and it is recommended for ducks. To make your own feed, give them cooked oatmeal with water for breakfast, scrambled eggs with water for lunch, and whole wheat bread with water for dinner, with some cut up greens. Gradually give them more and more greens, and at two weeks start giving them some grit.

Flying ducks:

If you have a flying breed of duck, use big scissors to clip the long feathers off one wing when they first grow, and after each molting after that. Without the long feathers the birds can't fly.

Differences between ducks and geese:

Geese can live with a small pond just like ducks, and heavy breeds need more water or they can't breed (just like ducks). They can also live just off grass pasture (if there is enough), and unlike ducks they don't need new grass. Six geese is the maximum population per acre of water surface. Geese are meat birds, and do well as watchdogs or guards, although they are very quiet. If you keep a goose for a long time, it can become too big for you to handle and can be dangerous, but a meat goose doesn't get big enough to pose a threat. Some geese become attached and some don't and will become very protective of eggs, nests, females, or whatever. Geese lay any time of the day, and their eggs taste like chicken eggs, but are bigger.

Differences between ducklings and goslings:

Use the same amount of floor space for goose chicks, but per 250-watt heat lamp there should be only twenty-five goslings. Feed them four times a day, with enough food to eat in 15 minutes. They need tender,

Gosling

green grass or weeds, along with a bit of duck food and some grit. At 5–6 weeks they can survive completely on a big enough pasture (1 acre per 20–40 geese), or you can add some grain. In winter give dried grass, alfalfa hay, corn fodder, grain, and whatever scraps you would feed chickens.

Choose a goose:

Depending on the purpose of the goose, some are for eggs and some are for meat. Dark breeds are harder to feather when butchering. It is hard to tell the difference between a male and a female. The only sure way is to catch the goose and lift it by the neck and legs. Lay it on its back on a table or over your knee with the tail away from you. Bend the tail downward and put some petroleum jelly on your finger. Put your index finger into the cloaca ½ inch. Move it around in a circle a few times to relax the muscle, then apply pressure directly below the vent to expose the organs. If it is not a female it is probably a male. The easy but less accurate way is to watch the flock for a broader head, longer neck, and more aggressive bird. A gander's (male bird's) call is deeper and stronger, and louder. Ganders are much more aggressive, and will protect their females, and an easy way to tell which is which is to watch during mating. An old goose does not have soft, yellow down on its legs. Don't eat a goose more than three years old.

Natural farming and geese:

Geese are natural weeders if you use them before your wanted plants come up. Put seven geese (over eight weeks old) per acre in the field before the weeds are tall and coarse. A fence around the geese should be 3 feet high, and let them clear out all the weeds until your sprouts start coming up.

Clipping wings:

Cut 5 inches off the long feathers of one wing with big scissors. This takes the bird off balance so it can't fly (that's why you don't cut both). Don't cut during molting or the bird could bleed to death, and don't cut the wing.

Carrying a goose:

Scoop the goose up with one arm under its body, but with its head facing towards your back. Pin the wings tightly with your arm. If you have the head facing forward it can bite your face.

Housing:

Geese need a yard with a house that is leak-proof. The yard should have 30–40 square feet, and the house 10 square feet per goose. Geese need a simple box feeder inside the house, and a water bucket or trough outside so that they don't get wet. The litter in the house should be chopped straw, wood shavings, peat moss or ground corncobs. During snow they only need to be watered once a day, and other times twice a day. If you make nests, they should be 2 feet square, and there should be one nest per three geese, lined with straw or other liner.

Feeding:

There is no commercial goose feed available, but chicken feed can be given if it has 15 percent protein. Geese also need oyster shells or grit at all times. A ration mixture usually includes 10% ground corn, 20% ground wheat, 10% wheat bran, 20% ground barley, 21% pulverized oats, 8% soybean oil meat, 2% dried whey, 6% dried alfalfa meal, 1% ground limestone, 1% dicalcium phosphate, and 1% iodized salt. If geese have pasture that is not alfalfa (they won't eat it), they don't need any extra feed.

▲ Size difference between ostrich, goose and chicken eggs

Eggs:

Gather eggs two times a day and clean like chicken eggs. If you are incubating the eggs, a goose can hatch 10–12 eggs in 29–31 days, but they need to be turned by hand 3 or 5 times a day if the setting goose does not. Never turn them an even number of times or they will lie on the same side every night. You can also incubate in an incubator if you follow the same turning rules. Whether under a goose or not, during the last half of the incubation period sprinkle the eggs with lukewarm water for 30 seconds every day to help with hatching. Remove the goslings from the nest as they hatch and keep in a warm place until they are at least three hours old. This prevents the goose deserting the nest.

Breeding:

A large gander will mate 2–3 geese, and a light gander will breed 4–5 females. Ganders will stay with the same females year after year. Breeding season is usually late fall or early winter, and ganders should be kept separate until that time. Geese lay bigger and better eggs the second year so some people wait until a goose is two years old before breeding.

Brooding:

Don't let birds brood in January because none of the eggs will hatch. To do this take all the eggs but one (a different one every time) every day. Geese would rather nest outside, such as in an old tire with straw in it, or in a small house (like a doghouse). When a goose starts laying, take all new eggs but two every day until the nest is full. Both the male and female geese care for the new goslings.

Plucking a live goose:

1. Catch the goose and hold it tightly by both feet. Turn it on its back with its head behind you.
2. Carefully remove only the breast feathers of the goose, without tearing skin or injuring it in any way.
3. If the geese were hatched early, you might be able to pluck them four times a year. A ½ pound of feathers per goose is a good yield.

DOGS

We long for an affection altogether ignorant of our faults. Heaven has

accorded this to us in the uncritical canine attachment.

~George Eliot

Choosing a dog breed:

First decide what the dog is for. For a farm usually you will want a dog that serves some purpose. Even a golden retriever is valuable . . . they can be a companion to children and to yourself, warn you of bears and other predators, and keep your kids from drowning in the pond. Dogs can be trained to do other things but each breed has been specially bred to do a certain task—they'll just do it better and they'll like doing it too. Dogs also grow to different sizes so be prepared to feed a St. Bernard an enormous amount of food. A guard dog is not as good with children—for instance a Rottweiler is not a good choice for a family dog, whereas a German shepherd might be a bit more relaxed but not a bad guard at the same time.

Working dog breeds:

Great Pyrenees: guards large livestock

Border collies: herds medium-sized livestock

Blue heeler: herds medium-sized livestock

Labrador retriever: game-retrieving

Foxhound: hunting

St. Bernard: pulling heavy carts

Newfoundland: pulling heavy carts

German shepherd: aggressive guarding

Husky: pulling a sled in snow

Samoyed: pulling a sled in snow

Collie: watching children

Golden retriever: man's best friend

Terrier: killing mice and hunting small animals

Judging a puppy's temperament:

An already grown dog is fairly easy to judge as far as temperament and obedience go, but a puppy will have a hidden personality. The best dog for you and the easiest to train is an obedient dog, not a dominant dog. The simplest way to find this out is to lay the puppy on its back and look in its eyes. If the puppy submits to you, lets you scratch his belly, and looks away from your gaze, then it will be obedient. Also, never pick the last one left of a litter—there's usually a reason that no one picked him.

Choosing breeding dogs:

Selling puppies will pay for a dog's food for a whole year. You will have to buy a male and female from different ancestry of the same breed, so you'll have to see a pedigree. Take them both to the vet before

Great Pyrenees ▸

▲ Border collies

◂ Blue heeler

Labrador retriever ▾

▲ Foxhounds

▲ St. Bernard

▲ German shepherd

◂ Newfoundland

Husky

Samoyed

Collie

Terrier

Golden retriever

you buy, know about the breed and plan to keep the dogs far past their breeding life. Even if you keep just a female dog, you can decide to pay to have her bred with a stud.

Shots:

A puppy you buy should have already had its first shots and been dewormed. By six months it will need two more shots and a rabies vaccine. Then once a year boosters and a twice a year deworming should keep the dog happy.You can buy the shots and administer them yourself, but rabies vaccines have to be administered by a vet.

Cleaning a dog:

To get rid of fleas, wash and comb your dog. Pour boiling water over two lemons and let stand overnight, then sponge onto dog. If sprayed by a skunk wash the dog in tomato juice.

Training a dog:

There is probably no one right way to train a dog, but there are a few fundamental rules that every training system follows.

The dog should be treated like a dog, not a child. People have a tendency to see dogs as people, but dogs have an animal psychology just like any other animal. This doesn't mean they aren't a member of the family, but it means you relate to the dog differently.

Do a bit of training every day, but keep it short and sweet. Five to ten minutes is long enough, and always end on a positive note. If the dog isn't doing what you asked, give a simple command that it will do, just so she finishes by being rewarded.

Every dog should be able to come, sit, stay, down, and heel. She should also know how to give you something, or leave something alone.

During a training session there should never be any negative aspects—no hitting or yelling. The only punishment is not getting a reward. If a treat is used, it should by tiny and not crumble so that it is very quickly eaten. Rewards have to be given with a half a second of doing the command.

Use a hand signal when giving a verbal command. Eventually the dog will learn to use the hand signal interchangeably with a verbal command. The signal should be simple: a closed fist for sit, flat palm facing towards the ground for down, etc.

Be consistent. Enforce a command if you give it, and don't give one unless you intend to make sure it happens. If you say sit, and she doesn't sit then you will need to get her attention and make sure she does it. Don't let her ignore you.

When the dog is doing something undesirable like chewing or nipping, you should use an authoritative voice and say "no," and immediately replace the behavior with something good. Give the sit command and praise for obedience, or give a chew toy that she can turn her attention to.

GOATS

If you're short of trouble, take a goat.

~Finnish saying

How many goats to have:

An average doe makes about 3 quarts of milk per day. one to 1 ½ quarts goes to her kid, leaving only 1 ½ quarts per day. If she is a new mom, she will give even less than that. The more does you have the better it is because they can feed their own babies, plus adopt orphaned babies of other animals, plus feed extra milk to other

▾ Tiny individual goat house

animals (such as chickens), plus make cream and butter and feed you milk. Have one goat per person in your family.

Buying a goat:

Find out what breed she is, the age, if she has been bred and to what kind of goat, and how many times. Find out how many kids she had if she has been bred, how long she milked, what it is like to milk her, how much milk she gives, how many teats she has, and if she jumps fences. Find out if she is hard to catch, if she bites or butts. Has she had a distemper shot, and will she wear a halter? See her parents, sisters or brothers, or other family members. If you buy a buck, remember that they stink, and they must be handled carefully. You will need to decide whether to keep him separate or with the herd (see below).

Housing and fences:

Goat fencing should be at least 4 ½ feet high, with ¼ acre per goat. Wrap trees with chicken wire so they can't strip bark, and make sure any fence does not have a gap wider than 8 inches. If the goat can't see through it they won't try to get out, but if you must have a rail fence make sure they can't squeeze through. Goats can unlock most standard latches with their tongues so a padlock may be necessary. If the goat does try to get out all the time, put a Y-shaped yoke on its head so it can't fit and soon it will give up trying (then take off the yoke). Housing can be any kind of sturdy shed in their pasture, or a 6x6 foot space per goat in the barn with clean hay for bedding.

Catching a goat and goat manners:

Goats are very curious, so pretend to examine something while showing him some grain. Not only will he want the grain, he will want to see what you are looking at and will hopefully walk closer. Dive for him and grab whatever you can, then ask for help. To find out the sex of the goat, a male kid pees from the middle of his belly, while a female kid

has to pee squatting backwards a bit. If you have a goat who is jumping on you, butting other goats, or stealing food, slap him on the nose and yell "no".

Feeding:

Each goat needs about 4–5 pounds of hay per day of mixed grass and legumes (such as alfalfa). Half a bale of hay twice a day for 10 goats should be adequate for their diet. Don't put it on the ground, and watch how much they eat. If they leave some it may be too coarse. Goats also need salt and water at all times. Goats won't lick a salt block enough so loose salt is better. Breeding bucks and does and pregnant does all have grain requirements (see below). Goats can also eat weeds, vegetables, and garden cuttings. Don't give them chicken feed, or milkweed, laurel, or azaleas.

General health care:

Lice: Feed goats cider vinegar for blood-sucking lice. Rub the goat with a cloth wet with vinegar for scale lice. You can also dust with ashes.

Chronic mastitis: Symptoms can include bitter-tasting milk and a few creamy chunks that get caught in your milk filter. To treat, milk the udder completely empty several times a day and use intramuscular antibiotics.

Hooves: Trim the hooves once a month or they will keep growing until they can't walk. Use a knife or get hoof nippers.

Worming: Worm three times a year, and one of those times all the goats should be done together. Does should be wormed 6–8 weeks before kidding. You can worm before putting them inside before winter, before breeding, or before letting out in the spring.

Vaccinations: Give shots just before turning them out in the spring, and at least 6–8 weeks before kidding.

How to treat a buck:

Whether you raise a buck yourself or buy an adult goat, always treat it gently and carefully so that the butting instinct is not awakened. Don't make a pet out of it and don't allow it to think you are a part of its herd. The best policy is to get him to lead, stand quietly for grooming, shots, and hoof trimming, and then leave him be.

Breeding:

Breed your does 149 days or 5 months before you want to have kids. For small farms, having kids about April 1st is ideal, so breed November 1st. Don't breed does that are less than 70 pounds because they will have health problems. They should reach that weight by 7 months old, but wait until at least 10 months old, and two years old is actually the best for them. You can do well with one buck per 50 does but 30 is easier. A buck should have his own sturdy pen and as the does goes into heat she should be put in with him until they breed. Then you will know the exact due date because does almost always kid exactly 149 days later. The alternative is to keep the buck in with the herd and allow breeding whenever. If she does not get pregnant, the buck is infertile and you can replace him quickly. Does are in heat when they spend time sniffing and wagging tails towards the buck pen, and will make more noise.

Pregnancy:

Some goats can be milked during pregnancy and have special diet requirements (see below). Six to eight

weeks before kidding vaccinate, worm, and trim hooves (unless they are organic, then don't vaccinate). Give blackstrap molasses in the last two months of pregnancy to prevent ketosis, which can be caused by stress, overfeeding, underfeeding and lack of exercise. Symptoms are dullness, disinterest in feed, teeth grinding, and wandering.

Kidding:

The doe will be close to delivering when the lips of the vulva swell and the opening gets longer and larger, she will have a full udder, and a thin mucous discharge. When she gets hollows under her tail, slightly above her vagina on the left and the right, she will give birth within two hours. When a doe is close to kidding you should check on her every morning, and if she's delivered you need to find the kids as soon as possible after kidding. To be on the safe side, keep the goats nearby to be away from predators and within your sight. A barn floor with dry bedding is ideal if the goat is familiar with the place. Give her clean, fresh water in a small bucket and lots of good hay. When the kid's nose is out, you can clean the nose and mouth to help breathing. Otherwise, leave her alone unless: the water breaks and two hours pass without kid is poking out, if she's a great pain and 30 minutes pass, or if she's totally exhausted and 15 minutes pass. To help out with a normally-presenting (head first) kid, pull gently downward with each contraction. To help with an abnormal position or if one of the situations above occurs, for instance if both feet are tucked under, head turned back, or rear-end first, first scrub your hands and arms well making sure the nails are very clean. Reach in very carefully, determine the kid's position and reposition it until its feet and nose are coming out together. After delivery give the doe a bucket of warm water with some molasses in it and let her deliver the afterbirth. She may eat it, or you can bury it.

Newborn kids:

Wipe the kid's face, and if the goat isn't doing a good job of drying the kid, then go ahead and dry it off. If you find a kid outside after birth bring it inside and wrap it up next to a heat source until it is warm. It isn't necessary to tie the cord, just dip the end in iodine or alcohol to kill bacteria. Goats usually have 2–3 kids, and they should be standing up right away. If they don't nurse within the first 15 minutes, help them to do it by holding the teat in its mouth so it can suck. If it doesn't suck, squirt milk in its mouth and then try to get it to suck again in 3–4 hours, or when it seems hungry. Some kids need help like this for three days. If it still doesn't suck after several hours, then you'll have to bottle-feed, but only a couple of times so it learns to suck—then try to get it back on the mother. Don't keep the kid in the house or away from the mother for more than 6 hours during this time or the mother will reject it. For 3–5 days after birth the kids and mother should be kept separate from the herd until they are strong and nursing well.

Feeding kids with the mother:

Let the kids stay with the mother until they are strong and eating solid food well. This usually takes two months for singles and twins and three months for triplets. Then separate the kids from the mother at night and milk her in the morning, then put her back with the kids. In a few days they will learn the routine. As the kids wean, keep milking her more and more so she will get used to holding more milk in her udder at a time and her teats will lengthen. You can either keep letting the mother feed the kids at night and only milk once a day, or you can gradually work toward milking her every twelve hours. If you let the mother nurse once a day, she will eventually wean them herself.

Feeding the kids with a bottle:

Give the mother grain and milk her. Save the colostrum and put it in a pop bottle with a lamb nipple on the end and feed it to the baby. If the kid is too weak to use the lamb nipple, use a human baby's nasal aspirator (or syringe). Never feed a kid cold milk. It is so important to get the colostrum into the baby—it is the difference between life and death. Feed every 2 hours on day 1, every 4 hours on day 2, and every 8 hours on day 3. Then feed morning, noon, and night for 10 days, and then morning and night after you milk. If you can't give the kid goat's milk the next best is cow's milk, but they can still get scours (or diarrhea) and die more easily (chilling and overfeeding can also cause it). Most likely you will have to butcher the kids if you don't have goat milk, or if you need the milk yourself.

When to not keep kids:

Besides the kids that aren't nursing and you can't get milk for, you may also get more billy goats than you need. Nanny goats are valuable and you will hardly ever need to kill them, but billy kids don't give milk and if you have too many then you get no milk for yourself. The best way is to decide how many you need before the

birth, then any extra take out of earshot and hit them over the back of the head with something heavy like a hammer. Butcher them like you would a rabbit. Or you can wait a little bit when they are big enough to have some more meat and then butcher them . . . generally in cool weather around November is the best because then you don't have to feed them all winter.

Dehorning:

Dehorning is done for safety reasons. If you have a goat that will be around children, or a ram for breeding that you need to handle, you might consider dehorning. If possible dehorn at 3 days old by burning. You can burn horns of as long as they are shorter than ¾ inch. Get an iron rod or a specially made dehorner with a base the same size as the horn. Heat an electric dehorner according to instructions, or heat the iron rod in a fire until it is very hot. Press the hot end to the horn button while a helper holds on to the goat so it can't move. Hold the rod there until a circle of copper-colored skin appears around the horn's base, about 10–20 seconds. Do it for the other horn and you're done. If the horns are too big to do this, you will have to use dehorning nippers, which will make the horns bleed.

Feeding does:

Start feeding the breeding does ¼ pound of grain per day starting on October 1st and increase a ¼ pound per week until by the start of November until each doe is getting 1 pound of grain per day. This will increase the chances of having twins and triplets. For healthy older does, taper off this grain starting December 15 (6 weeks after breeding), and then bring it back up again starting February 15 (6 weeks before kidding). First kidders and unhealthy animals can keep eating grain all the way through.A buck should taper off grain when he is in good condition.

Feeding milk goats:

Check the mother's udder twice a day after kidding to make sure it isn't too full. Her milk will come in 3–5 days after kidding, so during that time don't give her grain or the milk will come in too fast. What you feed your milk goat can change how the milk tastes—wild mustard, the cabbage family, wild garlic, wild onions, etc. If the milk tastes bitter try removing some of these high-odor foods from her diet. Goat milk may taste naturally "goaty," but you can fix this by putting a pan of baking soda in a pan in their feed trough. Keep it full of soda and in a few days the milk will taste sweet (any goat eating grain should eat baking soda to avoid a sickness from too much acid). Pregnant does who are producing only a little milk should stop milking two months before kidding, but you can keep heavy-producing does milking all the way through continuously. Pregnant milkers need high quality feed, but not too fattening.

Caring for milk equipment:

Stainless steal without seams is the best and most expensive container for milk. Food-grade plastic and glass will work also if they are seamless, but don't use commercial plastic milk jugs for storing milk. Rinse buckets, containers, and utensils in lukewarm water right after using or you will get milk deposits which are hard to remove. Wash by scrubbing thoroughly in warm, soapy water, then rinse again in scalding water and air-dry upside down. Any cloth used for straining should be rinsed and boiled directly after. Use homemade soap and rinse cloth with a little lye to sterilize it completely.

Steps to milking:

1. Clean the milking utensils in warm soapy water (bucket, strainer, storage container).
2. Put the goat in a stanchion (a frame that holds the goat by its neck). Brush the fur to get out loose hair and dirt, put down fresh bedding, and keep the long hair under the udder clipped.
3. Put some feed in the stanchion's trough. Milking is done every 12 hours, and the best time is in the early morning before they go eat. Always be on time or the goat will get over-full and be in pain.
4. If you want to, groom the animal by brushing and look it over for problems. Then wash your hands and dry them, and fill a bucket with water 120–130°F and a fuzzy cloth.
5. Wash her udder and teats. This helps the milk let down, removes dirt and bacteria, prevents milk from souring quickly, prevents a weird taste in milk, and makes the teats easier to grip. Wait 1 minute after washing to start milking.

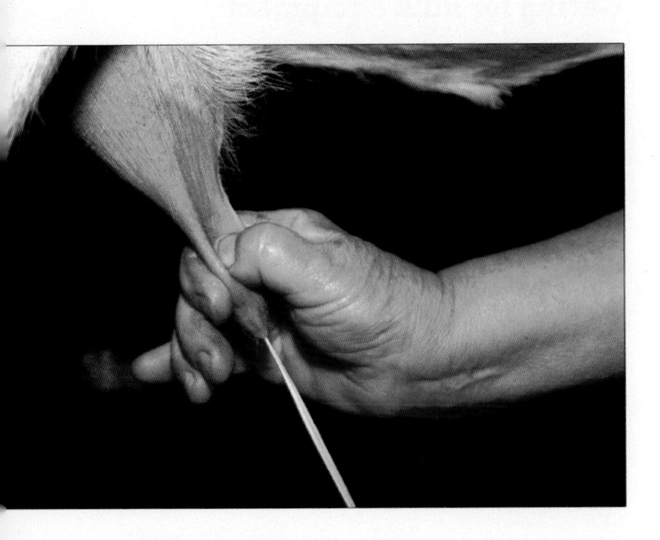

6. The goat needs to be happy and relaxed for the milk to let down. Put your thumb and forefinger around the teat near the top of the udder, pushing up slightly and allow the teat to fill with milk. Then close your hand around the teat and squeeze the milk out while pulling down, or stripping. Keep your hand away from the nipple hole or the milk will go all over. Squirt the first three squeezes into the ground because the first milk has more bacteria in it. Make sure you completely empty the udder or she will produce less and less until it is gone.
7. Sometimes the goat may need to poop while you are milking. Grab the bucket quickly and get out of the way.
8. Strain the milk. Use a regular kitchen strainer lined with several layers of clean fabric such as a dishcloth, muslin, or even a cloth diaper.
9. If you choose to pasteurize, put the milk into a double boiler on your stove. Use a thermometer to heat the milk to 161°F, and keep stirring so it is heated evenly. Maintain that temperature for 20 seconds, then quickly remove from heat and put the pot in very cold water, stirring constantly until it gets down to 60°.
10. If you don't pasteurize, put the storage container inside another container full of cold water. It should be chilled to 40°F within 1 hour. Store any milk in the coldest part of the fridge.

Milking problems:

The milk won't let down: If you have washed her udder with the warm water and waited a minute and the milk doesn't let down, try massaging either with the cloth or with bag balm, or gently pat the udder like a kid butting her. If she still won't let down, suck on the teat a little and the milk will come down.

Drying up milk: If you want to have a kid every year, you have to dry her up after ten months of milking. However it's not necessary and you can wait until she dries up by herself. To dry her up purposely, gradually milk less and less, leaving the udder with a little milk in it and milk less often. Watch carefully for mastitis.

Mastitis: The first sign is usually flaky, lumpy, or stringy milk. Check the milk at least once a week by squirting it into the cloth. Feel the udder for tumors, or large hard areas, or an abscess—a red, tender swelling of the whole side of the udder. The udder will feel warmer than usual, and it will be more difficult to milk. If it gets worse, the milk may turn yellow or even brown or pink from pus and blood. *Don't drink the milk!* Don't throw it somewhere that an animal can lick it, and wash your hands very well after touching the animal because it is infectious. Get antibiotics in her quickly. A goat who has had it once is susceptible to it. Having a bruised udder or drying up too quickly so that she gets bruised from being too full can make her likely to get mastitis.

Self-sucking: Goats can be flexible enough to suck their own milk. Other than butchering, you can use an Elizabethan collar or side-stick harness that can prevent her from reaching the udder.

Milk taste:

Besides baking soda to prevent the "goaty" taste in milk, you can prevent a distasteful flavor in milk by only using seamless stainless steel, food-grade plastic, or glass to store milk. Keep milk out of sunlight and fluorescent light, keep the barn clean, clean the udder, and keep your hands clean. Don't feed the goat strong-flavored foods (such as onions, garlic, cabbage, or turnips) closer than 7 hours until milking, and don't let her smell the buck or let the buck near the milk. Don't smoke around the milk.

Preserving milk:

Parts of the year goats don't produce milk. To can the milk fill jars to ½ inch from the top and process 10 minutes at 10 pounds. It won't taste good but works in cooking. To freeze you can use jars, plastic containers, or freezer bags. Don't fill completely to the top because the milk will expand, and before you use it completely thaw it and stir.

SHEEP

I would have been glad to have lived under my wood side, and to have kept a flock of sheep, rather than to have undertaken this government.

~Oliver Cromwell

What to look for when buying sheep:

Ram: Has good legs and feet. Does his job as a ram.

Ewe: Doesn't have mastitis. Udder is soft and in working condition.

Is not overweight or underweight.

Does not limp.

No hooves are hotter than the others and no green tinge (signs of foot rot).

Mouth is sound, gums have no anemia and are very red.

Eyes have no anemia, arteries around eyes are red.

No swelling or lumps under the chin.

Ask about sheep history.

If possible, ask if the sheep had rectal or uterine prolapse (lining protrudes from body), as it is genetic.

Teeth and age:

90 days: eight temporary incisors

6 percent of ewes 18 months: temporary teeth are longer and narrower, more space between teeth.

10 percent of ewes 18 months: first pair of permanent teeth poking out, with three lamb teeth on each side.

84 percent of ewes 18 months: two permanent teeth are fully out.

37 percent of ewes 2 ½ years: four permanent teeth with two lamb teeth on each side.

57 percent of ewes 2 ½ years: six permanent and one lamb tooth on each side.

28 percent of ewes 3 ½ years: six permanent and one lamb tooth on each side.

50 percent of ewes 3 ½ years: full eight permanent incisors.

5 percent of ewes 4 ½ years: six permanent and one lamb tooth on each side.

88 percent of ewes 4 ½ years: full eight permanent incisors.

A sheep with a "broken mouth" (teeth missing and crooked) is not worth it.

Sheep pasture and feeding:

A good pasture will feed five ewes and eight lambs per acre for the season in northern regions. It's a good idea to have 25-50 percent more than that for management. Legume-grass mixtures are good for sheep. Alfalfa-grass and trefoil-grass will re-grow in 15–20 days. Rotate their pasture every week. Put loose salt (not in blocks) under cover outside for them. Sheep need 1½ gallons of water a day, fresh and clean. Make sure that you clean the feeders before giving food, watch that each sheep is eating well, and that every ewe has 1½ feet of rack space. If you are using the wool you may need to keep the wool clean—don't let burdock, thistles, fleabane, Spanish nettle, or other burred plants grow (give them to chickens). The pasture should be divided in half so that you can rotate when the grass is eaten down. Rotation helps prevent worms.

Feeding adult sheep:

Pregnant ewes: Feed ewes in the first four months of pregnancy 3½–4½ pounds alfalfa or clover hay. In the last month of pregnancy give ½–1 pound grain also, such as whole oats or shelled corn.

Ewes after lambing: Give 1 pound of grain per head to those with one lamb, 1½–2 pounds to ones with twins. After two months, feed ½–1 pounds grain. When lambs are weaned, take away all grain suddenly, and only give a little hay for the week to help dry up the ewes.

Before breeding: During breeding season give rams 1 pound of grain a day. Put ewes on top quality hay, lush pasture, or give some grain.

When turning out to pasture: Every spring when you turn out the sheep to pasture wait until the grass is a few inches high and give them a big pile of hay so that their feed will change more gradually.

During winter: You can let your sheep eat from a harvested cornfield, give them roots from the garden, and 2–4 pounds of hay per day per sheep, or 10 bales of hay per winter. Plus give them 75 pounds of grain.

Creep feeding young lambs:

Start feeding lambs at two weeks even if they will be finished at pasture. Grind the feed at first, then feed whole. The creep feed should contain 15 percent protein, and 20–40 percent high-quality alfalfa hay given in a separate rack. A lamb will typically eat 1½–2 pounds of rations per day from 10–120 days old.

Keep these records on your sheep:

Name, sex, birth date, ear tag number, comments on birth (single, twin, etc.), name of sire and dam, body weight, date bred, birth weights and sex of lambs, space for comments, disease record (with space for symptoms and treatment). When an animal is removed from the flock, note the reason for it.

Housing:

During summer all sheep need is some basic shelter such as trees or a three-sided shed. Pregnant ewes need better shelter in the winter and during lambing to protect the babies. If you shear and the weather gets cold they may also need better shelter. Otherwise sheep can live in the open.

Health care:

Stress: Take particular care of sheep after stress, such as storms, transportation, or abrupt feed changes. They may stop eating or even have heart attacks.

Watch for: Labored breathing, dullness and listlessness, refusing to eat, lying down a lot, and going off alone. Keep a rectal thermometer. Normal temperature is from 100.9°F to 103.8°F, average is 102.3°F.

Enterotoxemia in lambs: Symptoms include tremors and convulsions, diarrhea, and collapse. Death usually occurs within two hours. It is caused by the bacteria *Clostridium perfringens*. Lambs get it from overeating and quick transition to rich foods. To prevent use gradual transitions, timed and balanced feed and vaccinate pregnant ewes. It cannot be cured once caught. It is prevented with *Clostridium perfringens*, Type C and D toxoid, administered by a vet during pregnancy and a booster at least four weeks before lambing.

Vibriosis: Causes abortions, dead and weak lambs. The most common cause of sheep abortions. Requires two vaccinations to ewes and lambs, then yearly boosters.

EAE: Caused by *Chlamydia psittaci*. Produces abortions, eye infections, lamb arthritis, epididymitis (fertility problems in males), pneumonia, and diarrhea. The vaccine is usually given with *Vibriosis bacterin*.

Leptospirosis: Causes abortions, anemia, and systemic disease. A killed bacterium is given to immunize sheep.

Tetanus (lockjaw): *Clostridium tetani* enters the body through open wounds and causes muscle spasms, stiffness, and other nervous system problems. Seventy-five percent of lambs die from it. A tetanus toxoid vaccine every 10 years prevents it.

Clostridial disease: Also called blackleg, malignant edema, and braxy, it is a soil bacteria found in the intestinal tract. Symptoms are lameness and swelling just beneath the skin (subcutaneous), and rapid death.

▾ **Great Pyrenees guarding herd of sheep**

Bluetongue: Also called sore muzzle, the virus is spread by the biting midge. Produces mouth ulcers, nasal discharge, crusty nostrils, and lameness. The vaccine is a live virus vaccine that must be introduced to a flock only if they have it because it will give it to them. Wear rubber gloves and follow directions.

Sore Mouth: Also called "contagious ecthyma," causes pox, or lesions on lips, nostrils, eyelids, mouth, teats, feet, etc. Spread through direct and indirect contact.

Predators:

Sheep are especially vulnerable to predators since they are not that smart and get scared very easily. Sheep have died from heart attacks when hearing thunder. A dog can chase a sheep to death without biting it, and of course any larger animal can and will eat a sheep. The prevention to this is a dog. Either get a border collie to herd the sheep when you need to (they also are great at watching children), or get a guard dog to live with the sheep in the field. One of the best guard dogs for this purpose is a Great Pyrenees. They stay in the field and attack sheep predators. The ideal way to raise one of these is to have an ewe nurse and raise the puppy. It will become part of the sheep family but keep its guarding instincts.

Breeding:

If the sheep is too fat, start a diet six weeks before breeding or else she is at

▲ **Lambs should nurse very soon after birth**

risk for pregnancy problems. Two weeks before breeding, "flush" the ewes by putting them on the richest pasture or giving them really good alfalfa hay, or give them 50 percent more grain. This will increase the chances of having twins or triplets. If the sheep breed is not "open-faced" (it has hair on its face), clip the wool away from the eyes, and also "tag" her, or clip the wool away from the vagina. Check feet and trim at the same time. The calendar that I have included has sheep lambing in February, so since sheep mating season is from September–December, breed in October. In 5 months she will lamb. You can buy a yearling ram every year to breed the ewes and then eat him, or you can keep a ram full time. Don't use one of your own rams born of your ewes or you will have problems because of in-breeding.

Lambing preparation:

Get your ewe into a lambing pen that is completely dry, protected from wind, with a room. Check your herd every 2 hours because all the sheep will give birth around the same time. Gather together iodine, warm water, soap, lubricant, and old clean towels or blankets

Lambing:

Note: Make sure there is a veterinarian or knowledgeable person available around the time of lambing if possible.

1. If the ewe's water broke and she has been straining for 20–30 minutes and no lamb is born, you will need to help.
2. If you haven't tagged her already, clip the dirty and excess wool from around the ewe's birth canal. Then wash her with warm, soapy water.
3. Wash your arms with warm, soapy water and apply an antiseptic lubricant. You might want to use a sterile obstetrical sleeve instead.
4. Enter the ewe and feel the lamb to see what's going on. Find out how many lambs there are, and which feet go with which legs before you start pulling.
5. Deliver the closest lamb first (if it's twins). Pull gently downward between the legs of the ewe, rotating gently if you need to free a shoulder.

What to do if it's a breech:

If you feel a tail instead of a head, call a vet. If you can't get the vet, get an assistant. Have the assistant hold the ewe up by the hind legs. Put on a rubber glove and push the lamb forward. Then pull its hind legs straight (they will be bent forward). Gently but quickly pull the lamb out the same as if it were head forward. Clear out its nostrils.

New lamb care:

Pinch or cut off the umbilical cord 4–5 inches from its belly. Dip the navel into iodine. If the lamb is really chilly, wrap it in a blanket or towel and bring it into the

▲ Non-electric clippers

house to get warm. Then get it back to the mother quickly to start nursing. It should nurse within 30 minutes of birth, and if it isn't interested, milk some colostrum into its mouth. If the mother isn't doing what she's supposed to, rub some placenta on the baby, especially around its anus. Even an orphaned lamb can be adopted by a ewe if you rub a new placenta on it.

Orphan lambs:

Bottled-fed lambs are called "bum lambs." If you have goat milk, use it. If not, buy a replacement with 30 percent fat, 20–24 percent protein, and less than 25 percent lactose. Orphan lambs occur somewhat frequently because a ewe will not recognize its lamb if it does not smell right. If you take the lamb in the house to get warm and bring it back, the ewe may have totally forgotten it. Feed every 4–6 hours until it is 3 weeks old, then every 8 hours for the next 2 months. At the first few feedings the lamb will not drink more than 2 ounces.

Milking:

Sheep's milk makes excellent cheese and makes more cheese per pound than cow's milk. It is also easier to make into cheese than goat's milk. You should get 1 quart–1 pint of milk per day. It also makes great yogurt and ice cream but not very good butter because it is naturally homogenized.

Shearing:

Shear during dry weather, and for a small flock use hand clippers. Put the sheep in a small pen or in a barn with a

◂ Shearing a sheep

clean floor or tarp nearby. Have someone hold the sheep but not by the wool. Put the sheep between your knees on its side with its head pointing towards your back. Cut off any dung locks (wool with poop in it), grease tags (knots of matted wool), and throw them out. Then clip the wool off, cutting as close to the body as you can (about ½ inch away), and it should come off all in one piece. Start with the head, clip across the body from one side, across the back and to the other side, zigzagging all the way to the rear. Don't cut the skin, and don't cut twice—if you don't get it with the first clip, forget it. Treat any cuts with disinfectant, check the hooves, and let her loose. Lay the fleece out and examine it for dirt, let it dry, and gather it up with the side that was against the sheep's skin on the outside. Don't store it with plastic, instead use paper or cardboard. When you are done, go have someone check you for ticks.

RABBITS

Practicality

Rabbits are an easy way for homesteaders to raise extra meat on the side, as they require very little work and don't eat very much. They are especially practical for urban homesteaders because rabbit hutches are legal in most places. Any rabbit breed can be eaten, but there are rabbits that are designed specifically for meat and are much larger. The nutritional value of rabbit is not as high as chicken or other staples of a meat diet, but since they are very self-reliant it doesn't hurt to have some around to eat sometimes. They eat hay, grass, rabbit feed, and vegetables, and can eat other kitchen scraps such as bread. They should have a constant supply of fresh water, and they should have a raised hutch, or cage, with a wire mesh floor big enough for droppings to fall through but not their feet, usually with

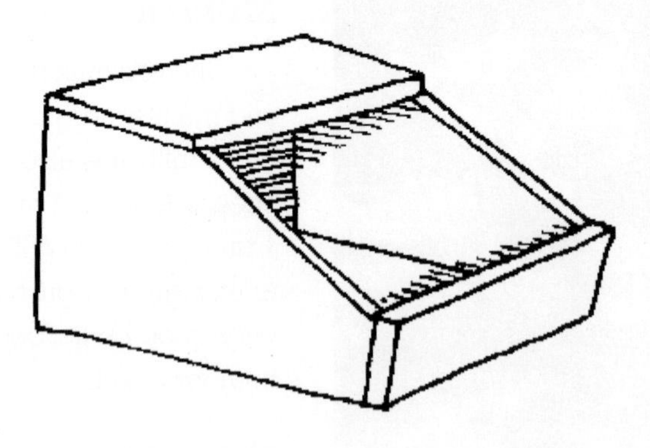

▲ **Rabbit nesting box**

½-inch squares. Their home should be kept clean and dry, sheltered from the weather.

How do I breed rabbits?

Rabbits breed very prolifically and easily, but may not always be the best mothers. The males and females should be kept separately and only put together for breeding for a short time under supervision. Once she is pregnant, she should be provided with a nesting box and soft bedding such as hay and down or other material. After the babies are weaned the males and females should be separated so that they will not breed too early. As they grow, you will probably have to separate them as they become very territorial.

RAISING WORMS

Worms have three lips, five hearts, a brain, and red wigglers are bisexual, although it takes two worms to make babies. They are useful for fishing, bird food, composting, and gardens. They tunnel and loosen the soil, reduce soil acidity, free plant nutrients into the soil, and make your topsoil deeper.

Getting worms:

You will need 2 pounds of worms per 1 pound of food waste you produce. There are many types of worms for different uses.

Garden worm: The kind you find in your backyard or garden, and turns white in water. Don't use these for your compost bin.

▾ Red worms

African nightcrawler: About 5 inches long but can grow much bigger, they like warm weather and die in cold air or water. They reproduce every two years.

Native nightcrawler: Found in your yard and garden, they require lots of soil, like temperatures under 50°F, and don't like to be disturbed. These are best for the garden but not the compost bin.

Red worms: Consume lots of garbage, reproduce every seven days, and has alternating red stripes, 1 ½–3 inches long. These are used for compost and garden and fishing.

Sewer worms: These are found in manure piles.

Worm bin:

If you are going to be putting your worms into your compost bin, make sure that the bin is not made of a fragrant wood such as redwood or cedar. A wood box will last 2–3 years. The bin should also be no deeper than 12 inches since worms stay near the surface. Never put manure into your worm bin or they will cook and die.

To feed the worms:

Besides your regular compost (organic materials like food scraps), add shredded newspaper, decaying leaves, peat moss, and a bit of dirt. Feed once or twice a week, varying where you put the food.

To harvest the soil:

Clear out half the area and put in new bedding (newspaper or peat moss or other materials) in the clean side. Move their food over also, and the worms will migrate over to the new spot. When they have moved, take out the old stuff for the garden.

5 | The Comforts of Life

MAKING TOOLS AND FARM EQUIPMENT

The ordinary arts we practice every day at home are of more importance to the soul than their simplicity might suggest.

~Thomas More

How do I make an awl?

Use oak or birch to carve out the handle. Attach a metal spike securely to it. Or, instead use a nail, file off its head, and make a wood handle.

▲ Awl

How do I make a helve (ax handle)?

1. Split a hickory log into four quarters.
2. Using a hand ax, hew one quarter into a square timber beam.
3. Make a pattern for a handle on a board. Make sure the helve will be the perfect size for your hands and grip.
4. Using the pattern, draw the helve onto the beam.
5. Hew the beam into the rough shape of your helve.
6. Put the unfinished helve onto the shaving horse and shape it more.
7. Use a spoke shave to shape it even more finely, until you are ready to smooth it.
8. Use a piece of glass to smooth it.

Levels and plumb bobs:

Levels and plumb bobs work on the same principle, and are used to make sure the thing you are building is straight. A plumb bob is simply a string with a weight on the end, and a level is a straight piece of wood with a small hole. In the hole there is a tiny string with a weight, and a line in the

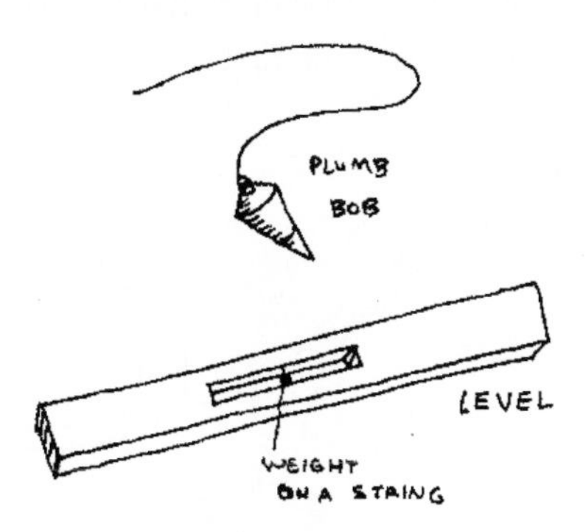

center. If the thing is level, the weight will line up with the center line.

How do I make a simple harrow?

1. Find an elm tree with a large, defined fork like a V, with both limbs as alike as possible.
2. Chop down the tree, cut the tree so that you have a Y, and hew it smooth.
3. Attach a crosspiece in the fork and bore holes evenly down it.
4. Drive in iron harrow teeth in the holes.
5. Bore a hole through the foot of stalk of the Y (or the long trunk sticking out of the top of the A), and attach a clevy (an iron U-shape with holes for a crosspin). The clevy is for hitching animals to the harrow.

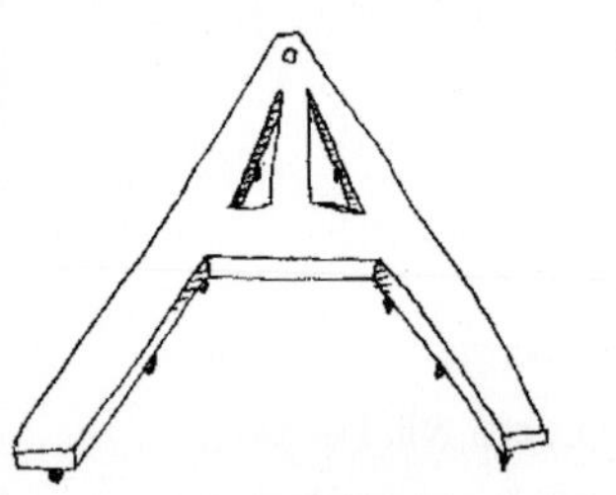

Simple harrow

How do I make a simple plow?

1. Cut a branch of a hardwood tree into a long handle that curves down until it is straight at one end.
2. Cut another branch of hardwood about ⅔ the size of the first one and carve it into a curved shape that is thicker at one end than the other.
3. Cut a slot into the angle of the larger piece and insert the thick end of the smaller piece of wood.
4. Attach a piece of steel or wood as a blade to the bottom of the vertical end of the large piece of wood.
5. The smaller piece of wood will have a skinny end sticking straight forward. Drill a hole in this and insert a rope for an animal to pull.

A: Hardwood handle

B: Hardwood piece inserted into slot in A

C: Hardwood piece with large slot through which A is inserted

D: Piece of steel inserted into C

E: Attached to oxen or horse

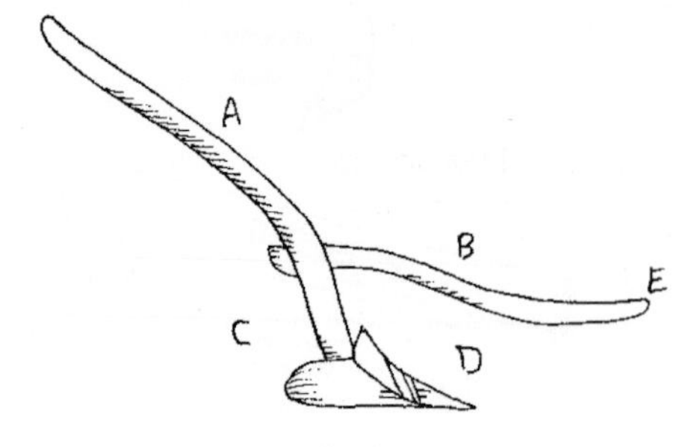

Simple plow

How do I make a rice paddy rake?

1. Drill a series of holes through piece A (the hardwood) and insert small hardwood pegs to create a comb for raking.
2. Use this when the paddy is plowed and flooded to remove weeds.

Rice paddy rake

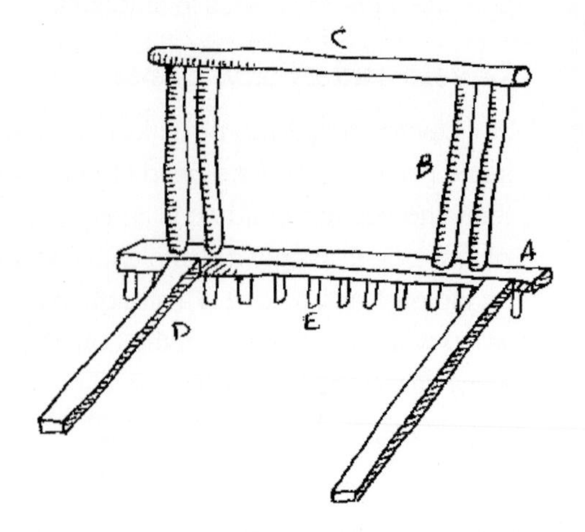

▲ **Lambs should nurse very soon after birth**

risk for pregnancy problems. Two weeks before breeding, "flush" the ewes by putting them on the richest pasture or giving them really good alfalfa hay, or give them 50 percent more grain. This will increase the chances of having twins or triplets. If the sheep breed is not "open-faced" (it has hair on its face), clip the wool away from the eyes, and also "tag" her, or clip the wool away from the vagina. Check feet and trim at the same time. The calendar that I have included has sheep lambing in February, so since sheep mating season is from September–December, breed in October. In 5 months she will lamb. You can buy a yearling ram every year to breed the ewes and then eat him, or you can keep a ram full time. Don't use one of your own rams born of your ewes or you will have problems because of in-breeding.

Lambing preparation:

Get your ewe into a lambing pen that is completely dry, protected from wind, with a room. Check your herd every 2 hours because all the sheep will give birth around the same time. Gather together iodine, warm water, soap, lubricant, and old clean towels or blankets

Lambing:

Note: Make sure there is a veterinarian or knowledgeable person available around the time of lambing if possible.

1. If the ewe's water broke and she has been straining for 20–30 minutes and no lamb is born, you will need to help.
2. If you haven't tagged her already, clip the dirty and excess wool from around the ewe's birth canal. Then wash her with warm, soapy water.
3. Wash your arms with warm, soapy water and apply an antiseptic lubricant. You might want to use a sterile obstetrical sleeve instead.
4. Enter the ewe and feel the lamb to see what's going on. Find out how many lambs there are, and which feet go with which legs before you start pulling.
5. Deliver the closest lamb first (if it's twins). Pull gently downward between the legs of the ewe, rotating gently if you need to free a shoulder.

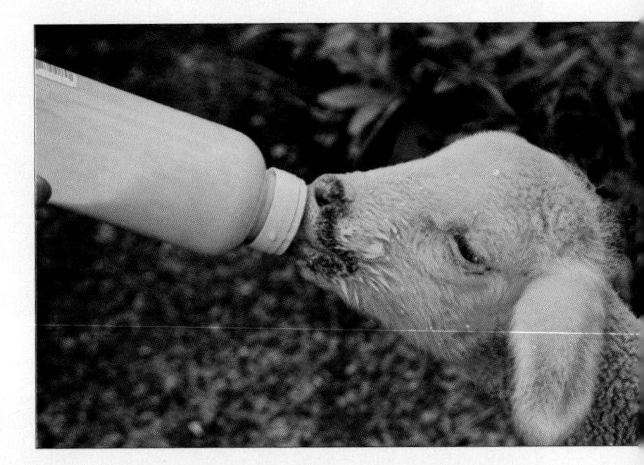

What to do if it's a breech:

If you feel a tail instead of a head, call a vet. If you can't get the vet, get an assistant. Have the assistant hold the ewe up by the hind legs. Put on a rubber glove and push the lamb forward. Then pull its hind legs straight (they will be bent forward). Gently but quickly pull the lamb out the same as if it were head forward. Clear out its nostrils.

New lamb care:

Pinch or cut off the umbilical cord 4–5 inches from its belly. Dip the navel into iodine. If the lamb is really chilly, wrap it in a blanket or towel and bring it into the

3. Harness to a water buffalo or ox.
   A: 150 x 5 x 7 cm hardwood
   B: Soft timber or bamboo
   C: Softwood handle
   D: Softwood
   E: Hardwood dowels for teeth

How do I make an ox yoke?

1. Measure the oxen for a yoke by standing them side-by-side 6 inches apart. As the oxen grow you will need to replace the yoke when their sides start to touch.
2. Cut a piece of hard maple and shape it roughly with an ax.
3. Cut two hickory saplings and when they are green and flexible, bend into a U-shape and tie securely.
4. Hang all three wood pieces from the ceiling to season and dry over the winter.
5. Use a drawshave and pocketknife to shape the main portion of the yoke, which will rest on the oxen's shoulders.
6. Use an iron poker heated in the fire until it is red-hot to burn holes into the wood, which will be close to the oxen's necks on either side.
7. Insert the ends of the hickory bows (the U-shapes) into the holes so that they can protrude downward and encircle the oxen's necks.
8. Fasten the hickory bows by drilling small holes through ends sticking out. Drive in a peg that is wider than the bow.
9. Put an iron bolt through the center of the yoke with a ring on the bottom surface to be attached to a cart tongue or log chain.

Ox yoke

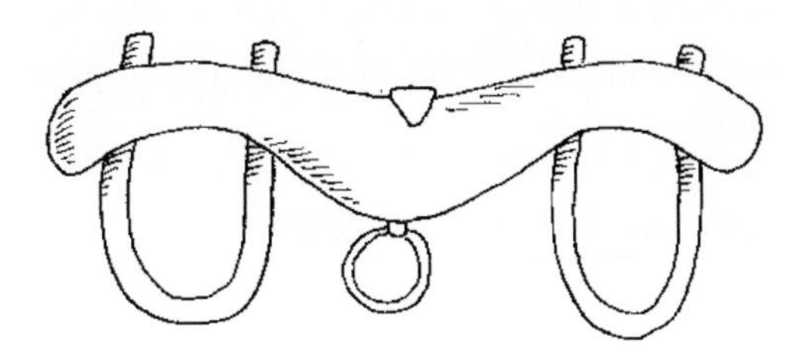

How do I hook an animal up to an implement?

One of the easiest ways is with a single tree. It is a single piece of wood with metal supports. Two chains or ropes are attached to each end, which go to the harness or ox yoke. A ring in the middle is tied to the implement.

How do I care for an ox yoke?

Every time you use the yoke, wipe it down with wisps of straw to remove all the moisture and dirt. Hang it up on the wall so it will stay clean.

SIMPLE HOME FORGING

Under a spreading chestnut tree
The village smithy stands:
The smith, a mighty man is he,
With large and sinewy hands;
And the muscles of his brawny arms
Are strong as iron bands.

~Henry Wadsworth Longfellow, "The Village Blacksmith"

What is forging?

Forging tools is beyond the scope of this book. However, making horseshoes is a possible skill, since the small homesteader would not need to rely on a farrier. Other useful items are nails, hinges, axes, rings for animal harnesses, horse bits, candle sticks and lanterns, chains, pokers and fireplace hooks, and other simple items which are better if metal.

What are forging safety rules?

Keep water nearby at all times; don't wear cloth gloves, only leather; don't try to catch anything; don't touch something if you're not sure it's cool.

Tools you will need:

Blacksmith cross peen (hammer): Get one weighing between 1–3 pounds. If you are small you will need a small hammer.

Anvil: A forging (or blacksmith) anvil weighs between 75–500 pounds, while a farrier's anvil weighs between 50–150 pounds. The smaller your hammer, the smaller you want your anvil. Some antique anvils are a cross between the forging and farrier's anvil. Some people substitute the anvil with a heavy metal railroad rail or even a large rock.

Forge: An apparatus that creates enough heat to soften metal, or an extremely hot fire. Also the place that a blacksmith works.

Vise: The most important tool is a blacksmith's leg vise (or post or "solid box"vice), and bigger is better. Anchor the vise to a heavy bench, or anchor it to a post in the ground.

▲ Blacksmith tools

▼ Farrier's anvil

Tongs: Tongs can be made on your own forge, or you can use channel-lock pliers.

Safety supplies: A really good welding hood or safety glasses with side shields, a leather apron, leather gloves, and cotton clothes (synthetic fibers melt to your skin).

How do I build a forge?

The old-style wood-burning forge consists of a brick or stone platform that

is wider on the top than on the bottom, so that you can stand close to the fire. On the top of the platform there is a trough that runs right through the fire so that things can be laid directly in the flame. A metal ring runs all around the top edge of the platform in order to hold in the rock or stones. A fire is built in the center of the platform through a hole in the front side on the ground, and the flame reaches up through a hole in the top of the platform. A pipe runs through the brick directly to the fire so that air can be blown through it with a bellows. Over the top of the platform is a large sheet metal hood that directs the smoke to a hole in the brick chimney wall so it can go outside. The inside of the chimney is 1 foot square.

How do I forge wrought iron?

1. Heat the metal you are shaping until it's orange, then use the hardy (on the anvil) to hammer the amount you need off of it if necessary, holding the end with your hand.
2. Reheat the metal to a cherry red and shape it by hammering. Make an edge of something using the hardy.
3. If the part you are holding gets too hot, hold it with tongs and plunge just that end in water. Keep shaping your metal.
4. When you are done, put the whole thing in water with the tongs, and then clamp it in a vise to sharpen it if necessary.

How do I make nails?

1. Heat a nail rod and hammer one end to a point with a forge hammer.
2. Dent the rod on the pointy end at naillength (how long you want your nail), by hammering it over a hardy on an anvil.
3. Slip the nail end of the rod through a nailheader and snap the rod off at the dent.
4. Using the forge hammer, pound the top end of the nail flat to make a head.
5. Dip the nail in cold water, and then dump it out of the nailheader.

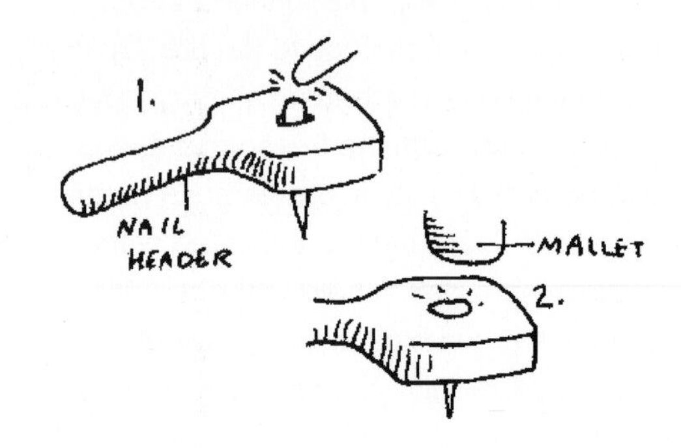

▲ **Handmade nails**

How do I make a horseshoe?

1. Use a straight piece of about $\frac{5}{16}$ inch by $\frac{3}{4}$ inch by 11-inch mild steel. If you have a large or small horse, adjust the size of the metal.

2. Mark the center of the steel with the hardy. Then make two more marks $1\frac{3}{8}$ inch on either side of the center on the flat side.
3. Heat the steel and those marks will create a bubble, which will help you find center. Only heat about $\frac{3}{4}$ of the steel so you can hold one end with tongs.
4. Lay the steel over the round end of your anvil. Use a series of hammer blows to bend the steel to a 90° angle with the center mark exactly at the outside corner of the angle.

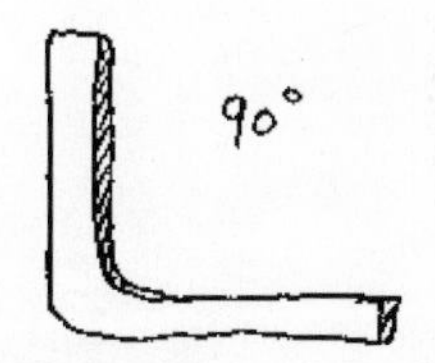

5. Round off the ends of the steel by hitting the corners against the flat side of the anvil.

▼ **Rounding off the ends**

Rounding and leveling the shoe

Making nail holes

6. Lay the end of the steel on the round end of the anvil again. Curve the straight end of the steel around to be the same roundness of the horse's hoof. If you have an already-made shoe it can be your pattern.
7. Use a 3/8-inch wrench to mark the spots where you will put the nail holes. This should be done when the metal is dark red, not red hot.
8. Use a forepunch (made for farriers to make nailholes) to make a dent in the marks only 1/3 of the way through. It will make a bulge on the outside of the shoe called a "frog eye" that needs to be tapped back in.
9. Use the forepunch to hammer in 2/3 of the way through. Then use a much

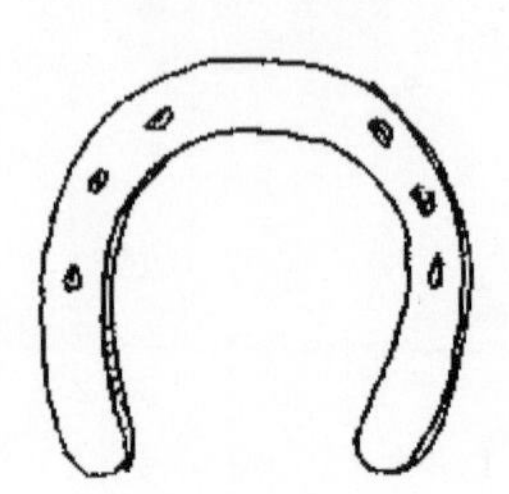

Measuring the number of holes to place

◂ **Fitting nails properly**

smaller forepunch, with a skinny head (also called a "stamp") to punch the nail hole through. When it cools go to the next hole.

10. Use the pritchel or hardy hole on the anvil to make sure the stamp goes completely through the shoe. The shoe should be black by now.
11. Only one half of the shoe is done. Go back and do the other half exactly the same way. Your goal is to make the two halves symmetrical.
12. Hammer the inside of the shoe to make it rounded. Clean out any holes that did not go through. This time, make sure that the holes are angled to be the same angle as the hoof wall.
13. Run around the outside edges to bevel them and round them as you did the frog eyes. Then level up the whole shoe and make sure it is a mirror image. You will also have to make nails for the shoe.

How do I make tongs?

1. Use a round metal rod and make a mark 1 inch from the end using the hardy. Then make another mark 1 inch below that.
2. Make two round dents in the metal on those two marks. Line one dent up with the corner of the anvil and hit it very had so that it twists around flat. Then

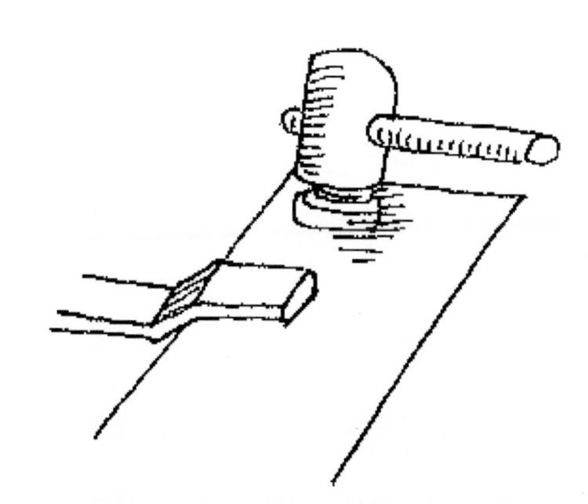

▴ **Hammering the tong end**

do the same for the other dent until the 1-inch space between is completely flat.

3. Bend the dent that was the lowest on the rod to a 45° angle.
4. Flatten the round part at the top of the rod. At this point the rod should look like one half of the tongs.
5. Use a tapered punch (a cone-shaped piece of metal) to drive a hole.
6. Cut off the handle of the tong to the length that you want. Use the extra pieces of rod to make a rivet.
7. To make the rivet, cut a small piece of rod and flatten the top. Insert it into the hole, and then flatten the other end of the rivet so it can't come out.

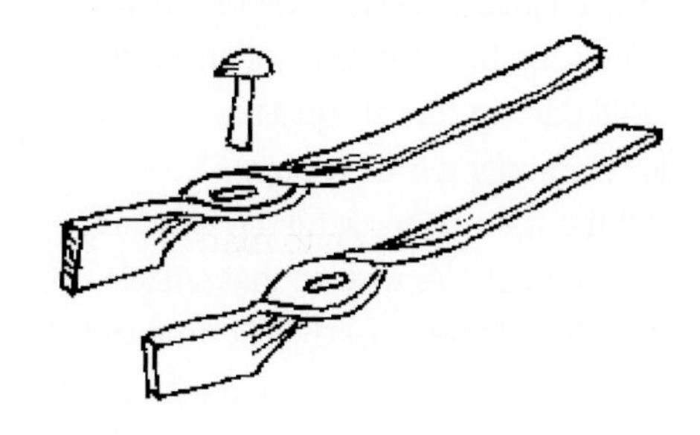

▴ **Fastening the rivet**

The finished tongs

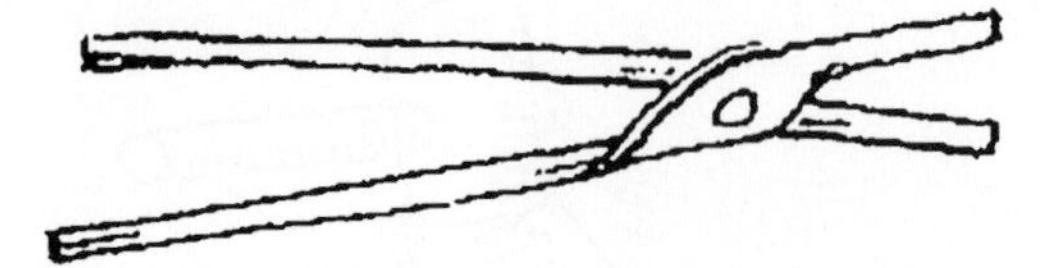

MAKING THINGS FROM WOOD

The carpenter stretches out a line. He marks it out with a pencil. He shapes it with planes. He marks it out with compasses, and shapes it by the figure of a man, with the beauty of a man, to reside in a house.

~ Isaiah

How do I make charcoal?

1. Get a big metal drum. Make five 2-inch holes in the bottom near the center of the drum.
2. Put three bricks or big rocks on the ground and set the drum on top with the open end up. Don't cover up the holes with the bricks.
3. Chop seasoned hardwood into big pieces about 5x5 inches. Only use broadleaf trees like maple, almond, hickory, etc. Start a fire in the drum with small kindling and when it is done start putting the larger pieces at the bottom, filling the bucket to the top.
4. When the fire is really hot and burning well, pile up dirt around the bottom, leaving only a 4-inch gap.
5. Put the lid on almost all the way, but not quite all the way so that smoke can escape. White smoke will come out, and when it starts to slow down, bang on the drum so the wood will settle and more smoke will come out. It will take about 4 hours to finish.
6. When the smoke turns to thin blue, close off all the air to the wood by shutting the lid tight, and pile up more dirt to cover the 4-inch gap at the bottom. If any air gets in, the charcoal will burn so be extra sure it is airtight. Let it cool.
7. After 24 hours dump out the charcoal.

How do I make charcoal in a traditional earth mound?

Instead of putting the wood into a drum, you can pile it up on some cleared ground or in a small pit, with the largest wood at the bottom. You can also insert a pole in the middle. Then cover it with

An old charcoal kiln

Charcoal

Pine resin

grasses and leaves, and on top of that put sand and dirt, leaving small openings all around the bottom for air to go in. Remove the pole, start the fire in the hole that is left, and watch the smoke come out the top. When the wood is done burning, cover the holes around the bottom and the hole in the top with more dirt.

How do I make pitch?

1. Pitch is somewhat like tar but it is from a tree. First collect pine resin by cutting diagonal cuts into a pine tree, and put a bucket at the base so the groove with the resin will run down into the bucket.
2. When all the resin has run out of the tree, chop down the tree and make it into charcoal.
3. Powder the charcoal by grinding it up, and bring the pine resin to a boil. If you want very liquid pitch then add less charcoal to the resin, but if you want a very strong pitch then add more.
4. The pitch is used to make wood things waterproof, such as buckets and Archimedes screws.

How do I make a log raft?

Find six to seven logs, no longer than 16 feet long and 1 foot wide. Fasten them together with three sticks attached across the logs at front, middle, and end. You can

Log raft on a river

▲ Dugout canoe

make a tent on one end of the raft, and oarlocks. You can use a long pole to push the boat along, or you can make oars.

How do I make a plank canoe?

Get plank of wood, 16 feet long and 2 ½ feet wide. Cut a bow (front end of a boat) on the plank on one end by shaping it to a point. Make or find a little stool that can be fastened to the back end of the plank, about 3 feet from the end. Make a double-ended or single paddle by hewing a piece of wood.

How do I make a dugout canoe?

Get an old-growth white pine tree. Build small, slow fires in the log and burn part of the center out, and then scrape the rest out with an adze. Leave cross sections to brace the sides.

What kinds of buckets are there?

A sap bucket's top is smaller than its bottom, and it has a bail (a half-circle wire for a handle). A water bucket has straight sides, and two of the staves (or pieces of wood that make up the side) are made long so that a pole can be put between them for a handle.

▼ Sap bucket with tap

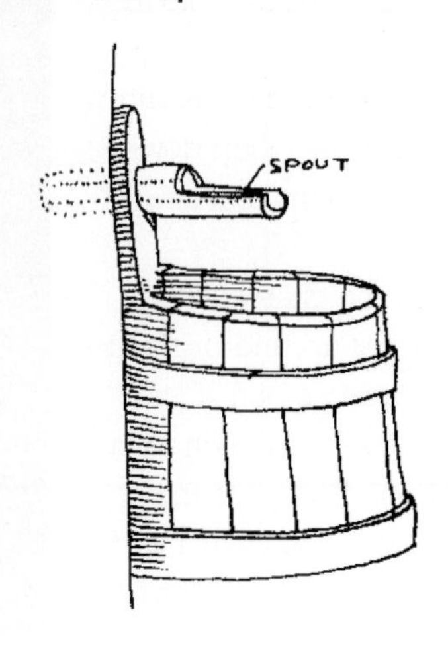

▲ A well-made bucket can last a century

How do I make any bucket or barrel?

1. Cut white oak into staves (the pieces that make the walls), all the same size. Don't make them taper or misshapen. Figure out how many you'll need beforehand.

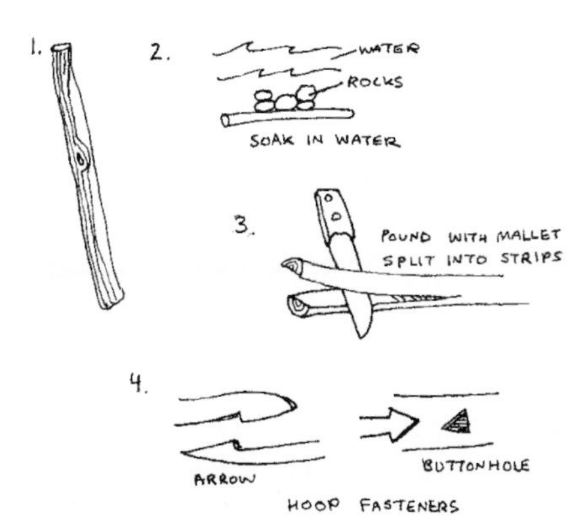

▲ Making a bucket hoop

2. Bevel the edges of the staves so that they fit in a circle.
3. Cut hickory poles in spring for making the hoops. Soak them in water for a while.
4. Split the hickory poles in half to make two strips for the top and bottom hoops, and then soak them in hot water until they bend.
5. Shave one strip big enough to bend all around the bucket or barrel with enough room to make a clasp.
6. On one end of the hoop, cut a bullet-shaped hole, with the point of the bullet pointing to the end of the wood.

▼ Wooden water bucket with iron handle

7. Shape the other end into an arrow shape, so that it will fit into the hole on the other end, but won't be able to pull out without pushing the hoop smaller.
8. Cut grooves around the staves where the top and bottom hoops will be, and slip the hoops on.
9. The bottom is made of a round piece of wood fitted closely inside, and fastened to the staves with small nails or other fasteners. Some bucket makers use no fasteners at all, and are able to fit the staves together so tightly with the hoops that nails are not necessary.
10. Seal with wax on the inside and linseed oil on the outside.

How do I make a round black ash basket with white ash handle?

1. This is one of the most traditional baskets to make. Cut down a smaller black ash, take off the bark and soak the log in water.
2. When it is well softened, take a big maul and hit the log as hard as you can all over it. Every year a tree grows a soft and hard ring. Pounding will loosen the hard ring so that you can peel it off in strips (called splines).
3. If you are making a fancy basket, lay the splines on a bench and smooth them with a plane. If you're not worried about looks, skip this part.
4. Lay out some splines like wheel spokes, with their middles in the center.
5. Weave the bottom in a circle around and around flat from the center, until it's as big as you want it. This is the bottom of the basket.
6. Put the bottom over a round chopping block or log and bend down all the splines that stick out, to make the walls.
7. Weave the walls as high as you want.
8. Bind the edge with white ash, wrapping it around and around
9. Make the handle of white ash. Cut a notch on each end of the handle to make a hook, stick them through the binding so the hook will catch on the weaving and won't pull out.

Basket materials and their preparation:

Brown willow: Gather older growth in late fall or winter. To prepare, boil 4–6 hours or soak 3–4 days and peel off bark. Soak ½–1 hour in warm water.

▾ Brown willow basket

Cattail stalks: Gather when fully grown in early fall. To prepare remove top, clean base, split in half, then in quarters. Soak 1–5 minutes in lukewarm water.

Green willow: Gather green first-year growth in spring. To prepare, soak ½–1 hour in warm water.

Hardwoods: Gather first-year growth in spring or fall. Use right away, peeled or with the peel, and split if you want. Do not soak.

Honeysuckle: Gather 1- to 2- year-old vines from late spring to early fall. To prepare, boil 3–4 hours then rub hard with a towel to remove bark. Soak 20 minutes in lukewarm water.

How do I make a birch basin for collecting maple sap for maple sugaring?

1. Cut down a white birch or another tree with soft wood, and split the tree in half.
2. Use an adze to hollow out a small section the size you want your basin, and then chop it off.
3. Hollow out the next one, and so on.

How do I make a shaved broom?

1. Take piece of birch 2 or more inches wide, and as tall as you want your broom to be.
2. Starting at the bottom, shave up toward the handle for how tall you want the bristles to be, until you have as many bristles as you need. Cut off the wood core.
3. Starting toward the handle, cut downward, but not close enough to cut off your lower bristles. This makes more bristles.
4. Fold the top bristles down on top of the bottom ones and tie them tight with braided cornhusks or some other handy thing.
5. Every time you use this broom, dip it in water before you use it, and it will last longer.

How do I make a birch twig broom?

1. Make a handle of birch with a bulge on the end.
2. Take a bunch of birch twigs and stuff them into a small iron hoop.
3. Drive the handle top-first through the middle of the bunch until the bulge stops it.

▲ Twig broom

▼ Straw broom woven tightly with thread

4. When the twigs give out, just take out the handle and get new twigs.

How do I grow broomcorn?

Grow broomcorn the same as corn—plant 4 inches apart in rows 30 inches apart, and thin out when they sprout. A 10-foot row will make one or two brooms.

How do I harvest broomcorn?

When the seeds are still green, bend the stalks 2–3 feet below the top tassel. Let them dry, still rooted in the garden, for a few days. Then cut off the brush with 6 inches of stalk. Pull off all the leaves, spread out on a clean surface in the sun for 3 weeks (bring in at night if it will rain). When the tassels spring back when you gently bend them, they are ready to be made into brooms.

How do I save broomcorn seed?

Wait until the seeds are fully ripe, and they will form into clusters that grow like corn tassels. Cut them, thresh out the seeds by scraping them off, and use like sorghum grain.

How do I make brooms from broomcorn?

1. Comb the seeds out, and trim the stalks, allowing 4–8 inches of stem plus the tassel length.
2. Make a broom handle out of straight young wood 4–5 feet long. Birch works well.
3. Soak the tassels in boiling water until they are softened. Take them out and when they are still hot and wet, and bind them super-tight with twine or wire around the handle. One way to do this is to hang a rope from the roof with a loop at the bottom. Wrap the middle of the rope around the tassels twice, put your foot in the loop, and step down. The rope will tighten, and you can wrap the twine around. Make sure

▾ Broomcorn brooms

to put the rope lower or higher on the broom than where you are putting your twine.

4. Trim the bottom of the broom at an angle that will help you sweep in corners.

MAKING HOME NECESSITIES

Most of the luxuries and many of the so-called comforts of life are not only not indispensable, but positive hindrances to the elevation of mankind.

~*Henry David Thoreau,* Walden

How do I prepare clay for pottery?

1. First wedge the clay. Make a ball and push it, then fold it, then push it and fold it. It is very much like kneading dough. Use your upper-body weight to push the clay, and work on a porous surface such as canvas or clean dry concrete. The goal is to get out all air bubbles and to mix it. Cut the clay in half to check for bubbles.
2. The clay is ready when it is the same consistency all the way through (no softer or harder spots), all the same color and no bubbles. Don't worry about sand or grog (dried clay particles) in the clay.

▾ **Wedging**

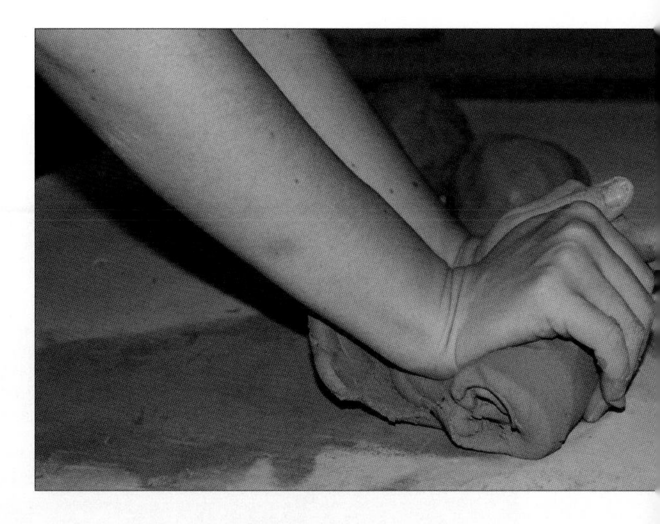

How do I throw on a wheel?

1. Make sure the wheel is dry and form the clay into a ball. Put a bat (a wooden or plastic disc made for potter's wheels) down on the wheel. Slam the ball down onto the wheel hard, as close to the center as you can. Make sure it is stuck to the wheel.
2. Dribble water onto the clay, making the surface very wet. Start the wheel spinning, and use your left hand to center the clay. Brace your left elbow against your left leg, and hold your left thumb in your right hand. Push against the clay with the heel of your hand.
3. Centering takes a lot of strength, but the goal is to form the clay into a perfectly circular shape right in the center of the wheel. You will be able to tell when it is centered because it will no longer wobble if you are holding your hand stiff against it.
4. In the same position, flatten the top of the clay with your right hand. The

The clay should be as centered as possible

trick to centering and flattening is just holding your arms in a stiff position, which puts pressure on the clay.

5. To open the clay, cup your left hand around the edge of the clay, and press into the top with your right hand's middle fingers, slightly right of the center. Remember throughout to keep the clay wet. Hold your fingers and hands steady or things will get crooked. You can use a small sponge with the right hand to help you.
6. Continue to shape the clay as it spins, drawing the edges up into walls until it is shaped as a bowl or cup or other container. This takes much practice. Always raise the walls with the right hand at the 4–5

Centering the clay

▲ **Press with middle fingers into the center**

o'clock position and brace the inside with the left hand.

7. Once the basic shape has been formed, many potters use a steel or wooden slip, which is just a simple oval used to scrape the outside clean of wet clay and remove imperfections. It is held in the right hand and pressed gently against the clay as it turns.
8. Use a stick or dowel to clean the base of the container by holding it with both hands. Put it on the right side, and keep it at an angle to cut away the clay that forms at the bottom that is too thick. Then hold the stick horizontally along the bottom to remove the excess clay from the wheel.
9. Soak up the water on the inside using a sponge (attach it to a long tool if you have to), and don't add any more water to the clay. Leave the pot on the bat until it is a little bit dry or it will stick to your hands or break.

▼ **Forming the shape**

▲ **These hands are not in the correct position**

10. Use a wire or fish line twisted around your fingers to cut the container from the bat after it has dried enough that it holds its shape (but is still moist). Then you can fire it.

How do I make a coil pot?

This is a very easy kind of pot to make. You simply roll out a very long, even rope of clay, and starting from the bottom of the pot you coil it around and around, filling out the bottom and working your way up the sides until it is the size you want. If you want to add more rope, you just pinch the ends together gently and smooth the clay over the attached seam. Once the pot is the right size, you use a bit of water and smooth and blend the coils together until you can't tell they were there. You will probably have to support the pot from the one side while you push with your other hand on the other side.

How do I pit fire my clay creations?

1. You can create an artistic pattern on the pot by tying materials to its surface. For instance: string soaked in salt water then dried, dry grass, dried seaweed, thin copper wire, wire mesh, steel wool or banana peels.
2. Make a fire pit by making a container 3–5 feet in diameter of metal or concrete, or simply dig a hole. Put the pit somewhere away from anything flammable such as in some sand or dirt. You can also create a couple of holes or small vents near the bottom of the pit for airflow.
3. Put a layer of sawdust 2–4 inches deep on the bottom of the pit. You can add rock salt, Epsom salts, or copper carbonate on top of the sawdust for more patterns on the pots.
4. Place the pots on top of the sawdust. The parts on the sawdust will probably turn black. Keep the pots toward the center of the pit where it will be hotter.
5. Add a layer of crumpled newspaper on top of the pots. Then put small kindling

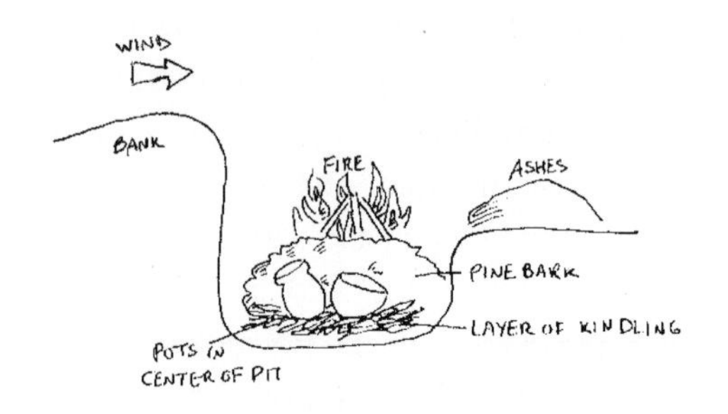

▲ **Setting up a pit firing**

▲ Glazed stoneware pottery that has been fired in a kiln

on top, followed by larger pieces of wood (hardwood is better but not essential).

6. Light the fire and keep adding more wood to keep it going strong. Do this gently and if you have to stir the wood, be careful not to break any pots. Put the last piece of wood on about 3–5 hours after starting the fire.
7. Let the fire burn down to ashes by itself and wait for it to completely cool, which can take a day or two. If you take them out hot they can crack easier. Sometimes even cool pots will crack—every firing you will have at least one cracked or shattered pot. You can tell if they fired properly if you flick them and they make a ringing sound.
8. Rinse off the ashes. Then you can polish them. These pots are not fired as hot as in a kiln and so may not hold water unless the inside is sealed.

How do I seal or waterproof my pots?

Fill the pot with milk and allow it to sour for four days. Dump it out and it will be fairly waterproof. If you use the pot regularly in cooking it will also seal it but you must soak it in water before each use, and you can't use detergents to wash it. If you are just going to carry water in it you can waterproof it the same as a bucket. Never put an empty pot on the fire because adding cool food to a hot pot can crack it.

How do I make a cornhusk bed?

Drive a nail into a board and sharpen the end of the nail with a file. Draw the husk through the nail to shred it in 1 inch or less strips. Let the shreds dry in the attic or loft. If you have enough, use only the papery, thin ones. Use heavy, tough, twill cloth for your ticking (or bed cover). Fill it with husks until it is as firm as you want. When you need to clean it, take out the husks and let them sun, then sew them up again. Instead of corn husks you can also use rice straw.

How do I make a feather bed?

It is nice to have a feather bed on top of your cornhusk bed. With the same heavy cloth for your ticking, fill it with goose feathers gathered throughout the year. It is very important to clean the feathers or you will get mites in your bed. Gather green flat cedar and simmer it in an enamel or glass pan, although stainless steel will work. If you simmer it gently it will make a green tea that is not too dark and not too light. If your feathers are very dirty, wash them with mild soap first, and then wash with the green tea. The tea protects the feathers and helps preserve them. You will have to change your feathers every fall though.

How do I make a feather pillow?

For a good pillow, use a mixture of ¼ small body feathers and ¾ down so that it will be both soft and provide support.

How long do candles burn?

¾" diameter x 4" tall: 2 hrs 20 min

⅞" diameter x 4" tall: 5 hrs

2" diameter x 9" tall: 63 hrs. (7 hrs an inch)

How do I make a wick?

Soak heavy cotton yarn for 12 hours in turpentine, limewater, and vinegar. Braid three strands together when it is dry. The wick must be the right size for your candle for it to burn well.

Wick:

Size	Candle diameter
Flat braided 15 ply	1"–2 ½"
Square braided 24 ply	3"–4"
Square braided 30 ply	more than 4"

How do I make a rolled beeswax candle?

Take a sheet of beeswax that is used to start new hives, about 16 inches long. On a warm day cut the top of the sheet off at a slant. Roll the sheet around the wick, starting with the wide end first. Trim the bottom flat.

How do I make a dipped beeswax candle?

Take two containers. Put melted beeswax in one, ice water in the other. Dip in wax, then ice water over and over until the candle is the size you want. Many people use paraffin to make candles but since this is made of petroleum it can only

▾ Rolled beeswax candle

▲ Dipped paraffin candles on a dipping frame

be bought at the store. It works the same way, but beeswax takes a while to melt and smells delicious.

To make a candle out of bayberry:

Note: Warning! Do NOT pick wild bayberry because it is endangered. Grow your own. Gather in the fall when it's ripe.

1. Put the berries in a copper kettle filled with water and boil. The fat will melt off and float to the top. Skim this off and put in another pot.
2. Melt this fat and skim off the stuff that comes to the top until the wax is a transparent green color.
3. Leave the melted wax on low heat so it will stay melted. You can mix in beeswax to make more wax. Take your wick, dip it in the wax, and let it harden,

▼ Bayberries

▲ Kerosene lantern

dip again, and so on until the candle is as fat as you want.

Lamp troubleshooting:

Warning: Remember to have lots of ventilation when using a kerosene lamp. The fumes are bad for you. An oil lamp can be used inside, while a kerosene lamp should not be.

Smoking lamp: Soak the wick in vinegar and dry it well.

Broken chimney: Tie a string soaked in kerosene around the bottom of a canning jar. Light the string and plunge the jar in cold water. When it has burned all the way around, the jar will crack where the string was. Rub the edge smooth with sandstone.

Tempering chimney: This will prevent the chimney from cracking. Put the chimney in cold water and gradually bring to a boil. Allow it to cool slowly.

▼ Oil lamp

▲ Walnut hulls

Wick: Soak cotton yarn or string for 12 hours in 1 cup water, 1 tablespoon salt and 2 tablespoons boric acid. Dry yarn and braid it.

Ran out of kerosene: You can use biodiesel from when you make soap or when making biodiesel for your vehicle. Or you should probably be using an oil lamp anyway

How do I make all-natural ink?

Brown: Mash butternut, peanut, hazelnut, chestnut, or walnut hulls in a towel with a hammer (or use tree bark off the ground); simmer 1 hour in water mixed with ½ teaspoon salt and ½ teaspoon vinegar.

Black: Add lampblack (burn a pine knot in a stove and hold a can over it to catch the smoke—the soot is the lampblack) to the brown ink.

Blue: Put blueberries in a strainer and crush over a bowl until all the juice is squished out. Mix ½ cup of blueberries, ½ teaspoon vinegar, and ½ teaspoon salt.

How do I make alcohol ink?

Fill a small baby-food-sized jar ¼ full with coloring powder. Then add rubbing alcohol to the top of the jar.

▲ Quill pen and handmade paper

▲ Making a stone inkwell

How do I make milk paint?

1 quart skim milk (room temperature)

1 ounce hydrated lime (do not use quick lime, it reacts with water)

Optional: 1–2 ½ pounds of chalk, glue or rice paste for filler

Put the lime in a bucket, and add enough milk to make a cream, stir it well, and then add the rest of the milk. Now add any coloring powder that you want and paint. Keep the paint until the milk goes sour. Paint takes 4 hours to dry and covers 280 square feet per gallon. Let it sit in the sun for 2 weeks during a dry season.

How do I make oil paint?

Mix linseed oil with a dry coloring powder using a paint spatula. It does not take very much powder (or pigment) to make paint because the paint is usually a very thick paste. Once it is well mixed, it needs to be ground to a finer consistency, which you can do with a glass muller (a hand grinding tool). If your pigment just turns into lumps, add a tiny bit of mineral oil to wet it. All of the materials used to make paint should be stored in airtightf glass containers.

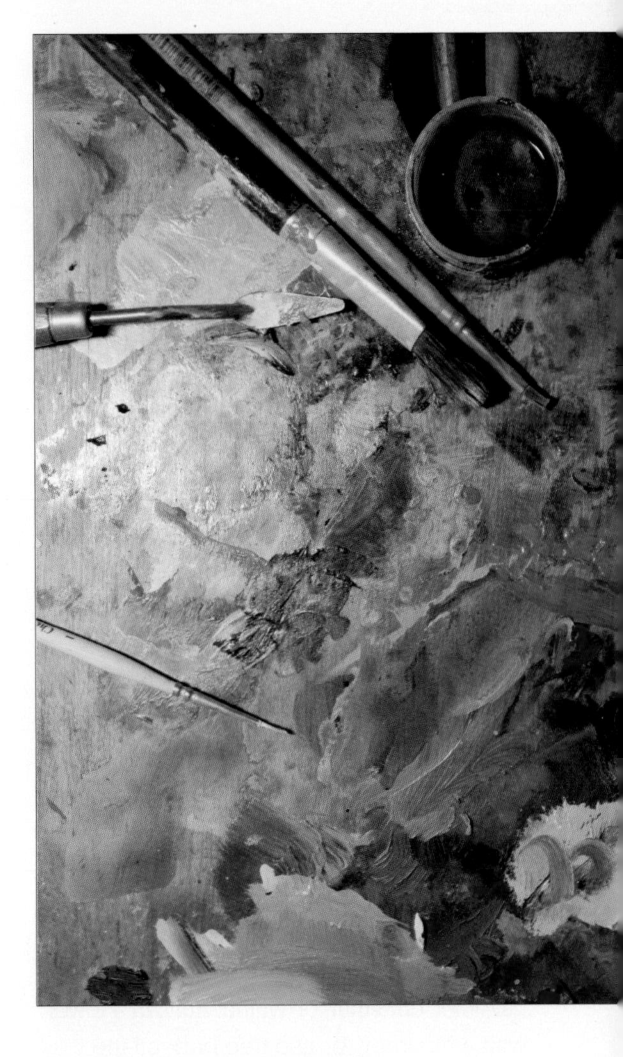

Old milk-painted barn with red coloring powder

How do I make gesso?

Gesso is a white or brown paint used to prepare a canvas to receive oil paint. Mix 3 parts glue with 1 part chalk and 1 part pigment in a metal container, put in a doubleboiler bath and heat, stirring well but slowly. Apply with a broad brush and let dry, then recoat.

How do I make whitewash?

Put 25 pounds hydrated lime (***not*** quick lime) in 10 gallons boiling water, cover, and let sit four days. Strain it, and then add a peck (8 quarts, or 2 gallons) of salt dissolved in warm water. Boil rice to a

▲ Coloring powders

thin paste and add to the lime. Dissolve ½ pound clear glue in water and add to lime. Mix well and let stand at least 3 days before using. Heat it to as hot as possible when you paint. Keep in a kettle.

What are coloring powders?

You will need to find paint powders that do not contain a binder. The Venetian powders are made with iron oxide (or rust) and may be somewhat difficult to find. If you are in a fix, you can make some using a red brick, or with red clay. Experiment with other minerals to create other colors.

How do I make glue out of flour?

To patch flour sacks, cover the patch with flour and water and iron it until it's dry. You can glue any paper with flour and water, and you can make items out of papier maché. Papier maché can be done with newspaper, water, and flour. To create a good consistency for your glue, all the flour should be completely mixed in to be smooth, and it should be able to be poured, but not so thin that it is drippy. White flour is better for artistic projects, but any flour will work.

How do I make paper from other paper?

1. Use a coat hanger and attach panty hose or screening over it to make a rectangular screened frame (or deckle) the size of the paper you are making.
2. Get paper and shred or rip it into small pieces into a bowl ¾ full of hot water.
3. Blend the paper until it is turned completely to pulp (called slurry). Add glue or starch to add strength if you want.
4. Pour the slurry into a shallow pan.
5. Scoop the deckle into the bottom of the shallow pan and lift straight up very slowly—count to 20 while you are lifting.

▼ Handmade paper with plant and other fibers

6. Let some of the water drip off. Place a clean dishtowel (or newspaper or felt) on a flat surface and flip the deckle paper-side-down on cloth so that paper comes off. You can lay flowers or other materials on the paper for texture.
8. Lay another cloth on top and squeeze out moisture with rolling pin.
9. When paper is mostly dry iron it on the hottest setting to dry and flatten it.
10. Put the extra slurry in plastic bag and freeze.

What are the benefits of hemp paper?

Unlike other plants, both the fiber and pulp of hemp can be made into paper. The paper made from the fibers (or bast) is rougher and more brittle. The paper made from the pulp (or hurd) is softer and thicker and better for everyday use. Some people use both to make a strong paper. Wood pulp paper has been treated with chemicals that make it have acid—in a hundred years this acid eats away at the paper and destroys precious books. Hemp paper is naturally acid-free and will last a very long time.

How do I make hemp paper slurry?

1. Get 12 ounces of retted hemp and soak it for 12 hours.
2. Put the hemp and water on low heat. Add some soda ash and simmer for 8 hours. It will still feel tough.
3. Put the hemp into the blender or grind it in a grinder until it turns fluffy, cloudy and soft. This is your hemp slurry. You can substitute this slurry for the paper-made slurry above.

What other plants can be used to make paper?

Just about any plant can be made into slurry. Sometimes if you add larger plant bits and leaves to the slurry you can make an artistic paper. You also do not necessarily need to use a square deckle. You could make spherical paper in a bowl if you wait long enough for it to dry.

How do I make brushes?

Brushes can be made using a green tree branch, and pounding the end of it so that the fibers split apart. Or you can tightly attach horsehair to a stick using string or wire.

How do I make pens?

Pens can be made using a quill (goose feathers are best), washing it thoroughly using the same method used for a feather bed, then cutting the end diagonally. Then split the end to provide a channel for the ink. It is also quite easy to find metal pen nibs that work the same way.

▲ **Making a quill pen**

All-purpose spray cleaner:

Mix 1 part white vinegar with 1 part water. You can also soak stains in lemon to remove them. This can be used on showers, tubs, countertops, hardwood floors, exterior of the toilet, etc.

Tough all-purpose cleaner:

Undiluted white vinegar can take out the toughest stains. Use on shower walls, toilet bowls, etc. Works best if you let it soak.

Fabric softener:

Add undiluted white vinegar. If you do hand laundry, this may not work. For those with washing machines, add to rinse cycle.

Or put all your cooking ashes into a large barrel and fill the barrel with rainwater. After about an hour the ashes will settle to the bottom and the water on top will become clear and soft. Scoop this water out and use it to wash the laundry.

Scouring powder:

Use baking soda or borax on a damp sponge. This works for bathtub rings, bathroom counters, kitchen sinks, toilet bowls, etc.

Glass cleaner:

Cornstarch polishes glass very well, but another recipe is 1 cup rubbing alcohol, 1 cup water, and 1 tablespoon of white vinegar.

Furniture polish:

Lemon oil can be used on all wood furniture, or mix 1 cup olive oil and ½ cup lemon juice.

Oven spill remover:

Sprinkle salt in the oven while it is still hot, and it will soak up spills.

Mildew remover:

Mix 2 teaspoons white vinegar in 1 quart warm water, then soak mold, then scrub off.

Laundry aid:

Mix natural soap with borax. This softens water and has fewer allergies.

Dishwasher soap:

Use straight borax.

To clear a drain:

Use a plunger or metal snake. Try it until your arms feel like they will fall off. If it still doesn't work, pour ½ cup vinegar, ½ cup baking soda, and 2 cups boiling water mixed together down the drain. Let stand 3–5 minutes, and then flush with hot water.

How do I use a rain gutter to make lye?

WARNING: ***Lye is extremely caustic. Don't touch it! Wear gloves, keep it away from any living thing, and don't splash while you're mixing. And DON'T breathe in the fumes. If you get it on your skin, pour vinegar on the burn to neutralize it. It is best to make soap outside if you can. Mark all containers that you store lye in or have used with lye (such as your spoons and pots) and lock them away from living beings so they are safe, and you know to only use those things for lye.***

Use a 10-foot length of plastic rain gutter with a cap on one end and a down spout connection on the other end. Fasten it to fence posts 4 feet off the ground. Fill the gutter with an even layer of wood ashes (do not pat down or compact it). Place a plastic bucket under the spout, and drill ***in the lid*** 6 holes in a circle 6 inches from the middle, each hole a ½ inch in diameter. Place a second bucket on top with matching holes in its bottom. Pack straw into the top bucket, and put a lid on it with a hole matching the diameter of the spout of the rain gutter. Do not use any metal in this setup because the lye will eat through it.

How do I use a barrel to make lye?

Get a small wooden barrel and make sure it's waterproof. Drill lots of holes in the bottom of the barrel and stand the barrel on blocks with enough room for a waterproof wood or glass container (not metal). Put a layer of gravel in the bottom of the barrel and then a layer of straw. Then put ashes into the barrel leaving at least a couple of inches at the top. Pour rainwater into the barrel.

Old-style lye leecher made of bark

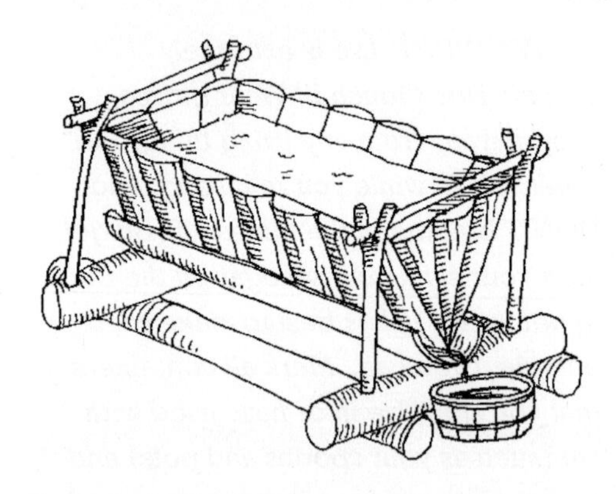

How does rainwater and ash make lye?

In either setup the rainwater runs through the ashes and then into the straw. The straw acts as a filter removing the contaminants and purifying the lye water. This lye water can then be made into lye (see below). Softwood ashes make soft soap, hardwood ashes make bar soap, and kelp makes very hard and durable soap.

How do I make lye water into lye?

Use an old iron or steel pot that you will never need for anything else and boil the lye water until it is concentrated. To test the concentration, put a whole egg with the shell on into the water and it should float on top. Destroy the egg when you are done, and don't let any of the liquid touch your skin or clothing.

What can I use lye for?

Lye is used to make soap, wash non-carpet floors, for sanitization, killing bacteria, making glycerin, and biodiesel. For safety precautions, use the same rules as for making soap.

Equipment and ingredients for making soap:

Large glass
Pyrex measuring cup
Enamel or stainless steel pot
Glass or stainless steel thermometer (such as candy thermometer)
Fat such as lard, or another kind of oil
Water and/or milk
Plain salt
Vinegar
Soap molds, non-metal, heat resistant with lid (you can make one out of a cardboard box lined with plastic or get a rubber container)
Plastic wrap or wax paper
Wooden spoons
Rubber gloves
Goggles
Long-sleeved shirt
Newspaper
Lye (find it in the drain cleaner section of your grocery store)
Old blanket
A scale to measure ingredient weights

How do I make simple soap?

1. Put the fats into a lye-resistant container such as enamel or stainless steel

(***not*** aluminum) and put in a glass or stainless steel thermometer. Don't let it touch the bottom. Heat 16 ounces of oil and fats to the proper temperature (see fat table below). If you want to add herbs, add it to the fat now (no more than 1–2 tablespoons per pound).

2. Put on eye protection and rubber gloves before dealing with the lye. Put 1 cup of water that is 70–75° into a heatproof container—a Pyrex glass measuring cup works well. Stir the water with a stainless steel, plastic, or wooden spoon while you slowly add 2.8 ounces of lye. The water will get very hot and cloudy, and will give off fumes that you must ***not*** inhale. Put the thermometer into the lye water until it reaches the proper temperature for the type of fats you are using (see fat table below). Always, ***always*** add the lye to the water—not the other way around. If you add water to lye, it will cause a violent eruption and get toxic lye everywhere.
3. As soon as the fat and the lye water have both reached within five degrees of the temperatures you want, put the fat pot into the sink or if you are outside, into a tub, then pour the lye water into it slowly while stirring rapidly (but not fast enough to splash). Also make sure you are pouring the lye into the fat, not the other way around.
4. Keep stirring the mixture until it "traces." At first it will be thin and watery, and gradually it gets thicker and more solid. Some soaps, such as those made with coconut oil have a very clear trace, while other soaps such as Castile soap (made with pure olive oil) it is hard to tell. Draw a line

▾ **Making soap**

in the soap with the spoon and if the line remains for a few seconds, it has traced. Stir constantly for the first 15 minutes, and then after that stir every 15 minutes, because it can take 90 minutes for it to trace and you will get tired.

5. Add up to 1 tablespoon of essential oil if desired, and stir for a few more minutes. This soap should remain scent-free, but essential oil will cover up fat smells that the soap may contain.
6. Pour it into molds as soon as you're sure the soap has traced and put the lid on. Wrap it in the old blanket so that the insulation will help in saponification. Allow it to harden for 48 hours and don't look at it because it may look gooey, but that's all right. If it gets stuck in the mold, freeze it and it will come out easily.
7. Wrap the bars tightly in plastic wrap or wax paper. Store the soap away and allow it to age for at least 3 weeks. This decreases the pH level and makes it milder. If air comes in contact with the soap, it will get a white crust called soda ash. Cut it away, or wash it off.
8. If you have any batches of soap that don't turn out, you can always save this soft soap for tanning leather.

A general simple soap recipe:

2 cans or 3 pounds of vegetable shortening

1 can or 12 ounces of lye

2 cups of water

Dissolve lye, melt shortening, and allow them both to cool until you can touch the outside of the pans but they are still hot. Pour lye into shortening and stir until the consistency of mashed potatoes. Pour into mold and let sit 24 hours uncovered. When it is firm, cut into bars and wrap. Let them age at least 3 weeks.

How do I make goat milk soap?

1. Dissolve ¼ cup honey in ½ cup hot water. Pour it into a large enamel pot (***not*** aluminum). Add 2 ½ cups cold milk.
2. Stir well with a wooden spoon and slowly add 6 ounces of lye (using all the safety precautions). It will start to get very hot.
3. When the lye solution has cooled to 70°F, heat 7 cups of rendered lard (or lard and olive oil) to 80° and slowly pour it into the lye.
4. Stir it constantly until it is like thick honey (this is called tracing—you will be able to make a trail in the liquid), and pour it into molds. Cover with blankets or newspapers so it will cool slowly. Let it sit 48 hours.
5. Take the soap out of the mold and let it cure for 6 weeks before using. Don't worry if the soap turns brown. That is just the sugars in the milk cooking. You can add the milk last to make the soap

Fat temperature table:

Type of Fat	Fat Temperature	Lye Temperature
Sweet rancid fat	97–100°F	75–80°F
Sweet lard/soft fats	80–85°F	70–75°F
Half lard/half tallow	100–110°F	80–85°F
Tallow	120–130°F	90–95°F
Vegetable fats	100–120°F	90–95°F

▲ **Goat milk soap**

lighter, or you can freeze the milk and add it to the traced soap for lighter soap as well.

Another milk soap recipe:

1 cup melted lard

1 cup melted coconut oil

1 cup milk (such as goat)

¼ cup Red Devil lye granules (not flakes or crystals)

¼ cup water

Dissolve the lye, add the lye water to the fat, and stir in the milk just as you normally would. It should trace in about 1 hour and 15 minutes. Leave in the mold for 2 days, remove and age for at least 3 weeks.

What kind of additives can I put in the soap?

Fragrance: Essential oils and herbs can all be used to make fragrance in the soap. Essential oil is the best; however, use caution. These smells have a powerful effect on people, and if you have a condition such as high blood pressure or are pregnant, you should watch what you are sniffing. Add ½–1 ounce of oil for every ¾ pounds (340 grams) of soap, when the soap has started tracing.

Coloring: Cinnamon, clove, vanilla, and nutmeg make brown soap. Cornmeal makes yellow soap. Paprika makes peach soap. Add herbs, rose petals, or oatmeal to make other colors. Essential oils also affect the color of the soap. For spices, add a bit of water before adding to the soap after it has started tracing, to make a thin paste. Then add the paste to the soap.

Skin care additives: Skin care is a major part of soap making and a variety of things can be added.

Pectin: Add to soap and it will prevent your liquid shampoo from separating.

Rosin: A fine powdered by-product left from distilling pine resin. It not only helps your soap bars keep their shape, it also produces lather. Mix with vegetable oil before adding to the soap.

Making glycerin:

Glycerin is a by-product of making biodiesel. You can buy it, but if you really want to make it, see the Biodiesel section. It is made like this: put vegetable oil into a container. In a separate container mix ethanol with lye (be ***careful!***), then add it to the oil. Stir just like soap (with blender on low or by hand), then instead of waiting for it to trace let it settle for about

Additives for soap:

Additive	What to Use	Result
Almond	Grind up blanched almonds into fine meal	Unclogs pores and absorbs oil, makes a good lather
Aloe vera	Use aloe vera gel	Heals abrasions and tones the skin
Apricot	Use fresh or dried	Softens skin and adds vitamins
Bran	Break out the husk of any grain	Mild abrasive cleans pores and dead skin
Calendula	Use fresh or dried petals and remove all seeds	Softens and soothes sensitive and dry skin
Carrot	Grate carrots	Adds vitamins
Clay	Add powdered clay	Absorbs oil, may be drying
Coffee	Use fresh unbrewed coffee grounds	Absorbs odors
Cornmeal	Add cornmeal	Absorbs oil and unclogs pores
Cucumber	Add grated cucumber	A mild cleanser and cleans pores
Elderberry	Distill elderberry flowers and add liquid	Softens and tones
Goat milk	Use fresh milk	Cleanses skin
Honey	Use raw honey	Softens and smoothes skin
Kelp	Use large leafy algae	Cleanses and adds vitamins
Lemon	Use the oil in the peel	Antibacterial and adds Vitamin C
Lettuce	Use fresh clean leaves	Antibacterial
Marshmallow	Use ***Althea officinalis***	Softens the skin
Oatmeal	Use long-cooking or rolled oats ground in a blender	Soothes sensitive and irritated skin
Rosemary	Use ***Rosemarinu officinalis***	A mild cleanser
Sage	Use ***Salvia officinalis***	Antibacterial and cleanser
Strawberry	Use fresh strawberries	Tightens and whitens the skin, adds Vitamin C

an hour. There will be two distinct layers, glycerin on top, and biodiesel on the bottom. Drain off the glycerin and you have liquid soap (you can then add it to the soap recipe you want to make it bar soap).

What is a feminine product?

For those who don't know, this is the polite term for tampons, pads, and other products used during a woman's menstrual period. Feminine products are often uncomfortable, ineffective, and costly, but there are alternatives to store-bought brands.

How do I use a sea sponge?

Safety rating: Sea sponges are used internally but unlike tampons they are safe to use all night. They are also soft. It is best to get farmed sponges made for this

▲ Sea sponges

purpose, rather than wild sea sponges, which does hurt the environment.

To make them: Get a large sea sponge, and divide it into strawberry-sized pieces that are rounded and oval shaped (not long and skinny like tampons). To insert them, have clean hands, wet the sponge, squeeze out excess water, and insert it. It should feel comfortable (if it doesn't, then trim down the size of it). Then just pull it out when it needs changing. To reuse the sponge, use one of the following cleaning mixtures. If the sponge isn't giving enough protection, also use a cloth panty-liner pad (see below).

Reusing sea sponges: Wash in 1 tablespoon baking soda and 1 cup warm water, or 1 tablespoon cider vinegar and 1 cup water, or ½ cup hydrogen peroxide and ½ cup water, or boil for 2 minutes. Air dry in sunlight and store in a cool dry place. They should last 1 year or maybe longer if you use several.

How do I make cloth pads?

Fabric: old receiving blankets or old towels (or both)

Light flow: 4 layers of flannel per pad

Medium flow: 6 layers of flannel per pad

Heavy flow: 2 pieces flannel, 2 pieces of terry (or substitute 4 flannel)

Cut the pieces 12 inches long and 4 inches wide, rounding the corners if you want. Layer all pieces together and stitch ¼ inch from the edge and trim. A person with heavy flow should have 4 light flow pads, 10 medium, and 12 heavy flow pads. Double up if you get leakage.

▲ Mullein

▼ Wisteria

Other options for feminine products:

There is also a product made of natural rubber called the Keeper or Diva Cup, which is supposed to last 10 years. You squeeze it, insert it, and dump it out as it fills up during your period. Some native tribes have a special hut where a woman can sit and do handiwork the entire week without getting up, but this is not really an option for the rest of us.

What can I substitute for roll toilet paper?

Newspaper: Crumble it in your hands to make it softer.

Paper: Treat just like newspaper, but you might have to crumble it for longer.

Cloth: Old scraps of cloth cut up into squares.

Mullein: Harvest the large velvety leaves of the mullein plant.

▲ Oak

▼ Persimmon

▲ Fig

▼ Corn silk

Your hand: In India most poor people use their left hand, and then wash it.

Sand: Some people dip their damp hand in a jar of sand and scrub a bit with it.

Wisteria: Commonly used as toilet paper, harvested for the leaves.

Persimmon: These tree leaves have been used as toilet paper.

Oak: Oak leaves are also a common toilet paper.

Fig: Fig leaves not only do the job, they cure hemorrhoids.

Straw: A handful of clean straw can easily be put into a composting toilet.

Corn: The corn silk is much softer, but if you don't have that even the ear leaves work.

Cedar insect repellant:

3 cups rubbing alcohol

1 ½ cups red cedar wood shavings

½ cup eucalyptus leaves

Mix together in a large bowl or jar. Cover and let stand 5 days, then strain and store the remaining liquid. Store tightly sealed, yields 2 cups of repellant. To use, use a spray bottle to spray a light coat on the skin.

Citronella repellant:

Mix aloe vera gel, essential oil of citronella, tea tree oil, and lavender oil. Stir it all together. The mixture will become opaque.

Garlic juice repellant:

Mix 1 part garlic juice to 5 parts water. Put into a spray bottle and put it on your skin before you go into the woods. Repels chiggers also.

How do I make yummy toothpaste?

¼ teaspoon peppermint oil

¼ teaspoon spearmint

¼ cup arrowroot

¼ cup powdered orrisroot

¼ cup water

1 teaspoon ground sage

Mix all dry ingredients in a bowl. Add water until it is a paste that is thick enough to spread but not to pour. Store in a tightly covered jar.

How do I make tooth powder?

Mix 3 parts baking soda with 1 part salt, and add a few drops of peppermint oil. Dip your brush in and scrub thoroughly. Tooth brushing is less about the paste and more about the brushing. Scrub at least 3 minutes twice a day.

How do I make mouthwash?

2 cups water

3 teaspoons fresh parsley

2 teaspoons whole cloves

2 teaspoons ground cinnamon

2 teaspoons peppermint extract

Boil water and remove from heat, then add all ingredients and let steep for 10–15 minutes. Strain, put into a tightly covered container and store it in the fridge.

What if I don't have a toothbrush or toothpaste?

If you don't have a toothbrush or toothpaste, you can use soap, salt, or baking soda and a green twig. Chew the end of the twig until it kind of has bristles. You can also drink an infusion of green tea after meals to prevent tooth decay.

Deodorant:

Alfalfa: Eat some alfalfa, which may help neutralize odor.

Apple cider vinegar: Coat underarm with vinegar, which kills the bacteria. White vinegar may also work.

Baby powder: Use in areas of heavy perspiration.

Baking soda: Apply powder to dry armpits.

Cornstarch: Use alone or mix with baking soda.

Parsley: Eat some parsley, which may help neutralize odor.

Radish: Juice 24 radishes and add ¼ teaspoon of glycerin. Put it in a spray bottle and use as a deodorant.

Rosemary: Put 8–10 drops essential oil in water and apply as deodorant.

Sage: Use 1–½ teaspoons of dried sage, steep 10 minutes, and drink in small doses throughout the day. Or blend fresh sage in tomato juice and drink.

Tea: Put 2 drops essential oil into 1 ounce water and apply as deodorant.

FIBER ARTS AND CLOTHING

I say beware of all enterprises that require new clothes, and not rather a new wearer of clothes.

~Henry David Thoreau, Walden

How do I make the braid to make a braided rug?

Use any kind of fabric strips; scraps work well. You will need three strips 8–9 feet long, which can be made by sewing thin strips (1 ½–2 inches wide) together into one long piece. Sew the three strips together at the ends, and fasten them to something stable. Braid the strips together until 4 feet are braided, then make three

Braided rug made of scrap fabric

more long strips and sew them to the ends of the braided strips. If you try to braid strips any longer than 8 feet they will become tangled.

How do I sew a braided rug?

Once the strips have been braided into a very long braid, the braids must be sewn together flat into a shape. The rug can be oval, round, rectangular, heart-shaped, square, etc. To sew, lace through alternate loops of adjoining braids. When it starts to curve stitch through two loops at a time and skip one.

▲ Sewing the braid

Linen fiber statistics:

1 bushel flax = 56 pounds
1 acre hemp = 1,500 pounds

How do I prepare flax to make linen?

1. Pull the flax out of the ground when it begins to turn yellow at the base. Leave it in the field a day or two to dry.
2. Remove the seeds by pulling through a ripple (a plank with wood or wire teeth sticking up like a rake) and save for herbal remedies.
3. Soak the stalks in water to rot the outer husk, or bundle into a tipi shape and keep damp until they are somewhat rotted. This process is called "retting" and they must be gathered before they are too rotten, but still soft enough to beat and crush the fibers out.
4. Beat and crush the flax in a flax breaker, until the outer shell comes off and only the fibers are left. If you don't have these tools, beat them over the back of a chair with any dull tool.
5. Pull handfuls of fibers through a hetchel, until the finer part of the husk is combed out and it is all silky and fine. Any refuse after the first hatchling can be used to spin into bags and

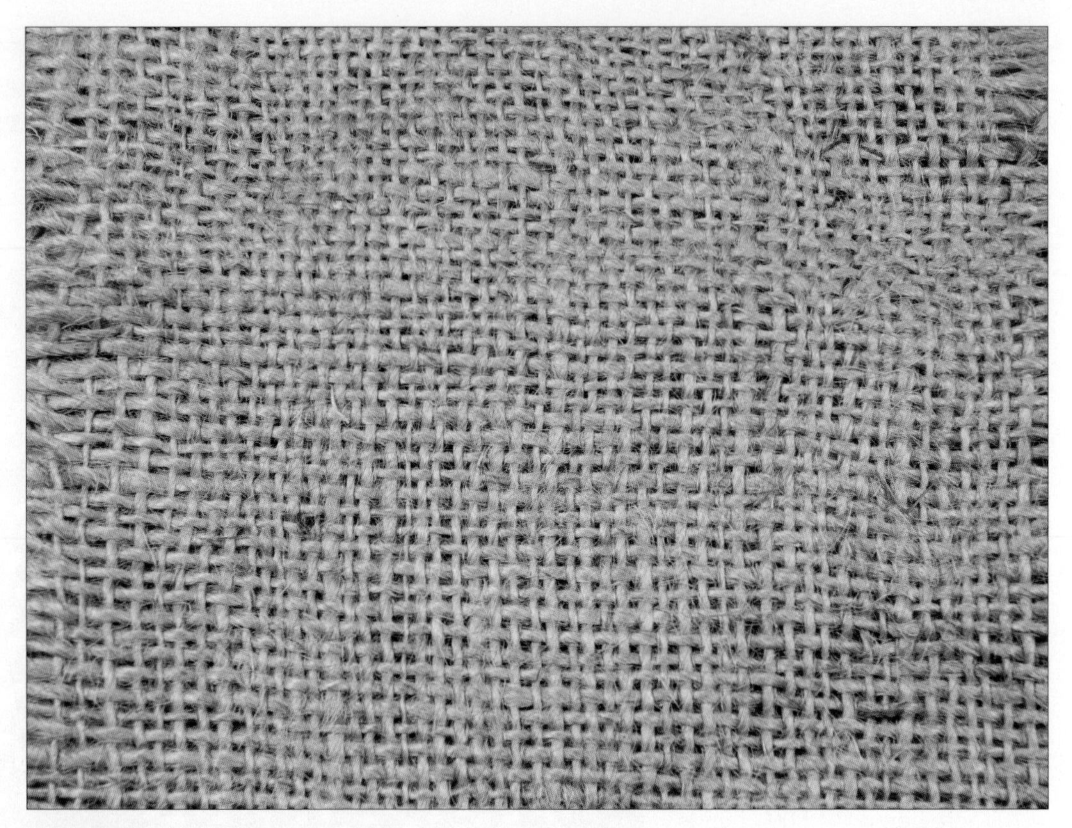

▲ **Undyed linen**

tarps. Any refuse after that can be twisted into ropes.

6. Comb the flax with flax cards, much like you would wool, by pulling the cards with the flax between them, away from each other.
7. Spin the flax into thread with a flax wheel or drop spindle.

▼ **Flax breaker**

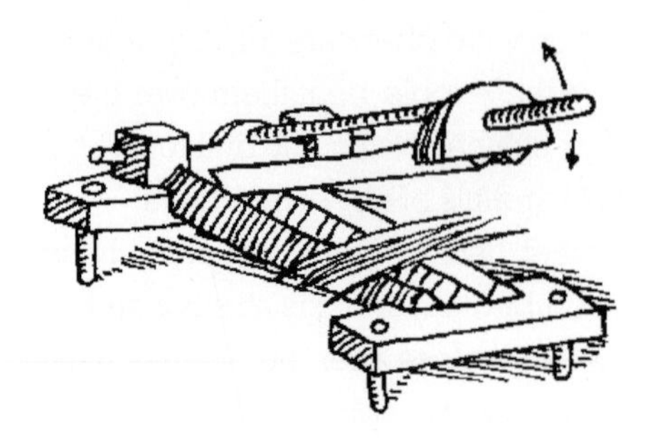

How do I spin flax?

It is possible to drop-spin flax, but of course a wheel is much easier. Keep your fingers moist and spin the thread as you would wool, into a "skein" or thread that has been looped into a loose coil. Soak and boil the skeins in rainwater, wash with soap, and rinse out thoroughly. Then dry and wind on a loom bobbin.

▼ **Hetchel**

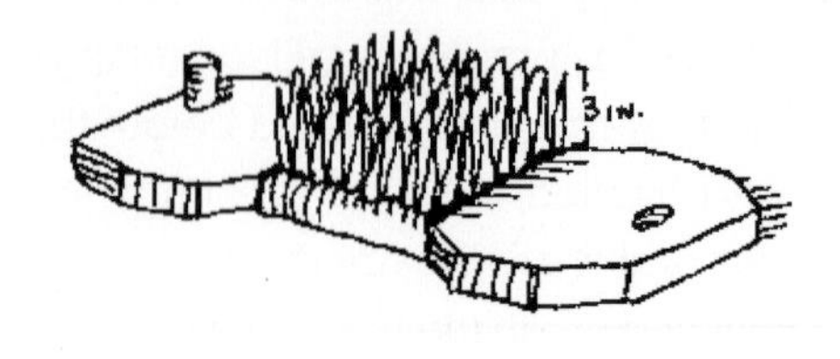

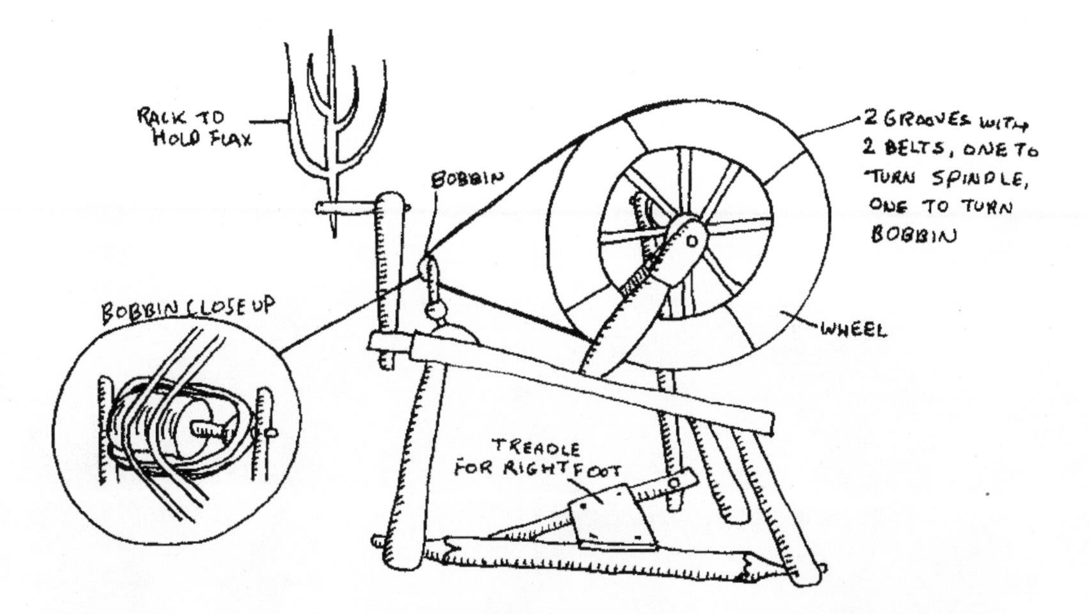

▲ Flax wheel

How do I make nettle or hemp cloth?

Stinging nettles at one time were used more than flax to make a linen-like fabric. Nettles do not necessarily have to be replanted and can sometimes be harvested several times a year because they grow so fast. They are prepared just like flax

▼ Hemp

▲ Nettle

except that the tipi rotting method is never used—it is always soaked in water to soften it. It can make fine linen or very strong course sailcloth and sacking. It is stronger than flax and so makes very good fishing nets. Hemp on the other hand is the very strongest and is also very soft. It is also prepared just like flax, and both hemp and nettle can be spun on a flax wheel or with a drop spindle. However, according to U.S. federal law hemp cannot legally be grown without strict security (because it is a plant related to marijuana—although you could never smoke it). But if you live in Canada and other countries there are many places that allow you to grow "industrial" hemp.

Cotton fiber statistics:

1 acre cotton = 500 pounds

How do I grow cotton?

Cotton is usually planted in rows and irrigated with furrows down the rows. The field is slightly sloped so that the water runs down the entire field through the furrows. Drip irrigation works very well for those who need to conserve water. Cotton needs warm weather, lots of water, and rich soil. Ten days after planting the seedlings will appear, 35 days after planting the flower buds will form, and when the flower falls off the cotton boll forms. The cotton boll holds the fluffy cotton fiber and when

▾ **Cotton bolls**

it is mature the boll bursts open exposing the fiber. About 180 days after planting the cotton can be picked. Cotton does not do well with weeds, which need to be pulled right away, and they are vulnerable to more than 30 kinds of pests that eat the plant and the boll. Ladybugs, spiders, and ants all help cotton, and planting crops nearby which attract the pests away from the cotton can also help, but an organic grower will get a lot less cotton than nonorganic growers.

How do I harvest cotton?

Harvesting cotton by hand is actually better because you will get more fiber out of the boll. You may also have to harvest several times as the cotton ripens, in order to get all of the cotton. You simply pull the cotton out of the boll and put it into a sack.

How do I prepare the cotton?

Eli Whitney invented the cotton gin in 1793, and it was the first machine that was able to remove the seeds from the cotton fiber. It consisted of a rotating cylinder that pulls the cotton fiber through a slotted opening so that the seeds fall out. However, you can't buy these hand versions anymore because they are all in museums. The small homesteader usually has to pull the seeds out by hand (unless you can build your own cotton gin), and card the fiber with the same cards that you would use for wool. Some also spin the cotton directly from the boll by holding onto the seed, instead of making rolags, because picking out the seeds is very tedious.

How do I spin the cotton?

1. Once the cotton is carded into squares, it is rolled into tight rolags by using a thin stick. The cotton is wrapped tightly around the stick by rolling it, and then the stick is removed, leaving a tight roll of cotton.
2. So that the rolags (the rolls of cotton) don't unroll, wrap them tightly in paper with a rubber band around them. You will also need a leader, which is a piece of previously spun cotton yarn. How do you get a leader if you've never spun before? Find someone who does spin, or you can buy sewing thread or fine crochet cotton. The leader should be 24 inches long. Fold it in half and knot the ends together. Wind the knotted end twice around the base of the spindle shaft, leaving out a bit of the loop above the shaft. Pull the other end up through the loop and pull it gently until the thread is tight on the shaft.
3. Now you can spin with either a drop spindle, a regular spinning wheel, or with a charkha. A charkha is the best wheel to spin cotton on not only because it works very well, but also because it was especially designed by Gandhi to spin his cotton clothes.

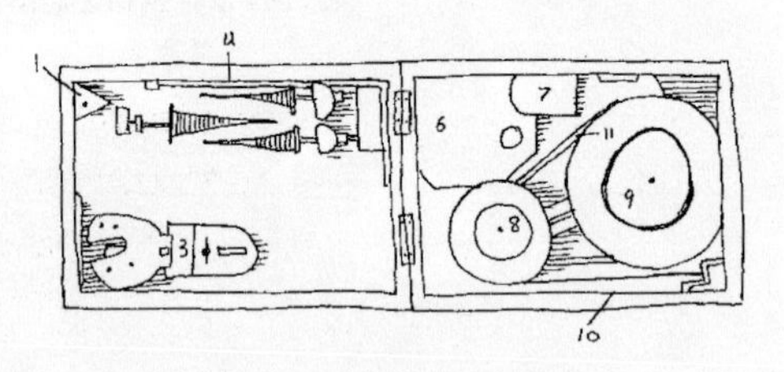

▲ **Charkha**

1. Wooden block for handle
2. Tensioning eye for skeinwinder (flips up vertically)
3. "Mousetrap" spindle bearing assembly
4. Arms for the skeinwinder

    5. Spindles
    6. Center block for skeinwinder
    7. Metal handle used for turning the drivewheels
    8. Small drivewheel that runs the driveband
    9. Large drivewheel where handle is mounted
    10. Wooden handle to hold the charkha steady while you work
    11. Driveband
4. If you bought a new charkha it may have come with instructions. First you have to set up the spindle. On the opposite side from the wheels there is a spindleholder which folds up and is attached to the wheels with a cord. It is in the proper position when it is straight up, so you have to adjust the cord tension until it is.
5. The spindle pulley must rotate between the arms of the spindleholder without touching either one. There should also be either a long, thin, parchment-like piece, or a leather washer that should go right against the pulley so that the pulley doesn't rub the spindleholder. You can also use a braided strip of dried cornhusk, which sometimes lasts longer. The other washers you must clean and oil after each use. Lock the spindle in with a screw or wedge.
6. The drive should turn easily but firmly. You have to make sure the cord is not too tight, but also not too loose or the cord will slip. First remove the main drive cord, but don't let it touch the oiled axle. Then move the smaller wheel, but don't move it too far away from the edge of the case on your spinning side.
7. Once you've adjusted the wheel to where you need, put the main drive cord back on by putting it around the metal pulley under the small wheel (don't let it touch the oily axle), part-way around the big wheel, and then turn the big wheel to get the cord all the way around it. If you can put it on without turning the wheel then the cord is too loose. If you need to make a new cord, make sure to make the knot as small and as strong as possible, and coat the cord with beeswax.
8. Sit down on the floor cross-legged and place the charkha to your right pointing forward. This should put the spindle in front of you pointing left. A briefcase charkha would have the big wheel parallel to your hip so your right hand can easily access it. Hold the charkha steady with your left foot, either on the case handle or the footrest (on a book charkha).
9. As you turn the big wheel clockwise with your right hand, you pull the cotton thread or yarn out with your left hand at an angle—your hand should be at a level a little higher than your stomach. Don't pull the yarn too quickly—you should try to draw it out at about the same speed that you are turning the wheel. Pull it out until your arm is extended as far as it will go.
10. Keep the yarn tight and lift your yarn hand to about the level of your eyes. Reverse the direction you are turning the wheel so that you are turning counterclockwise, and at the same time the yarn will begin to unwind from the spindle tip.
11. Make sure the yarn is still tight, and reverse direction again so that you are turning clockwise. This will wind the yarn tightly on the spindle. As the yarn gets near the tip of the spindle

move your hand back down to an angle so that it is a little higher than your stomach. The yarn will not wind as tightly as you do this, but that's all right. The process of drawing, unwinding, and winding takes a practiced spinner only a few seconds.

How do I add more cotton as I spin?

1. Eventually you will have to add more cotton to the end of your yarn, or if your yarn breaks. Hold the rolag in your left hand with your palm turned slightly upward. You should join the rolag at least one foot away from the spindle, so make sure that you think ahead and prepare to join when the end of your yarn is still at least 1 foot long. Lay the last few inches of yarn on top of the rolag in your left hand and hold it very lightly with your thumb.
2. Turn the wheel and at the same time pull the rolag very slowly away from the spindle, letting the spun yarn slide out from under your thumb. Use your index finger to push the thread up slightly so that it makes contact with the fibers in the rolag and the yarn will draw it in. You should then continue spinning as you did in the beginning. A spindle should hold about 100 yards of thick yarn.

How do I finish off the cotton yarn?

1. Experts are able to make a completely consistent yarn that is strong enough to finish without checking. The last bit of yarn is spun very tightly so that it won't unravel. A beginner usually has to draw out the yarn to arm's length with an incomplete twist.
2. Draw out any thicker or looser portions of this section of yarn by pulling carefully and adding a little twist, to make sure it is completely even.
3. Pinch the yarn at the tip and hold it there as you turn the wheel a few more times to give it the final strengthening twists. The yarn should shorten slightly.

Cotton spinning problems:

My yarn falls apart when starting to pull the rolag: Twist more and don't pull the rolag out so far so quickly.

I can't pull the unspun fiber even when I'm not turning the big wheel: It is twisted too much. Let go of everything, hold the yarn up in the air, and allow it to unwind a little. Pull apart the clogged fibers. Next time make sure you pinch the yarn tightly enough to prevent the unspun fibers from twisting.

When I start a new rolag I get a big lump: Don't spin the big wheel so fast because it is causing you to get too much twist in the yarn. The lumps are called slubs, and you can stop and loosen the slubs as you go.

I got a nice length of yarn but it broke before I could wind it: You had a thin spot in the strand, which is caused by twisting more in one area than the other areas.

I can't wind onto the spindle without the drive cord slipping: Your spindle is full and is too heavy for the drive cord to turn the spindle. Get an empty spindle.

How do I prepare wool?

1. Pull the strands apart and let the seeds fall out onto newspaper. If you want to sort the wool, separate the softest and finest strands from the courser ones.

▲ Shearing a sheep

2. Put it in warm water at least 95° with very mild soap that is completely dissolved. Soak 10 minutes, drain it, and soak it again.
3. Drain it and rinse by soaking it in clear water 10 minutes, drain it, soak it again, and drain it.
4. Hang it in the shade or spread it on paper. Then you can dye the wool if you want.

How do I get a spinning wheel?

It is impossible to make your own spinning wheel unless you are a master craftsman. However, you can make a drop spindle and they are good for beginners, and it is much cheaper. A Saxon wheel is

▼ Wool quality

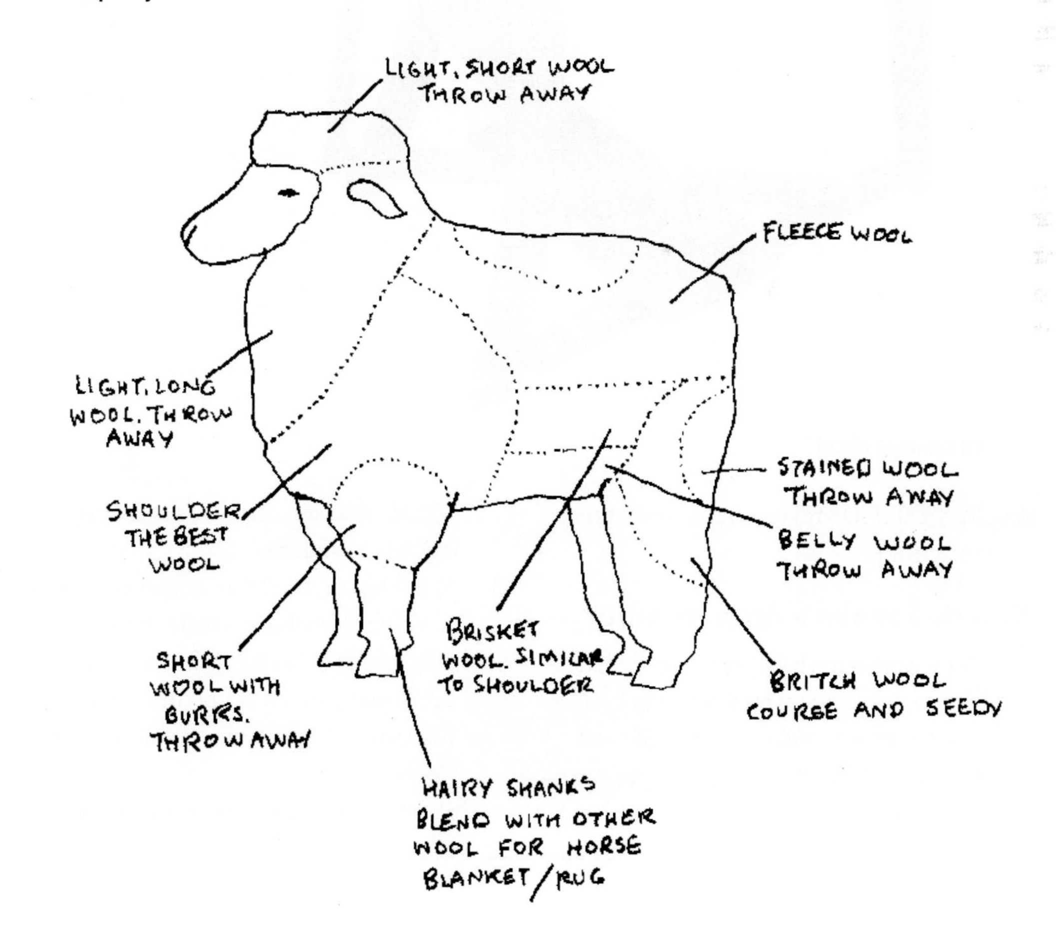

▲ **Saxon spinning wheel**

usually $300 USD and up, and is sometimes difficult to find.

How do I make a drop spindle?

1. Get a wooden wheel, or circle, with a hole in the exact center. It has to be the exact center or else your spindle will wobble, and it has to be 2–3 inches in diameter.
2. Place a 4-inch dowel in the hole so that it sticks out 1 inch out of the bottom of your wooden disc. This is called a low-whorl spindle.
3. At the top of the dowel screw in a metal hook, such as those used for cups. It only needs to be a big enough hook or circle that yarn can slide through easily. You can also make one out of wire or wood.
4. Sharpen the bottom end of the dowel, either with a knife or pencil sharpener. This allows it to spin easily.

▲ **Wool card and drop spindle**

How do I spin with a drop spindle?

1. Tease the fleece with your hands to loosen it (and to get particles out if it is unwashed).
2. Card the wool using two wool cards. Lay pieces of wool on one card, put the other card on top in the opposite direction, and pull them away from each other. Stroke several times until both cards have the same amount of wool. Transfer the wool from one to the other by turning one card around and pulling it across the other. Stroke it again several times until the wool is fluffy. Put the entire fleece on one card and lightly use one to pull it off the other. Take the wool in your hands and roll into a rolag (long roll of even fiber).
3. You will need a leader, which is previously spun yarn. How do you get

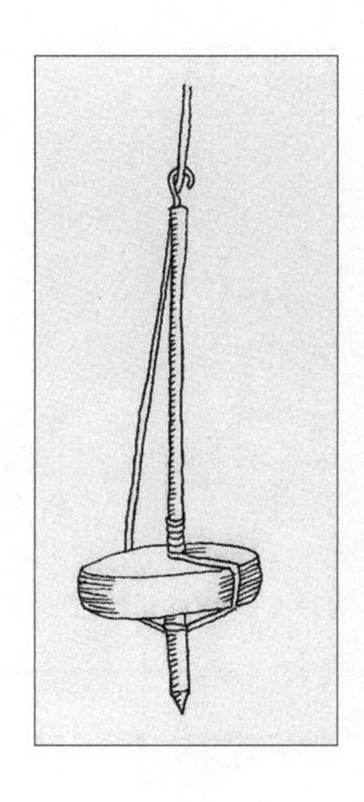

◂ **Winding the leader on a drop spindle**

a leader if you've never spun before? Find someone who does spin, or you can buy wool yarn or wool thread. The leader should be 24 inches long. Fold it in half and knot the ends together. Wind the knotted end twice around the base of the spindle shaft, leaving out a bit of the loop above the shaft. Pull the other end up through the loop and pull it gently until the thread is tight on the shaft. Unravel two inches of the end of the yarn.

4. Unravel two inches of the rolag and lay the yarn on it. Hold this part together with your left hand while you start the spindle spinning with your right.
5. When the ends are twisted together, use your right hand to pinch the spinning yarn 1 inch below the left hand, so that anything above that hand will not twist. Keep holding it until the yarn below is as tight as you want.
6. As you pinch, pull out more fibers with your left hand, moving your right hand up as the yarn reaches the required thickness. When the spindle slows, spin it with your right hand. The trick is to make the yarn even and thin.
7. Don't spin to the end of the rolag; instead add another rolag (see #4). When the spindle reaches the floor, wrap the yarn in a figure eight between your thumb and pinkie fingers. Take the yarn off the top of the spindle, and wrap the yarn on your hand around the base to make a cone shape. Then start again, using the end of the yarn to connect it.

What problems could I have drop spinning?

Yarn comes apart: Add more wool to your twist. To connect your ends back together, untwist both ends, loosen the

▾ **Spinning on a wheel**

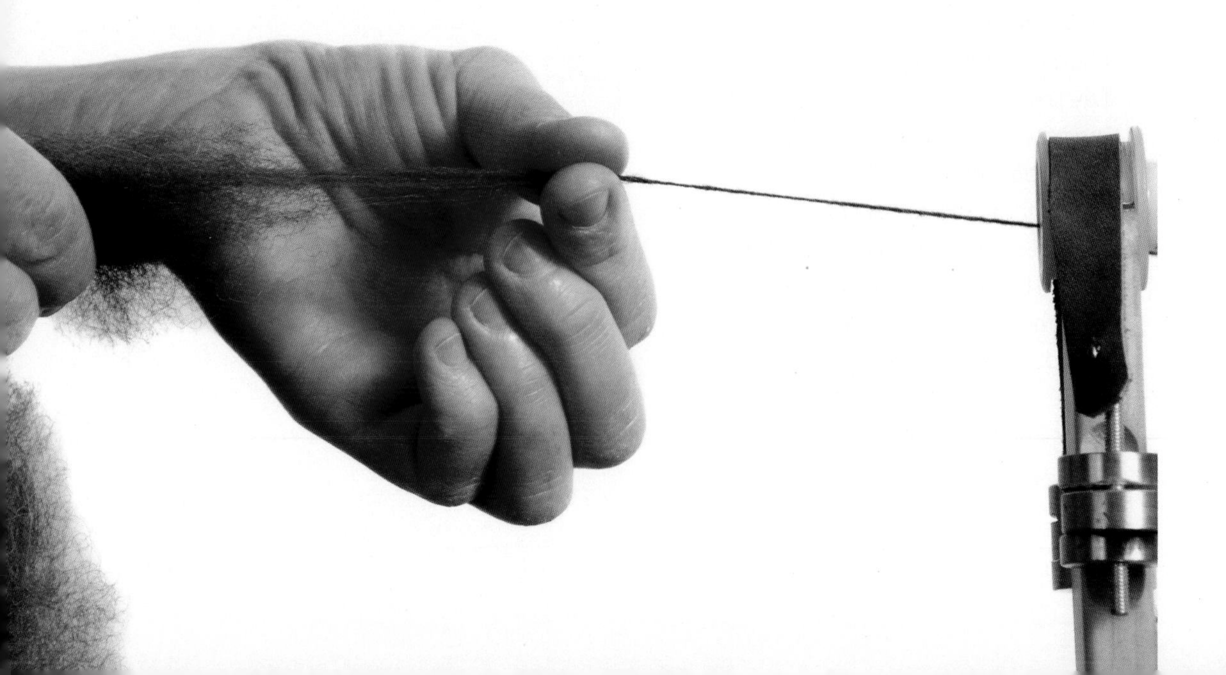

fibers, lay one side on top of the other, and twist them back up again.

The spindle runs away and you get a fiber mass: Stop the spindle, untwist your mess, and twist them back up again.

Wool soft spots and thin spots: Raw wool often has irregularities known as slubs where there are thick spots and thin spots. Remove them by pinching and untwisting and save them to make novelty yarn.

How do I dye yarn, fabric, or thread a solid color?

1. Collect three gallons of rainwater for every pound of fiber. When using grass fiber soak in water until it is soft. For silk and wool, put it in the water with enough strong soap to make it sudsy, and simmer for 30 minutes. For cotton and linen, add ½ cup washing soda with the soap, and boil for 45 minutes. Allow to cool and rinse in more rainwater completely.
2. Put 4 gallons of lukewarm rainwater in an enamel or stainless steel pot and put 4 ounces alum in it for animal fibers, 6 ounces for plant fibers. Let it dissolve, and then immerse the fiber in it. Slowly bring to boil, and boil for 1 hour. Alum is a natural mordant (a mordant fixes the colors of the dye into the molecules of the fibers), and it can be bought from the grocery store since it is used for pickling. The mordant also determines what color you will get. Alum should never be ingested and you should wear protective gloves.
3. Prepare the dried plants by shredding leaves and blossoms, cutting up twigs bark, and roots, and crushing nutshells. Take out all other particles that are not part of the dyeing, and then wrap the stuff in cheesecloth. Put it in an enamel or stainless steel pot with enough water to cover it, and then cook according to the natural dye's directions.
4. For every pound of fiber in the pot, add 4 gallons of water. Cook it until the water is a rich color, adding more water occasionally to keep the dyestuff covered and turning the fiber with a wooden spoon.
5. Allow the fibers to cool, and then rinse them until the water runs clear. Gently squeeze it out and hang it out to dry.

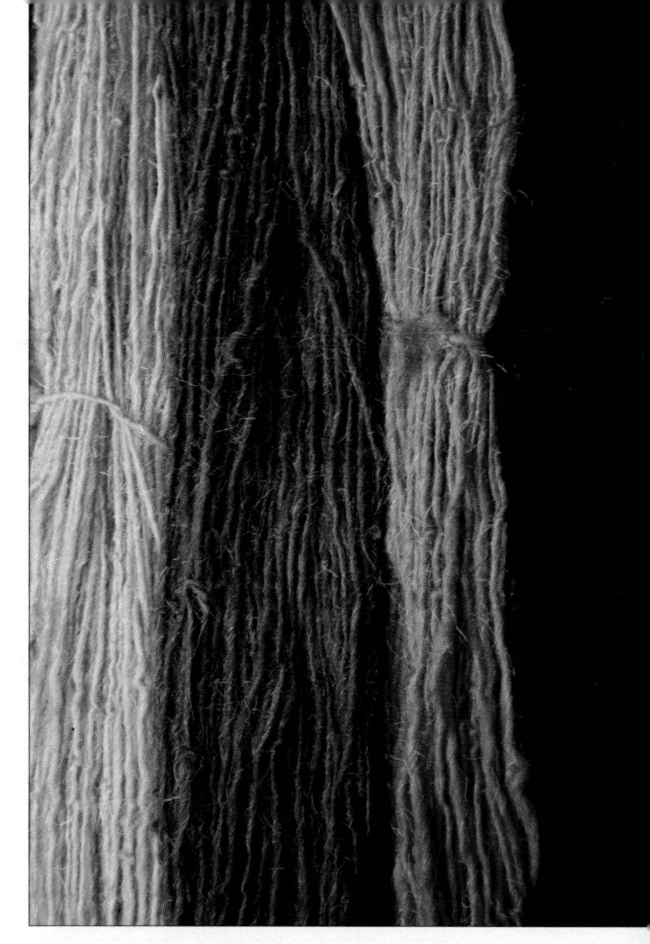

▲ Naturally dyed spun fiber

Natural dyes: (*Note: These amounts are for 1 pound of fiber and 4 gallons of water. Colors may vary—colors listed here indicate only the color family the plant creates.*)

Material	Color	Gathering	Preparing	Dying
Acorn	Brown	7 pounds off the ground	Soak overnight	Boil 2 ½ hours, then simmer with fiber 1 hour
Beets	Red	2 pounds		Boil 1 hour, then simmer with fiber 30 minutes
Blueberries	Blue	16 quarts fully ripe berries		Boil 30 minutes with 1 cup of vinegar, then simmer with fiber 30 minutes
Butterfly weed	Yellow	1 bushel blossoms in full bloom	Soak 1 hour	Boil 1 hour, then simmer with fiber 30 minutes
Celery leaves	Yellow	2 pounds fresh green leaves		Boil 1 hour, then simmer with fiber 30 minutes
Cherry fruit	Pink	16 quarts fully ripe berries		Boil 1 hour, then simmer with fiber 30 minutes
Cherry root	Blue	2 pounds		Boil 30 minutes with 1 cup of vinegar, then simmer with fiber 30 minutes
Coreopsis	Yellow	2 bushels blossoms in full bloom	Soak 39 minutes to 1 hour	Simmer with fiber 30 minutes
Dandelion root	Red	2 pounds		Boil 1 hour, then simmer with fiber 30 minutes
Elderberries	Lavender	16 quarts fully ripe berries		Boil 30 minutes with 1 cup of vinegar, then simmer with fiber 30 minutes
Goldenrod	Yellow	2 pounds fully blooming blossoms and stems		Simmer 30 minutes, then simmer with fiber 30 minutes
Grapes	Blue	16 quarts fully ripe grapes		Boil 1 hour, then simmer with fiber 30 minutes
Grass	Green	2 pounds fresh green grass		Boil 1 hour, then simmer with fiber 30 minutes
Juniper berries	Brown	16 quarts fully ripe berries		Boil 30 minutes with 1 cup vinegar, then simmer with fiber 30 minutes
Lily of the valley	Light green	2 pounds fresh green leaves		Simmer 1 hour, then simmer with fiber 20 minutes

Continued...

Material	Color	Gathering	Preparing	Dying
Madder	Red	2 pounds madder root		Boil 1 hour, then simmer with fiber 30 minutes
Marigold	Brown	2 bushels fully blooming blossoms		Simmer 1 hour, then simmer with fiber 30 minutes
Onion skin	Yellow	2 pounds dry outer skins		Simmer 20 minutes but do not overcook, then simmer with fiber 20 minutes
Red onion skin	Red	2 pounds dry outer skins		Simmer 20 minutes but do not overcook, then simmer with fiber 20 minutes
Pokeberry	Yellow	16 quarts fully ripe berries		Boil 30 minutes with 1 cup of vinegar, then simmer with fiber 30 minutes
Privet	Gray	1 pound fresh green leaves		Simmer 30 minutes to 1 hour, then simmer with fiber 20 minutes
Queen Anne's lace	Green	1 bushel fully blooming blossoms and stems		Simmer 30 minutes, then simmer with fiber 30 minutes
Raspberries (red)	Pink	16 quarts fully ripe berries		Boil 30 minutes with 1 cup of vinegar, then simmer with fiber 30 minutes
Red cabbage	Blue	2 pounds leaves		Boil 1 hour, simmer with fiber 30 min.
Red cedar root	Purple	2 pounds red cedar roots		Boil 1 hour, then simmer with fiber 30 minutes
Rhododendron	Green	3 pounds fresh green leaves	Soak leaves overnight	Boil 1 hour, then simmer with fiber 30 minutes
Rose hips	Red	2 pounds rose hips		Boil 1 hour, then simmer with fiber 30 minutes
Sassafras	Orange	2 pounds fresh green leaves		Boil 1 hour, then simmer with fiber 30 minutes
Spinach	Green	2 pounds fresh green leaves		Boil 1 hour, then simmer with fiber 30 minutes

Continued...

Material	Color	Gathering	Preparing	Dying
Strawberries	Pink	16 quarts fully ripe berries		Boil 30 minutes with 1 cup of vinegar, then simmer with fiber 30 minutes
Sumac	Black	2 pounds fresh green leaves		Boil 1 hour, then simmer with fiber 30 minutes
Tea (no mordant)	Light brown	Gather ½ pound dried leaves		Cover with boiling water and steep for 15 minutes, then simmer with fiber 20 minutes
Virginia creeper	Peach	2 pounds of all parts of Virginia creeper		Boil 1 hour, then simmer with fiber 30 minutes
Walnut	Brown	7 pounds of walnut hulls off the ground	Soak overnight	Boil 2 ½ hours, then simmer with fiber 1 hour

How does a loom work?

In its most basic form a loom is a frame which is used to make any kind of woven material including tapestries, fabric, carpets, mats, belts, beadwork, etc. Schoolchildren make looms from a picture frame with nails or notches on both ends, which hold the "warp." The warp is the tight, stationary thread through which you weave. Looms range from simple placemat looms, tabletop looms, wall-sized belt looms, or looms the size of an entire room. The more complicated looms have "heddles" which hold intervals of the warp-thread up to create alternating "sheds" (the space between the warp when it is lifted) so that the shuttle (a tool with the weaving thread, called weft, wrapped around it) can be thrown through the warp without the tedious job of going in and out of the threads.

◂ **Large loom with heddles**

▾ **Ancient warp-weighted loom**

How do I make a loom?

Building a loom with treadles or levers to switch between your warp threads is very complicated and beyond most people's carpentry skills. However, there is an ancient kind of loom called a warp-weighted loom, which is fairly easy to build. It consists of a large frame that is leaned against the wall, and the warp is only tied at the top. They are held at the bottom by heavy weights. Instead of mechanical heddles, the weighted loom uses string heddles that are moved with a heddle rod. You can make the loom as wide as you want, but do not make it so tall that you can't reach it when standing up.

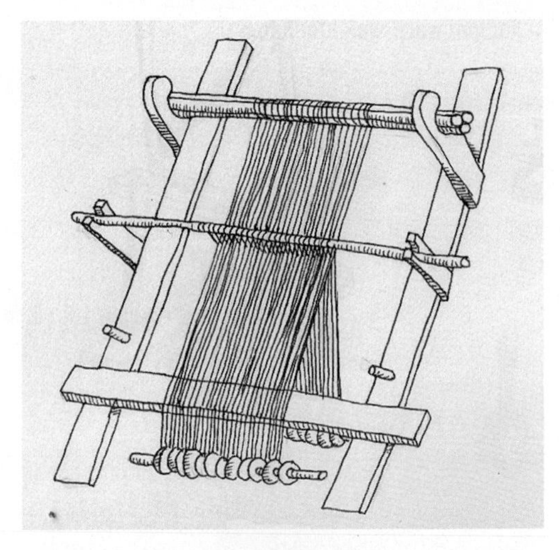

▲ **Parts of a warp-weighted loom**

How do I make the warp-weighted loom?

1. Sand two 8-foot 2 x 4 beams and lay them side-by-side. These are the vertical support beams.
2. First decide if you want to be able to remove the horizontal top dowel that holds the warp. The ancient loom pictured has a stationary frame, while the modern version has brackets, which hold the top beam in place. If you would like to be able to remove the top beam and rotate it to make long items, cut two 12-inch pieces of 2x4 and shape them with a curve large enough for the top beam to drop into. These are your support brackets.
3. Measure 15 inches down from the top of the vertical support beams and make a pencil mark. Center the base of support bracket on this mark and attach with a wooden peg of screw. You don't have to worry too much about angling these brackets since the loom itself angles against the wall and the curve allows the top beam to stay stationary. Repeat for the other bracket. This will create two curved holders which will hold the top beam dowels.
4. To make the top beam, you can either lathe a 4-inch cylinder as wide as you want your loom to be, or you can get a 1½-inch dowel and measure 4 inches from each end. Mark those spots with a pencil and then mark holes equally all down the dowel 2 inches apart and ¼ inch in diameter. Get two more 1½-inch dowels and fasten them to the first dowel (with the holes) either with pegs, wood glue, small screws, or even lashing with leather bands. The dowel with the holes needs to have the holes clear, and if you were to hold the three dowels together with the holed dowel on the bottom, the holes should be horizontal. The three dowels put together should form a triangular shape, which will be the top beam. Place the top beam into your brackets.
6. Now you will need to make brackets for the heddle rod. Get a 1x6 board and cut two 12-inch lengths. On one end of each piece cut a notch 2 inches deep and 2 inches wide in the center.

Lean the loom (the vertical beams) up against the wall exactly as you will be using them.

7. Fasten these brackets to each vertical support beam on the side, so that the notch is sticking straight out. Put them at a height that is comfortable for your weaving. Make sure that the two brackets are exactly the same height.
8. Get a 1½-inch dowel for the heddle rod and cut it a couple inches longer than your top beam dowels. Set it either in the heddle rod bracket's notches or on top of the heddle rod brackets, next to the vertical supports. Both positions are used during weaving. You can also tie leather thongs 1 inch apart down the length of the dowel, which will prevent the threads from sliding together.
9. Use a 1x6 board to cut a length the same width as the loom. This is your shed rod and it is not moveable. It is fastened 2 ½ feet from the bottom of the vertical beams as a crosspiece. Unless you decided to permanently attach your top beam, this is the only piece of wood that is attached to both vertical beams and holds them together. If you have a moveable top beam, you could also fasten another beam at the top behind the top beams at the very back for greater strength.

Heddle support bracket

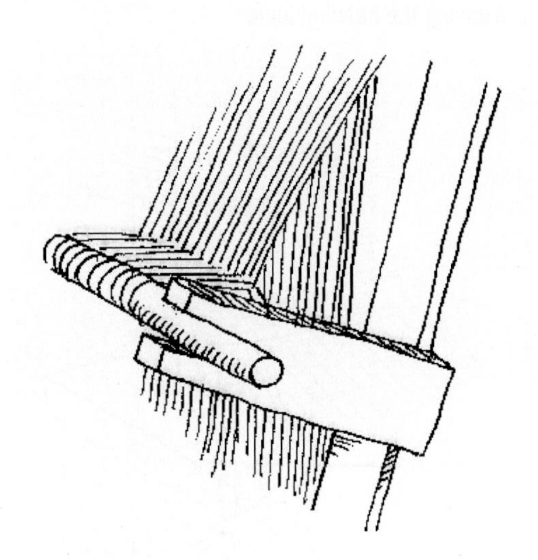

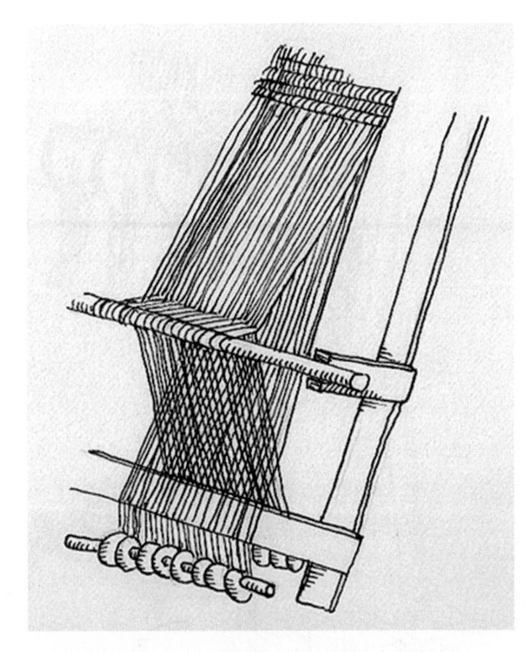

Pulling the warp with the heddle rod

How do I thread the loom?

1. First you have to make a heading band. Cut a warp thread as wide as the width of the item you are going to weave (but no wider than the loom of course).
2. Cut 23 more warp threads the same length as the first. Traditionally half the

Rigid heddle

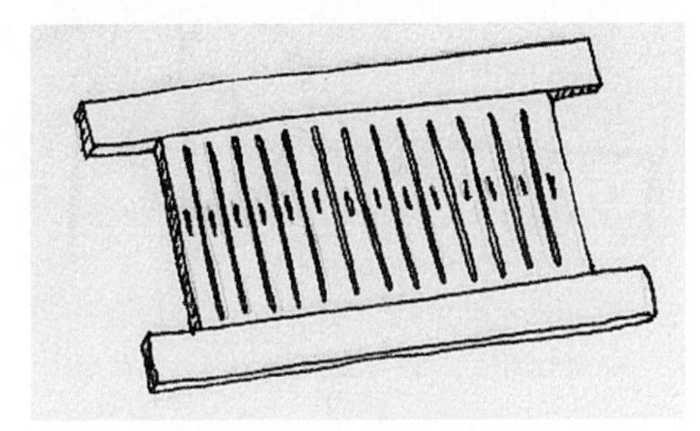

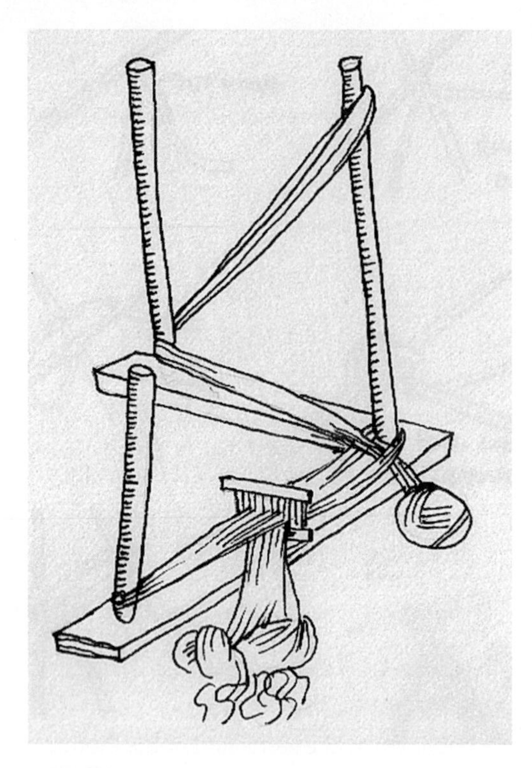

▲ **Warping frame**

threads are a dark color, and half are a light color.

3. Get a rigid heddle about 4 inches wide. You can buy these or you could make one by cutting a rectangular piece of heavy cardboard or thin wood and cutting thin notches in it. It is like a comb but enclosed on all sides. In the center of each tooth of the comb put a hole called an "eye."
4. Make a warping frame by getting two boards 4 feet long. Fasten them together to make a right angle.
5. Get three 1 ½ inch dowels, one of them 2 feet long and the others 4 feet long. Lay the right angle boards flat on the floor and fasten one of the 4-foot dowels straight from the center of the corner of the right angle.
6. Fasten the other 4-foot dowel at one end of the right angle, and the 2-foot dowel at the other end of the right angle 6 inches from the ends. Cut a deep notch a few inches from the top of both of the taller dowels.
7. Take the warp threads and thread them through the rigid heddle with the threads of one color through the notches and the threads of the other color through the eyes.
8. Gather the threads at one end, untangle them, and tie them to the other end to make a loop. The loop should fit around the smaller dowel and the dowel in corner of the right angle.
9. Begin weaving by getting a long ball of yarn with enough thread on it to make your entire warp (that's a big ball of yarn). Using the rigid heddle to separate the strands, pull the end of the ball of yarn through the gap in the thread (the shed), around the bottom of the far dowel (it has nothing on it at this point), and then up to the notch at the top of the dowel in the right angle and tie it there.
10. The other end is now going to a ball of yarn that you can't pull through the gap. Instead you pull a loop of yarn

▼ **Weaving the heading band**

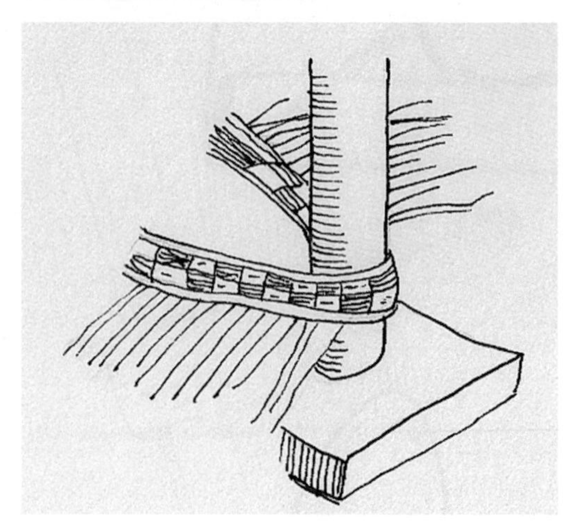

through the ***opposite*** threads than the ones you did previously, and again pull it around the far dowel and loop it around the notch at the top of the angle dowel. This sounds more complicated than it is . . . you are just weaving, but without a loom.

11. Switch the threads with the heddle again, grab another loop and follow the same path and repeat. As you weave you will have to rotate the heading band around the dowels because the spot that you are working on needs to be facing down towards the far dowel.
12. As you rotate the heading band the loops around the angle dowel will make it impossible to turn it. Remove the loops and divide them into two sections (it doesn't matter if they are mixed up). Tie each section into a very loose knot or ball to keep them untangled and out of the way.
13. When the heading band has been woven all the way around, take it off the warping frame and tie each end to the loom top beam, leaving the knots hanging down.
14. Using strong thread sew the heading band through the holes of the top beam. There are traditional ways to do this with a stitch sort of like a buttonhole stitch, but anything will work if it is secure.

▾ **Heading band stitch**

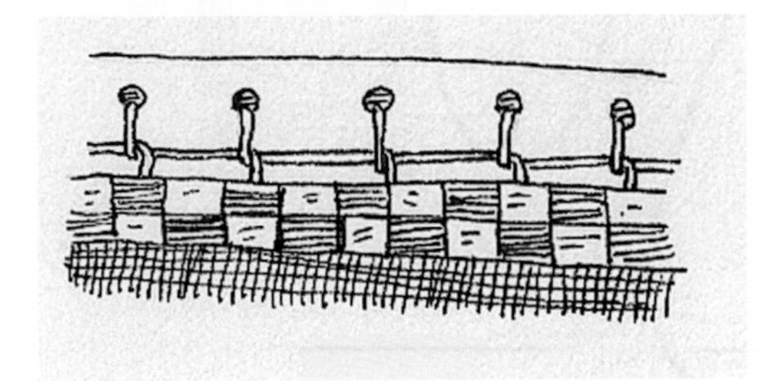

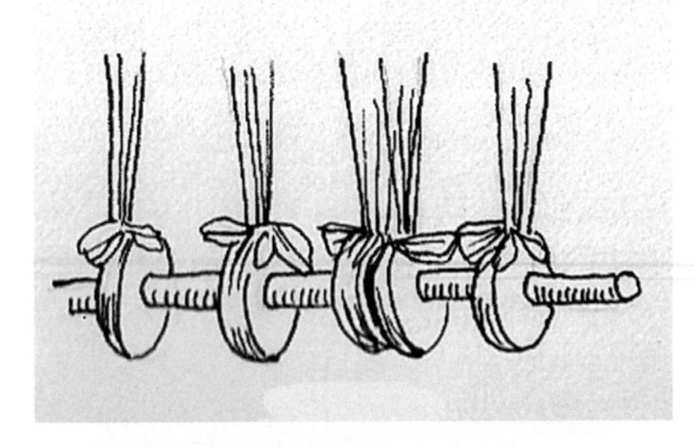

▴ **Weights made of concrete**

15. Let all the knots or balls hang down towards the wall. Untie all the knots of the warp, and carefully separate and untangle them. Pull half of the threads over the bottom shed rod, being careful to pull only every other thread (don't skip over two). Be careful not to bump them. You should now have two sections of threads, one hanging over the front of the bottom shed rod and one hanging in the back near the wall.
16. Carefully pull the warp threads tight and cut the loops at the bottom so that all the threads are the same length. Now you have to attach the threads to the weights. You must have enough weights to attach to 1-inch sections of threads for the entire warp, and they all have to be 12–16 ounces (and all the same), and they can't touch the floor. You can use soda cans full of sand, bags of sand or rocks, concrete disks, fired clay disks, or other materials. If you use disks with a large circle you can put a stick or dowel through them so they don't swing all over.
17. Divide the warp into 1 inch sections and tie each section to a weight.
18. Using cotton cord or thread, you must tie the back warp (the threads hanging closest to the wall) to the heddle rod

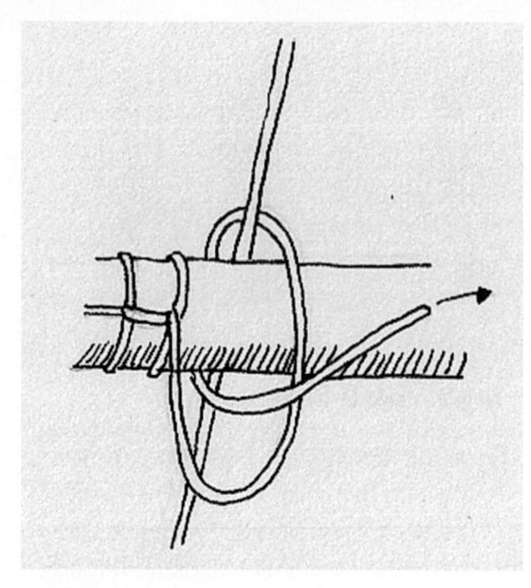

▲ **Heddle knot**

(the dowel near the middle of the loom). Put the heddle rod into its notch in the brackets.

19. Get a large ball of string or yarn. First tie the string around one end of the heddle rod. Pull one of the back warp threads through (the ones nearest the wall) until you have a large enough space behind it to put your hand. Loop the string around that warp thread, leaving enough loop to keep the warp sitting in that spot that leaves you enough room to put your hand, bring it under the heddle rod and up through the loop again and pull it tight (still leaving the loop to hold the warp).
20. Keep doing this until all the back warp threads have been brought through and tied with the heddle string. You will want to make each string loop the same length so that all of the warp will be pulled forward the same distance. If you have leather thongs around the

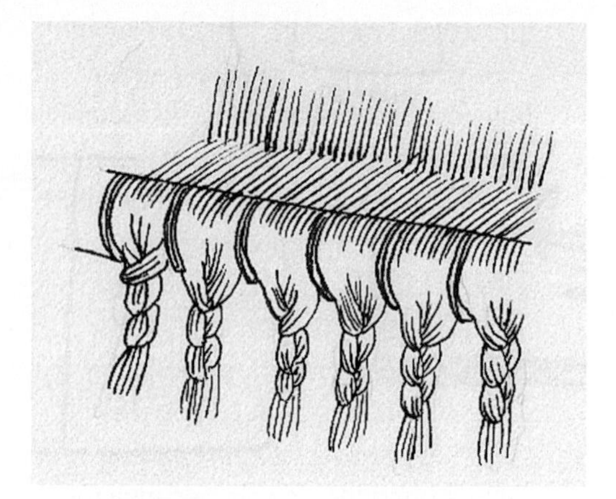

▲ **String heddles**

heddle rod it will prevent the string from bunching up towards the center.

21. To make patterned cloth you can use two or more heddle rods and tie different intervals of the warp so that you can add different colors for each heddle rod.
22. The front warp that is sitting across the shed rod should be held with cord so that it does not get tangled. Using more of the string or yarn, tie a chain with a loop around each warp thread. First tie a loop at one end, bring the end up through the loop, make a loop around the warp thread, go up through the previous loop, and loop around the

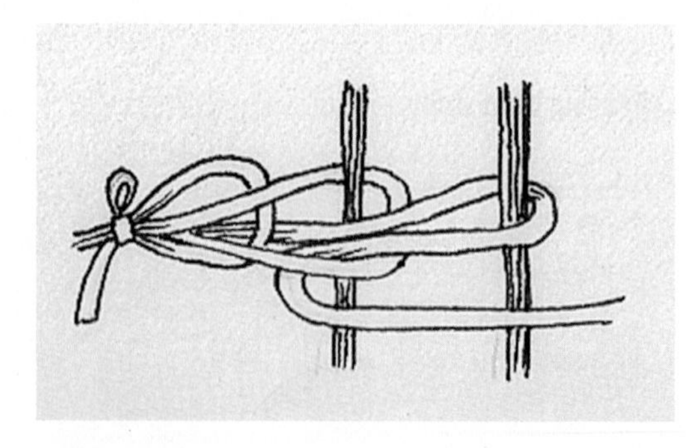

▲ **Warp chain**

▲ **Shuttle**

next warp thread, and so on around all the threads.

How do I weave with a warp-weighted loom?

1. Once you have gotten the warp set up on the loom, weaving is fairly simple. You will need a shuttle, a stick or a store-bought shuttle with the thread that you are going to weave with wrapped around it. This thread is called the weft.
2. You will also need a sword beater. This can be a yardstick or ruler, or a traditional sword-shaped piece of wood that is used to beat or push the weft tightly into place.
3. Hold one end of the weft with your left hand and feed the shuttle through the shed (the gap between the warp threads held by the heddle rod and the threads hanging over the shed rod) to your right.
4. Use the beater to push the weft thread up as far as it will go, but not so tight that the ends pull in past the outer warp threads.
5. Move the heddle rod from the notches and set it on top of the brackets right next to the vertical support beams. This will create another shed, but this time the warp threads leaning on the shed rod will be in front. Use your right hand to feed the shuttle through the shed.
6. Put the heddle rod in the notches and feed the shuttle through with your left hand. Beat it with your sword beater or yardstick anytime that the threads seem looser.
7. The goal is an even, tight weave, with straight outer edges. Most beginning weavers have a tendency to create hourglass-shaped weaving. This can be avoided by estimating how much the

thread will tighten when it is beaten, and leaving a small amount of slack on the end. This skill only develops with practice.

8. As the fabric grows longer turn the top beam over, wrap the fabric around it, and tighten the weights at the bottom until you run out of warp or the fabric is long enough.
9. Fringe or hem the bottom to prevent unraveling and wash it to shrink it.

What is clothing made of?

Wool, flax, hemp, nettle, cotton, basswood bark, llama fur, rabbit fur, goat fur, and silk are all materials which can be used to create fabric. Individual fibers from raw materials can be spun into thread, and then woven together on a loom. Or, for thicker and warmer cloth, the fibers can be spun into yarn and made into sweaters through crotchet and knitting. Or, if you have scraps of already-made cloth, you can make patchwork, in which the small pieces are sewn together to make a big piece of fabric. These fabrics can all be dyed using natural colorants to create whatever colors you want.

How do I make my own clothes?

Clothing is a complicated process involving patterns and measurements. A cheaper and more resourceful way is to get clothes secondhand, unless you are weaving your own cloth. The following are simple instructions for making clothes, but you can also buy patterns to use instead of designing your own.

The measurements you need for making your own clothes are as follows:

Circumference of: chest, waist, hips, biceps, elbow, forearm, wrist, thigh, knee, calf, ankle, head.

Measure from: nape of neck to shoulder, top of shoulder to underarm, top of shoulder to elbow, top of shoulder to wrist with arm bent, inside arm to elbow, inside arm to wrist with arm straight, elbow to elbow with arms outstretched, wrist to wrist with arm outstretched.

Measure: width across front and back of shoulders, width across front of chest, width across back of chest, point to point on chest, top of shoulder to point of chest, forehead to nape of neck, from ear to ear, over top of head.

How do I make a clothing pattern?

Making clothes that actually look good without a pattern bought at the store is beyond most people's ability. You can buy very easy patterns from many stores. However, making ugly clothes without a pattern is quite simple and easy. You can create clothes patterns out of newspaper, and a dressmaker's dummy is extremely helpful. Use the measurements of yourself to create the patterns, and then pin them together on the dummy in order to see if they are properly fitted together. Always remember to leave ¼ inch all around the pattern for a seam allowance.

How do I sew?

Knotting the thread: After you thread the needle you should knot the other end. Make a loop around your finger, and then use your thumb to roll the knot until it is all twisted up. Then pull it by running your thumb and index finger down the thread until it is tight.

Running stitch: The basic stitch for all sewing is the running stitch. To make the best quality items you want each stitch to be as small and as close together and

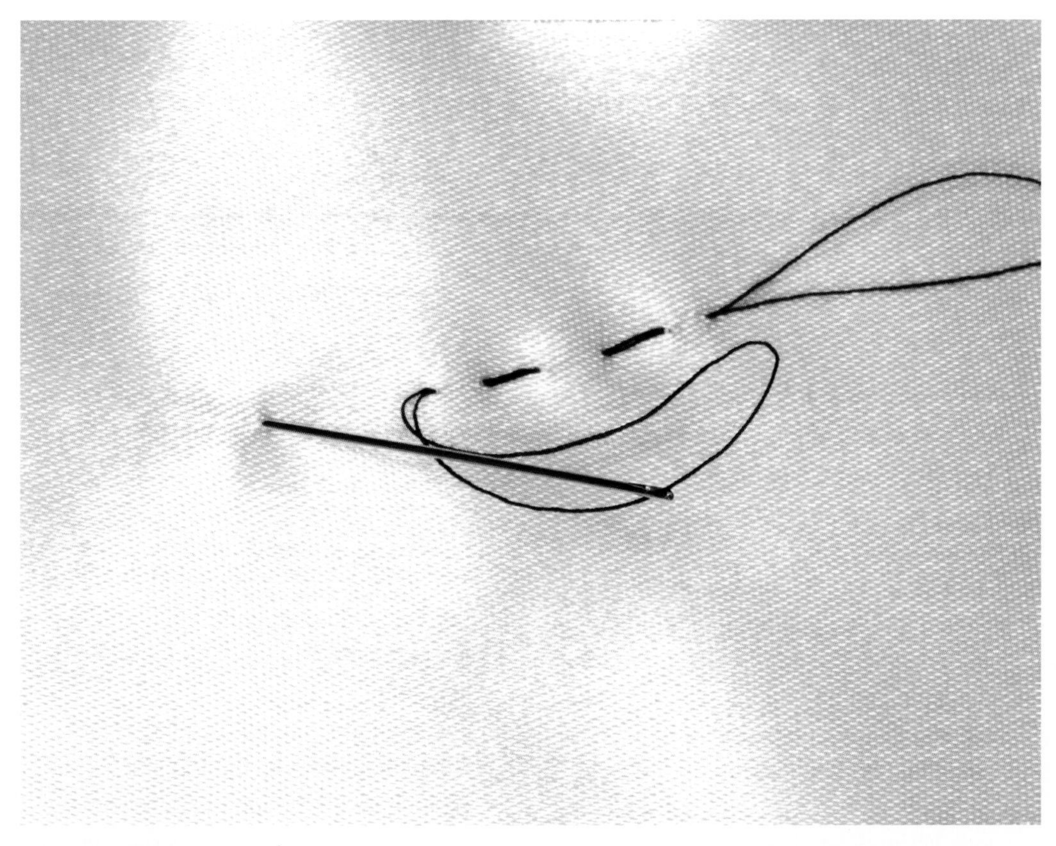

▲ **Running stitch**

as straight as possible. This is easy with a sewing machine, but takes much practice and patience when sewing by hand. Basically you go in and out with the needle. A thimble makes this easier, especially with several layers of thick fabric.

Buttonhole stitch: A buttonhole is a slit in the fabric that a button must fit through. Starting at one end, you whip around the edge of the slit with the threads very close together so that the entire edge is covered in loops of thread with tiny knots along the top. You go in one side, around the edge, in the other side. But before you tighten the loop you just made, you stick your needle back through it. When you tighten it, your thread should be caught along the edge of the buttonhole in a half-knot.

Hem stitch: Sometimes you want the hem to be easily removable because you might want to adjust it later. Instead of a tight running stitch, you can make a looser diagonal stitch. Start right above the hem, dive diagonally down through the hem, and then right up above the hem again.

▼ **Buttonhole stitch**

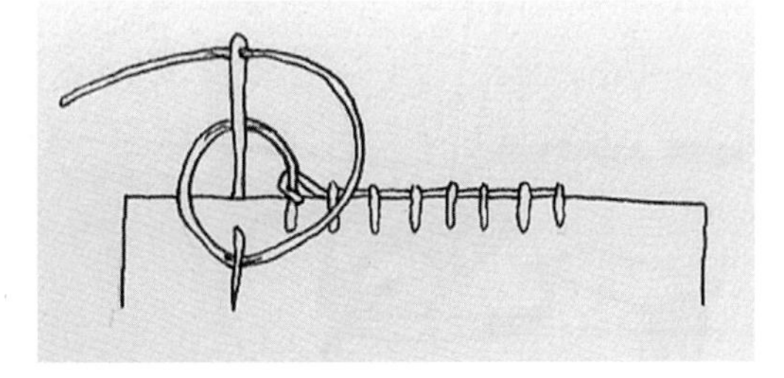

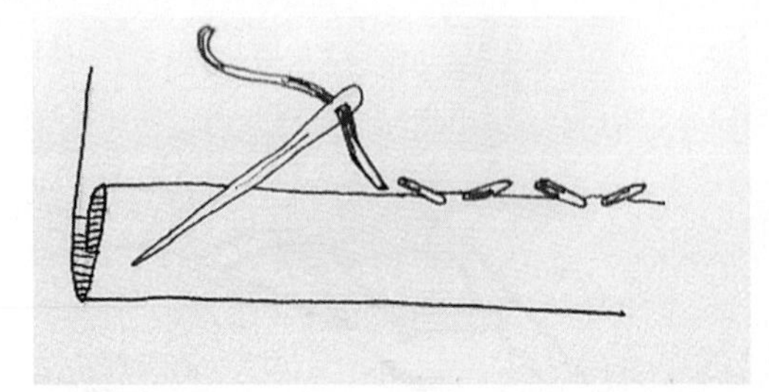

▲ Hem stitch

How do I make a dress or skirt?

1. Calculate how much fabric you'll need by multiplying the length of the skirt by 2 and adding ½ a yard.
2. Fold the material in half, selvage (finished edge of fabric) to selvage.
3. Fold in half again, selvage to selvage. The fabric should now be a quarter of its original size.
4. Lay the fabric on a flat surface, with the selvages on the left side.
5. Divide your waist measurement by four. Write down this number.

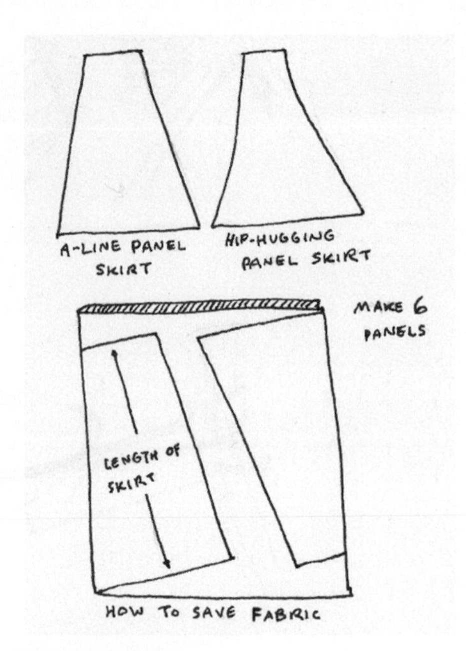

▲ Panel skirt pattern

6. On the top right corner of your fabric draw a curve from the top side to the right side, enclosing the corner of the fabric. Make the line as long as the

▼ Full skirt pattern

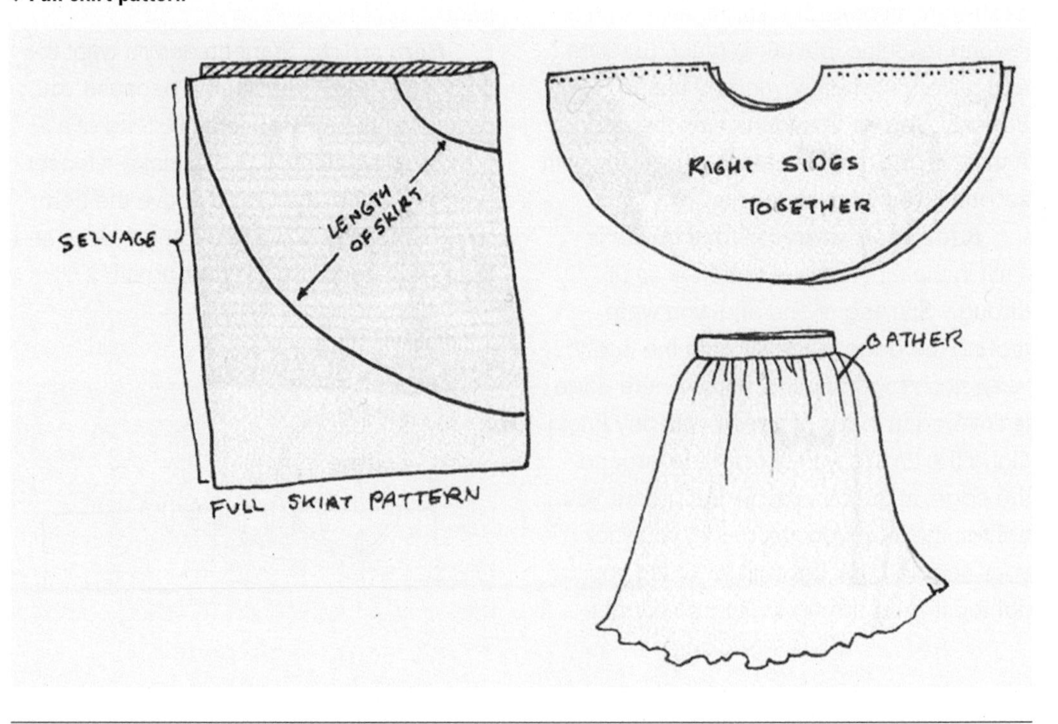

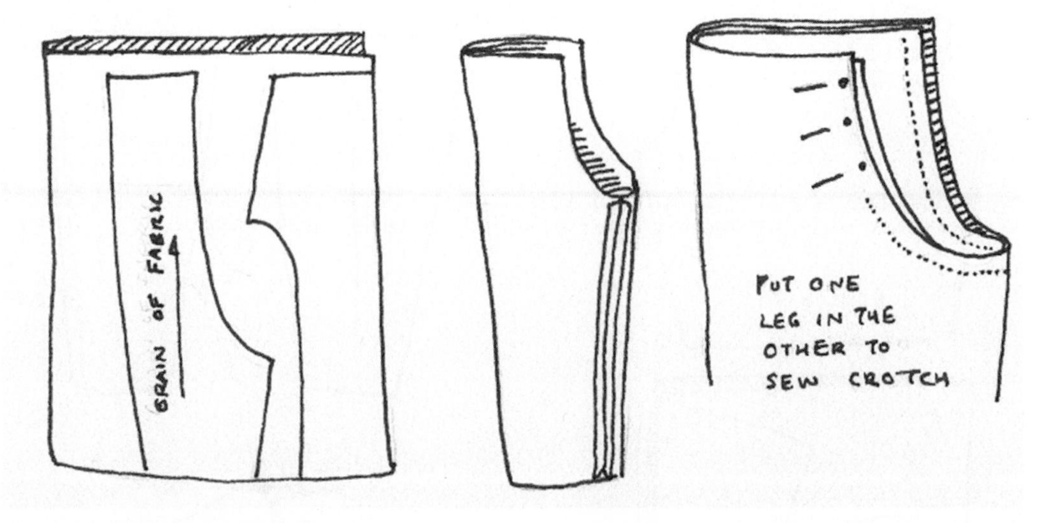

▲ **Pants or shorts pattern**

number you wrote down. This is going to be the waist.

7. How many inches long do you want the skirt to be? Write down this number.
8. Draw a curve from the top left corner to the bottom right corner. It should be as many inches away from the waist line as the number you just wrote down. This is your skirt bottom.
9. Cut on the two lines, leaving two half circles.
10. Stitch the two half circles together, making sure to sew the right (the side with the print) sides together. On one seam, leave 9 inches from the waist unsewn. This gap will be for getting into the skirt.
11. Press the seams and hem (sew the edge on the bottom) the skirt.
12. Hem the gap you left, all around it.
13. Cut a strip 3 inches wide and as long as your waist measurement from the scraps of fabric.
14. Sew the strip to the waist of the skirt, making a casing over the top for a waistband, with the ends ending on each side of the gap you left in the seam. Sew a button and buttonhole on the waistband at the gap.

How do I make pants or shorts?

Pants are usually made of four pieces, all the same, that are sewn together and a button is made at the waist so that someone can get into them. The pattern piece is basically one leg and it curves in at the crotch. The two front pieces are sewn together so that the two crotch sides meet in the middle.

How do I make a shirt?

A T-shirt is made of four pieces, two for the sleeves, and two for the front and back chest area. The front and back chest area pieces are different from each other because the neck hole in the front is a different shape from that of the back. The sleeves can be made in a variety of shapes, but the end that will be attached to your shoulder is wider than the part that goes around your arm.

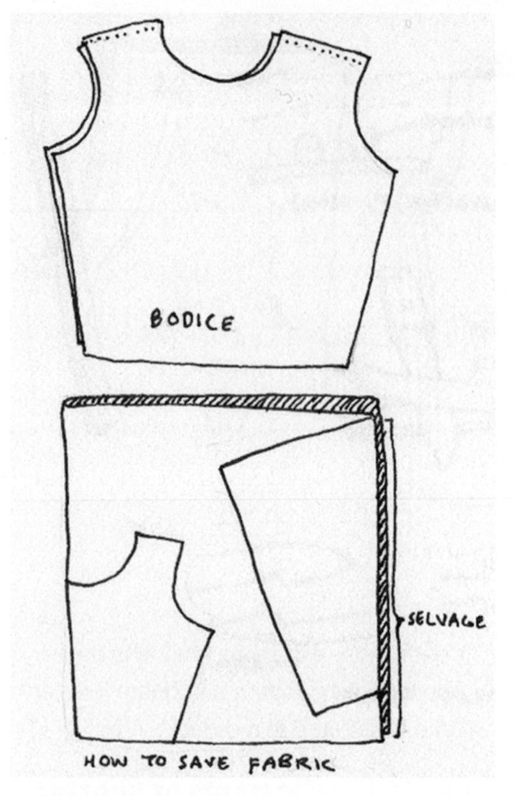

▲ **Blouse or shirt pattern**

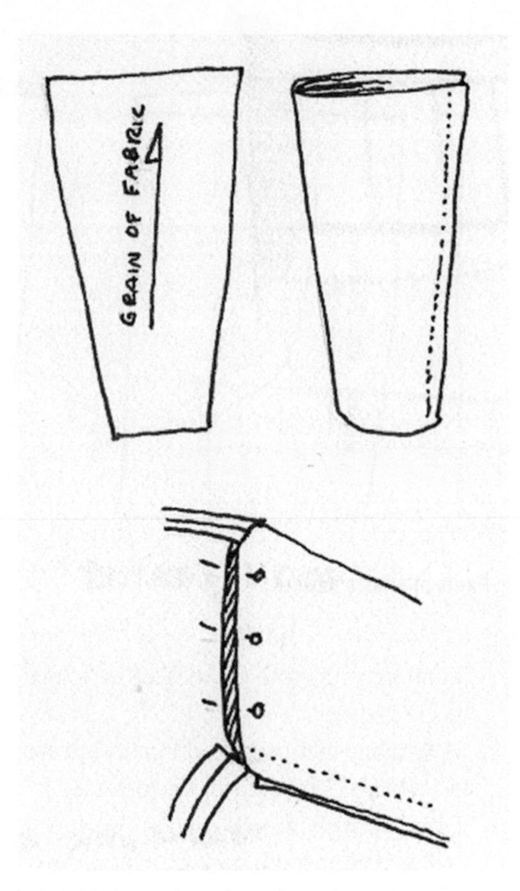

▲ **Sleeves**

How do I make a sleeve?

1. To find how wide the top of the sleeve panel should be, multiply the depth of your armhole by two and add 1 inch. Write this number down.
2. How long do you want the sleeve? Write this number down.
3. What is the circumference (the length around) of your arm where the end of the sleeve is? Write this number down.
4. Draw a four-sided panel on the fabric, with the grain of the fabric running down the length of it.
5. The top side should be the first number, the sides should be the second, and the bottom should be the third.
6. Fold the panel in half with the right sides together. Sew the seam and press.
7. Attach the top edge to the armhole by pinning the top of the sleeve to the shoulder seam, and the sleeve seam to the side seam of the garment. Make sure it is on correctly and sew.

▼ **Cape pattern**

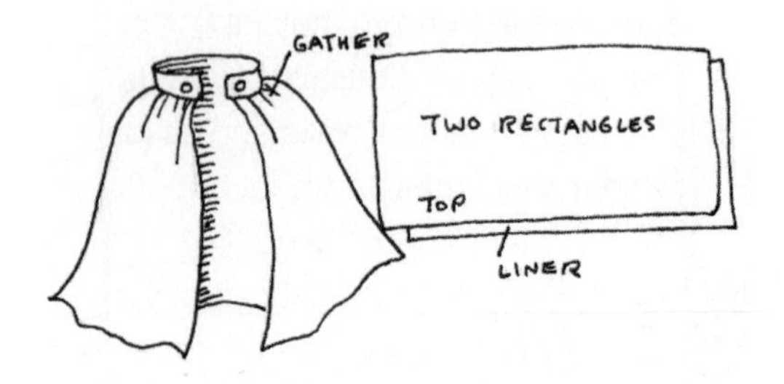

▲ Underneath patch

What cloth terminology should I know?

Right and a wrong side: When you use a ready-made pattern, they will refer to sewing wrong and right sides together. The wrong side is the side of the cloth that will be worn inside; the right side is the side that will be seen on the outside. If you make your own cloth and dye it yourself, you may not have a right or wrong side.

Pattern: If you buy cloth and it has a pattern, you probably will have a right and wrong side. Cloth with stripes is not usually a good choice for beginning sewers because to make it look good, you have to match the stripes to each other. The pattern may also bleed in the wash, so beware.

Grain: The grain is the direction that the weave in the cloth is going. There are two directions that are at right angles to each other, and when you lay the pattern on the fabric, you want the grain to be going the same direction. For example, you would want the grain in your sleeve to be running straight down your arm from your shoulder to your wrist, and not diagonally. This makes the fabric stronger.

Selvage: On manufactured cloth, the edge of the fabric is woven thicker to be stronger so that it does not fray. It is a good idea to use the selvage on hemlines and sleeve ends because those are high-wear areas.

What kind of patches can I use to mend clothes?

Underneath patch: Match the straight grain of both patches and pin the

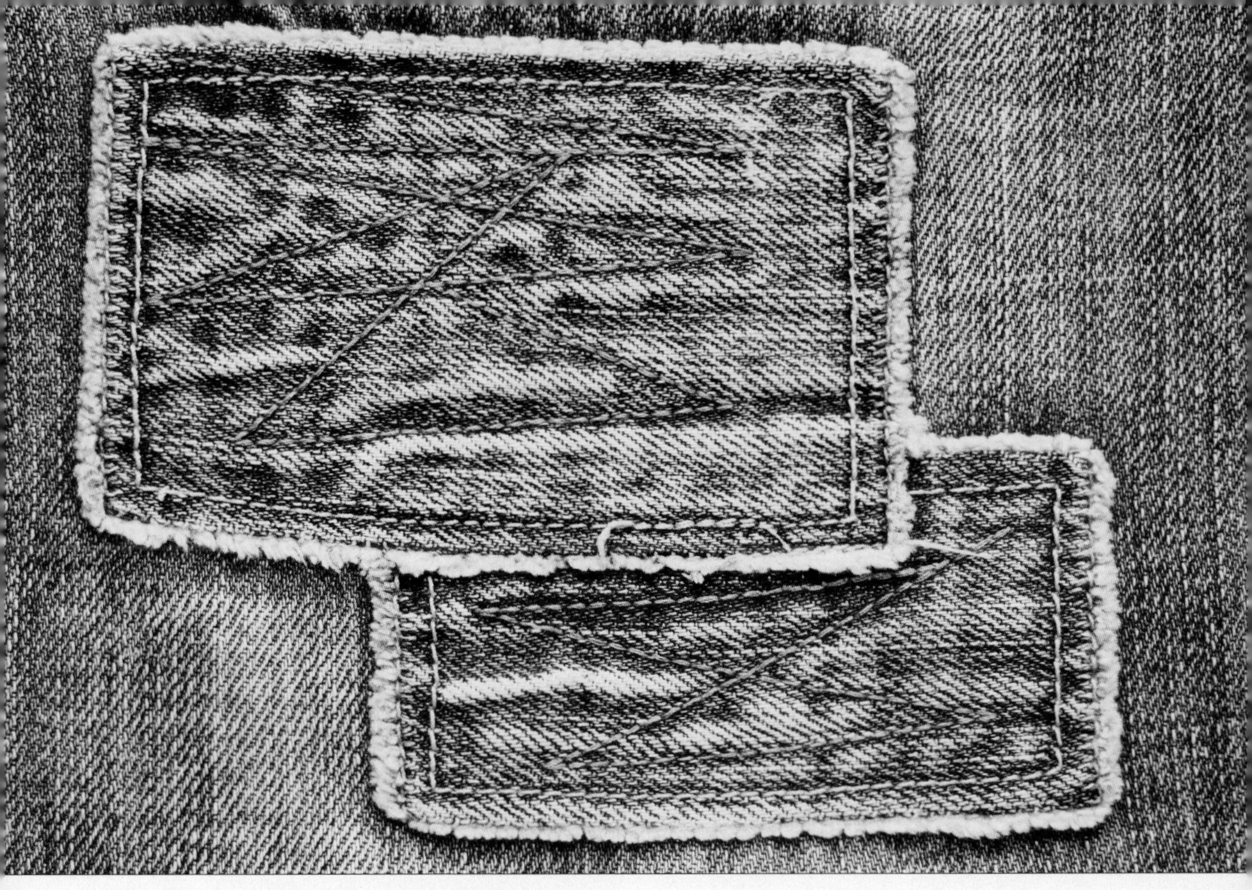

▲ Torn edges don't always have to be turned under

patch under the tear. Turn under the torn edges, and sew to the patch underneath. Turn to the wrong side and trim patch ½ inch from the stitching. Use this patch for irregular tears.

Applied patch: Use an iron-on patch or appliqué a patch on (make a shape and use a buttonhole-type stitch). Use this patch for elbows, knees, and the seat of your pants.

Insert patch: Draw a square on the right side of fabric with chalk and cut it out around the tear. Cut a patch to exactly match the fabric and two inches larger than your hole. Pin it on behind the hole. Cut ¼-inch slashes in the corners of the hole, fold the edges under and iron. Sew all around the edges. Use this patch for plaid, stripes, and designs.

How do I knit and crochet?

Knitting and crochet are two different but similar ways of knotting yarn to make a variety of useful items. Like anything else in this book, it is an art in itself, and most easily explained with pictures. For the most basic knitting and crochet, choose a yarn that is about as thick as your needle or crochet hook for the best results, and avoid synthetic fibers. Beginning knitters can most easily knit squares and rectangles, which can then be sewn into other objects such as scarves and blankets. Crochet lends itself more easily to round shapes than knitting does. Once you have figured out the basic stitch with either of these crafts, you can experiment with more complicated projects such as socks, hats, and sweaters.

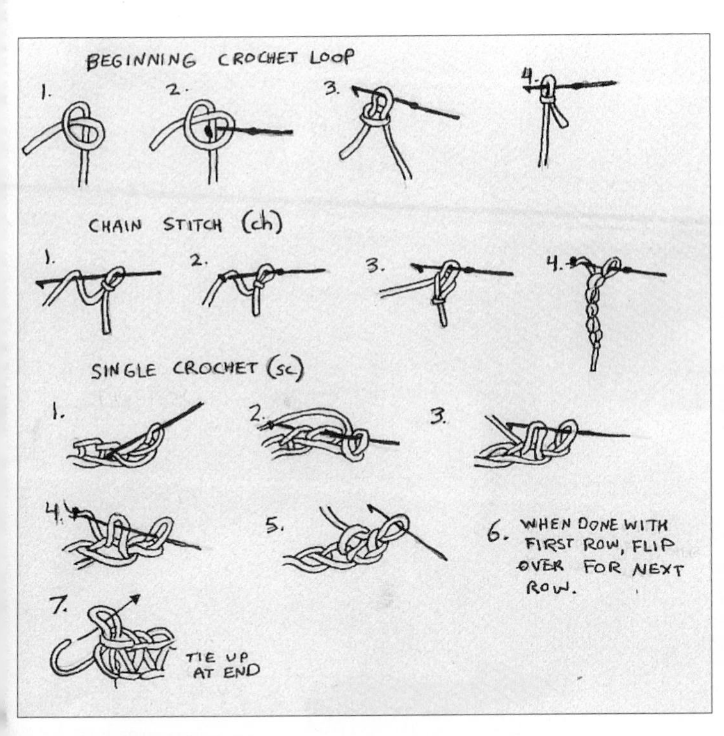

▲ Simple crochet

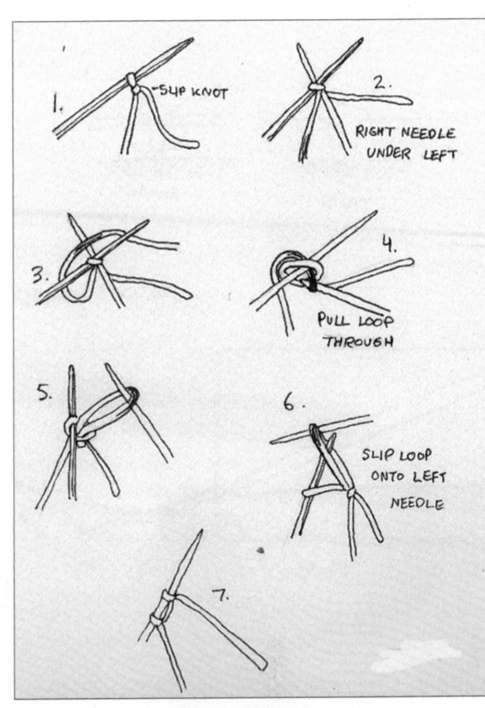

▲ How to cast on knitting needles

▼ Hold needles closely together

▲ Insert left needle into top loop of right needle

▼ Use yarn wound on left finger to wrap around right needle

▲ **Pull the new loop of yarn on right down, held open with left needle, and scoop loop off left needle at the same time**

Why leather?

Whether you are vegetarian or not, if an animal dies (or you eat it) you may want to utilize the resources available from the animal. Leather is an extremely valuable material—shoes, tools, clothing items, bags, and harnesses all need to be made of leather.

What kinds of methods are there?

The skins of animals can all be treated in a variety of ways in order to preserve and protect them (as well as remove any remaining fleshy material). In the old days a hide was dried with the fur still on it in preparation to being sold. If a hide was to be tanned, the hair was scraped off, and then preserved with a tanning solution (usually brains in the olden days). It is ideal to use rainwater because it is the softest and will prevent spots on the leather. Always use wood, plastic, or ceramic containers, and when you put the hide to soak in a tanning solution, hang it over a wood stick so that it has as few wrinkles as possible.

How do I dry a hide?

1. Make a small square frame about the size of the hide. Bore or burn holes a few inches apart all the way around it.
2. Use basswood bark or buckskin thongs to stretch the hide taught inside the frame. Hang to dry.

How do I use a scraping horse?

1. Make a scraping horse with rounded board (a log cut in half) elevated three

▲ **Hide in a drying frame**

feet high at one end, and resting on the ground at the other.

2. Place the hide on the rounded surface of the horse with the hair side down.
3. Stand at the tall end of the horse and use a dull scraper to scrape ***away*** from yourself (not towards). This is done to remove fat and flesh from the skin. (Wear a protective apron or you will ruin your clothes with grease.)

How do I remove the hair from the hide?

1. Use a large plastic can or barrel and fill with 10 gallons water and 1½ pounds of unslaked lime or a bucket of wood

ashes. Stir until dissolved and immerse the fur in it.

2. Cover and let stand for 48 hours, and when the hair starts to pull away it is done. Don't leave it too long or the hide will be ruined.
3. Use the scraping knife to remove the hair and throw the limewater in the garden. Rinse the hide with water and a large amount of baking soda, and then soak for 2 hours.
4. Either work the hide to make it soft for buckskin, or finish the hide by tanning (see below).

How do I tan the hide with soap and brains?

1. When making soap, save any that comes out unsatisfactorily. Soak the hides in the soft soap until the hair can be slipped off with the very thin outer skin.
2. Pound a whole brain into the inner surface of the skin to firm it.
3. Soak the hide in alum (see tanning solutions below).
4. Wash the skins very thoroughly to get the alum out, and then soak for 24 hours in water. Pull it out and squeeze all the liquid out.
5. Use the oak bark solution (see below) to soak the hide as a finisher. Do not soak it for as long, only for a week or two. Pull it out and squeeze all the liquid out.
6. Stretch, pull, and work the skin until it is broken in and soft, then hang it in the shade for two days. You can finish the hide, or if you leave it this way it is rawhide.

How do I tan the hide with salt?

1. Lay the hide out flat in a dry place. Cover with a thick layer of salt for at least two days. Either salt it when it is still warm off the animal, or wash in warm water.
2. Scrape the hide with a sharp edge such as a butter knife to remove membranes, fat, etc. Scrape in the direction from head to tail, and then use a scrub brush to clean.
3. Wash all the salt off, then soak the hide in a tanning solution (see below) for the specified length of time.
4. Stretch and pull and work the skin until it is broken in and soft, then hang it in the shade for two days. You can finish the hide, or if you leave it this way it is rawhide.

Tanning solutions:

Alum: Four gallons water per 1 pound powdered alum and 1-pound salt. Soak 1 week.

Oak bark: Gather bark when the sap starts to rise in trees so it is easiest to peel. Any time of the year works; however, the oldest trees have the most tannin. Use double the weight of the hide. Boil cracked acorns and ground oak bark in water. Strain, and then add an equal amount of water to it. Continue this until the plastic can or barrel is full, and let hide soak 1 month.

Brain: Boil the brain until it is a paste. Apply evenly to the hide and let it sit for 3 days. (Note: With the advent of mad cow disease, only use brain if this is your own animal because similar diseases exist in wild animals too.)

How do I finish the leather?

1. Dry the leather by laying it out flat and cover it with old dry clothes. Walk all over the surface to press out the moisture.

2. When the leather is still damp, work neat's-foot oil into the grain side only.
3. Hang up and allow to dry.
4. For tougher leather, work a mixture of tall and neat's-foot oil (50/50) into the grain side. Rub with sawdust to get excess grease out.
5. For softer leather, dampen the leather, and rub in more oil. Stretch it out more, work in more oil and allow to dry.

Making catgut:

Note: Catgut is a very strong and elastic material that is used for sutures (stitches in wounds), strings for musical instruments, bow strings, and lacings for clothing and bags.

1. Clean the intestines of a sheep and pull or scrape off the fat carefully.
2. Soak them in water for a few hours, then use a blunt knife to scrape off the external membrane, or the outside layer of tissue.
3. Throw 1 gallon of wood ashes into 3 gallons of boiling water, then allow it to cool. Put the intestines in this solution overnight.
4. Flatten the intestines into an even sheet by stretching and drawing it out if necessary.
5. Bring 5 gallons water to boil, and add 1 pound of powdered alum and 2½ pounds salt. Stir until it is all dissolved, and let cool to a lukewarm temperature.
6. Add intestines to the alum mixture and let soak for 3 days, stirring several times a day.
7. Drain out the water and wash the intestines in clean water. Squeeze them out as much as you can. Cut into very thin ribbons, the size depending on what they will be used for. If you are making lacings then they can be smoked like buckskin.

Leather dyes:

Dark blue: Mix urine with indigo and soak the skin in it. Then boil with alum. Or boil elderberries, wash the skins in the liquid, then boil the elderberries with alum and dip the skins in the liquid once or twice. Allow to dry.

Sky blue: Steep indigo in boiling water, and allow to cool to a lukewarm temperature. Spread it over the skin.

Green: Mash buckthorn berries into a thick paste and spread on the skin. Boil the skin in alum-water.

Yellow: Smear the skin with aloe and linseed oil, then rinse off.

Light orange: Smear with fustic berry paste and boil in alum-water.

Dark orange: Smear with turmeric paste and boil in alum-water.

How do I make hard-soled moccasins?

1. Put your foot on a piece of heavy leather or buckskin and trace it. Cut out this shape, flip it over and make a reverse one and cut it out—they are the soles of your moccasins.
2. Get a piece of soft leather and draw a pattern which is curved into a triangular shape with the tall point at the same spot your big toe will be. It should be 1 inch longer than the sole of the moccasin. Cut it out, flip it over and make a reverse one.
3. Draw a T-shape in the center of the flat edge of the triangular soft leather and cut it out. Do the same for the other moccasin.
4. If you want to embroider the top of the moccasin do so now.

5. Starting at the toe, stitch the sole to the upper soft leather until you get to the heel, then do the other side starting at the toe. Repeat for the other shoe.
6. Cut two little rectangles out of soft leather to be the tongues of the moccasins. Sew them into the T-shape, leaving the sides open.
7. Sew up the back of the moccasins.
8. Punch holes with an awl for lacing and use leather laces to tie them up.

Moccasin pattern

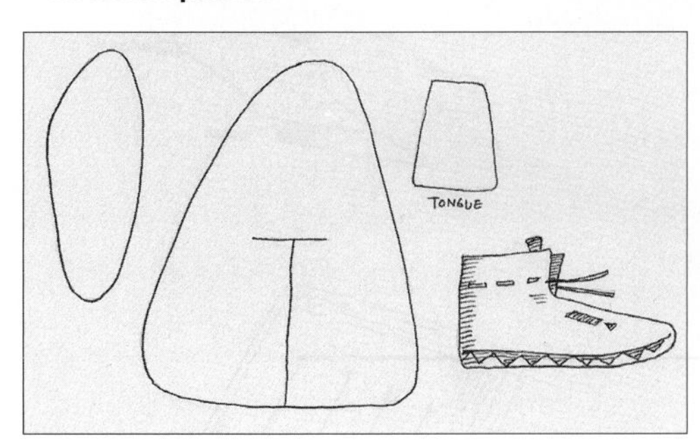

9. When wearing the moccasins, sometimes the bottoms can get slippery in winter, but these moccasins don't wear out as quickly as other styles do. Make sure to wear good socks with them.

How do I make a braided straw hat?

1. Cut oat or wheat straw when it has changed from green to yellow but not too ripe so that it is brittle.
2. Soak the straw in water and then sort them into sizes. Remove the joints and leaves and tie them loosely into bundles or put them in separate containers. Immerse them in water again.
3. Braid the straw with 7–11 strands of straw, adding more straws into the braid as you get closer to the ends, so that it makes a long continuous braid.

▲ **Beaded moccasin**

▼ **Braided straw hat**

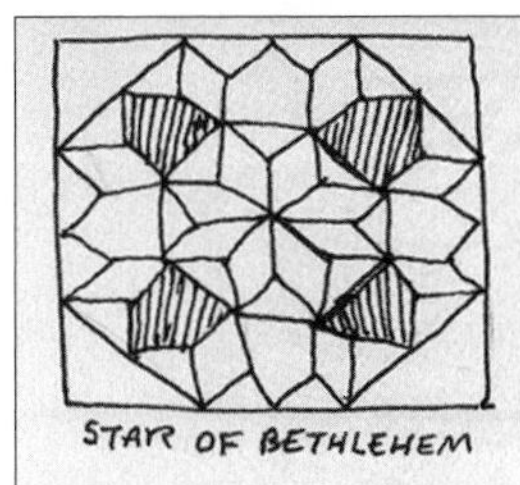

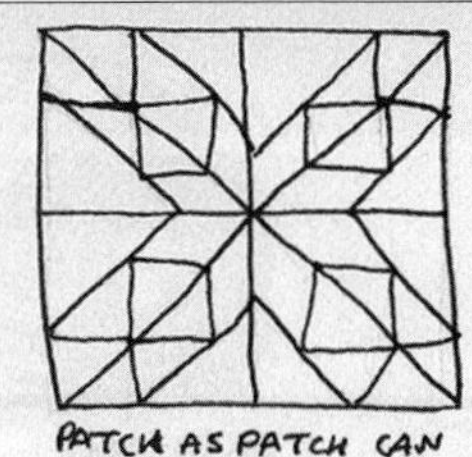

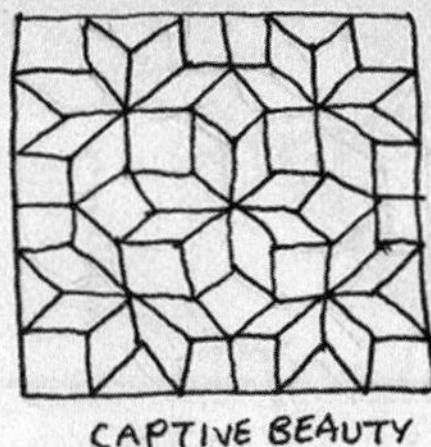

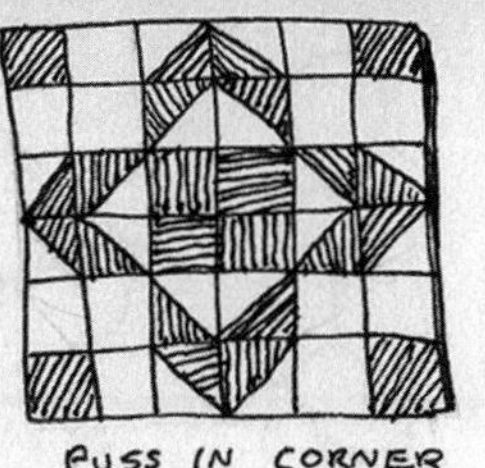

▲ **Traditional quilt patterns**

4. Coil the braid into the shape of a hat by sewing as you would a braided rug. Allow it to dry flat.

How can I make a quilting frame?

A quilting frame can be made by fastening four boards together, making a frame that is a little bigger that your quilt will be. Cover each side with some cloth wrapped around the board, so you can pin the quilt to the frame. Put legs on the frame or support with boxes or something handy.

How do I make a quilt?

1. Find some fabric, large enough to make the top of the quilt (big enough to fit what you want to cover, like your bed), the bottom of the quilt, and the border.
2. Cut the fabric into the appropriate shapes for the design you are quilting.
3. Sew the shapes together for the top of the quilt, sew the border, and create the bottom of the quilt also. Sew the top and the bottom together on three sides only.
4. Put batting or filler for the middle of the quilt (which provides warmth), and sew the last side of the quilt (the open side) closed. You can either use a whip stitch (a stitch which goes in one side, over the edge and into the other side, "whipping" around the edge), or you can turn the edges under and do a regular stitch.
5. Now the quilt needs to be "quilted" or tied. Quilting means that you stitch a pattern into the entire face of the quilt. Some quilting designs are quite detailed, but you can do whatever is necessary. Tying means that yarn or thread is pulled through and back again, then tied on the top. A button can also be used to fasten the thread.

How do I make a crazy quilt?

A crazy quilt involves using irregular pieces of fabric and sewing them together however they will fit. This is usually done when you have a lot of fabric scraps and don't know what to do with them. Just sew enough together to make a large square, and then add four border pieces, which can be fitted together with a diagonal seam, sort of like a picture frame.

How do I make a charm quilt?

A charm quilt is made of all the same shaped piece (example: all squares or all diamonds). The simplest way is to make a square 4x4 inches with a seam allowance, decide how big your quilt will be and decide how many squares will be needed to fill it, and sew them together. The easiest way to stitch them is to make a long

▲ **A form of crazy quilt which uses scraps to create squares**

▼ **Charm quilt**

strip of squares, then sew all the strips together. Then add your border pieces.

How do I put on the quilt back and batting?

The back of your quilt can be one large piece of fabric of any material, although it is nice to use longwearing soft fabric for bed quilts. The batting can be made of any material that you have, although unspun wool or cotton would be the traditional filling. Feathers are also a good filling but do not last as long or wear as well.

How do I make a feather quilt?

Sew the feathers into either several smaller bags, or one big bag the size of the quilt, made of cheesecloth or a very light, airy fabric. Lay the bag or bags on top of the bottom layer of the quilt, and then put down the top layer and sew quilting or tie it. For down quilts tying may be easier, but having a few stitched areas will last longer.

DOING ODD JOBS AT HOME

The best things in life are nearest: Breath in your nostrils, light in your eyes, flowers at your feet, duties at your hand, the path of right just before you. Then do not grasp at the stars, but do life's plain, common work as it comes, certain that daily duties and daily bread are the sweetest things in life.

~ *Robert Louis Stevenson*

What to do with leftovers:

Use the juice from all your canned foods as soup juice—freeze it until it's ready to use.

Use the juice from cooking vegetables as soup juice.

Canned Liquid Equivalents:

1 can condensed milk = 15 ounces, 1 ⅓ cups

1 can evaporated milk = 6 ounces, ⅔ cup

1 can frozen juice concentrate = 6 ounces, ¾ cup

Flour substitutes:

Note: When using dark flour use twice as much baking powder.

1 cup self-rising flour = 1 cup all-purpose flour, 1 ¼ teaspoon baking powder, pinch of salt

1 cup white flour = 1 ⅜ cups barley flour

1 cup white flour = 1 cup corn flour

1 cup white flour = ⅞ cup corn meal

1 cup white flour = ⅜ cup potato flour

1 cup white flour = ⅞ cup rice flour

1 cup white flour = 1 cup rye meal

1 cup white flour = 1½ cup ground rolled oats or 1 cup oat flour

1 cup white flour = 1 cup whole wheat flour

1 cup whole wheat flour = ⅞ cup amaranth

1 cup whole wheat flour = ⅞ cup garbanzo bean or chickpea

1 cup whole wheat flour = ¾ cup corn flour

1 cup whole wheat flour = 1 cup cornmeal

1 cup whole wheat flour = ¾ cup millet flour or oat flour

1 cup whole wheat flour = ⅝ cup potato flour

1 cup whole wheat flour = ¾ cup potato starch

1 cup whole wheat flour = ⅞ cup rice flour

1 cup whole wheat flour = ¾ cup soy flour

Removing cream from goat's milk:

Goat cream is harder to remove than cow cream. Cow cream will just rise to the top if you let it sit. If you put goat milk out for 24 hours some of the cream will rise but not all of it. To get it all you will need a cream separator specially made for goat's milk. The best type for this is a glass container with a hand crank at the top. Run the milk through the separator and save the cream. The rest is your skimmed milk. You can freeze cream for later if you thaw it completely before using.

Canned milk:

Pour the fresh milk into clean jars up to ½ inch from the top of the jar. Put on the lids and process in a pressure cooker for 10 minutes at 10 pounds of pressure or 60 minutes in a water bath canner.

Make curds and whey:

Let the milk sit out in a covered enamel roaster, a canning kettle, heavy crockery, or a stainless steel bowl, until it gets completely sour or "clabbered." Don't use aluminum! Keep the temperature at 75-85°, on a warming shelf on your wood stove or other warm place as if you were letting yeast bread rise, but not too hot. When the curds separate from the whey it has "set up." It will feel like jelly, and it will form a single large curd floating in the whey. If you need clabbered milk sooner, add 1 tablespoon vinegar or lemon juice per 1 cup of milk. Goat milk takes longer, as many as 5 days.

▾ **Dazey butter churn**

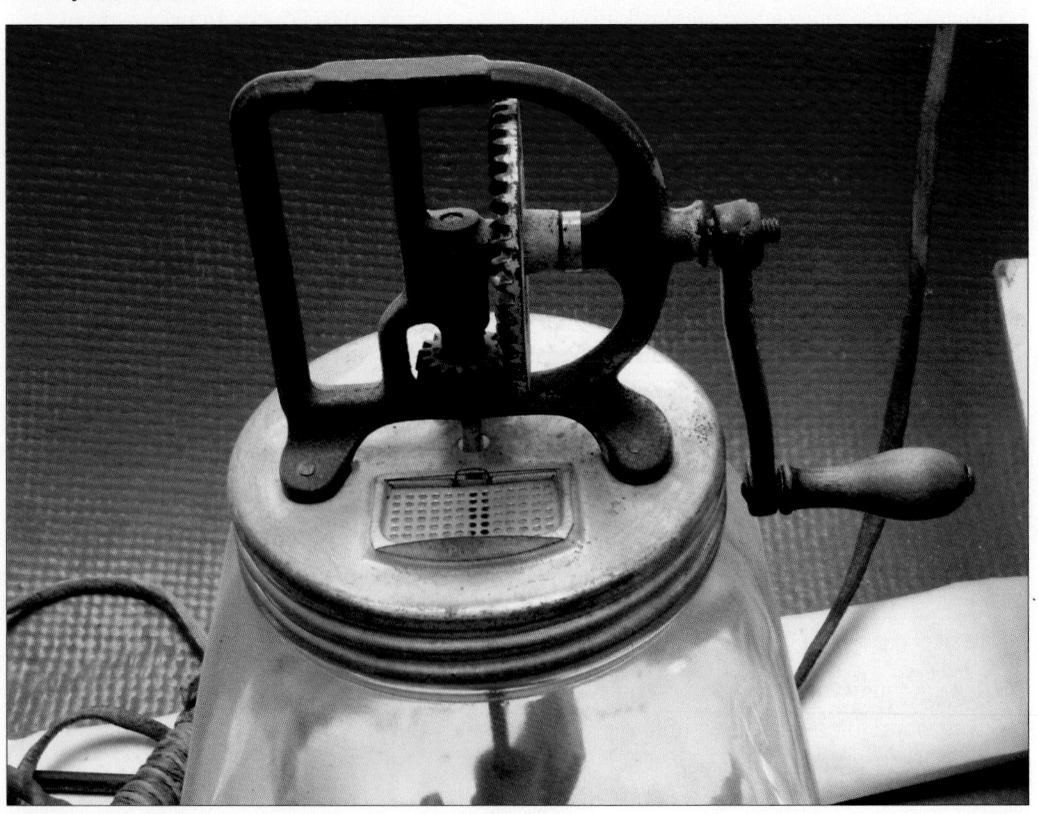

Churning cream into butter:

1. You will need some kind of churn to make butter. Either use a traditional churn or make your own using a quart jar with a lid.
2. The cream must be 60°, or about room temperature, and put it in the churn or jar so it is half full.
3. Churn steadily and rhythmically, not too fast and not too slow (if you are using a jar, simply slosh the cream back and forth). After 20–30 minutes butter clumps will form.
4. Remove the butter from the milk (the milk is buttermilk) and put it in a bowl of cold water. Roll it into a ball, then flatten it and repeat just like kneading bread. Change the water frequently when it gets cloudy, and as the milk is rinsed away the butter will start to feel waxy.
5. Keep working the butter until the water stays clear, then take it out and squeeze out the water. Mix in flavoring (such as honey, salt, etc.) if you want, then pat dry with a clean towel and press into a mold or form into a cake. Cover and put in the fridge or freeze for later.

Making yogurt:

1. Use pasteurized milk so the good bacteria don't have to fight bad bacteria. You can make yogurt right after pasteurizing if you cool it to 110° (no hotter or it will kill the acidophilus bacteria).
2. While the milk is still on the heat, add 1 heaping teaspoon of cultured starter per quart of milk. Stir gently, then remove from heat and pour into clean warm jars.
3. Put the milk in a warm place and keep it warm. Some easy ways to do this is put the jars on a heating pad and cover with a towel, or placing them in an insulated cooler filled with 100° water, or utilizing your oven.
4. Once it's warm don't disturb it. Six to eight hours later it should have thickened and then it is ready. Add ½ cup powdered milk to it if you want it to be thicker and mix in flavors such as fruits, honey, or vanilla. Refrigerate immediately.

Yogurt starter:

Most people use plain yogurt from the store that has live cultures. Make sure it is live—some yogurt is not. Make sure it is well stirred, and keep it separate from the starter you will be using. Save ½ cup of your first batch and use it as your next start within 5 days. If your yogurt gets a weird taste or smell, scrap your starter and begin again. In the old days people hoped to get yogurt in the air by chance and saved it for years.

Types of cheese:

Soft: non-ripe fresh cheese such as cottage, cream, pot, ricotta, gjetost, and Neufchatel. Brie, Camembert, double and triple creams ripen naturally in a few days.

Semi-soft: ripened using specially developed molds and bacteria, includes basic Swiss, Brick, Muenster, Liederkranz, Limburger, Port Salut, Roquefort, and Gorgonzola.

Firm: includes cheddar, Chesire, Lancashire, Caciocavallo, Swiss Emmenthal, Gruyere, and Jarlsburg.

Hard: firm cheeses that are matured longer to make them grainier, including Asiago, Parmesan, Romano, and Sapsago.

▲ Straining cottage cheese

Making cheese:

You can mix cheese with seasonings, or you can smoke it. You will get a clear fluid that pours off the curds which is called "whey." In food value, 1 ¼ gallons of whey equals 1 pound of ground barley. Don't ever used galvanized containers for making cheese.

Cottage cheese:

1. 1 quart of milk will make 1 cup of cottage cheese. After setting (or souring) the milk (see above), cut the curd using a big, clean knife that is long enough to reach the bottom of the pan. Make squares 1 inch wide.
2. Stir the curd very slowly and gently to stop it from clumping, but without breaking the curd.
3. Next you will heat the curd. If you like softer cheese, heat below 110°F. If you like firmer cheese, heat up to 120°. Heat it very slowly to your desired temperature, stirring gently.
4. Pinch a little bit of the curd between your thumb and index finger and if a tiny bit sticks to the ball of your thumb then it is ready.
5. Strain the whey in a cheesecloth-lined colander or other strainer to drain away the whey. If the milk was very sour you can rinse the curd in cold water and strain a second time.
6. Put salt on the curd, knead it in, and then put the cheese in a bag and hang it to drain for an hour.
7. Store the cheese covered in the fridge for about a week. This is dry cottage cheese. You can add cream before serving, but if you add it now it will not keep very long.

Making Colby cheese:

What you need:

Stainless steel or unchipped enamel pots and utensils. No aluminum.

Non-iodized salt such as kosher and canning salt.

Cheese press. You can make one out of a coffee can and wood, or buy one.

Cheese wax. It is reusable so get good quality wax. Do not use paraffin or beeswax.

Cheesecloth is not the same as the stuff at the grocery store. It is 100 percent cotton muslin-type cloth. You could also use old pillowcases. Use bleached shoelaces to hang the cheese.

1. First you need rennet. Use the biggest stomach of a suckling lamb (a lamb that has only eaten mother's milk). Add salt to the stomach, cut it into strips, and dry it like jerky. The equivalent of 2 drops of liquid rennet is 1 square inch of stomach.
2. Thistles and nettles can be used to make rennet. Harvest thistles when the flowers are brown, but before they produce down. Air-dry and store in jars until you make cheese. Use 5 heaping teaspoons of powdered plant per 1 gallon of milk. Put it in a mortar and pour a small amount of warm water or whey just to cover and let soak 5 minutes. Pound with the pestle for 5 minutes, let soak again, and repeat until it has soaked four times. It should turn dark brown. Strain it and it is ready to add to the milk. Don't add too much or it will taste bad.
3. Next you need culture, or good bacteria. In the old days yogurt and cheese bacteria were caught in the air by chance. You can try to get your own by souring your milk and hoping it turns into good cheese. Otherwise you will have to buy a cheese culture, or get cultured buttermilk from the store (it can also be used for yogurt). Add ¼ cup buttermilk to 1 quart of milk and let it sit 24 hours at room temperature. Add that quart to 3–5 gallons of milk that is 86–90° F and let it ripen for 1–2 hours.
4. Pasteurize 12 gallons of milk, then quickly cool to 88–90° F, the quicker the better. Add 2 cups of your cultured milk and let it sit for 30 minutes.
5. Dilute 2 drops liquid rennet or 1 square of dried rennet into 1 cup of water. Mix the rennet into the milk for 3 minutes and then let it sit for another 30 minutes.
6. Push your finger into the milk at a 45° angle. If the curd breaks in a clean, straight line then it is ready. Your finger should not have anything stuck to it. Cut the curd using a big, clean knife that is long enough to reach the pan. Make squares 1 inch wide.
7. Let sit 15 minutes, and then set it up like a double boiler, such as a pail put into a canning kettle with the water halfway up the sides. Heat it very slowly, taking about 30 minutes to get to 100°. You can also just fill the double boiler with hot water from the tap so you can control the temperature easier.
8. Drain off the whey and stir up the curd so it is not one block of curd. Add salt to taste, about 3–4 tablespoons per 10 pounds of curd, and mix thoroughly.
9. Sterilize cheesecloth by boiling and put the curd into it. Don't use a coffee can as a cheese press. Buy or make a real cheese press so it doesn't have lead in it. Put the curd into the press while it is still hot.

▲ Tiny aged cheeses

10. For 30 minutes use 5 pounds of pressure, and gradually increase to 20–25 pounds. Leave it overnight, or 12–16 hours.
11. Soak a cloth in vinegar and wrap the cheese in it, or dip it in cheese wax. If you use vinegar you will have to resoak it periodically. These procedures prevent mold. Then store the cheese at or below 55° F. This is the ideal temperature so that in 3 months you will have reasonably cured cheese. Turn the cheese once a day for the first month. You can put the cheese in a turned off chest freezer in the basement so mice don't get to it. One and a half years of curing is better.

Other cheeses:

Monterey: At step 7 leave the curd to cook at 100°F for 1 hour and 45 minutes.

Cheddar: At the end of step 7 the curd should not stick when pressed together. Test by sticking it to something hot and if it pulls away with fine threads 1/8 inch long then it is ready. Continue to step 8 and drain the whey, but then put the curd evenly into a roasting pan with a hole at one end over a warm stove (don't let it cool) and tilt it slightly so the whey can drain off. Create channels in the curd for the whey to run through. Stir it occasionally until it breaks apart very easily, and when you put it on a hot surface the strings extend ½ inch long when you pull away. Then complete step 8, but the salt should be 2–3 ounces per 5 gallons of milk that you started with, and make sure it completely dissolves. Continue on with step 9.

Quick cheese:

1. Use 3 quarts fresh milk at 98.6°F, add ½ teaspoon liquid rennet and stir

gently. It will start to thicken, so add 1 quart of boiling water and continue stirring.

2. The curd will separate from the whey. Put a reed cheese basket into a colander and dip the curd into it. Press down on it, and then turn the basket upside down, dumping out all the curd at once (onto a clean cloth). Put the curd back into the basket with the bottom side up and press.
3. Leave it in the basket for 36 hours, in a warm dry place, and take the curd out of the basket 2 times a day and turn it over.
4. When it is firm enough to keep its shape, take it out of the basket and put it on a plate. Sprinkle salt on the top and keep turning it twice a day, sprinkling salt on the top and sides each time. Change the plate if it gets wet.
5. When no more moisture comes out, put it in a stone crock and put it in the cellar. If it gets moldy wipe it with salt water. Let it cure for 6 months for the best taste.

Methods of leavening bread:

Cream of tartar:

When you make your own grape juice, the tartar (or argol) will gather at the bottom after it has been sitting several months. Pulverize the flakes to make cream of tartar. The scientific term is potassium hydrogen tartrate.

Baking powder:

12½ ounces cream of tartar
5½ ounces bicarbonate of soda
3½ ounces cornstarch

Mix together very well and let it dry if necessary. When using baking powder put it in last and stir as little as possible so the bubbles will produce more rising in the bread. This is single-acting baking powder—double-acting contains aluminum. Cream of tartar is a by-product of making grape juice. Soda is too hard to make at home—it involves making ammonium salts, which are mixed with carbon dioxide, which is produced by burning lime, etc. Soda can be used without cream of tartar also.

Baking powder equivalents:

1 teaspoon double-acting baking powder (usually the kind you buy at the store) =

2 tablespoons homemade (single-acting) baking powder

½ teaspoon soda plus 1 cup sour milk or buttermilk

½ teaspoon soda plus 1 tablespoon vinegar plus 1 cup milk

½ teaspoon soda plus 1 tablespoon lemon juice plus 1 cup milk

Baking powder ratio:

Cake with eggs: 1 teaspoon baking powder per 1 cup of flour

Biscuits, muffins, and waffles: 2 teaspoons baking powder per 1 cup flour

Buckwheat and whole grain with no eggs: ¾ tablespoon baking powder per 1 cup flour

Wood ash:

Add a teaspoon of wood ash to an acidic food such as real sour cream, buttermilk from slightly sour milk or yogurt. It must be used very quickly because the bubbles don't last long.

Sourdough starter from yeast:

1. Find a large-mouthed container with a loose lid that can hold 5–6 cups. An old cookie jar with smooth insides or an old ice bucket works well. If it has a screw-on lid, poke a small hole in it. Never use a metal container or metal utensils to stir it!

2. Mix together 2 cups wheat flour, 1 cup water, 1 cup cooled potato water, 1 package bread yeast (or 1 tablespoon yeast). Mix well and set in a warm area until there are no more signs of fermentation. It will be separated and have no more bubbles.
3. Stir well, remove 1 cup of it, and stir in a cup of flour (rye, black rye, or semolina if you want—never "self-rising" flour though) and a cup of water.
4. You can use it after 2–3 days, but it will taste like sourdough if you leave it weeks or months.

Sourdough starter from the air:

Mix 1 cup flour with 1 cup water or unpasteurized milk in a non-metal container. If the milk is pasteurized, let it stand 24 hours first. Let the mixture sit on the counter 2–3 days covered with a light cloth. If it ever looks moldy (turns pinkish or gray) or smells like mold throw it out. If it foams a lot, it is good yeast and you can continue to feed it.

Caring for sourdough starter:

Leave it on the counter, and use it even if it gets flat. Don't put it in the fridge, but you can put it in the freezer in 1-cup portions. Sourdough can't spoil because it's fermented. Feed it once a week by adding a spoonful of water and a spoonful of flour.

Yeast:

Yeast is a living organism that must be stored in a cool place. For short term put it in the fridge, for long term put it in the freezer. Cultured yeast is the stuff you buy at the store, and wild yeast is what you get in sourdough starter. Use it sooner than 1 year, and test it before using by putting it in warm water with a little sugar. If it bubbles it is still good and your bread will rise.

How to make bread:

1. Put your yeast into a bowl. Mix your yeast with 80–90°F water, milk, fruit juice or potato juice (¼ cup water per 2½ teaspoons dry yeast). How much liquid determines how much bread you'll have—it will become 1½ times as big as your liquid. Mixing does not harm yeast.
2. Add ½ teaspoon sugar, fruit juice, honey, molasses, or maple syrup if you want. Sugar feeds the yeast, and they can find their own natural sugars but giving extra helps them. Liquid sugars make the bread moister.
3. Mix in melted shortening, either lard, butter, bacon grease, margarine, olive oil, vegetable oil, or mashed banana or avocado.
4. Salt is optional, but if you want to you can add ¼ teaspoon of salt per 2 cups of flour.
5. You can now add any other items, but they should make up no more than a quarter of the volume of the rest of the ingredients. Fruits, nuts, sauces, vegetables, etc. all add wonderful flavor to bread.
6. Add your flour. For light bread, ¾ of the flour should be wheat flour, because it contains gluten which will help your bread rise. For heavy bread, ½ the flour should be wheat. The other ¼ to ½ can be any other kind of grain flour that you want. There should be 3 measures of flour per 1 measure of liquid. If you are making biscuits or other drop batters, it should be 2 measures of flour per 1 measure of liquid. If you are using sourdough starter, only add half the flour right now, and you will add the other half later.

▲ **Mixing and kneading dough**

7. Knead the dough thoroughly until it is smooth. To knead, place the dough on a clean table covered in flour and push it with your hands, fold it over, push it, turn it, push it, fold it over, and on and on. Kneading is a natural thing that most people can do without much practice. If it is too sticky, add a bit more flour by dipping your hands in flour and kneading, but don't add too much. It should turn into a smooth rounded glob of dough after 15 minutes or so, but if it is too dry you can't add more water.
8. If you are using sourdough starter, set it in a warm place and let it rise at least a few hours or overnight, until the batter is light and yeasty. Then mix in the other half of the flour (this is called the sponge method). If you are using yeast, simply knead and let rise.
9. Take a big enough bowl that it will fit double the dough that you have. Lightly grease it and put the dough in. Roll the dough around so that it also gets some of the oil on it, and then cover with a clean cloth.
10. Heat the oven on low until it is warm inside, then turn it off and set the rising bowl inside. Or if you have a woodstove, set the bowl near to it. If it is summer and it is very hot, set it in a very warm place. Let it rise for about an hour or until it is near double the original size.
11. Poke the dough gently with two fingers. If the hole stays instead of springing back, it is finished rising. Punch the

▲ **Dough rising**

dough in the middle as far as it will go, then punch all over the top of the dough. Turn it over, cover, and let it rise again.

12. Prepare the bread pans by greasing them, then shape your dough into loaves. For a conventional loaf pan, fill the pan ½ full for light bread (white), and ¾ full for heavy bread. If you patch on pieces of dough because you misjudged, there will be lines or seams in the bread. You can also form rolls and treat them the same way.
13. Let the loaves rise one more time, until they are almost double in size. If the bubbles are big and it looks spongy, your bread may fall. If it seems too close to falling, punch it down and let it rise again.
14. Bake higher than 350°F for 30–60 minutes. The bread is done when the crust is brown and it sounds hollow when you tap on it. Pull it out and gently dump it out of the bread pan onto a cooling rack, and wrap with a dishtowel.
15. Serve right away, or store in a sealed plastic bag. Extra loaves can be frozen and will last a long time.

Extracting vegetable oils:

Any oil extracted from a plant is a vegetable oil. This includes almond oil, avocado oil, castor oil, coconut oil, hazelnut oil, olive oil, wheat germ oil, soy oil, sunflower oil, canola oil, peanut oil, sesame oil, etc.—any nut, grain, bean, seed or olive. The oil is 100 percent pure fat. Only sesame seeds and olives can be pressed without being heated. This is called "cold pressing" and is the most nutritious. Other foods need to be heated before pressing. There are several modern ways to extract the oil by pressing, either hydraulically or by expeller.

▲ **Freshly baked bread**

The other way is with a solvent, which is definitely bad for your health. Most store bought oil has been refined to the point that its age and taste are gone. Homemade oils are rich and flavorful, and it is easy to tell when they have gone rancid (store-bought and processed oils are more difficult to tell). When you purchase olive oil at the store it is labeled Olive Oil, Pure Olive Oil, Virgin or Extra-Virgin. Virgin means that no chemicals were used in extracting the oil. Regular olive oil has some virgin oil mixed in, but isn't very high on flavor and has had solvents in it. Extra-virgin has exceptional taste.

Home production of oil:

With olives it is just a matter of crushing the juice out and letting the oil rise to the top. With other foods first crush them and press the oil out, then boil the pulp in water and more oil will rise to the surface and can be skimmed off. There are no home-size presses for this so either a fruit press can be used or a homemade press can be devised for this purpose.

To make mother of vinegar:

Keep a mixture of half vinegar and half cider at 80°F for a few days. The thin scum on the top is mother of vinegar. Or you can do it the hard way and mix rain water with late-run maple sap. Add sugar-pot scrapings, sorghum settlings, discolored honey, and other sugar leftovers. Let it sit at a warm temperature and it should form a thick, hard covering over the surface of the sap.

To make vinegar the hard way:

1. Take sweet apple cider, uncooked with no preservatives, and fill a gallon glass jug to the neck.
2. Buy an airlock for the jug, or make one from a corncob. Take a piece of corncob that will fit in the jug's neck and burn or punch out the middle. Insert a grape vine or sumac into the cob lengthwise. Put the end of a piece of rubber tubing over the grape or sumac and put the other end in a jar of water. As the juice ferments the carbon dioxide will come out the tube and bubble the water.
3. At room temperature you should wait 4–6 weeks for the cider to ferment. When the bubbling stops, pour half the cider into another jug. Then add mother of vinegar to each by putting a little on a dry corncob and floating it inside, or add already-made vinegar to it (1 part vinegar to 4 parts cider). Cover the jugs by tying a cloth on over it.
4. Try to keep the jugs at 70–80°F. Ordinarily it will take 3–9 months. When it is vinegar, dilute it before you use it.
5. To store it, strain the vinegar through cheesecloth and store in a cool dry place in bottles.

▾ **Apple cider vinegar**

To test your vinegar to make sure it is strong enough (titration):

1. Mix a small amount of baking soda in water in a small jar. There should be enough soda that some of it settles to the bottom.
2. Steam a head of cabbage in a little water and keep the juice (make sure it is very purple liquid). Pour this juice in another jar.
3. Add a few ounces of water to two drinking glasses, making sure they're equal to each other. Using an eyedropper, add enough cabbage juice to each of the glasses to make them purple (put the same amount in each).
5. Rinse the dropper, and then put seven drops of five-grain store-bought vinegar into one of the glasses of purple water.
6. Rinse the dropper and add seven drops of the homemade vinegar to the other glass.
7. Rinse the dropper, and then add 20 drops of baking soda water to the

◂ **Herbal vinegar**

store-bought vinegar mixture. This will turn the water blue.

8. Add baking soda one drop at a time to the homemade vinegar mixture, counting each drop, until it is the same color blue as the store-bought one.
9. To calculate the acidity, divide the number of drops you used by four. Thirty drops divided by 4 = 7.5% acidity.

To dilute the vinegar:

The following formula uses 7.5 as the example homemade vinegar acidity. Most recipes use 5 percent acidity, so that's probably what you'll dilute it to.

1. Subtract 5 from 7.5 (your current acidity) to find the difference: 7.5 - 5 = 2.5

2. Multiply the answer by your total amount of vinegar (in ounces). 1 quart = 32 oz.

32 x 2.5 = 80

3. Divide the answer by 5. 80 / 5 = 16

4. The answer is how much water you add to the vinegar. So in this case you would add 16 ounces of water to it.

To pasteurize the vinegar:

Pasteurizing prevents you from using the vinegar as a starter, but it also keeps it from being cloudy because no mother forms. Put the bottles (loosely corked or unsealed) into a pan filled with cold water. Heat the water gradually until the vinegar is about 145°F and keep it there 30 minutes, then cool.

Herbal vinegar:

Collect some aromatic herbs. You can use the leaves, stalks, flowers, fruits, roots, and even nuts. Cut them up very finely, and fill any jar up to a thumb's width from the top. Pour room temperature vinegar into the jar up to the top—apple cider vinegar is best. Cover it with a plastic lid, several layers of plastic wrap or wax paper held on with a rubber band, or a cork. Don't use metal as the vinegar will corrode it. Label it with the name and date and in six weeks it will be ready to use. The fun part about this is that not only is the vinegar useful, it can make a great gift because the jar can be beautiful. To use, pour on beans and grains as a condiment, in salad dressing, in cooked greens, stir fry, soup, or any recipe that calls for vinegar. You could even use it for cleaning. After 6 weeks you can decant the vinegar into a better jar so it does not get stronger.

The best tasting herbs:

Apple mint (*Mentha sp.*) leaves, stalks

Bee balm (*Monarda didyma*) flowers, leaves, stalks

Bergamot (*Monarda sp.*) flowers, leaves, stalks

Burdock (*Arctium lappa*) roots

Catnip (*Nepeta cataria*) leaves, stalks

Chicory (*Cichorium intybus*) leaves, roots

Chives and especially chive blossoms

Dandelion (*Taraxacum off.*) flower buds, leaves, roots

Dill (*Anethum graveolens*) herb, seeds

Fennel (*Foeniculum vulgare*) herb, seeds

Garlic (*Allium sativum*) bulbs, greens, flowers

Garlic mustard (*Alliaria officinalis*) leaves and roots

Goldenrod (*Solidago sp.*) flowers

Ginger (*Zingiber off.*) and wild ginger (*Asarum canadensis*) roots

Lavender (*Lavendula sp.*) flowers, leaves

Mugwort (*Artemisia vulgaris*) new growth leaves and roots

Orange mint (*Mentha sp.*) leaves, stalks

Orange peel, organic only

Peppermint (*Mentha piperata* and etc.) leaves, stalks

Perilla (Shiso) (Agastache) leaves, stalks

Rosemary (*Rosmarinus off.*) leaves, stalks

Spearmint (*Mentha spicata*) leaves, stalks

Thyme (*Thymus sp.*) leaves, stalks

White pine (*Pinus strobus*) needles

Yarrow (*Achilllea millifolium*) flowers and leaves

Herbs for use as a calcium supplement (take vinegar 2–4 tablespoons daily):

Amaranth (*Amaranthus retroflexus*) leaves

Cabbage leaves

Chickweed (*Stellaria media*) whole herb

Comfrey (Symphytum officinalis) leaves

Cronewort/mugwort (*Artemisia vulgaris*) young leaves

Dandelion (Taraxacum off.) leaves and root

Kale leaves

Lambsquarter (*Chenopodium album*) leaves

Mallow (*Malva neglecta*) leaves

Mint leaves of all sorts, especially sage, motherwort, lemon balm, lavender, peppermint

Nettle (*Urtica dioica*) leaves

Parsley (*Petroselinum sativum*) leaves

Plantain (*Plantago majus*) leaves

Raspberry (*Rubus species*) leaves

Red clover (*Trifolium pratense*) blossoms

Violet (*Viola odorata*) leaves

Yellow dock (Rumex crispus and other species) roots

What is pectin?

Pectin is used to make jelly or jam. It turns your juice into a thicker substance. Many people buy Certo Fruit Pectin in packets and add it to the jam, but you don't need to do that if you have apples.

To make pectin from apples:

1. Pick lots of really tart apples early in the season. They should be as crisp, sour, and small as possible. The best varieties are Northern Spy or Russel, and wild apples are perfect. MacIntoshes work if they are very small, but you definitely want to find apples that are the opposite of sweet, mealy Red Delicious apples. The reason small is better is there will be more core per apple.
2. Cut the whole apples into 1 inch pieces, leaving the core and the skin. Throw out the stems and leaves, and wash the rest.
3. Put the pieces into a large sauce pan or kettle (not aluminum). Add 1 pint water per 1 pound of apples. Bring to a rapid boil for 15 minutes, then cover and turn off the heat.
4. Pour the juice through one thickness of cheesecloth—don't squeeze the pulp. Some people use a Foley food mill to sieve it. Then put it back in the pot and add 1 pint water per pound.
5. Simmer on low heat for 15 minutes, turn off, and strain. Then let cool for 15–30 minutes or until barely warm. Squeeze the pulp to get the last juice out. This is your pectin.

How to make jam:

1. To use the natural pectin in the fruit, first clean the fruit and taste it for tartness. If it is not very tart, add some lemon juice for extra acid.
2. Clean the fruit, mash it through a jelly bag. To make a jelly bag, take unbleached muslin and soak it. Wring out it and attach it firmly to a bowl and mash the fruit through it into the bowl. If you want clear jelly don't squeeze the bag.
3. Add water to the fruit and put it in a saucepan. Bring to boil and simmer until fruit is soft, stirring occasionally.
4. You can test to see if the jam is ready to set by removing a small amount of it and putting it on a plate. Stick it in the

Pour hot jam into hot jars

freezer for a few minutes. If it gels, it is ready.

4. Sterilize the jars as you would for canning, then put the jam in hot and seal them. The recommended processing time is 5 minutes at sea level, 10 minutes at 1,000 feet, and 15 minutes above 6,000 feet.

Good ingredients for jam:

apple
apricot
blackberry and raspberry
blueberry and cranberry
cherry
citrus
grape
peach
pear
quince
strawberry

Apple jam recipe:

1. Prepare the apples as you would to make pectin. Cut up enough apples to make 5 pounds, add 1 cup of water, and simmer (you may have to add more or less water for a better simmer).
2. Sit frequently until soft. Sieve the pulp in cheesecloth or the Foley food mill.
3. In your jelly pot measure in 10 cups of pulp and 2 ½ cups of sugar.
4. Cook the pulp quickly over medium-high heat (220°F), stirring constantly. When it is done it should be boiling fast, will have a sheen to it and will have reduced in volume. This is tricky . . . if you stop the boiling too soon it will not set, but too late it will be rubbery.
5. Fill your canning jars and seal them. This is supposed to make 11 cups of jam.

How to harvest maple syrup:

1. Before the first thaw in March, chop down a maple tree and hollow out the log into a canoe.
2. Make white birch–bark basins for each tree to hold the sap, or you can use a bucket with a handle.
3. In March or April, when it freezes at night and thaws in the daytime, test the maple trees to see if the sap is flowing, by cutting a small gash in it on the side with the most or the biggest limbs. The best trees are sugar, black, red, or Norway maples with new growth. Some will be running and some won't.
4. When a tree is running, set a birch basin under the gash and drive a hardwood chip deep into the gash. The sap will drip from each corner of the chip and down into the basin. Rabbits and squirrels will try to eat the buckets so guard them. If you use buckets, you will have to make a hollowed-out wooden tap and hang the buckets from the tap. There are also buckets and taps sold that are designed for this purpose.
5. The sap is best when there are icicles from the chip, there is a west wind, and it is sunny. Some trees are better than others, and some years the sap is worse. You will probably have to collect the sap in the morning and afternoon. Pour the basins into a bucket (if you are not already using buckets), or use a sap yoke, then dump it in the canoes. Each tree will yield 6–12 gallons, but it takes 20–35 gallons to make one gallon of syrup.
6. Make a long fire and hang a row of kettles over it (or you can do it one pot at a time). Take some sap from the canoes and put it in the kettles. Watch the kettles so the fire is constant and the syrup doesn't boil over. You will have to stir the syrup with long-handled spoons. As it boils down, add more syrup to the pots.
7. The sap is usually done when its temperature is 7 degrees hotter the boiling point of water. Use a candy thermometer to measure this. Don't let it scorch or it will ruin the syrup.
8. When it is syrup, use a wooden paddle to pour some of it on the snow to cool it, and eat it. If it is as sweet as you want, it is done. Pour some of the syrup into molds of birch, hollow canoes, or

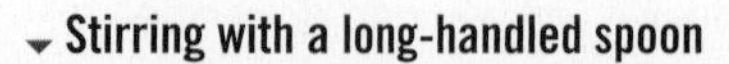
▲ Kettles of syrup

▼ Stirring with a long-handled spoon

reeds to make hard blocks or hard candy. Use the canoes as canoes when done.

9. Stop tapping when the snow melts, the ground thaws, and buds start to swell, or your sap will not taste good. You can make sugar by pounding your hardened syrup to make it fine.

How do I store and care for an ax?

When you leave an ax outside, find a log lying along

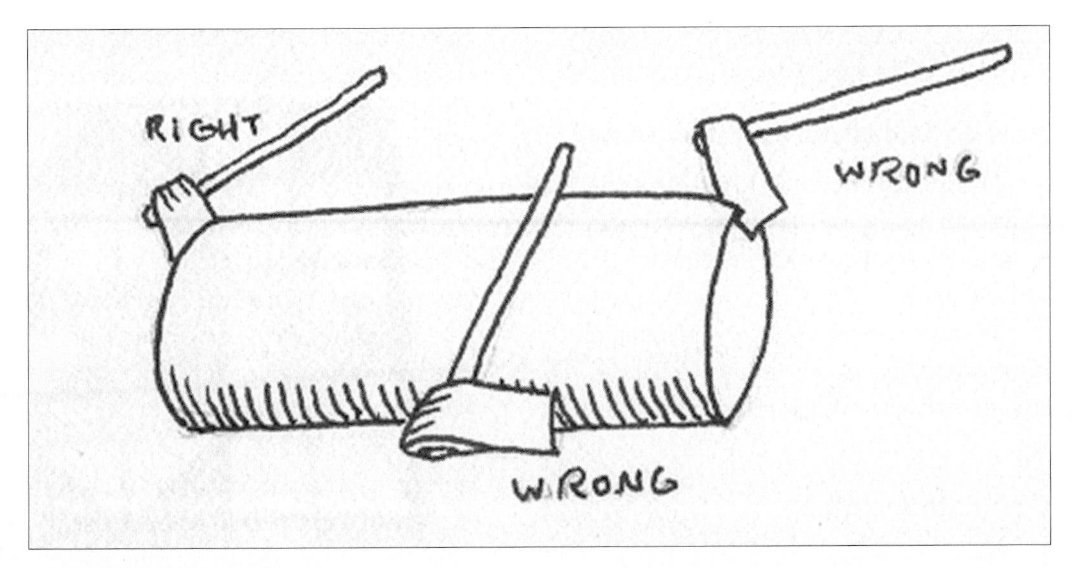

▲ Storing an ax

its side, and drive the ax into one end so that the handle is over the log. Or you can lean the blade against the wall. In very cold weather heat the ax to make it less brittle. When you are done for the day, rub it carefully with grease. When using a grindstone, roll the edge of the ax on the

▼ An old grindstone

grindstone, instead of holding it steady. If you try to hold it steady, you won't be able to.

How do I sharpen or grind an ax?

Ax for splitting wood: Make the edge sharp, but don't grind the sides flat. Instead leave a bulge so that it will split the wood while it cuts it.

Broadax: Grind it only on the side that is away from the wood, so that it will travel straight down. Leave the edge to the log flat.

Grindstone rules:

1. Never leave your grindstone in the trough—it softens the stone.
2. Don't leave the stone outside, unless you have to. When you do have to, put a board over the top to shelter it.
3. To clean the grindstone grind a piece of ice on it.

How do I care for a bucksaw?

Never leave the wire tight while you're not using it or the weather will snap your blade. The teeth on a bucksaw are not straight—they alternate angling to the right and left. There also is a "raker" tooth that cleans the sawdust out of the way as you saw. There is a tool called a blade setter that is placed over one tooth at a time. First you set all the teeth going one direction, then flip the saw over and set all the teeth that go the other direction.

How do I sharpen a chainsaw?

1. Ensure that you can see what you are doing. Use a stand or tree to arrange the bar (the long piece that the chain is on) vertically. Sharpen your saw every 1–2 times you fill your gas tank.
2. Adjust the tension of the chain so the teeth won't be loose, and then put gloves on. Make sure the switch is off and it is unplugged. Mark with chalk where you start.
3. Remember when sharpening to only file when pushing forward—don't drag the file backwards across the tooth. After each tooth is done dust it off on your pants or with a cloth. Don't use a regular round file—always use a chainsaw-sharpening file, which looks similar but is different.
4. On a new chain there is a diagonal line across the rear part of the top outside surface of each tooth. If yours doesn't have this, draw a line across the back of whatever you have leaned your saw up against, to help guide your angle as your file.
5. Hold the top of the bar with your left hand to brace it, and file with your right. Each left-hand tooth is shaped like a hook with a flat top. The round file goes into the hook and is drawn up and out to sharpen the corner between the hook and the flat top. To do this, draw the file in a straight line parallel to the guide (either on the saw, or the line you drew) and pull slightly towards you (and the cutting edge) so you are sharpening the edge and not the bottom of the hook (or throat).

Sharpening a chainsaw

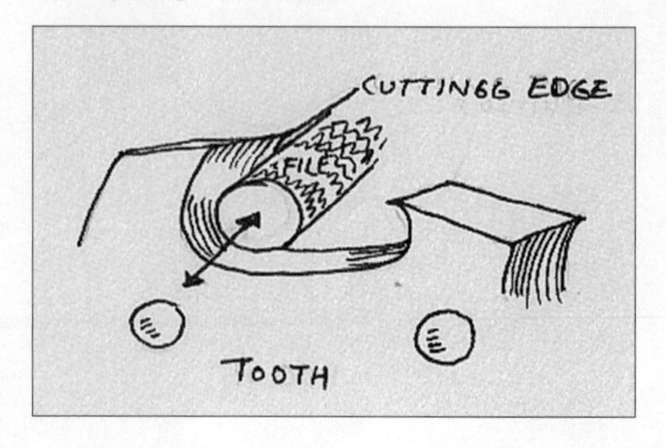

6. Sharpen the tooth 4–5 strokes until it is sharp enough to peel a bit off your fingernail. Repeat for each tooth all the way around to the chalk mark you made.
7. Switch hands so that your right is bracing the bar and your left hand holds the file. Now file the right-hand teeth the same way you filed the left-hand. The right-hand teeth are smaller than the left ones, but work the same way.
8. File the same number of times on each tooth, and only draw the file one direction—out. If some teeth are shorter use fewer strokes on those and more on the others. If the teeth are rock damaged, completely file off the point with 10–12 strokes being careful to keep the right angle, and then sharpen if necessary.

How do I set the teeth on a regular saw?

Use a hammer and an iron with a beveled edge. Lay the saw on the iron, and then hit the first tooth with your hammer, then the third, then the fifth, and so on. Then flip the saw over and hit the second tooth, the fourth, and so on. If you have a small saw, use a smaller tool.

How do I sharpen a chisel?

Sharpen a chisel by putting it in a jig and grinding against the grindstone. After sharpening, you can hone it with a piece of petrified wood.

Pocket knife rules:

1. Keep it clean, dry, and sharp always.
2. Don't use it on stuff that will dull it or break it.

The essential hand tools

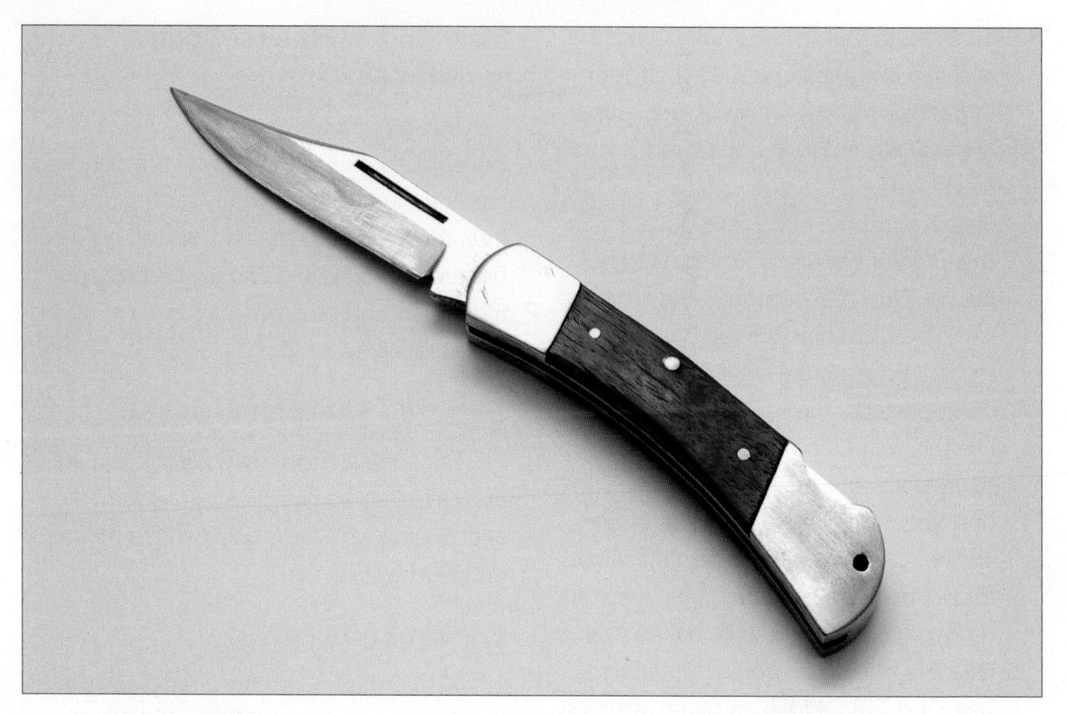

▲ **A versatile pocket knife**

3. Don't put it on the ground.
4. Don't put it in the fire.
5. Wipe the blade clean and close it after using it.
6. Oil the joints sometimes.

How do I sharpen a knife?

Run the side of the blade along a sharpening stone towards you as if you were cutting the stone. Then do it on the other side. Keep switching sides until the edge is sharp, and then wipe the blade.

How do I grind a scythe?

You have to be very careful when sharpening a scythe. Hold it diagonally over the grindstone, moving it in one direction: from tip to heel diagonally. If you don't do it this way it won't cut. A scythe has a ridge on one side, so you can follow it. You can also do this by hand with a corncob-shaped stone, but be very careful.

How do I wash clothes by hand?

1. The reason clothes used to be washed once a week (on Monday) was the enormous amount of work it takes. With this method it is only practical to change your clothes once a week.
2. Over a large fire or on the woodstove heat two large kettles of water to a boil. While it is heating separate your colors and whites and scrub on the bad stains with your hands or stiff brush.
3. Soften the water of one kettle by adding a small amount of lye or ashes. You can skip this step if you are washing with rainwater.
4. In the other kettle put the clothes in to boil for about five minutes, stirring frequently.

5. Pour the rainwater or lye-softened water into the washtub and put the boiled clothes in with it.
6. Use soft soap or strong lye soap to scrub the clothes with a washboard.
7. When washing quilts or heavy bedding, put them instead into a barrel and stomp on them with bare feet instead of scrubbing by hand.
8. Rinse (see next section).

What hand-washing devices are available?

There are two main non-electric hand-washing devices available today. There is a Wonder Washer which is a hand-turned sphere but which only holds about 2 pounds, and there is the James Washer which can hold 16 gallons of water and do much more laundry. You would scrub the clothes with the devices, and then use the regular rinsing techniques (see below).

Rinsing:

1. Use three large laundry tubs or other tubs that hold 20–30 gallons. You can also set these on a small table to bring them to a nice height (about 28 inches) so you don't have to bend over.
2. Get two toilet plungers (plumber's helpers), brand new and a different color than your regular toilet plunger

(or make a stick with a similar shape on the end).

3. Fill one laundry tub with cold water. As the clothing comes out of the washing device squeeze out a little of the soap and put it into the cold-water tub to soak.
4. When you are done scrubbing one whole load (such as all your colors), use a plunger to plunge each piece of clothing until the soap is mostly out.
5. Hang a wooden bar between two trees or some other support. You can cover it with a cotton cloth to keep clothes clean if you want. Loop one article of clothing around it, and then twist it to wring it out. Use an empty tub to catch the water so you don't waste it.
6. Fill the last tub with half cold water, half boiling water, and 4 teaspoons of salt if it is wintertime. The salt will keep the clothes from freezing to clotheslines. Put the wrung-out clothes into that tub and plunge some more.
7. Wring out the clothes and hang them to dry.
8. You will have to change the water and wipe out the tubs if your clothes are really dirty or if you do many loads, depending on the load size about every three loads.

How do I conserve water and energy during the laundry process?

Four regular loads of laundry will use about 55 gallons of water to wash and rinse, and will probably take 2½ hours. This is why the empty washtub for recovering water is important. If you use this water and put it into your graywater recycling system it can be used for the garden.

6 | Health and Family

BEAUTY AND HEALTH

Plainness has its peculiar temptations quite as much as beauty.

—*George Eliot*

How do I routinely care for my skin?

The recommended skin care routine consists of cleansing, toning, and moisturizing, in that order. For cleansing, homemade soap can be used. For toning, acne treatments can be used, or a mild juice such as cucumber. For moisturizing, homemade lotion, olive oil, or honey may be used. It is very important to drink at least eight glasses of water a day, and eat fruit, tomatoes, whole grain bread, oatmeal, vegetable oils, carrots and broccoli for your skin. Eating salt, spinach, and shellfish can cause acne—and don't worry about chocolate, it won't.

How do I care for homemade skin care products?

Most homemade herbal mixtures have a shelf life of 2-3 months if stored in a cool, dark place. Mixtures with fruits in them do not last as long and should be thrown out after a week. To preserve longer, add .5–5% grapefruit seed extract to the mixture. Glycerin and rosewater also act as a preservative. Also remember that essential oils and herbal ingredients can be very powerful—watch out for allergies.

Hair and scalp moisturizer:

Shampoo your hair. Then cut an avocado in half and scoop out the pulp. Mash it up in a bowl and add a little water if necessary. Smooth it over your face and rub it into your scalp. Rub the inside of the avocado peel over any dry spots on your body. Wait half an hour, and then take a shower.

Soybean moisturizer for dry hair:

Mix 1 teaspoon soybean oil with 2 teaspoons castor oil. Warm on a low heat. Massage into scalp and hair, and wrap your hair in a hot towel for 15 minutes. Shampoo and rinse.

Control dandruff:

Mix ½ cup water with ½ cup white vinegar. Apply directly to the scalp before shampooing twice a week.

Soft water shampoo:

This is the same as fabric softener. Put all your cooking ashes into a large barrel and fill the barrel with rainwater. After about an hour the ashes will settle to the bottom and the water on top will become clear and soft. Scoop this water out and use it to wash your hair (use some homemade soap with it).

Dry shampoo:

Mix 2 tablespoons corn meal, 1 tablespoon ground almonds and 1 ½ tablespoons orrisroot. Massage 1 teaspoon into your scalp and brush through hair, then repeat. Rinse out if you want.

Honey locust shampoo:

Boil the fruit of the honey locust in a large pot of soft water. Allow the liquid to cool and either put it in a bottle for washing with later, or dip your hair into the pot and scrub your hair. Then rinse if you would like. Some women leave it in, as it smells very nice.

Conditioner:

Mix ½ cup olive oil, ½ cup vegetable oil, and ½ cup honey in a pot and heat until it just starts boiling, and then remove quickly from heat. Let it cool, then put in a spray bottle. To use, spray ends of hair, then wrap your head in a warm, wet towel for an hour. Then rinse and dry. Store in a cool dry place.

Rosemary Conditioner:

Mix ½ cup dried rosemary leaves and ½ cup olive oil. Heat until warm, then strain. Work it into your scalp and all the way to the ends of your hair. Wrap your hair in plastic wrap and a towel for 15 minutes. Wash and rinse twice to remove the oil. You can use this twice a month.

Chamomile hair shine:

Make one cup of chamomile tea and set aside to cool. Shampoo your hair and lightly condition, and then pour the tea over your hair. Work into the hair from scalp to ends and rinse quickly under cool water. Towel dry your hair gently.

Hair gel:

Put 2 tablespoons flax seeds and 1 cup water in a pot and bring to boil. Take from heat and let sit 15-20 minutes. Strain and allow to cool, and then add a few drops of essential oil if you want. Place in a glass container and store.

Hair herbs:

You can mix these herbs with oil or your regular rinse, or you can make them into a tea and work it into your hair.

Burdock: root prevents dandruff

Catmint: leaves encourage hair growth and sooth scalp

Flannel mullein: lightens hair

Goosegrass: tonic and cleansing, helps prevent dandruff

Henna: red hair dye and conditioner

Horsetail: non-fertile stems and branches strengthen the hair

Lavender: antiseptic, antibiotic, stimulates hair growth, degreaser

Lime: flowers clean and soften

Marigold: lightens hair color

Nasturtium: stimulates hair growth

Parsley: enriches hair color and gives a shine

Rosemary: tonic and conditioner, shines and adds body and darkens hair

Rhubarb: yellow hair dye

Sage: tonic and conditioner, darkens hair

Southernwood: encourages hair growth and prevents dandruff

Stinging nettle: tonic and conditioner, prevents dandruff

Witch hazel: leaves and bark are astringent and cleanse oily hair

Lip gloss:

1 tablespoon shredded beeswax

1 tablespoon essential oil of your choice

Beet powder, raspberry, blackberry juice for coloring

Mix beeswax, oil and coloring and heat to 160°. Pour into containers while hot and let it set until completely cool.

How do I make aftershave?

2 cups rubbing alcohol

1 tablespoon glycerin

1 tablespoon dried lavender

1 teaspoon dried rosemary

1 teaspoon ground cloves

Mix in a bowl and put into a jar. Cover and refrigerate 3-4 days, shaking occasionally to mix it up again. Then strain the liquid out and keep it refrigerated. It will keep 1-2 months.

How do I make leg wax?

2 cups sugar

¼ cup lemon juice

¼ cup water

Mix together and boil about 10-15 minutes until 250°F or until it forms a soft ball. Pour into jars. This is supposed to be similar to 'Moom Wax', but unlike wax it can be washed off. This makes about three jars. Heat before using in the microwave for 10-20 seconds, or heat by putting the jar in a bowl of very hot water. I recommend the water method or you can really burn yourself. Powder the area to be waxed with baby powder or other homemade powder, the spread it on in the direction of the hair growth. Place a piece of fabric on it and press it down, and the pull it off very quickly AGAINST the direction of hair growth. If the wax is not hot enough it won't work. To make a thicker mixture you can also add ¼ cup vegetable glycerin. When you are done wash off with warm water and put lotion on. Don't use anywhere around your eyes.

Herbal facemask:

Mix 1 tablespoon honey, 1 egg, 1 teaspoon dried chamomile flowers, 1 teaspoon finely chopped fresh mint in a small bowl. Apply to face and neck and let dry 10-15 minutes. Rinse off with warm water.

How do I moisturize regular skin?

Moisten your face with water and cover it with slightly warmed honey. Or mix a cup of lavender, lemon verbena, violet, honeysuckle, mint, carnation or geranium leaves or petals in a pint of boiling water in a glass or stainless steel saucepan. Simmer for 2 minutes, and then add 1 cup of cider vinegar. Store in a clean jar and tightly seal with a lid. Wait two weeks, and then strain through a clean cloth.

How do I moisturize normal to dry skin?

Moisturize with cucumber, strawberry, bell pepper, and grape or cabbage juice.

How do I moisturize oily skin?

Moisturize with grapefruit juice.

How do I moisturize dry skin?

Dilute the juice of borage leaves in an equal volume of water to make lotion, or apply fresh aloe gel. Or dilute the juice of borage leaves in an equal volume of water to make lotion. If you are allergic to most things, you can apply a small about of olive oil with your fingertips, massage it into the skin, and then put a damp hot face cloth on until it cools. Then rinse with cool water.

Almond and rose lotion:

¾ cup apricot or almond oil

⅓ cup coconut oil or cocoa butter

1 teaspoon lanolin

½ ounce grated beeswax

⅔ cup rosewater

⅓ cup aloe vera gel

1–2 drops rose oil

Melt almond oil, coconut oil, lanolin and beeswax over low heat and cool to room temperature. Mix in rosewater, aloe vera, and rose oil and whip until it is creamy. Store in a covered jar.

Orange lotion:

½ ounce melted cocoa butter

1 ounce warm olive oil

1 ounce orange juice

2 drops essential oil (orange flower is ideal)

Mix together until light and fluffy. If it separates mix together again. Keep in the fridge.

Oatmeal lotion:

½ cup juice of almonds

1 ½ cup oat flour soaked in water

1 ½ cup baking soda

½ cup beeswax

Warm beeswax and mix all together until creamy. Keep in a jar and store in the fridge.

How do I avoid acne?

Acne, irritations, and pimples can all be prevented through a regular skin care routine and a healthy diet. Also avoid touching your face with your hands, and avoid spicy food, fabrics or plants you are allergic to, or caffeine. Drink lots of water.

What herbal remedies can I use for acne and other irritations?

Angelica: Apply a cream of angelica leaves for skin irritations.

Apple: Mix 1 grated medium apple with 5 tablespoons of honey, smooth over skin and leave it there for 10 minutes. Rinse off with cool water.

Burdock: Boil 1 quart of water, reduce to simmer and add 4 teaspoons dried burdock root cut into small pieces. Cover and let simmer 7 minutes, remove from heat and allow to steep for 2 hours. Use as a wash.

Cabbage and witch hazel: Mix 250 g fresh cabbage leaves and 1 cup distilled witch hazel in a blender, strain, and add two drops of lemon juice oil. Use night and morning for acne.

Chicory: Chop finely and apply directly to irritated skin.

Garlic: Rub a fresh clove of garlic on the area.

Lavender: Mix a drop of lavender oil with a drop of tea tree oil and apply to blemish.

Marshmallow: Apply a poultice of the root or a paste of the powdered root of marshmallow mixed with water for skin inflammations.

Milkweed: Wash warts in milkweed juice.

Oak: Boil oak bark in water to make strong tea and use to wash blisters.

Pot marigold: Apply cream made from petals of pot marigold for inflammation, dry skin, and sunburns.

What types of allergens are there?

Animals: People are not just allergic to fur, they are also allergic to skin oils, dander, saliva, and urine which have proteins that float in the air causing sneezing, coughing, itchy eyes, etc.

Chamomile: If you are allergic to pollen, you will also be allergic to chamomile. Many soaps and teas contain chamomile and they can cause reactions such as itchy rash.

Chemicals: Chemicals don't create a true allergic reaction but many people have similar symptoms from paints, carpets, plastics, perfumes, smoke, etc.

Dust mite: A microscopic organism that lives in the dust in your house, and creates the same symptoms as pollen and asthma.

Lanolin: A common skin-contact allergen, it is the natural oil found in sheep's wool.

Pollen: People with pollen allergies often develop all-year allergies such as dust mites, are also allergic to chamomile, and get itchy eyes, runny nose, sneezing and coughing.

Rue: Can cause skin irritation if it contacts your skin.

How do I avoid allergies?

Pollen: Pollen is hard to avoid, but be careful in the morning or in a windy place to minimize reactions.

Mold: Keep lawns mowed and leaves raked, and clean mold from the house. Wear a dust mask, and use a dehumidifier in most places in the home.

Dust mite: Remove carpets, blinds, down blankets and pillows, forced air heat, dogs, cats, closets, anything that gathers dust. Vacuum often with a high-efficiency particle air (HEPA) filter. Dust surfaces with a damp cloth, and wash clothes with water hotter than 130°F.

How do I treat allergies?

Avoidance is the best way, because most anti-allergy drugs cause numerous side effects. Use air filters in cooling or heating units, or buy a stand-alone air filter. Your vacuum can have a significant effect on your allergies by either aggravating them (if the filtration is poor) or really helping them (if you have a HEPA certified vacuum). Children raised on farms and in the outdoors with exposure to animals often have fewer allergies than other children.

What is the food pyramid?

Every day a human being needs 10 servings of bread, cereal, rice or pasta, 8 servings of fruits and vegetables, and 4 servings of dairy and meat. The dairy and meat can be substituted by eating a good vegetarian protein diet, but the bread and fruit and vegetables are important no matter what. The serving size for each is about the size of your fist.

What kinds of vegetarians are there?

Raw food vegetarian: These people only eat raw plants that grow above the ground, especially fruits and nuts.

Vegan vegetarian: They keep a fairly strict diet that contains absolutely no animal products.

Ovo-lacto vegetarian: They don't eat meat products, but they do eat non-killing animal products such as eggs and milk.

Meat-eating vegetarians: Although this may seem contradictory, they are mostly vegetarian but may allow fish and poultry.

Vegetarian food pyramid

Vegetarian (and healthy) food substitutes:

Boiling: steam, bake or sauté in water or tomato juice

Bouillon: tamari and miso

Calcium: apricot, beans, broccoli, carob, collard, cornbread, dandelion, kale, lentils, mustard greens, navy beans, nuts, pinto beans, prunes, raisins, sesame seeds, soybeans, tahini, tofu

Fats and oils: banana, avocado, nuts, nut oil, tahini

Milk or egg for holding recipe together: bread crumbs soaked in rice or soymilk

Meat: tofu seasoned with tamari, water, sage, basil, and oregano

Marinade: lemon juice or tamari

Salad dressing: lemon juice, tamari and herbs

Milk: soy, rice, almond milk, or for cereal cook in apple juice

Mayonnaise: lemon juice in nut butter

Healthy food pyramid ▸

White sauce: fry whole wheat flour dry until toasted, remove from heat, blend with water then warm and stir until thickened, season as desired

Sugar: substitute 1 cup frozen apple juice concentrate for 1 ½ cup sugar, or use honey

Casserole cheese: bread crumbs, herbs and tamari

Salt: tamari and miso, or herbs

Benefits of vegetarianism:

1. Vegetarians don't kill animals.
2. It doesn't promote inhumane animal raising practices.
3. There are no weird hormones in the food.
4. There are less bacteria and diseases to combat.
5. It is healthier for people if they are careful in how they are eating.
6. It doesn't waste valuable land with large grazing fields.

Eating a healthy omnivore diet:

Some modern diets advocate eating tons and tons of protein (i.e. meat). Some people advocate total vegetarianism. Most people eat some meat and some vegetables. There are some benefits to eating meat. It is more difficult with a vegetarian diet to get enough iron, and very difficult for a vegan to get enough vitamins B12 and D. Many vegetarians have lower mineral content in their blood and need to take a mineral supplement. Doctors recommend that you avoid red meats, and stick to poultry and fish, which can help lower cholesterol and decrease your chances of colon cancer. Meat eaters tend to eat fewer vegetables and eat more sodium, so the best diet may be "meat-eating vegetarianism."

What are the unhealthiest foods to eat?

White flour and white flour products, mustard, pepper, vinegar, salt, condiments, salted and canned meats, gravy, fried and greasy food, pastry, very hot or cold food, cane sugar and processed sugar products, pickles, caffeine, brandied fruits, and white rice.

How much water should I drink?

Everyone needs 6–8 glasses of water a day. A glass does not mean a tall, huge glass; it means a small glass (but not a shot glass). Water cleanses your body, clears your skin, and prevents infection.

HERBAL REMEDIES

Staying inside the house breeds a sort of insanity always. Every house is in this sense a hospital. A night and a forenoon is as much confinement to those wards as I can stand—and then I must go outdoors.

~Henry David Thoreau's journal, 1856

Extracting essential oils from plants through distillation:

Warning!!! Never boil herbal remedies in metal pots or mix them with drugs . . . doing so changes the chemistry of the herb and can make it dangerous. The proper method is to pour the boiling water over the herbs as you would tea.

1. Use fresh or dried herbs and put it in some kind of perforated container, such as a sieve that will not let all the herbs drop through.
2. Fill a pot with only enough water that when it boils it will not reach the herbs in the sieve. Put the sieve into the pot and turn on the heat. If you have a still (for water purification or for ethanol) then use it to siphon off the steam. Otherwise cover the pot with a lid.
3. Boil the water steadily so that the boiling water does not touch the herbs.
4. If you used a lid to collect the steam, remove the lid and collect the liquid off of it. The best method is with a still. The steam carrying the oils will be deposited into a container and the oil will rise to the top of the water. This oil is the essential oil.

Cold pressing essential oils from citrus:

1. Roll the peel over something sharp to break it up, or grate the peel.
2. Press the peel with something heavy, such as a fruit press or your own method. Put the oils and juice into a container.
3. Let the oil rise to the top of the liquid and separate the two.

Making an infusion:

Making an infusion is the same as making tea. Bring water to a boil and pour over the herbs, usually 1 cup water per 1–2 teaspoons of herb. Let it steep for 15 minutes, and then strain it into a glass jar. It will keep 3 days.

Making a decoction:

Put 1 or more ounces of the herb into 1 quart of water and cover with a lid. Bring the mixture to a boil and simmer for two minutes. Steep for 15 minutes, then strain.

Making an ointment or salve:

Mix a decoction of the herb with olive oil and simmer until the water has completely evaporated. Add enough beeswax to give it the firmer consistency that you want. Then add a drop of tincture of benzoin per ounce of fat to preserve the ointment.

Making a non-alcoholic tincture:

1. A tincture is longer lasting than decoctions or infusions. Put the amount of herb that you need into a glass jar.

2. Most tinctures are made with 80–100 proof vodka or rum. However, you can use vinegar. Slowly pour in enough vinegar to entirely cover the herbs, plus 1–2 inches.
3. Seal the jar tightly so that no liquid can leak or evaporate, and put it in a dark place or paper bag.
4. Shake the jar every day for a couple of weeks. Then pour the tincture through cheesecloth into another glass jar (dark colored is better). Squeeze the herbs to get the last liquid out.
5. Close the lid tightly and label. It can last up to 2 years. The average dosage of tincture is 1 teaspoon, 1–3 times a day, diluted in tea, juice or water.

Herbs

The following is a list of common medicinal herbs. Some of these may be found in the wild, but some may be endangered. There are also many similar-looking plants, so it is best to grow them yourself, choosing the seeds based on the Latin names, or buy them from a reputable source. Don't go looking for them in the wild unless it is an emergency.

Aloe (***Aloe vera L.***)
Angelica (***Angelica atropurpurea***)
Agrimony (***Agrimonia parviflora***)
Barberry (***Berberis vulgaris***)
Bee balm/bergamot (***Monarda didyma***)
Black elder/elder/elderberry/elderflower (***Sambucus nigra***)
Borage (***Borago officinalis***)
Burdock (***Arctium lappa***)
Chamomile (***Matricaria chamomilla***)
Chickweed (***Stellaria media***)
Chicory (***Cichorium intybus***)
Eucalyptus (***Eucalyptus globulus***)
Fennel (***Foeniculum vulgare***)
Feverfew (***Tanacetum parthenium***)
Goldenseal (***Hydrastis canadensis***)
Hyssop (***Hyssopus officinalis***)
Jewelweed/impatiens (***Impatiens aurea***)
Lady's Mantle (***Alchemilla vulgaris***)
Marshmallow (***Althaea officinalis***)
Milkweed (***Asclepias syriaca L.***)
Mugwort (***Artemisia vulgaris***)

▲ **Aloe**

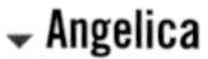

▼ **Angelica**

Mullein (***Verbascum thapsus***)
Sorrel/sheep sorrel (***Rumex acetolla***)
Stinging nettle/nettle (***Urtica dioica***)
Vervain/verbena (***Verbena officinalis***)
Watercress (***Nasturtium officinale)***
Witch hazel (***Hamamelis virginiana***)
Wormwood (***Artemisia absinthium***)
Yarrow (***Achillea millefolium***)

◂ Agrimony

▴ Barberry

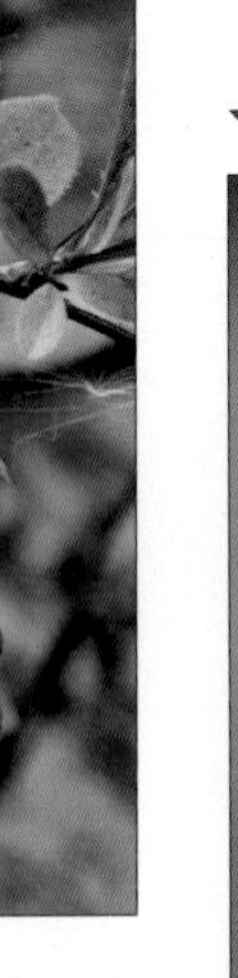

▾ Bee Balm

Black Elder ▾

▲ Elderberry

▼ Borage

Burdock ▼

Chamomile ▲

Chickweed/stellaria ▼

Chicory ▼

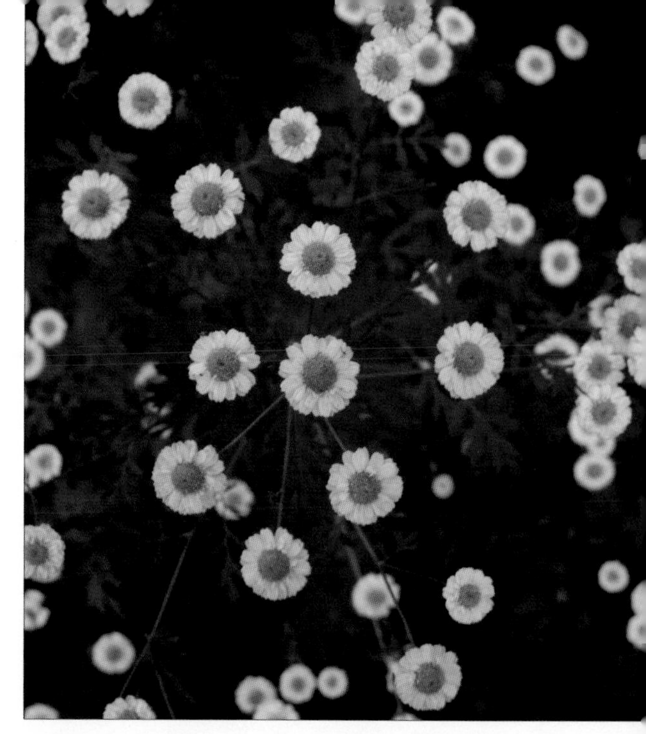

Feverfew ▲

◂ Eucalyptus

Fennel seeds ▾

▲ Goldenseal

Hyssop ▲

Hyssop ▲

◂ Jewelweed

▲ Wormwood

◂ Yarrow

Arthritis/Rheumatism/Gout

Angelica: Dilute up to 10 drops of angelica oil (from the root) in 2 tablespoons of almond or sunflower oil for arthritic or rheumatic pains and massage into joints.

Almond or sunflower: Dilute 5–10 drops celery oil in 1 ½ tablespoons almond or sunflower oil. Massage into joints.

Cabbage: Strip out the central rib of a cabbage leaf, beat it gently to soften it, and bind to arthritic joints with a bandage.

Celery: For gout in feet and toes, soak feet in a bowl of warm water with 15 drops celery oil. Or liquefy a whole fresh celery plant and drink the juice for rheumatoid arthritis.

▲ Watercress

Witch hazel ▼

▲ Sorrel

Vervain ▲

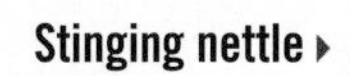

Stinging nettle ▸

▲ Marshmallow

Milkweed ▲

Mullein ▸

▾ Mugwort

Mullein: Make an infusion of the leaves, and strain very well to get the plant fibers out. Apply as poultice to arthritic joints.

Asthma

Cabbage: Take ¾ tablespoon syrup from decoction of cabbage leaves.

Cranberries: Cook cranberries in water until it turns into mash. Take 2–3 teaspoons per day.

Elderflower: Elderberry flowers taken as an infusion or tincture with yarrow and peppermint helps clear congestion and cleanse the body.

Licorice: Put ½-teaspoon licorice root in hot water and steep 10 minutes. Strain and drink 1–2 cups per day.

Mullein: Make an infusion of the leaves, and strain very well to get the plant fibers out. Drink it like tea.

Stinging nettle: Make a tincture of nettle and drink.

Watercress: Eat lots of watercress every day.

Yarrow: For mild asthma, use fresh yarrow in boiling water and inhale the steam.

Athlete's Foot

Apple cider vinegar: Soak in 50/50 mixture of apple cider vinegar for 10 minutes daily for up to 10 days. Or coat the area with the vinegar. Any vinegar will work.

Baking soda: Soak in a mixture of baking soda and water 30 minutes per day.

Garlic: Rub a clove of raw garlic on or crush the garlic and apply as a poultice for 30 minutes. Then wash with water. Doing this once per day for 7 days should clear it up.

Ginger: Boil 1 cup of water and add 1 ounce of chopped fresh ginger. Simmer 20 minutes, let it cool and apply to feet once a day.

Rubbing alcohol: Apply rubbing alcohol on the infected area and allow it to dry.

Canker Sore

What is a canker sore?

Unlike a cold sore (which is a form of herpes), a canker sore happens inside your mouth and is a small ulcer, or pocket of fluid. It has a white or gray base and a red border. They heal in 1–2 weeks.

Aloe: Use aloe juice as a mouth rinse several times a day.

Baking soda: Make the baking soda into a paste by adding a bit of water and apply directly to the sore several times a day. You can also mix it with water and use as a mouthwash.

Burdock: Boil 1 quart of water, reduce to simmer, and add 4 teaspoons dried burdock root cut into small pieces. Cover and let simmer 7 minutes, remove from heat, and allow to steep for 2 hours. Drink 2 cups per day on an empty stomach.

Goldenseal: Mix ¼ teaspoon salt, ½ teaspoon goldenseal powder, and 1 cup warm water. Use as a mouthwash several times a day.

Onion: Apply raw onion to the sore.

Plum: Use 2 tablespoons of fresh plum juice as a mouth rinse for several minutes, or apply as a compress directly on the sore.

Salt: Mix 2 tablespoons salt in a tall glass of water and use as a mouthwash 3–4 times per day.

Sorrel: Take a fresh piece and apply it on the sore until it is wet.

Tea: Put a tea bag on the sore and hold it in your mouth as long as you can.

Colds

Bee balm: Steep 1 teaspoon of dried bee balm leaves or 2 teaspoons fresh bee balm leaves in a covered cup of hot water for four minutes. Sip the tea three times per day.

Borage: For feverish cold take an infusion of borage leaves.

Burdock: For feverish colds with sore throat and cough take a decoction of burdock seeds.

Cayenne pepper: Add a teaspoon of pepper to a glass of water and drink as soon as you feel cold symptoms.

Elderberry: Elderberry flowers taken as an infusion or tincture with yarrow and peppermint helps clear congestion and cleanse the body.

Eucalyptus and yarrow: Mix yarrow oil with eucalyptus, peppermint, hyssop, or thyme oil. Dilute 20 drops oil in 1½ tablespoons of almond or sunflower oil and mix with other herbs.

Lemon: To make a nasal rinse that kills the virus, mix 1 tablespoon fresh lemon juice and a pinch of salt in a glass of warm water. Plug one nostril and inhale the juice into the other, hold it as long as you can; and then blow it out. Repeat in other nostril. Just salt water may also work.

Mullein: Make an infusion of the leaves, and strain very well to get the plant fibers out.

Olive leaf: Drink the extract of olive leaf.

Stinging nettle: Make a tincture of nettle and drink.

Vervain: Mix 3 tablespoons blue vervain (also known as New Jersey Tea), 2 tablespoons fennel seed, and lots of peppermint leaves. Put in teapot and pour boiling water over it and seep 10 minutes. Strain and add some honey. Drink three cups a day. It works to expel mucus from the chest and lungs.

Congestion

Agrimony: Gargle an infusion of agrimony leaves/aerial parts (***A. eupatoria***) for nasal mucus.

Elderflower: Elderberry flowers taken as an infusion or tincture with yarrow and peppermint helps clear congestion and cleanse the body.

Mustard: For chest congestion, sift 1 tablespoon dry muster and ¼ cup flour in a bowl (for children use 6 tablespoons flour). Slowly add enough lukewarm water to make a paste. Spread the plaster on a piece of muslin big enough to cover the chest, and then cover with another piece of muslin. Make sure the skin is dry and there is no allergic reaction. Remove when chest starts to turn red (usually 10–20 minutes), and no longer than 30 minutes at a time. Then rub petroleum jelly on to keep the heat in.

Stinging nettle: Make a tincture of nettle and drink.

Yarrow: For upper respiratory phlegm, drink an infusion made of yarrow flowers.

Constipation

Apples: Eat lots of apples.

Apple cider vinegar: Take 2 teaspoons twice a day with a glass of water.

Beets: Eat two small beets in the morning.

Carrot: Drink carrot juice several times a day.

Dates: Soak 6 dates in a glass of hot water, let it cool, and drink the water and eat the dates.

Cranberry: Drink cranberry juice to prevent bacteria growing in the mouth.

Hydrogen peroxide: Use it as a mouthwash until the infection goes away, and soak your toothbrush in it.

Lemon juice: Squeeze the juice of one lemon into a glass of warm water. Use as a mouthwash for one minute after brushing your teeth.

Sage: Pour 1 cup of boiling water over 1 tablespoon of sage leaves. Cover and steep for 15 minutes, strain, and add 2 teaspoons of sea salt. Use twice a day after brushing your teeth and keep it in the fridge.

Tea tree: Add one drop of tea tree oil on your toothbrush over your toothpaste and brush.

Gum Disease

Gum disease is also known as periodontal disease. After having gingivitis for a while without treating it, the plaque may grow beneath the gum line. Gradually the bacteria in the plaque breaks down and destroys the tissue and bone supporting the tooth, leaving pockets of empty space which can get infected, but you may not even feel anything. Then the tooth becomes loose and has to be removed.

Baking soda: Make a paste of baking soda and water and apply gently to the gums.

Pot marigold: Use a mouthwash of an infusion of petals of pot marigold.

Salt: Use a warm saltwater rinse and swish it in your mouth before spitting it out.

Tea: Hold a wet tea bag up against the gum abrasion or even a canker sore.

Hay Fever

Mullein: Make an infusion of the leaves, and strain very well to get the plant fibers out. Drink like tea.

Olive leaf: Drink the extract of olive leaf.

Yarrow: Harvest fresh yarrow flowers and put in boiling water. Inhale the steam.

Headache

Agrimony: Apply a poultice of leaves of agrimony (***A. eupatoria***).

Apple cider vinegar: Add a few tablespoons to a pan of boiling water and inhale the fumes for five minutes. Lie down for 15–20 minutes and the headache should be gone. Repeat if necessary.

Cayenne pepper: Mix ½ teaspoon of cayenne pepper in a glass of water and drink slowly.

Celery: Soak the seeds in hot water, strain, and drink slowly.

Feverfew: Chew some feverfew leaves.

Ginger: Eat fresh or powdered ginger in a meal, or put grated fresh ginger in a drink.

Honey: Take 2 tablespoons of honey with each meal. Or boil equal parts honey and apple cider vinegar and inhale the steam.

Mullein: Make a tincture of flowers and drink.

Orange: Some headaches are caused by high altitudes, which vitamin C can help. Eat oranges or drink orange juice.

Peppermint: Drink 1–2 cups of tea, or put a few drops of essential oil on your forehead.

Water: Soak your feet in a pan of very warm water, adding more hot water as it cools off.

Hemorrhoids

Aloe: Apply aloe vera gel directly on the anus to relieve pain and burning.

Cayenne: To stop bleeding hemorrhoids, drink ½ teaspoon of cayenne in a glass of water every day for several days until the hemorrhoids are gone. Drink 1 or 2 cups per week for prevention.

Mullein: Make an infusion of the leaves, and strain very well to get the plant fibers out. Apply directly as a poultice.

Witch hazel: Put an extract of witch hazel on a wad of cotton; secure it to the hemorrhoids before bed and go to sleep. Use until the hemorrhoids disappear.

High Blood Pressure

Apple cider vinegar: Take if the cause of your high blood pressure is eating too much protein and too little acid. Take 2 teaspoons apple cider vinegar in a glass of water drops the pressure from 20–40 points in 30 minutes.

Beets: Slice a raw beet and coat it in fresh lemon juice. Let it sit in the fridge overnight and eat one slice per day.

Cayenne: Mix 1 teaspoon cayenne powder and 1 cup hot water and drink. If this is too strong, buy gelatin capsules and fill them with cayenne powder. Swallow two a day.

High Cholesterol

Garlic: Eat fresh garlic cloves to reduce levels.

Oolong tea: Use an infusion of oolong tea after eating a heavy meal to help lower cholesterol.

Infection

Cloves: Eat fresh cloves to prevent infection. Drink juice from fresh cloves to stop infection.

Honey: Apply honey to a wound or eat for resistance to infection.

Olive leaf: Drink the extract of an olive leaf.

Tea tree: Put 1 ½ tablespoons tea tree oil in a cup of warm water and use to cleanse wounds daily.

Insomnia

Chamomile: Drink chamomile tea just before bedtime.

Milk: Drink a glass of warm milk with a bit of ginger in it.

Oats: Take doses of ½ teaspoon of the fluid extract from oats. Other foods with the same properties include rice, tomatoes, bananas, barley, sweet corn, soy nuts, cottage cheese, chicken, pumpkin, and turkey, so eat plenty of those.

Lice

Lice are a common childhood calamity with many remedies, most of them old wives' tales. Suffocation with mayonnaise and washing all pillows in hot water will not completely get rid of an infestation of lice. The only sure cure is to purchase a lice comb with metal teeth (not plastic, or the eggs will stick to the teeth), and to the comb hair from scalp to tip thoroughly every day. As you comb, pull the comb to the tip of the hair over a clean white surface such as a bowl or bathroom sink. You will pull out the lice as you go, which will either get stuck in the comb or drop into the bowl. Comb every day for 14 days, even if you see no lice, as the adults may have laid eggs and after a few days you would see juvenile lice. After 14 days of clear hair, you are cured. Washing hair with the below solution will greatly deter lice and make them easier to comb out.

Essential oil rinse prevention: Add 5 drops each of lavender, eucalyptus, rosemary, peppermints and tea tree essential oils to 1 ounce sunflower oil, olive oil, or other oil. Then add 5 drops of the oil to a cup of water. Wash hair in this solution, then comb through hair thoroughly. This can be used to both prevent and help get rid of lice.

Rash

How do I treat a rash?

1. Find out what is causing the rash. Even if you can't, follow these rules: if it is wet, keep it dry, and if it is dry, keep it moist, and never scratch.
2. Clean the rash like a wound, using iodine, garlic, salt water, bee honey, or sphagnum moss (a natural source of iodine).

Chicory: Mix finely chopped chicory with clay and apply as a poultice. Change the dressing frequently (for a dry rash).

Goldenseal: Mix 4 teaspoons dry clay, 1 teaspoon goldenseal powder and 1 teaspoon comfrey root powder (for a wet rash).

Oatmeal: Put oatmeal in a stocking, knot the open end and fill a container by pouring water through the stocking. Soak the rash in the oatmeal water (for a dry rash).

Sore Throat

(See Cough and Congestion)

Agrimony: Gargle an infusion of agrimony aerial parts/leaves (***A. eupatoria***).

Cayenne: Add ½ teaspoon cayenne pepper to 1 cup boiling water, stir well, and gargle while mixture is still very warm. You can also add 4 parts echinacea, 1 part garlic, and 2 parts peppermint per 1 part cayenne.

Chamomile: Add 1–2 teaspoons dried chamomile blossoms to a pint of boiling water. Steep and drink 1 cup every 3–4 hours.

Garlic: Mix 1 inch of fresh ginger root (grated), ½ of a fresh lemon (sliced, with peel), 1 mashed clove of garlic, and 2 cups water in a pot. Bring to boil, and then simmer on low for 20 minutes. Strain into a mug and add lots of honey.

Ginger: Remove the skin from a small ginger root and slice it into thin disks. Put them in a pot of water and bring to boil. The water will turn a yellowish or tan color. Add 3 tablespoons of honey and sip the hot tea slowly.

Goldenseal: Boil 1 pint of water and add ½ teaspoon powdered goldenseal root. Drink while hot.

Hyssop: Put 2 teaspoons dried hyssop in 1 cup boiling water and cover for 10 minutes. Strain and drink hot.

Marshmallow: Put 1–2 teaspoons root bark in a cup of hot water and drink several cups per day.

Lemon: Mix 3 tablespoons lemon juice with 1 cup honey. Add 1-tablespoon comfrey, chamomile, or rosemary. Slowly stir in ¼ cup warm water. Store in a covered jar in the fridge, and take 1–2 tablespoons for a cough.

Thyme: Lightly crush thyme leaves and place in a cup. Fill with water that is just cooler than boiling and cover for five minutes. Remove leaves and drink or gargle.

Sores and Bruises

Burdock root: For skin sores, apply a poultice of the root of burdock.

Burdock leaves: For bruises, apply a poultice of burdock leaves.

Cabbage: Break a cabbage leaf so that the juice is released, heat it up, and apply directly to a bruise.

Calendula: Apply a cream or tincture to a bruise.

Elderflower: Elderberry flowers taken as an infusion or tincture with yarrow and peppermint and applied as a poultice.

Lady's mantle: For general sores, wash in an infusion of lady's mantle.

Mustard: Mix 2 parts of ground mustard seed with 1 part honey and 1 part finely chopped onion, apply to a bruise, and cover with bandage.

Onion: Cut up an onion, dip in apple cider vinegar, and rub on a bruise right after it occurs. Or make a poultice of roasted onion and apply to bruise.

Vinegar: Apply a compress soaked in vinegar to bruises.

Witch hazel: Make a tincture of witch hazel and apply as a cool compress to bruises.

Wormwood: For bruises, soak a compress pad in an infusion of the aerial parts of wormwood (***A. absinthium***) and apply to bruise.

Stress and Anxiety

Borage: Take a ¾ tablespoon dose of tincture of fresh borage leaves three times a day, or drink 2 ½ teaspoons of the juice of fresh borage leaves three times a day.

Pot marigold: Add 5–10 drops of pot marigold essential oil to bath water and take a bath.

Oats: Take doses of ½ teaspoon of the fluid extract from oats.

Teeth Grinding

Some people grind their teeth together involuntarily when they sleep. They may not even notice they are doing it until someone sleeping next to them wakes up because of it. It can wear your teeth down, give you headaches and a tired jaw and wreck your gums. It can be caused by having a deficiency in pantothenic acid (a B vitamin) and a calcium deficiency. Grinding your teeth is also called bruxism. In some cases bruxism can also be caused by a parasite, so if you know you are not deficient you should go get tested and get an herbal parasite cleansing.

Treatment:

Relax before bed by taking a warm bath and listen to soothing music. To give yourself pantothenic acid, eat avocados, mushrooms, yogurt, salmon, and sunflower seeds. It is also found in meat, beans, and grain. You can help boost your calcium by eating milk, yogurt, cheese, broccoli, spinach, tofu, soybeans, and lentils.

Tired Eyes

Lavender: Add one drop of lavender oil to 2 ½ cups of water and mix well. Dip a cotton cloth in the liquid, squeeze out the excess liquid and place it over the eyes for tired and strained eyes.

Tea: Use a poultice of black tea bags on tired eyes.

Ulcers

Banana: Eat bananas every day to strengthen the stomach lining.

Cabbage: Eat fresh raw cabbage to strengthen the stomach lining.

Cayenne: Combine 1 teaspoon cayenne pepper, 1 cup of hot water, and drink.

Honey: Eat lots of raw honey to kill harmful bacteria.

Vomiting

(See also Nausea)

Vomiting can be caused by food poisoning, the flu, viruses, motion sickness, anxiety, pregnancy, medications, etc. Usually vomiting does not last very long. If you have vomiting from an injury, have severe pain, fever over 105° F, blood in the vomit, or if it lasts longer than 24 hours, then see a doctor.

Treatment:

Drink small amounts of clear liquid such as soup, tea, 7-Up, Sprite, or water. It is important to drink very little or vomiting will start again. When vomiting has stopped, try eating toast or crackers.

Warts

Aloe: Soak a small cotton pad with aloe gel and fasten to the wart. Every three hours add more gel to the pad and change the pad daily.

Apple cider vinegar: Soak the wart in warm water for 20 minutes and then dry it. Apply the vinegar and leave it on for 10 minutes. Wash it off and dry it.

Ashes: Put wood ashes on the wart.

Baking soda: Rub baking soda paste on the wart three times per day.

Dandelion: Break open the stem of the dandelion and rub the milky white juice on the wart 2–3 times per day until the wart is gone.

Garlic: Mash fresh garlic and rub on warts.

Grapefruit: Make an extract of grapefruit seeds and apply a drop directly on the wart. Cover it with an adhesive strip. Reapply the drop twice a day and the wart should fall off in a week.

Lemon: Rub lemon juice on the wart 2–3 times per day.

Milkweed: Rub milkweed juice on the wart 2–3 times a day.

Potato: Rub the wart with raw potato pieces every day. Radishes also work for this.

Worms

Garlic: To prevent and treat worms in humans and animals, eat fresh cloves of garlic. A large cat or small dog only needs one large clove per month.

ILLNESS AND THIRD-WORLD DISEASES

The greatest wealth is health.

~Virgil

Bronchitis

Bronchitis is when the bronchial tubes, or the air passages to the lungs, become inflamed. If you get it you will start coughing and wheezing, and you might have a fever and a sore chest. Bronchitis is highly contagious. It can develop by having a virus or bacteria, smoking, or even inhaling dust.

Treatment:

Aloe: Use aloe gel in a steam inhalant for bronchial congestion.

Cabbage: Take ¾ tablespoon syrup from decoction of cabbage leaves for bronchitis.

Licorice: Use ½ teaspoon powdered licorice root in a decoction in 1 cup water. Take 3 cups per day.

Marshmallow: Use an infusion of marshmallow leaves for bronchial disorders.

Stinging nettle: Make a tincture of nettle and drink.

Chicken Pox

Most people get chicken pox when they are children. First you will get flat red spots that turn into pimples, which then blister, get a crust, and scab up. These spots are extremely itchy, and they will continue to appear and blister for 3–4 days. You might also get a mild fever, lose your appetite and be kind of lethargic, and if you get it once you probably won't ever have it again. It is highly contagious, and it may incubate for a couple of weeks before the spots show up. The scabs may stay on for as long as 20 days.

Treatment:

Most children go through chicken pox easily. However, if the fever rises to a dangerous level it can turn into encephalitis, but this is rare. Don't give a child aspirin for fever (or for any reason)—just treat symptoms with regular remedies. Cut the child's nails so they can't scratch, put them in loose clothes, and cover the spots with baking soda paste or oatmeal lotion to reduce itching.

Oat meal lotion: Mix ½ cup almond juice, 1 ½ cup oat flour soaked in water, 1 ½ cup baking soda, and a ½ cup beeswax and dab on chicken pox.

Cholera

A rare disease that has become mostly eliminated in modern countries, cholera is spread through polluted water or polluted raw fruits and vegetables. The bacterium damages intestinal lining and causes such bad diarrhea that you may lose up to 4 gallons of liquid per day. If you have it you will get abdominal pain and diarrhea, severe thirst, and sometimes vomiting. Dehydration can quickly lead to death.

Treatment:

Replenish body fluids as much as possible with clean water with an electrolyte solution either by drinking or intramuscularly. The patient can be given a broad spectrum antibiotic such as tetracycline or terramycin.

Honey: Mix fruit juice with ½ teaspoon of honey or corn syrup and a pinch of table salt to replace electrolytes and prevent dehydration.

Cold Sores

Many people get a cold sore and don't realize that it is actually a type of herpes. It shows up as a sore on or near the mouth or lips, which forms an itchy welt that blisters and then oozes. Finally it will form a scab and eventually disappear. It spreads from person to person through contact with the lesion, saliva, feces, urine, eye discharge, or germs on objects that the person has touched. It usually reoccurs once you've had it, sometimes simply by exposing the spot to sunlight. A cold sore is different from a canker sore because a cold sore occurs outside the mouth, while a canker sore happens inside.

Treatment:

You can use cold sore ointments, and when you have a lesion avoid highly acidic foods. Plain yogurt with live cultures is highly effective at fighting the virus—eat lots of it. Don't touch the spot with your hands, and wash your hands constantly. Especially avoid rubbing your eyes because it can spread and become even nastier.

Aloe: Use aloe juice as a mouth rinse several times a day.

Baking soda: Make the baking soda into a paste by adding a bit of water and apply directly to the sore several times a day. You can also mix it with water and use as a mouthwash.

Burdock: Boil 1 quart of water, reduce to simmer and add 4 teaspoons dried burdock root cut into small pieces. Cover and let simmer 7 minutes, remove from heat and allow to steep for 2 hours. Drink 2 cups per day on an empty stomach.

Goldenseal: Mix ¼ teaspoon salt, ½ teaspoon goldenseal powder, and 1 cup warm water. Use as a mouthwash several times a day.

Onion: Apply raw onion to the sore.

Plum: Use 2 tablespoons of fresh plum juice as a mouth rinse for several minutes, or apply as a compress directly on the sore.

Salt: Mix 2 tablespoons salt in a tall glass of water and use as a mouthwash 3–4 times per day.

Sorrel: Take a fresh piece and apply it on the sore until it is wet.

Tea: Put a tea bag on the sore and hold it on as long as you can.

Conjunctivitis

Conjunctivitis is an eye infection that usually starts in one eye, making it bloodshot, runny, and itchy. The eye will have a discharge and get gummed up, and will be sensitive to light. It is highly contagious, and is caused by viruses, bacteria, allergens, parasites, fungus, allergies, a blocked tear duct, or by silver nitrate drops given to babies in the hospital at birth.

Treatment:

First make sure that it doesn't spread by using separate towels, blankets, pillows, napkins, etc. Wash your hands after touching your eyes, and wash your clothes, towels, and blankets frequently. Eliminate any irritants such as smoke, dust, or allergies, and rest inside because sunlight may be too bright. Use plain water to wash out the eye frequently, or use herbal eyewashes. It will last from 2 days to 2 weeks.

Aloe: Soak a cloth in aloe vera juice and place as a compress on eyes or use as an eyewash.

Baking soda: Mix ¼ teaspoon baking soda in ½ cup water and use as an eyewash.

Barberry: Mix ½ teaspoon powdered barberry root bark in one cup of water, boil 15–30 minutes, and use as a compress or eyewash.

Chamomile: Mix 1 teaspoon dried chamomile flowers in 1 cup boiling water. Steep for 5 minutes, strain, and allow it to cool. Use as a compress or eyewash.

Elderberry: Elderberry flowers as an infusion or tincture and use as an eyewash.

Goldenseal: Mix 2 teaspoons of goldenseal in a cup of boiled water and use as a warm compress or as eye drops. When using eye drops use 2–3 drops three times per day.

Honey: Put a drop of honey into the eye or mix 3 tablespoons of honey in 2 cups boiling water. Stir until dissolved, allow to cool and use as an eyewash several times per day.

Potato: Grate a potato and place on the eye or make it into a poultice and apply for 15 minutes every day for three days.

Yarrow: Steep 1 teaspoon of yarrow in a cup of water for 5–10 minutes, allow to cool and use as a compress.

Croup

Croup is a virus that usually shows itself at night, and is highly contagious from person to person. It may only last half an hour, but it will probably reoccur. Your child will have a sharp, hoarse, barking cough; wheezing; and may have difficulty breathing. It becomes dangerous if the child is having enough trouble breathing that the chest starts to suck in but she can't get a breath. If your child starts turning blue or drooling a lot then you've waited too long to go to the hospital. It usually occurs in babies and children.

Treatment:

Stop giving the victim dairy products to reduce mucus, and watch the child constantly in case of another attack. Steam or a humidifier can help the child breathe and clear the nasal passages. If it is moist and cool outside wrap them up and take them out, or turn the shower on hot full blast to fill the bathroom with steam. Watch carefully for breathing problems, and because croup makes the child susceptible to pneumonia watch for symptoms of chest pain and a very congested cough. About 5 days after recovery the victim might get an ear infection. If your child gets croup it is best to go to the hospital.

Cryptosporidium

Cryptosporidium is a relatively new bacterium to humans. You will get watery diarrhea, and possibly abdominal cramps, nausea, a low fever, dehydration, and weight loss. It lives in a person's (or animal's) intestine and because the symptoms are pretty generic, only a test by a doctor can determine that it is cryptosporidium causing it. It is spread by putting something in your mouth that has had contact with infected animal or human feces. It happens more frequently in families, day cares, and nursing homes, and among people who have worked with animals with diarrhea. Or you can get it by touching soil that has been contaminated or eating or drinking food that is contaminated. You can even get it by swallowing a small amount of water in a chlorinated swimming pool. The most important prevention is to wash your hands. It is not killed by bleach or disinfectants, but temperatures over 160°F or drying clothes in a dryer will kill the bacteria.

Treatment:

The illness only lasts a few days to 2 weeks in a healthy person. Some may recover but then get worse. Even a well person may still carry the bacteria. Anyone with an illness or diabetes, a pregnant woman, the elderly, or infants can all be at risk because of the dehydration caused by the extended diarrhea. If this is watched carefully then the infected person should be fine. A person with immunity problems such as HIV, cancer patients on chemotherapy, and transplant patients may be in extreme danger.

Honey: Mix fruit juice with ½ teaspoon of honey or corn syrup and a pinch of table salt to replace electrolytes and prevent dehydration.

Diabetes

Diabetes is a disorder in which the body either does not make enough insulin or becomes resistant to insulin, which prevents blood sugar from getting into the cells. Type 1 diabetics have a complete destruction of the cells in the pancreas which make insulin, and Type 2 diabetics (the most common type) is when the cells of the body becomes insensitive to insulin (usually from obesity). The normal blood sugar range is 80–95. Evidence suggests that a healthy, natural diet and exercise prevents diabetes.

Symptoms:

If you get Type 1 diabetes you may suddenly be very thirsty, urinate frequently, be very hungry, lose weight suddenly, and feel very tired. If you get Type 2 diabetes you will gradually start to feel tired, get dry and itchy skin, feel numb or tingly in your hands and feet, urinate more often, and be hungrier. You also might get infections more and cuts won't heal quickly, and you might get blurry vision. The symptoms vary—they might all appear, or only one.

Diet:

Avoid any food that might cause your blood sugar to rise suddenly, which means avoiding carbohydrates like sugar, beans, starchy vegetables like potatoes, fruit, some dairy products, and bread. You can eat any non-starchy vegetable, meat, eggs, raw cheese, raw butter, or whole milk yogurt. It also helps to eat 6–8 small meals a day instead of 3 big meals because the blood sugar level will stay more even. Fish oil is very good for diabetics, so eating cold-water fish such as salmon three times a week can have a reversing effect.

Exercise:

Diabetics need long and strenuous exercise (on a homestead that is easy). But check your blood sugar every 10 minutes when exercising to make sure blood sugar doesn't get too low—if it does, eat a carbohydrate. Anaerobic exercise such as weightlifting lowers blood sugar more than aerobic exercise (like running).

Treatment:

Flaxseed: Eat ground whole flaxseeds or flaxseed oil in food.

Borage: Extract the oil and take ¼ teaspoon per day.

Douves

Found especially in the tropics, they are parasitic worms that can get into you through your skin. If you swallow them they infiltrate the blood and cause severe sickness and often death. The only prevention is to not drink stagnant or polluted water and don't even touch bad water. Liquor does not purify the water either. Only boiling to a very high temperature may kill the parasites. Douves can only be treated at a hospital. If you suspect douves get to a doctor as soon as possible.

Dysentery

Usually caused by drinking contaminated water, you will get painful diarrhea with blood and weakness. Eat and drink a lot, and as with any serious diarrhea prevent dehydration by drinking lots of liquids with electrolyte solution. Boiled rice is recommended for recovery.

Treatment:

Honey: Mix fruit juice with ½ teaspoon of honey or corn syrup and a pinch of table salt to replace electrolytes and prevent dehydration.

Tea: Take a strong infusion of black tea—2 teaspoons per cup of boiling water, without milk or sugar.

Encephalitis

This is inflammation of the brain and it is very serious. It is usually a complication of some other disease, such as the measles, and can only be prevented through vaccinations. The symptoms include fever, drowsiness and a headache at first, and usually during or right after having another disease or virus. If left alone it can cause brain damage or even a coma, so go directly to the hospital!

Epiglottitis

There is a little flap of cartilage that is behind the tongue and in front of the entrance to the voice box called the epiglottis. Its job is to close the passage to the windpipe when you swallow food or liquid, so that you don't choke. Epiglottitis is when this little flap gets inflamed, usually from *Haemophilus influenzae* (Hib), a type of flu. You will get a low-pitched cough, have difficulty breathing and swallowing, and you might drool. You will have a muffled voice, and sometimes your tongue will stick out and you will have a fever. It typically happens during winter, and hardly ever to children under 2 years old.

Treatment:

There is no home treatment—if you see any of these symptoms go to the hospital because breathing can quickly stop. Antibiotics are the only cure. Keep the victim upright and leaning forward, with the mouth open and the tongue sticking out until you get help.

Fifth Disease

On the first day you will have a very red, flushed face. The next day, a lacy rash will appear on the arms and legs. On the third day, the rash will spread to the crevices of the body such as between the toes and fingers. For the next 2–3 weeks anytime you are exposed to heat the rash may reappear. It mostly happens to 2–12 year olds in early spring, and is not dangerous unless you are pregnant or have immunity problems.

Giardia

Giardia parasites are found in contaminated drinking water. Giardia symptoms will occur 7–10 days after drinking the water, and result in diarrhea, gas, cramps, and perhaps nausea, tiredness, and weight loss for over 24 hours. The only cure is antibiotics.

Hand-foot-mouth Disease and Herpangina

Both of these diseases are caused by touching feces and not washing your hands properly so that the bacteria are ingested. With hand-foot-mouth you will get a fever, lose your appetite, and in 2–3 days will get lesions in your mouth, then your fingers and maybe your feet and buttocks. Sometimes the lesions spread to the arms and legs, and rarely they spread to the face. The mouth lesions usually blister, which makes the mouth and throat sore. With herpangina, the blisters in the mouth will be grayish white and then fill with fluid, but they don't spread anywhere else. The fever will also be higher (100–104°F), and sometimes you will throw up and have diarrhea. Babies and small children are most susceptible, but parasites are usually not dangerous.

Treatment:

Treat the symptoms with herbal remedies, and feed soft foods. With herpangina, the only complication that can occur is if the victim gets convulsions—if so, go immediately to the hospital.

Influenza

There are many kinds of influenza (often called "the flu"), but they are all characterized by a sudden fever between 100–104°F, shivering, lethargy, a dry cough, diarrhea, vomiting, and sometimes achiness. However, not all flu makes you vomit, but they all make you feel very tired. In a baby under 6 months old or an elderly person the flu can be dangerous and you should seek a doctor's help.

Treatment:

Anytime someone's fever is over 102°F they are in a danger zone—keep the fever down with acetaminophen (don't give aspirin to a baby ***ever!***) and cool baths and cloths. Give the victim lots of clear fluids and treat symptoms with herbal remedies.

Bee balm: Steep 2 teaspoons fresh bee balm leaves or 1 teaspoon dried bee balm leaves in a covered cup of hot water for four minutes. Sip the tea three times a day.

Cayenne: Take a teaspoon of pepper in a glass of water as soon as you feel flu symptoms.

Elderberry: Elderberry flowers as an infusion or tincture with yarrow and peppermint helps clear congestion and cleanse the body.

Lyme Disease

Lyme disease is spread by a tiny deer tick. After biting a deer or mouse or other woodsy animals, it then carries the disease to you by biting you. This is why it is so important to check for ticks after being around animals or in the woods, because it takes a while for them to infect you. If you remove them quickly you have less chance of getting it. Lyme disease only exists in certain parts of North America, and only from May 1st to November 30th. Up to a month after you are bitten you will have a bull's-eye rash around the bite, and sometimes other rashes on other parts of the body. You will feel tired, feverish and chilly, achy, and have headaches and swollen glands near the bite. Sometimes you will get conjunctivitis, have mood and behavior changes, and the testicles may swell up. What is scary about Lyme disease is that it never goes away—weeks or even years later you might have joint pain and heart abnormalities.

Treatment:

The only cure is antibiotics—many times Lyme disease is misdiagnosed so you need to be on the lookout for ticks. If you discover a tick and the symptoms show up later the doctor may be able to cure it because the doctor usually doesn't think of testing specifically for Lyme disease. If you don't know you were bitten many doctors might think it is something else.

Measles

Measles happens usually in winter or spring and you can only get it once. You will get little grainy white spots on the insides of your cheeks which may bleed, and a dull red rash will appear on your forehead. The rash will spread to behind your ears and then downward, making you look all red. For a couple of days you will have a fever, runny nose, watery eyes, and a dry cough, and sometimes diarrhea and swollen glands.

Treatment:

Measles itself is not dangerous unless it develops complications. If the cough gets very severe, the victim has convulsions, pneumonia symptoms, encephalitis, an ear infection, or the fever rises again after it has already dropped then you should see a doctor. Only the MMR immunization prevents measles. Isolate anyone who has it, and soak him or her in a warm bath. Sometimes the eyes are sensitive to light and dim lights can make them more comfortable. Use herbal remedies for the symptoms, and give extra fluids. For some small children and elderly people measles can become very dangerous so watch for complications carefully.

Meningitis

Meningitis is a very dangerous situation in which the membranes around the brain and/or the spinal cord become inflamed. It is usually caused by bacteria and sometimes a virus. The symptoms are fever, drowsiness, loss of appetite and vomiting, a stiff neck, sensitivity to light, blurred vision, and signs of a neurological disorder. A baby will have a high-pitched cry and the soft spot will bulge out of the head.

Treatment:

Only an Hib immunization can prevent meningitis, but if you get it once you will never have it again. If a virus caused it then it should eventually pass and the symptoms can be treated, but if it was bacteria then the victim must be hospitalized and treated with antibiotics. Babies usually get it from bacteria in a day care, and this can be very dangerous because the bacterial form can cause permanent brain damage and even death.

Mumps

Mumps is a virus that causes the glands on the sides of your jaw just in front of the ear (either one or both) to swell up, ear pain, painful chewing, pain when eating acidic or sour foods, and swelling of the salivary glands. Sometimes you will just have a very vague pain, but will also have a fever and lose your appetite, while about 30 percent of the people don't have any symptoms at all. It usually happens in late winter or spring. Mumps itself isn't dangerous unless it develops complications. If the victim starts to have the signs of meningitis or encephalitis then go to the doctor.

Treatment:

Treat the symptoms of fever and pain with regular remedies. A cool compress can help the rash feel better, and avoid acidic and sour foods. The complications can be rare, but if an adult male gets the mumps he is more likely to have problems and it can be dangerous.

Pneumonia

Pneumonia is when the lungs get inflamed, usually when in recovery from some other illness such as a cold. After apparently getting better you will suddenly get worse. You will get a high fever, your cough will become congested, and you might have abdominal pain and bloating. Your breathing will become rapid, wheezy, raspy, and difficult, and you even might turn slightly bluish.

Treatment:

Infants are high risk because the danger that they will stop breathing. If the baby has difficulty breathing, turns a bluish color or seems very sick then go to the hospital. Other cases can be treated at home with antibiotics. Usually the cough hurts so bad that the victim does not want to move, and so bed rest is essential.

Stinging nettle: Make a tincture of nettle and drink.

Rabies

Some animals may carry the rabies virus if they have not been vaccinated. You can catch it by being bitten by an infected animal. Any time that you have been bitten by an animal and are not completely certain that it has been vaccinated you must go to the doctor and get tested. Treat the bite with first aid, and the doctor will give you a rabies antiserum, a tetanus booster, and/or a human rabies immune globulin. If you get infected and don't get treatment, you will have a tingling, burning, or cold sensation at the site of the bite, and you will start to have flu-like symptoms with a 101–102°F fever. Your heartbeat will get more rapid, your pupils will be dilated, breathing will become shallow, and you may drool, sweat, and have watery eyes. Two to ten days later you will become anxious and restless, have problems seeing, the muscles in your face will be weak, and the fever will rise above 103°F. Often the victim will become afraid of water and the drool will start to become the classic frothy "foaming at the mouth." About 3 days later the victim will become paralyzed.

Treatment:

Without treatment, rabies is fatal. If you get the shots before you start to have symptoms then you will usually be all right, but once symptoms occur the mortality rate is still high. It is important to immunize your pets because they can carry it without showing any symptoms.

Roseola Infantum

A virus that happens mostly to babies and young children. First the child will become irritable and lose their appetite. They will develop a fever of 102–105°F, and sometimes get a runny nose and convulsions. On the third or fourth day the fever will drop and the child will seem better, but will soon develop faint pink spots on the neck, upper arms, and sometimes the face and legs. These spots will turn white if you apply pressure. It is not dangerous unless the fever lasts a long time, gets very high, or if the child gets convulsions.

Treatment:

Treat the symptoms with herbal remedies and give lots of fluids.

Rubella

Also called German measles, rubella only happens if you haven't been immunized with the MMR vaccine. It usually happens in late winter or early spring, and causes small, flat pinkish spots on the face, which spread to the body and sometimes to the roof of the mouth. Sometimes the neck glands will swell up and there will be a mild fever. Once the rash starts, it can last a few hours to 4–5 days. It is only dangerous to a non-immunized pregnant woman. There are hardly ever complications.

Scurvy

Scurvy is a disease resulting from not getting enough vitamin C. You become weak and get joint pain. Then you get raised red spots around the hair follicles of your legs, buttocks, arms, and back. Black and blue marks might appear on your skin because of internal hemorrhaging. After 5 months the skin forms small lesions, gums bleed, and tissues start to become weak and spongy, including the teeth.

Treatment:

Rose hips: Gather the "hips" or the part of the flower that remains when the petals fall off. Use 2–3 teaspoons of hips per cup, steep 10 minutes, and drink.

Spruce: Use spruce, pine, or fir, but spruce tastes best. Gather bright green buds (not all, just a few here and there from near the bottom), 1 teaspoon per cup.

Sheep or wood sorrel: Drink a tea made from the leaves.

Strep Throat and Scarlet Fever

Scarlet fever happens most often to children ages 3–12 any time of the year. They are both caused by bacteria, which in babies under 6 months only give them a fever and a mild sore throat. However, older children will get a high fever, a red pussy throat, swollen tonsils and glands, and abdominal pain. Scarlet fever acts very similarly to strep but the victim will also get a bright red rash on their face, groin, and underarms, which makes the skin rough and peeling. The rash will then spread to the rest of the body.

Treatment:

It is very important to avoid anyone who might have the *Streptococcus* bacteria. If you or your children have been exposed, wash your hands and gargle hot salt water to kill any bacteria in the throat. To treat the victim, watch the fever and see a doctor if it does not drop in 2 days. Give lots of fluids and treat the symptoms. Antibiotics may be given to prevent any complications; the infection can spread to ears, sinuses, lungs, brain, kidneys, and skin.

Tetanus

Also known as lockjaw, tetanus is caused by bacteria that enters most often through a puncture wound, but also through a burn, deep scrape, or an unhealed umbilical cord. The wound will spasm and the muscles around it will tighten, and you will have involuntary muscle spasms, which will make you arch your back, lock your jaw, and twist your neck. You may have convulsions, a rapid heartbeat, sweating, and a mild fever. It is only prevented through immunization (DPT).

Treatment:

Get to the hospital immediately if someone who has not been immunized has received a wound from something metal. Usually tetanus is caused by getting cut or punctured by some old, metal, rusty object, such as stepping on a nail. Or a newborn can get it by having their cord infected—if this happens go immediately to the ER. They will give the victim shots, muscle relaxants, antibiotics, and a respirator. Without any help the victim can freeze up and be unable to breathe or eat, will get ulcers or pneumonia, have an abnormal heart rate or even a blood clot in the lung.

Typhus

Typhus is a very rare disease during modern times but in the past has been a major concern. Typhus is highly contagious and is spread by being unsanitary. You can get it from a person or through contaminated food or water. Symptoms begin suddenly with headache, loss of appetite, and vomiting, followed by fever of 104°F, weakness, bloody diarrhea, and delirium.

Treatment:

Isolate the patient, and any bodily fluids should be contained and burned or sterilized with a huge amount of borax. Give the patient lots of electrolyte fluids, or salt and baking soda if nothing else is available. The antibiotics administered for this are tetracycline or Terramycin, both broad-spectrum. Recovery can take two weeks, but complications usually arise such as gastrointestinal bleeding or intestina rupturing, which make it life threatening.

Honey: Mix fruit juice with ½ teaspoon of honey or corn syrup and a pinch of table salt to replace electrolytes and prevent dehydration.

Viruses

People get a variety of nonspecific viruses, mostly in the summer, which cause fevers, loss of appetite, and diarrhea. Infants may also get many kinds of rashes. Only good hygiene prevents them and they are never dangerous unless the fever gets higher than 105°F or dehydration results from the diarrhea.

Treatment:

Treat the symptoms and give extra fluids to keep the victim hydrated.

Whooping Cough

Caused by bacteria, whooping cough starts out with cold symptoms such as a dry cough, a low fever and irritability. One to two weeks later, the coughing will begin to be in explosive bursts with no time to take a breath, and it will bring up thick mucus. Sometimes the coughing will be so violent that your eyes will bulge, and your tongue will stick out. You will sometimes have pale or red skin, sweat a lot, and may vomit. An infant may stop breathing and may get a hernia. This is the worst stage . . . gradually the coughing and vomiting will lessen and the appetite and mood will improve.

Treatment:

Infants should be taken to the hospital because of the danger that they might stop breathing, and they will be given antibiotics. They will also give an infant oxygen, suction the mucus, and humidify their air, all of which lessen the danger. Sometimes the victims must be fed intravenously. For a mild case, give small frequent feedings, lots of fluids, and humidify the air. The victim will never get it again, but there is a huge danger of complications including ear infection, pneumonia, and convulsions, and sometimes death.

FIRST AID

Warning: The conditions in this section are more serious and require more immediate attention than the one before. If possible they usually require a doctor immediately—these solutions are only for when that is absolutely not possible. And even after you have taken care of the problem, get to a doctor as soon as you can so he or she can make sure things are all right. These solutions are for the homestead, and I did not include very serious situations that require CPR. Get a Red Cross First Aid book and study it.

Abscess

An abscess is a pocket of pus somewhere in your body. One of the more common types is a dental abscess. This is when a tooth or gum infection (toothache or gingivitis) gets a pocket of pus. It causes pain and swelling, and can spread to the jaw, face, mouth, neck, etc. If it gets too big, it can stop you from breathing or give you a fever and kill you.

Treatment:

Infections should be treated by a doctor or dentist with antibiotics. If this is impossible, sterilize a scalpel, needle, or fishhook with a match, boiling water, or heat. It will hurt but will provide immediate relief. Poke a hole in the abscess and drain the fluid. Clean out the wound with hydrogen peroxide or antibacterial wash.

Back Injury

Back pain is caused by over-straining muscles, usually from lifting something too heavy for you. The pain can be mild or it can be so bad that you won't be able to bend over and you have to lie down. A ruptured, or herniated, disc is when the cartilage discs between the vertebrae break. This can be caused by heavy lifting also. Sitting, bending, coughing, or lifting a leg can cause pain and legs might tingle. It is best to seek medical care, but you can treat them the same as a muscle strain. If movement does ***not*** make pain worse, then the pain could actually be a kidney stone or infection, and you need to go see a doctor.

Treatment:

Have the patient lie on their back with a pillow under their knees, or on their side with a pillow between their legs for a couple of days. Give them a pain reliever with meals to help alleviate pain, and use ice packs on the back. After 2 days, resume gentle activity and after 2 weeks they should be fine.

Bee Sting

What do I do first?

About 4 percent of people are allergic to bee stings, and you can become allergic at any time, even if you have been stung before. First find the sting and pull out the stinger as fast as possible because it will keep pumping venom into the wound. The easiest way is to scratch it out with your fingernail but you might need tweezers. Then if the person is allergic and they know it they may have a bee sting kit nearby, and you just have to follow the instructions. An allergic person will have a headache, muscle cramps, difficulty breathing or swallowing, nausea, fever, drowsiness, sweating, and may lose consciousness. If this happens call 911. If not, wash the sting with soap and water and put ice on it.

Treatment:

Jewelweed: Break the stem and repeatedly apply the juice to a sting.

Black Widow

These are big, black, shiny spiders, and on the underside of their abdomen they have a red or yellow hourglass shape. Kids and smaller animals (such as dogs) are more at risk from these bites. Adults may not even realize that they were bitten, but a child should be checked by a doctor.

What do I do if a child has been bitten?

1. Stay calm. The faster the heart rate, the faster the venom will spread.
2. Apply something very cold, like ice, directly on the wound, and if it is on a limb, ice down the entire limb. This slows the blood to reduce spreading.
3. Do ***not*** use heat, herbal stimulants, alcohol, or put any kind of cream on the wound.
4. If you are close to a doctor, you will receive an antivenin made of horse serum. Some people are allergic to the antivenin, and if so, you will just be helped to fight the poison on your own.

Blister

Note: A blister is completely different from a boil. A boil is an infected bubble filled with pus, while a blister is a defense mechanism filled with lymph fluid to help heal the area.

A blister is a bubble of pus that develops on the skin. It is actually best to leave the blister alone because there is a perfect environment inside for healing. But, if you must drain the blister, apply a warm compress, which will bring the blister closer to the surface of the skin. Then use a sterile instrument such as a knife or needle to poke a hole in the bubble. Let the pus drain and clean it out with soap and water. Cover it and check it often for infection, because there is a much greater risk if you popped it. Disinfect it with hydrogen peroxide or some other disinfectant.

Broken Bone

What are the signs of a broken bone?

1. It looks deformed, or is bruised or swollen.
2. The part can't be used normally.
3. Bone parts are poking out, the victim hears bones grating, or they heard a snap or pop.
4. Injured part is cold, numb, painful, tender, discolored, or swollen.

How do I set a broken bone?

1. After a bone is broken you should first immobilize it. Just don't move it, and if you have to, move it as little as possible.
2. Check to see if the area below the break is starting to become numb, cool, pale, or swollen. Make sure the victim isn't showing signs of shock. If these things are happening then a blood vessel may have been cut and the victim is bleeding internally. Try to control the internal bleeding and treat for shock.
3. Basically to splint the break you have to feel the fracture, and then pull the bones far enough apart to set them in the right place. Of course the goal is to fit them together as closely as possible to their original form. If you are less than 2 hours from a hospital don't realign the bone—let a professional do it. But if the bones are left for more than 2 hours permanent damage can result, so go ahead and attempt it. Realignment can damage nerves and blood vessels, and do more harm than good. An amateur should also not attempt a wrist or shoulder because the risk of damage is even greater.
4. To realign the break, hold the limb below the fracture. Gently pull the limb along the long axis of the bone. Basically the limb stays in the same place, you are just pulling the broken off piece straight out from it, gently. This is called providing traction.
5. As you are pulling, move it gently and carefully in line with the upper part. Release and splint it.

How do I splint a broken bone?

A splint can be made of anything stiff and straight—rolled up newspaper, sticks, a broomstick, a pole, etc. The goal is to hold the broken limb as stiff and still as possible after it is set so it can heal at the break. Put the sticks on either side of the broken limb, and tie it around several times or wrap with cloth tightly (but not so tight that the circulation is cut off). This holds the limb straight against the sticks. For something like a broken shinbone you have to provide support at the bottom of the foot to hold the break together. This support is called traction, and you would basically make a frame around the leg, under the foot, and all the way to the hip or stomach so that the break remains stationary.

Broken Filling or Crown

How could I break a filling or crown?

You can break a tooth or filling, or dislodge a crown or cap by biting hard foods such as candy, nuts, or ice cubes, or chewing really sticky food such as taffy and caramel.

How do I make a temporary filling?

Just fill it with a small amount of temporary filling such as Tempanol or Cavit (also available with a toothache kit) using a dental instrument, knife blade, or popsicle stick. Have the patient bit down to form it to the shape of their bite and then remove any excess material. It will harden. Soft wax may also be used.

How do I fix a crown?

This is only a temporary solution. Clean out any dry cement with a dental tool or knife. Put a thin layer of temporary filling putty, denture adhesive, or in a fix you can use very thick paste of water and flour inside the crown. Align the crown to the tooth and have the person bite down to cement it together.

Burns

(See Sunburn)

What kinds of burns are there?

First-degree: A superficial burn which only burns the first layer of skin. The skin gets slightly red and swollen—such as a light sunburn.

Second-degree: Damages the second layer of skin. Blisters form, the skin is red and mottled, and the pain is much worse. Examples are a bad sunburn or bumping your hand against a hot pan.

Third-degree: Damages all layers of skin. Charred, black areas, or dry white areas form. It is at risk for serious infection.

How do I treat first-and second-degree burns?

1. If the burn is from tar, wax, or grease on the skin, don't try to remove it.
2. Flush by running large amounts of cool water over it for no more than 10 minutes.
3. Cover the burn with a sterile dressing—don't put creams on unless it is just a sunburn.
4. Don't pop blisters unless you can clean them out right away. A sunburn may blister, and tiny ones are relatively harmless.
5. If you get a big blister, treat like any blister.
6. If the victim is having trouble breathing, the burn is very large, or it happened from chemicals, an explosion or electricity, call an ambulance. Don't let the person go into shock.

How do I treat a third-degree burn?

Don't touch the burn, don't put ice on it, don't clean it, and don't use any ointment on a severe burn. Go to the hospital immediately. The only time you would get a victim wet is if they are still burning and water is the only way to get the fire out.

Herbal remedies (first-degree burns only):

Aloe: Cover well with aloe vera gel.

Honey: Apply honey and cover area with gauze or cloth.

Milk: Soak in whole milk for 15 minutes by putting a soaked cloth on top. Repeat with a new cloth every few hours.

Chiggers and Ticks

How do I remove a chigger?

A chigger is a type of mite. Lay a piece of tape over it, and then pull it up. The chigger will stick to it.

How do I remove a tick?

Use tweezers to grab the head of the tick and pull. Make sure you get the whole thing, not just the body. A more dangerous way is to light a match nearby. The heat will make the tick let go. To kill a tick you need to remove the head, or squash it.

Cuts

Note: Severe cuts require first aid. A severe wound is one that is bleeding a whole lot and doesn't stop. Before cleaning anything you need to stop the bleeding first.

What is a cut and how do I treat it?

A cut is not a wound, it is simply when the surface of the skin has been torn. Clean it first (use the following remedies if you can), then bandage it with an adhesive bandage or a clean cloth.

Agrimony: Use an infusion of agrimony leaves (***A. eupatoria***) to wash the cut, sore, or wound.

Aloe: Apply fresh aloe gel.

Cabbage: Strip out the central rib of a cabbage leaf, beat it gently to soften it, bind to wound with a bandage.

Chicory: Mix finely chopped chicory with clay and apply as a poultice. Change the dressing frequently.

Lanolin and marshmallow: Melt 50 g lanolin, 50 g beeswax, 300 g soft paraffin together. Heat 100 g powdered marshmallow root in the mixture for 1 hour over a water bath. When cool, stir in 100 g powdered slippery elm bark.

Stinging nettle: Use nettle tea compress or finely powdered dried nettles directly on the cut.

Tea: Soak a pad in weak green tea and use as a compress.

Yarrow: Wrap a poultice of fresh, washed yarrow leaves on cuts and grazes.

Frost Bite

Frostbite is when extreme cold causes your tissues to freeze and die. Light frostbite just makes your skin a dull white color. Deep frostbite kills tissue below the skin and makes it solid and immovable. Your face, feet, and hands are most likely to get frostbite. If you are with others check each other often, and if you are alone put your mittens over your nose and chin often to warm them. If you do get frostbite, do not use fire or snow to warm the area. Rub it gently with warm water, then dry it and put it next to bare warm skin. Once you have thawed the area, use an herbal remedy.

Treatment:

Aloe: Rub aloe gel or fresh juice on the area several times per day.

Banana: Cover the area with the inside of a banana peel.

Hypothermia

Hypothermia kills more people than gunfire, animal attacks, bites, wounds, or drowning. It can be caused by going into shock after an accident, getting too cold for too long, or getting wet in cold water for even a short amount of time. It happens slowly—at first the victim can walk and talk, but slowly becomes pale, starts to shake, and makes bad decisions or doesn't make sense when they say something.

How do I treat hypothermia?

1. Take the victim's temperature—it will be below 98.6°F.
2. Get the person warm quickly. Strip off wet clothes, get them next to a heat source (such as a fire), and wrap them in a warm blanket or sleeping bag. If you have no heat source, get inside the sleeping bag or blanket with them.
3. If the person is dazed or unconscious, don't give them hot drinks.

Insect Bites

Note: Some insect bites are quite dangerous, such as those from the brown recluse, or hobo spiders. For a serious

bite please use first aid and watch for signs of swelling, fever, and shock.

Aloe: Apply fresh aloe gel to the bite.

Tea: Place damp green or black tea leaves on the bite.

Wormwood: Apply a compress that is soaked in an infusion of the aerial parts of wormwood (***A. absinthium***).

Nosebleed

Nosebleeds happen all the time. They are caused by the membranes lining the inside of the nose becoming irritated and bleeding. This can be caused by dry air, allergies, and colds. Hang your head down and pinch the bones on each side (***don't*** put your head back).

Treatment:

Yarrow: Put a yarrow leaf into your nostril to stop it.

Poison Plants

What plants should I watch out for?

Don't touch, eat, or mess with plants with leaves in groups of three and ***any*** mushrooms (except the ones at the grocery store). You should also never eat any plant that you don't know what it is. The only time you possibly could is if you follow a careful testing process, starting with the juice only of a tiny bit of the wild food.

Treatment:

Jewelweed: Break the stem and apply the juice to a sting repeatedly. Or simmer a small amount in vegetable oil (not olive oil) for 10–15 minutes. Strain and add beeswax to make an ointment.

Scorpion

What do I do if a scorpion stings me?

1. Stay calm, and relax. The faster the heart rate, the faster the venom will spread.
2. Apply something very cold, like ice, directly on the wound, and if it is on a limb, ice down the entire limb. This slows the blood to reduce spreading.
3. Do ***not*** use heat, herbal stimulants, alcohol, or put any kind of cream on the wound.
4. Some scorpions are small and will not kill you. Others have very potent poison and could kill you. Seek a doctor if someone is stung by a scorpion, especially a child.

Shock

A person can go into shock because they've lost too much blood or other bodily fluids, from allergic reactions, infections, heart attacks, not enough oxygen, injuries, water accidents, and even being very afraid or seeing blood. The person might faint, the skin will be pale, bluish or cold, moist or clammy, and the pupils may be dilated. They will be weak, their breathing might be shallow or difficult or irregular, their pulse might be very fast, and they may throw up or feel extremely thirsty. As shock gets worse they may become unresponsive and have a vacant expression. Their eyes might look sunken and their skin may be spotty. You should get to a doctor if possible.

How do I treat shock?

1. If shock was caused by some accident then your first priority is to make sure they are breathing and not bleeding. The victim needs to lie down and you should keep them comfortable with blankets or shade or other help.
2. If the person has any bones broken then they should lie flat, with no part of the body elevated. If they have a head injury or are having trouble breathing then you should elevate the

head and shoulders only. If the person is bleeding from the mouth or vomiting then they should lie on their side. In any other situation only the feet should be elevated.

3. If the person is semi-conscious or completely out and may vomit, then don't give them anything to drink. Otherwise you should mix 2 pinches of salt and 1 pinch baking soda in an 8 ounce drinking glass of water that is room temperature (not too hot or cold). Have them drink it slowly over 30 minutes, by drinking half over a 15-minute period. Do this three times if they are still conscious and not throwing up. Children should drink smaller amounts, ¼ cup per 15 minutes, and babies should drink ⅛ cup per 15 minutes.

Sprains

A sprain is a common homestead injury. If you fall or turn your ankle, you can get a sprain (wear good boots to prevent this). Be careful that the sprain is not actually a fracture. A sprain is swollen, tender, and painful and may bruise. A fracture will have all these characteristics but it may also be deformed, make a grating sound, or the person heard a pop or snap. If you suspect a fracture, the only way to know is with an Xray. Splint the area and keep it immobile.

How do I treat a sprain?

1. Immerse the sprain immediately in cold water or ice packs to reduce pain and swelling, for 20 minutes 3–4 times per day, and take a pain reliever if you have it. Elevate the limb, keep it rested, and don't move it.
2. Wrap it in gauze or put on a big sock and wrap in an elastic bandage for support. Do not wrap it tightly or it will hurt worse.
3. After 72 hours you can soak it in Epsom salts, or use an herbal remedy. Rest is very important for healing—if you don't rest, it may never heal. You can tape or splint it so that it is immobilized so you can move around a bit.

Cabbage: Strip out the central rib of a cabbage leaf, beat it gently to soften it, bind to sprained joints with a bandage.

Sunburn

Aloe: Find an aloe plant and rub its juice on the burn.

Chicory: Mix finely chopped chicory with clay and apply as a poultice. Change the dressing frequently.

Elderflowers: Distilled elderberry flowers washed on the burn soothes the pain.

Honey: Apply honey and cover area with gauze or cloth.

Prickly pear: In the desert, make a hole in a prickly pear with a stick to get the juice. Use the juice.

Milk: Soak in whole milk for 15 minutes by putting a soaked cloth on top. Repeat with a new cloth every few hours.

Tea: Use a weak infusion of black tea to cool sunburn.

Tea: Make a decoction of 3–4 tea bags, and add 2 cups fresh mint leaves and 4 cups water. Strain into a jar and let it cool. Then dab onto sunburn with a cotton ball or washcloth.

Vinegar: Mix 2 tablespoons vinegar and water and dab on the burned area, or equal parts vinegar and olive or vegetable oil.

Tick

How do I pull out a tick?

Use fine tweezers to grab the tick at its head, not its body. The body will come off if you do the opposite. Some people hold a lighted match up near the tick and the heat makes it let go. This works but you can get burned. Wash the spot, and then clean with an antibacterial liquid.

Tooth Knocked Out

What if the tooth is just knocked out of place?

As in any case, see a dentist. If that isn't possible, push the tooth back into place and reposition it with steady, gentle pressure. Gently bite on a piece of gauze to hold it in place. A dentist can splint the tooth.

What if the tooth is knocked completely out?

1. If you put the tooth back within 30 minutes the body will usually heal the tooth back to itself. The ligaments will reattach (a root canal will still be necessary). If you wait longer than 30 minutes the tooth will not be accepted.
2. Find the tooth either on the ground or in the victim's mouth. If the mouth is bleeding, have them bite down on a gauze pad or a non-herbal tea bag.
3. Examine the tooth to make sure it is not broken. Hold the tooth by the crown (the part of the tooth that sticks out in the mouth), and clean gently with sterile saline, disinfected water or milk. Do ***not*** touch the thin, white layer of soft tissue covering the root, or else it will not reattach.
4. Put the tooth in the socket and gently press it into place. Have the victim bite down lightly on a piece of gauze to hold it in and go see a dentist.

Toothache

A toothache is when the nerve in the tooth (called the dental pulp) gets inflamed, either from decay from a cavity, the tooth is fractured, or other infections. If an infection is allowed to spread it can spread through the root and into the jaw causing an abscess. Symptoms include some pain in a tooth (or several teeth) and sensitivity to cold or hot or pressure.

Treatment:

1. Find the painful tooth and check for a cavity or a fracture. Clean out any food with a toothbrush, toothpick, or dental pick.
2. Put a tiny piece of cotton ball or small cloth soaked in a topical anesthetic such as eugenol or benzocaine solution right into the cavity. Use small tweezers or a toothpick to put the cotton in.
3. Cover the cotton with temporary filling material, dental wax, or softened wax from a candle (melt some wax then let it cool until pliable).
4. Take a pain reliever (but do not put aspirin next to the tooth), and get to a dentist. If it's going to be a while you will need to change the cotton.

Cloves: Put two drops of clove essential oil on a small cotton pad and put it next to an aching tooth in your cheek to reduce the pain.

Wound

(See also Cuts)

Note: A cut is different from a wound in that a wound is bigger. Always apply first aid before cleaning a wound, and stop the bleeding FIRST.

How do I treat a wound?

1. A wound is usually caused by some accident. Assess the situation. Is the

wound combined with other injuries, such as a broken bone, head injury, or internal injury? If so, do not move the victim unless they are in danger.

2. Talk to the victim if they are conscious. Find out where it hurts, how much pain they are in, what happened, etc.
3. Is the wound dirty? Is blood flowing or spurting from the wound? A small wound that is not bleeding very much should be cleaned and bandaged. A larger wound that is bleeding more needs to have the bleeding stopped first.
4. Apply pressure to the wound with a gauze pad (or your clean hand or a clean cloth . . . but ***don't*** use your hand if you don't know the person—wear a latex glove if possible) firmly until the bleeding stops. This could take as much as 30 minutes. If this is a large wound, soak sterile gauze with Betadine or another disinfectant solution and apply pressure. If you apply a clean cloth or bandage to stop the bleeding, do not remove it even if it becomes soaked with blood. Just add more layers. This is very important: If the wound is on the abdomen, chest cavity, or head then direct pressure will not work and can cause more harm than good. Just cover these wounds and hold the bandage there without pressure. For a head wound only use very light pressure unless you're absolutely certain it's not a fracture.
5. If direct pressure does not stop the bleeding, elevate the limb to slow the flow of blood to the area. If that still does not work, you can also use pressure points on the body, which squeezes the artery supplying blood to the wound. Simply press with your hand or fingers on the spot that is closest to the wound and between the wound and the heart. If you will be able to get to a hospital within 6 hours, use more dry sterile gauze and fasten with adhesive tape. If you won't be able to get to the hospital in that time, go on to step 6.
6. If the wound is over 1 inch long or the edges of the skin do not come together, the wound needs to be sutured. A smaller wound can be closed using butterfly adhesive strips.
7. Bandage loosely to help circulation and prevent pain and gangrene. Apply a cleaning agent such as tincture of benzoin on each side of the wound and let it dry for 30 seconds. Apply the tape to one side, close the wound gently so the skin edges touch, then tape the other side. The tape should stick out 1 inch on each side. Add more strips ¼ inch apart until the whole wound is closed.
8. Put a non adhesive dressing such as Telfa or Adaptic over the wound, wrap with absorbent gauze, and then cover with elastic bandage or rolled gauze to hold in place. Don't wrap it too tightly. You can also use sterilized cloth.
9. Keep the wound clean and dry, leaving the bandage off as much as possible to fresh air and sunlight. You may need to change the dressing twice a day.
10. Watch for pain, redness, fever, pus, or swelling, which are signs of infection. If you suspect infection, soak the wound in Betadine and give the patient antibiotics for 10 days (no less). If you have no antibiotics, soak the area in hot water with Epsom salts and go to the doctor.
11. If the person is in pain, give a pain reliever 1–2 days after the injury.

Give children Ibuprofen or Acetaminophen.

What if the wound is infected?

1. If you have no antibiotics or cleansing solutions such as benzoin, place a warm, moist compress soaked in an antibacterial solution right on the wound, and hold it there 30 minutes, making sure it stays warm. Do this 4 times a day.
2. Drain it using a sterile instrument (anything works as long as it has been heated to kill germs), then dress and bandage it and drink lots of water.
3. If it is very infected, expose the wound to flies for one whole day. Then cover it and check it every day for maggots.
4. Once maggots have developed, keep the wound covered but check it every day to watch for when they have eaten all the dead tissue out. If you suddenly have more pain and bright red blood, then you have waited too long and they are eating healthy tissue.
5. Wash the maggots out with sterile water or fresh urine. Then check it every 4 hours for several days to make sure all maggots are gone. Bandage and treat like a normal wound.

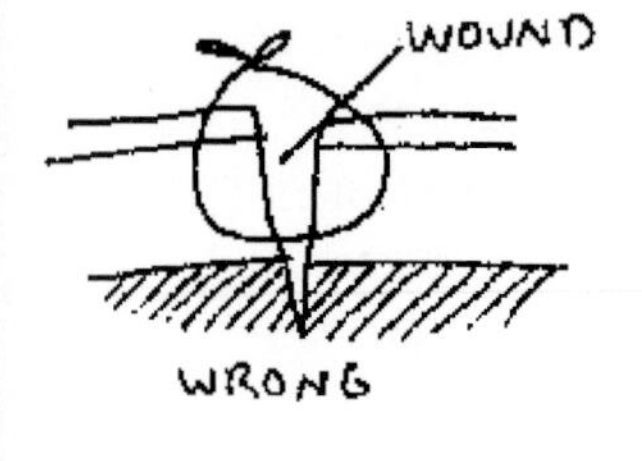

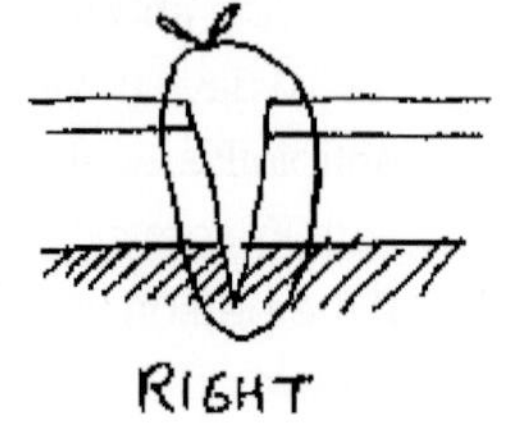

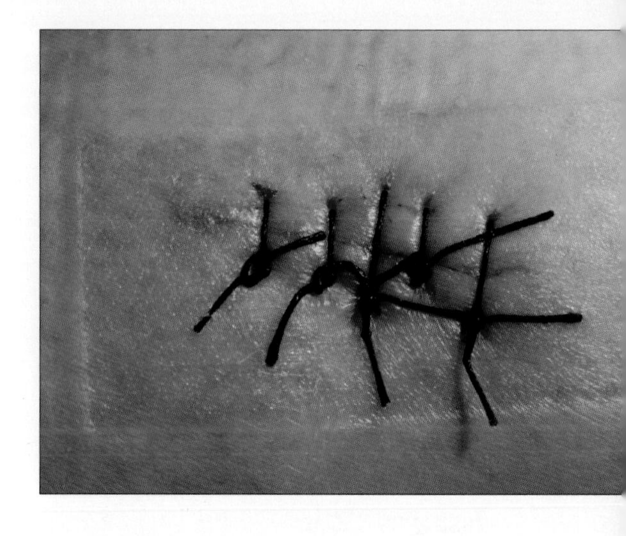

How do I give myself stitches or suture a wound?

1. You will need stitches if you see muscle or bone. Flush the wound with clean water for three minutes, then use iodine or other disinfectant to kill the germs.
2. Pick a needle that is less than 1 ½ inches long. Hold it over a flame for 1 minute to sterilize it, let it cool, and thread it with very thin silk or nylon thread. Don't use cotton because it collects more germs. The best material is catgut, thin strips of sterilized sheep intestine.
3. Start sewing as far away from the cut as the wound is deep. Push the needle through the cut and out the other side, pulling the flesh taught. Make sure the stitch goes under the cut, so that you don't see the thread inside it. The thread should not go through the wound at all. Tie a knot and cut the thread and repeat ¼ inch away.
4. Keep the stitches clean and dry and after 4 weeks snip the threads and pull them out.

BABIES AND CHILDREN ON THE HOMESTEAD

. . . a little child, born yesterday,
A thing on mother's milk and kisses fed . . .

~"Hymn to Mercury" (one of the Homeric Hymns), translated from the Greek by Percy Bysshe Shelley

Factors that determine your due date:

Your cycle length, its regularity, whether or not this is your first child, and your overall health and habits all determine the length of your pregnancy. In a 40-week system, you will date from the start of your last period, and assume that you ovulate on day 14. But if yours is longer you will need to know when you ovulate, by using the basal body temperature method.

Determining your due date:

If your cycle is 28 days and regular, then you will probably have your baby at almost exactly at 40 weeks.

If your cycle is longer than 28 days, but still regular, and you know when you ovulate using basal body temperature, then take the ovulation date, subtract 7 days, then add 9 months.

If this is your first child, you can on average expect to be a week later than your due date.

If it is your first child and your cycle is longer or irregular, you can be 1–2 weeks later than your due date.

The importance of the date:

You will never know when your actual due date will be, and it is only important if suddenly the baby is coming too soon, or if it is overdue. Knowing a general date will help determine if your baby is coming too early, because if you go to the hospital they will need to know. Preemies need special care because they are at risk for respiratory, heart and kidney problems. If you know when you conceived and you have gone 2 weeks over the date, you may need to help jog your labor, although this is not really necessary unless you feel that the baby really needs to come out—most of the time they come when they are ready.

The following discomforts are experienced during pregnancy:

Fainting, dizziness, lightheadedness, shortness of breath, allergies, dry nose, leg cramps, lower back pain, ligament pain, sciatica, carpal tunnel syndrome, bleeding gums, swelling, edema, headaches, migraines, heartburn, itchy skin, sleeping difficulty, varicose veins, constipation, hemorrhoids, groin pain, Braxton Hicks, and stretch marks.

What can help?

Herbal remedies that are safe for pregnancy

Drink lots of fluids

Stay calm and breathe slowly and deeply

Eat a banana every day

Take a walk and gently stretch or do yoga every day

Use pillows to support yourself while sleeping

Bend using your knees

Brush your teeth well every day

Take your prenatal vitamins

Eat green vegetables every day

Put lotion on your belly and thighs frequently

Take very warm showers

Keep out of stressful situations—don't get stressed

Herbs that can cause miscarriage:

Note: The culinary herbs mentioned here are only harmful if taken in a tincture or tea form because the dose is too large. Used as a seasoning in food should be fine.

Aloe (***Aloe spp.***)

American Mandrake (***Podophyllum peltatum***)

Barberry (***Berberis vulgaris***)

Bloodroot (***Sanguinaria canadensis***)

Black cohosh (***Cimicifuga racemosa***)—can be taken in the last month to stimulate labor

Blue Cohosh (***Caulophyllum thalictroides***)

Buckthorn (***Rhamnus cathartica***)

Calamus (***Acorus calamus***)

Cascara sagrada (***Rhamnus purshiana***)

Cayenne (***Capsicum frutescens***)

Celandine (***Chelidonium majus***)

Cinchona (***Cinchona L.***)

Cotton-root bark (***Gossypium herbaceum***)

Docks (***Rumex crispus***)

Ephedra (***Ephedra vulgaris***)

Juniper (***Juniperus communis***)

Fennel (***Foeniculum vulgare***)

Feverfew (***Chrysanthemum parthenium***)

Flaxseed (***Linum usitatissimum***)

Frankincense (***Boswellia carterii***)

Goldenseal (***Hydrastis canadensis***)

Juniper (***Juniperus communis***)

Lavender (***Lavendula officinalis***)

Licorice (***Glycyrrhiza glabra***)

Male fern (***Dryopteris filix-mas***)

Marjoram (***Oreganum vulgare***)

Mayapple (***Podophyllum peltatum***)

Meadow saffron (***Colchicum autumnale***)

Mistletoe (***Viscum album***)

Myrrh (***Commiphora myrrha***)

Oregano (***Origanum vulgare***)

Passionflower (***Passiflora incarnate***)

Pennyroyal (***Mentha pulegium***)

Rhubarb (***Rheum spp.***)

Rosemary (***Rosmarinus officinalis***)

Rue (***Ruta graveolens***)

Sage (***Salvia officinalis***)

Senna (***Cassia acutifolia***)

Tansy (***Tanacetum vulgare***)

Thuja (***Thuja occidentalis***)

Thyme (***Thymus vulgaris***)

Wild cherry (***Prunus serotina***)

Wormwood (***Artemesia spp.***)

Yarrow (***Achillea millefolium***)

Delivering a baby alone

Note: It is dangerous to deliver a baby alone. Get a midwife if you can, or at least have someone there with some experience; anyone is better than no one. You will need:

A big mirror (or one big enough to see the baby)

A nasal aspirator

A bucket or big bowl

Many clean blankets

Baby hat

Clean towels

A thick string or embroidery floss to clamp cord

Delivering a baby (for a mom alone, but useful for helpers too):

1. Don't take any painkiller or anesthetic. Drugs not only befuddle you, they also increase the chance of your child having breathing problems.
2. Your labor will start slow, with increasingly frequent regular pains that last longer and longer. During this time it is important to keep moving now and then, and keep semi-upright. Don't lie down flat, as the baby can move into a position that makes it more difficult to deliver.
3. During this time keep drinking juice, have some small snack, and try to stay relaxed. When you have contractions every 3–4 minutes that last 50–60 seconds then you are at the end of the first stage of labor.
4. At some point your water may break and you will be tired, irritable, feel sick, have a backache, and/or feel uncomfortable. Hopefully if you are not alone your helpers can make you feel more relaxed. During a contraction remember to breathe deeply and slowly and relax your muscles.
5. Finally you will feel an overwhelming urge to push. Sit down propped up against a wall, or pillows, and pull your knees up with your hands. Push only during contractions while counting to 10, then take a deep breath and relax. Do it again at the next contraction. A mirror placed in front of you can help you see when the baby is crowning and what is happening. If you have a helper, they should massage the perineum with warm olive oil and keep the area red or pink. This helps prevent tearing.
6. Pushing a baby out feels exactly the same as pushing out a bowel movement. During pushing you will be in a semi-conscious state, pretty much unable to concentrate on anything but pushing. If you are alone, use the mirror to watch what is happening.
7. At first, during a contraction the baby's head will peek out and go back in, but then it will start to become visible all the time and you will feel a stretching and burning. You must stop pushing and instead pant and relax during contractions. A helper can support the perineum with a clean washcloth without applying pressure so that the baby will come out slowly and gently. This will help the baby's head to slide out gently and prevent tearing. As the head emerges, have a helper support it with their hand, or you can try to do it. Continue panting until both the shoulders are out.
8. Wipe the baby's face with a clean washcloth and make sure there is no cord around the neck. If there is, gently unwind it—***don't pull it or it will tighten.*** If there is a membrane around the baby's face, tear it off. The umbilical cord will provide oxygen to your baby for 5 minutes after birth, unless your baby is blue.
9. If the cord is wrapped very tightly around the baby and is too short to be unwound and the baby can't make it out, it has to be cut and clamped (with the thick string or embroidery floss tied tightly) and the baby delivered quickly. The baby may be suffering from a lack of oxygen.
10. One shoulder will emerge, then the other, and the body will slip out quickly—be ready to catch it! If

several contractions pass without a shoulder coming, you or a helper may need to slip in two fingers and hook them under the armpit. The helper should rotate the baby's shoulder ***counterclockwise*** (turning to the left, towards the mom's thigh) while pulling out. The proper position for the baby is face down, and then it will turn towards the thigh.

11. When the baby is out, support the shoulder and neck with one hand without choking the baby, and hold the ankles with the other hand. Be careful because she will be slippery. Raise her up and hold her head lower than her feet. She should breathe and cry. If she doesn't, clear her mouth of mucus. Either use the nasal aspirator, or scoop it out with your fingers. If she isn't breathing yet, smack her on the bottom or the feet or rub her back. The baby doesn't necessarily need to cry, as long as she is pink and breathing.
12. Once she is breathing, wipe her off with a clean towel or blanket, and put her naked body directly against your naked chest. Put a blanket over her (but make sure her nose is out). Putting a hat on her will keep her warm also. ***If she doesn't breathe, you need to perform resuscitation!!! Study this before you are even near going into labor.***
13. Deliver the afterbirth into the bucket or bowl. This will have been going on while cleaning off the baby, and can take minutes or hours. If it is taking a while, go ahead and nurse the baby.
14. Now try to nurse her. This will keep her warm and help get all the mucus out. She will be really tired but she may

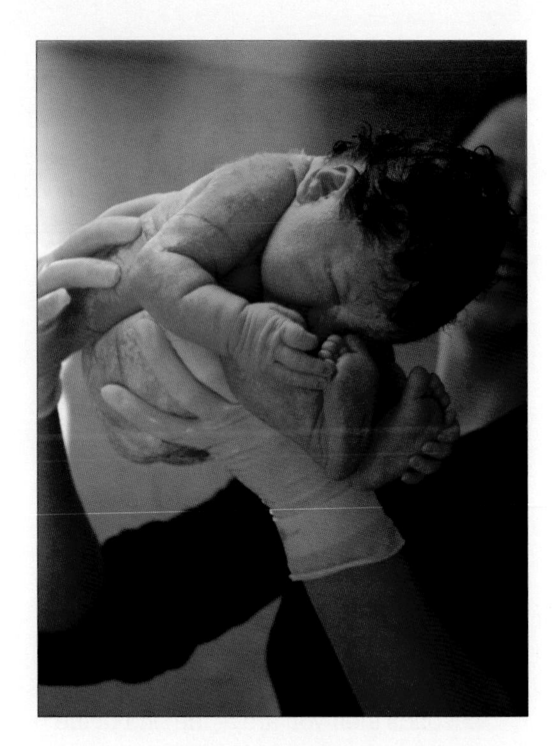

want to do it anyway. It is also good for the mother because the womb will start to contract and stop any bleeding.
15. Tie the cord when it stops pulsating and becomes limp. It is healthier for the baby to have the cord tied this late (rather than immediately which is the practice in hospitals), because it prevents anemia. Tie it about 1 inch from the baby's tummy with a sterile cloth or cord and cut.
16. If you begin to hemorrhage, lie down flat on your back at the end of the bed and elevate your feet (tilting the bed is the best way). The uterus should be massaged to keep it hard like a grapefruit, and a cold towel (dip in cold water and wring out) put on the abdomen. Put pressure on the perineum with towels and your hand. The baby should keep nursing. Even when things are normal, check

your abdomen every five minutes to make sure the uterus is still a hard ball.

17. To a helper: If the mother shows signs of vacant eyes, dilated pupils, pale and clammy skin, a faint and rapid pulse, shallow and irregular breathing, dizziness, or vomiting she may be going into shock. Keep her warm, dim the lights, talk softly and calmly to her, and continue the steps above—if possible go to a hospital!
18. Save the placenta to show a doctor. Check the amount of bleeding, it should be about 1–2 cups—use sanitary, sterilized pads or cloths. Make sure you and baby are warm, and you may wish to clean up, go to the washroom, etc. Don't contaminate the birth canal or you risk serious infection.

Baby resuscitation:

This is very important! You should know this first!

Pull the infant's lower jaw back and breathe gently into the mouth and nose with very small puffs at 20 puffs a minute. Try to clear out any excess mucus in the mouth.

Apgar score:

Hospitals and midwives use a score, which is given to a baby 1 minute and 5 minutes after birth. A score of 7–10 is normal, while 4–7 might require some resuscitative measures, and an Apgar of 3 or below requires immediate resuscitation.

Heart Rate

2 points: normal, above 100 beats per minute

1 point: below 100 beats per minute

0 points: absent, no pulse

Breathing

2 points: normal rate and effort

1 point: slow or irregular breathing

0 points: absent, no breathing

Grimace

2 points: pulls away, sneezes, or coughs with stimulation

1 point: facial movement only (grimace) with stimulation

0 points: absent, no response

Activity

2 points: active, spontaneous movement

1 point: arms and legs flexed with little movement

0 points: no movement, floppy tone

Appearance

2 points: normal color all over, hands and feet pink

1 point: normal color, hands and feet bluish

0 points: bluish-gray or pale all over

Everyone knows that breastfeeding is best. Here is why:

1. The milk is perfectly formulated for your individual baby.
2. It is the most digestible thing for a baby.
3. The milk has no allergens to babies.
4. Your baby won't have constipation or diarrhea (that yellow mustard is not diarrhea).
5. There is less risk of diaper rash.
6. Babies are less often overweight.
7. A mother's breast is exactly right for your baby's mouth.
8. It's easy (after it's established) and free.

9. It helps you heal after pregnancy better, and to lose weight.
10. The milk has tons of immunities for your baby.

You should not breastfeed if:

You have a severe debilitating illness or are extremely underweight.

You have a serious infection such as AIDS or hepatitis B.

You are taking regular medication such as an antithyroid.

You are taking any drugs, alcohol, or tobacco.

You've had serious surgery on your breast that might prevent your baby from getting nourished.

When to bottle-feed:

You've tried for at least 2 months and are still in pain, or your baby is not getting nourished due to an improper latch.

Your baby has a cleft lip or palate.

Your baby has a metabolic disorder such as phenylketonuria.

How to breastfeed:

1. You should try nursing soon after the baby is born. If you are at the hospital don't let the nurse stick a soother in the baby's mouth, or a finger.
2. You are most likely not making milk, just colostrum. This colostrum is the most important thing your baby will ever eat. Make sure you ask for help from anyone who's ever breastfed.
3. You can hold the baby in any way that is comfortable to you, whether you are lying down on your side, with the baby through your armpit, across your lap, or facing you. You will need many pillows of different shapes and sizes.
4. You should not lean forward; you should relax and lean back. The more relaxed you are, the more milk your baby will get. Tickle the baby's lips to get her to open her mouth.
5. To tell if your baby has a good latch, she must have as much of the areola (the dark circle around your nipple) in her mouth as possible. Her tongue should be under the nipple, and she should have a steady motion and you should hear a swallowing or gulping sound and little sighs.
6. It is recommended to feed from both sides, by first emptying one side, and then going to the other for a few minutes to make sure the baby is full. This keeps you from being lopsided. However, it is not necessary, and small babies may not want to do so, so do whatever is best for both of you.
7. Feed on demand. This means whenever your baby wants to nurse, do it. Their stomachs are very tiny, and until about 6 months they should have as much milk as they want. After

6 months they can start solids and you can start scheduling a bit.

Problems in breastfeeding:

Note: It is important that your baby keep feeding even if you are uncomfortable. Your baby should never be losing weight. If you have problems, contact your local La Leche League or a midwife can help.

My baby can't latch on and my nipples are bleeding: This usually happens for 6–8 weeks after birth. Babies have to learn how to nurse too. Put lanolin on your nipples before and after nursing, and let them air-dry thoroughly. Be very patient. If things get really bad, pump your milk, and use your finger and a straw to feed your baby until you are both more coordinated.

I have so much milk that my breasts are huge and sore and have red spots on them: This is called engorgement. Use very hot cloths on your breasts to sooth them, and pump out the extra milk. If you don't have a pump, express the milk by hand. You can keep that milk for later. Sometimes if you are really engorged, the baby can't get any milk. It is then absolutely necessary to get that milk out and give it to the baby, or she will lose weight. Engorgement shouldn't last more than a week. The best remedy for engorgement is to steam a cabbage leaf and break the veins to release the juice. Put the still-warm leaf on the breast and it will slow your milk production down immediately. If you're in a hurry you don't even really need to steam the leaves. Just break them a bit and put them inside your bra.

I leak all over the place: For the first 5–6 months you will have leakage, and not always small leakage, sometimes big leakage. Use a cotton nursing pad inside your bra to soak it up.

My baby eats all the time, even every hour: Feed your baby whenever she wants. She's growing at an amazing rate and is not going to be a baby for a long time. When she is done, burp her and try to feed her more, and sometimes that will fill her up for longer.

I don't have enough milk: If you are feeding on demand, your body will compensate to supply enough to feed your baby. It is pretty rare for a mother to be unable to provide enough milk, and you should see a doctor or midwife.

There are a couple of herbal remedies you can take to help increase your supply:

Anise: Crush 1–2 teaspoons of seeds just before using. Pour 1 cup of water over the seeds, cover and let stand for 5–10 minutes. Drink 1 cup 2–3 times per day to increase milk supply. Do not take if you cannot take estrogen. *Warning: Do not use Japanese star anise as it is associated with illnesses.*

Dill: Use 2 teaspoons of raw dill seed on your food at lunch and dinner, or steep 2 teaspoons of dill seed in 1 cup warp water for 10–15 minutes. Take ½ cup 3 times per day. Or steep ½ cup dill seed in water overnight, boil until very dark, strain, and drink 1 cup per day.

Food allergies:

Some babies have allergies to certain foods that are passed on through breast milk. This often takes the form of gas or stomachache, and can cause long crying spells. Keep track of what you eat and when, and if the baby has reactions within 24 hours, then avoid that food. These allergies don't last longer than 4–6 months.

Dairy products: often produces colicky symptoms in a baby.

Caffeine: cola, chocolate, and tea can sometimes produce a reaction. Some babies also won't sleep as well.

Grains and nuts: especially wheat, corn, oats, and peanuts.

Spicy food: your milk may take on the taste of garlic and peppers, or the baby might get gas.

Gassy food: broccoli, onions, Brussels sprouts, green (or bell) peppers, cauliflower, and cabbage almost always produce an upset baby. Any foods that gives you gas may also give your baby gas.

Acidic food: orange juice and citrus fruits may upset baby.

How do I recover after pregnancy?

Emotional health:

It could take at least 6 months for the pregnancy hormones to completely go away, at which point you will probably feel more "normal." You will still have breastfeeding hormones, but they are not as extreme as the pregnancy ones that you previously had. The key is to focus on your baby and don't get over tired. Postpartum depression is a big problem and most women get it to some extent. If you feel depressed at all, talk to someone right away.

Physical health:

Don't start exercising until your midwife or doctor says it's all right, but if you feel the need, then don't start until at least 3 months afterwards. In fact, you should be on bed rest the first 2 weeks, and then moving around normally. After the third month, you can start walking for exercise gently, and doing a few stomach crunches. Yoga is excellent for toning and stretching your sore body and ligaments. But remember, for at least 6 months don't do much more than focus on babies and housework, even if you feel restless. You are still in recovery from a year of physical stress. Of course this is great advice, but with more than one child this may well be impossible.

First foods:

First feed rice cereal, then progress to barley cereal, then applesauce, pears, peaches, carrots, squash, and sweet potatoes. Do not give your baby bananas, strawberries, spinach, chocolate, nuts of any kind, eggs, honey, salt, or dairy products until she is at least a year old. The exception with dairy products is if your child has very few allergies, you can introduce plain yogurt at about 8 months, and some cheese around 9 months. Also avoid stringy foods, which are a choking hazard, and chop everything into tiny pieces. Don't use seasoned, processed, or fried foods.

Calendar of solid food:

6 months: very liquid rice cereal

7 months: thicker rice and barley cereal, applesauce, and pear sauce

7–9 months: avocados, peaches, carrots, squash, mashed potatoes, teething biscuits (make sure they turn into mush immediately on contact with water—some can be choking hazards), pear and apple juice, and soymilk.

9–12 months: beans, poultry, bagels, rice cakes, yogurt, tofu, noodles, peas, yams, oatmeal.

12–18 months: eggs, peanut butter, fish, broccoli, spinach, cauliflower, melon, mango, kiwi, papaya, apricot, grape (halves), strawberries, tomatoes, pasta, graham crackers, wheat cereal, honey, pancakes, muffins.

Signs of food allergies:

Runny nose, sneezing, wheezing, watery eyes, sore throat, cough, persistent ear infection, congestion, face rash, hives, hands and feet swelling, dark circles under eyes, puffy eyelids, tongue soreness, diarrhea, constipation, gas, vomiting, and poor weight gain. Some more subtle signs are crankiness, anxiety, night waking, crying, headaches, sore muscles, irritability, and hyperactivity.

When to wean:

The amount of time you breastfeed is directly linked to the IQ of your child. The longer you nurse the higher her IQ. However, this is only really relevant when your child is younger (for example, if you stop breastfeeding at 2 months rather than 8 months). The general recommended time is between 1 year and 1 ½ years. Some babies want to stop nursing sooner, other babies want to go longer, or have to go longer because of food allergies. If your baby doesn't want to breastfeed for a whole year, she's not going to be less intelligent than other children.

How to wean:

Wean when your baby is ready; don't force it. The most natural way is to skip the feedings that aren't as important to your baby, and replace the breast comfort with some other comfort, like stories or cuddle time or hugs. Don't replace the breast with a bottle—if you have to, go to a cup. Make her feel like a cup is special and introduce a cup of water very early, at 6 months if possible, and she will like it more. Gradually you will find that you only nurse once or twice a day, and soon that will be gone too. It is a bad idea to wean to a bottle, which is much more difficult to wean from and eventually can cause dental problems.

Choosing disposables over cloth diapers:

Sometimes it is wonderful to have disposables ready in case of emergency. They are efficient, handy, and clean. Some babies prefer cloth because they have an allergic reaction to the chemical in the disposables. Other babies prefer disposables because they get rashes from the wet cloth diapers. Disposables stay dry next to the skin, but cloth is free and doesn't have chemicals.

Cloth diapers are cheaper and environmentally friendly, but also make more laundry for you. They are not really any less convenient because most people don't realize that you are supposed to dump poop into the toilet even from a disposable diaper. Human waste is not supposed to go into a landfill. It also takes a disposable 500 years to break down after it gets to the dump. Isn't it a strange thought to know that your baby's diaper will be around to see your great-great-great-great-great-great grandchildren?

What you need for cloth diapers:

5 diaper wraps or pants
24–60 cotton cloth diapers
Baby wipes (cloth or disposable)
Diaper rash ointment
Changing space
Diaper bucket

Folding a big square diaper:

Cloth diapers are large in order to make layers. Almost all of them (especially the ones you buy) are made to fold in thirds. However, if you are using the old-fashioned square kind then this is the easiest way:

1. Lay the square diaper out flat, and fold it in half to form a triangle.
2. Lay the baby on the triangle with her back in the center of the longest edge and the top point between her legs.
3. Fold the point between her legs up to her belly button, and then fold the other two corners to the middle on top of the point.
4. Fasten with a safety pin in the middle, or a much safer way is to use a clip made of rubber that has three ends that just stretch around and clip onto the cloth (one brand is Snappi Fasteners).
5. Then slip on the waterproof cover.

Making pre-folded diapers:

Diaper fabric sizes:

1 yard of 45-inch flannel will make 2 regular-sized diapers

11½ inch by 14½ inch: newborn–6 months

13 ½ inch by 18 ½ inch: 6 months–2 years

15½ inch by 21½ inch: toddler

Soaker pad fabric sizes:

4 inch by 13½ inch: newborn–6 months

5 inch by 17½ inch: 6 months–2 years

6 inch by 20½ inch: toddler size

To make the soaker pads, fold the layers of the fabric together and stitch around ¼ inch from the edge all the way around. Zigzag stitch to encase the raw edges. Take the piece of flannel for the diaper and center the soaker pad on it. Stitch down each side of the soaker pad to secure it. Put the top layer of flannel right sides together onto the layer with the soaker pad. Sew all around leaving an opening to turn it. Turn right side out, making sure the corners are poked out well. Sew the opening by tucking the edges inside and sewing. Stitch two lines in the center on either edge of the soaker pad to fasten them together.

To make diaper pants out of a sweater:

Get a sweater and wash in very hot water and dry it in the dryer so it will shrink. Measure the baby's hips. The sweater waistband is going to be the baby's waistband. Cut an equal-sided triangle with the waistband at the top the same width as your baby's measurements. Stitch the waistband ends together to form a circle. Then pull the last corner of the triangle up to the waistband to form leg holes and stitch partway down to be tight enough to comfortably go around the leg but not too tight. Cut off the cuffs of the sleeves of the sweater and sew in the leg holes to form cuffs for the legs.

To crotchet diaper pants:

Measure baby's waist and hips including the diaper and write the numbers down. Crotchet a triangle the same length as the measurement on all sides. Then stitch together just like the sweater pants, and create two ribbings and stitch onto the leg holes. Crotchet a chain twice as long as the length and thread through the waistband to make a tie to hold it on.

To knit very easy wool diaper pants:

This kind of diaper pant is also called a soaker, although it doesn't really soak anything. They need to be made out of 100% wool yarn, so that they will repel water. To make them, you cast on the number of stitches for the size of your baby (see below), and knit a triangle. The simplest way is to purl the whole thing, and at the start of each row take in one stitch. When you have finished the triangle, you fold the tip of it up to the waistband and stitch it together leaving two holes for the legs. Then crochet a long chain to use to tighten the waistband. Just use the crochet hook to pull the cord through at intervals and tie it in the front.

Preemie: cast on 50 stitches (10")
Newborn: cast on 70 stitches (12")
Medium: cast on 90 stitches (16")
Toddler: cast on 110 (20")
Super: cast on 130 (24")

Emergency diapers:

If you don't have any diapers and you need one right away, fold a flannel or cotton receiving blanket down and cover with the diaper wrap or plastic pants. If you don't have plastic pants, put on some tight baby thermal underwear or other tight bottoms. And if it really needs to be waterproof, take a plastic bag or grocery bag and use tape and scissors to fashion it into a waterproof cover. Other alternatives to diapers are cotton t-shirts, dishcloths, pillowcases and linens, towels, etc.

Washing diapers:

Experts say that you must wash them in hot water or the baby will get diaper rash. It really depends on the baby. If your child has sensitive skin, then slather her with diaper cream or petroleum jelly after drying her bottom and it will protect her from irritation. Hot water is better just to help get them clean, but if that's not possible it's all right. Dry them outside in the sun and they will bleach white.

Baby powder:

2 tablespoons crumbled dried chamomile flowers

¼ cup cornstarch

1 tablespoon orrisroot

½ teaspoon alum

Mix together and grind up further with a mortar and pestle. Sift and store in a powder shaker. Put on baby's bottom to prevent too much moisture.

Diaper cream:

Use aloe vera cream, or mix 1 cup olive oil with ¾ cup chopped herbs (cottonwood buds, pussy willow bark, chickweed, yarrow, and plantain are all good). Heat in a steel, enamel, or ceramic pan, and simmer on low for half an hour, stirring occasionally. Cool slightly and strain out the chunks. Put back in pan and add ¼ cup grated beeswax. Stir gently on very low heat. Drop some on a plate and put it in the freezer. If it hardens in a few minutes then the mixture is done—if it doesn't, then add more beeswax. If you can't spread it on your skin easily, add more oil.

Anti-fungal wipes:

1/2 cup distilled water

¼ cup vinegar

¼ cup aloe vera gel

1 tablespoon calendula oil

1 drop lavender essential oil

(If diaper rash is really bad leave out vinegar.)

Diaper rash treatment:

First wash the baby's bottom in warm water to make it completely clean, and then dry it thoroughly. For a wet rash use a goldenseal rash powder. For a dry rash use zinc cream or homemade cream.

Goldenseal: Mix 4 teaspoons dry clay, 1 teaspoon goldenseal powder, and 1 teaspoon comfrey root powder.

What is attachment parenting?

Attachment parenting is the opposite of many modern parenting philosophies in a lot of ways. Basically the belief is that you cannot spoil a baby, and you should be close to your baby as much as possible, since this is what most babies want. Often this includes baby slings, nursing on demand, baby sleeping in the parent's room, etc. Babies who are attended to promptly and with care, especially during the first 6 months, cry less later on than those babies who were left to "cry it out". This means that your 1-year-old will be less demanding if you stay attuned to her needs (this doesn't mean running every time you hear a squawk, but going promptly when you know she needs to be fed or comforted), than if you let her cry out her frustrations.

Making a western baby sling:

You will need:

2½ yards of 36–45-inch-wide fabric

2 metal or plastic rings

For a large baby use 45-inch fabric (no bigger). For a very small baby use 36-inch fabric. Hem all around the edge of the rectangle. Fold one end of the fabric into a fan shape 4–6 inches wide for about 10 inches down. Sew this securely by sewing over and over it back and forth. Put that end through both rings and fold it over and sew 3–4 inches down so that the fabric is holding the rings. Lay the sling flat with the ringed end at the top. Grab the bottom end (the end with no rings), and pull it through both rings. Then fold that end over the nearest ring, and down through the second

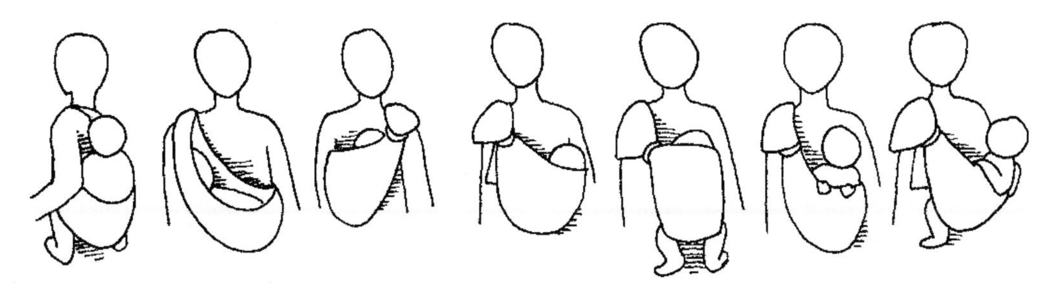

▲ Ways to wear a baby sling

ring. Put your left arm and head through the sling, laying the rings on your shoulder. Put the baby in the sling and adjust the rings so that they are sitting in front of your shoulder against your chest. The weight of the baby will pull it tight.

To wear the baby sling:

Lay the sling flat with the ringed end at the top. Grab the bottom end (the end with no rings), and pull it through both rings. Then fold that end over the nearest ring and down through the second ring. Put your left arm and head through the sling, laying the rings on your shoulder. Put the baby in the sling and adjust the rings so that they are sitting in front of your shoulder against your chest. The weight of the baby will pull it tight.

HOME EDUCATION

We who are engaged in the sacred cause of education are entitled to look upon all parents as having given hostages to our cause.

~Horace Mann, father of the public school movement

What is homeschooling?

When a child's near relative teaches that child at home, they are homeschooling. It is also called home teaching, home education, natural learning, home learning, etc., depending on the approach and philosophy of the individual family. Children who learn at home can be any age, and are generally more courteous than average children, are able to socialize well with any age group, score higher on all standardized tests, are well-rounded and diverse, live in families that have better unity, have more spiritual growth, and are more experienced than children who learn at public school. Though this is not always true, it is true of most homeschoolers, and has been proven by several studies.

What about people whose kids just don't go to school?

Homeschooling isn't a giant vacation. Even though homeschooled kids go to school for about four hours a day, they do more work during that time than is done in a public school within an eight-hour period. This is why public school kids have to do homework. Some people keep their kids home and call it homeschooling, but if they have no books in their house then it is not education, it is educational theft. There is a huge difference between unschoolers and those who simply have their kids at home. While the first group has filled its children's lives with educational resources, the other group has not and does not invest the time or energy in nurturing learning.

How do I grade my kids?

Some homeschoolers grade their children to track their progress, although it really isn't necessary. Generally this is done for high school. Here is the point scale:

A:	93–100	4 points
B:	85–92	3 points
C:	75–84	2 points
D:	70–74	1 point
F:	below	0 points

This is a fairly common grading system, but it is higher than what is used in many public high schools. In British Columbia's high schools, 86 and higher is considered an A. Unschooling homeschoolers have found that simply having a library available in their house and providing real-life experience, without actively sitting down to "do school," has been just as effective as correspondence school and their children still get higher scores than public schooled children. Grading may simply be a way to help you feel better.

Isn't homeschooling difficult?

The first years of homeschooling are the toughest, and some families with several children start homeschooling just one and add children as they go. You will probably feel overwhelmed, under qualified, ignorant, socially isolated, judged, and self-critical. If that doesn't deter you, it is also very common for total strangers to give you unwanted advice and criticism if they find out you homeschool. However, teaching your children at home is becoming more and more accepted by our society and an unprecedented amount of peopled are doing it too, which means you can find lots of support. The overwhelming majority of homeschoolers are Christian families, but even people who are not religious can find others who believe as they do. It does take discipline to be a homeschooling parent— instead of packing your kids to the bus every day, you will have to dedicate several hours a day to actively participating in your children's learning process, and they will be home the rest of the day instead of locked up at school, so you will have to find ways to provide activities for them as well. Most mothers find that while at first they have to really push the daily routine, their children soon begin to motivate themselves and start pushing the mom, becoming more independent and more fun to be around, and it becomes a rewarding lifestyle in itself.

What if one of my kids doesn't want to?

Some children, especially older ones, don't want to leave school and their friends to spend all day with their mother, which is understandable. Some children may find that adjusting to a slower social atmosphere to be absolutely intolerable (in their eyes). Or sometimes a child talented in a sports or drama program at school can't find the same opportunities staying at home. Sending your children to school is not a crime. If your kids are in a good school and the child seems to be thriving and doing well academically, it may not be necessary to homeschool. On the other hand, if your child just doesn't want to deal with the change, it may be a good idea to go through a trial period—show your child they don't have to lose their friends, and just try out homeschooling to see its benefits. Homeschooling is about actively working with your kids on their education, and forcing them to do it is counterproductive.

What about socialization?

How many adults spend an average workday constantly interacting with other adults? How many of those adults who do have a very social job feel energized and uplifted by their interaction? I am not sure where people get the idea that children require a full school day's worth of socializing with children their own age. I believe that this may actually hurt a child's development . . . how many adults only talk to people their exact same age? When children focus on academics for only a few hours a day, they have plenty of time to interact with others for the remainder of their time. Most homeschoolers regularly socialize with people of all ages, including the elderly, and sometimes volunteer as part of school. This kind of interaction is healthier, creates adults with excellent people skills, as well the obvious fact that it is closer to real life because it ***is*** real life.

What are my goals in homeschooling?

1. What are you preparing your children for?
2. Who is ultimately responsible for your children?
3. What are your religious beliefs?
4. What made you decide to home school?

Am I capable of homeschooling?

Anyone that can read can teach. Public school teachers read what to say to students out of books—if you look at a teacher's manual their entire conversation is written out. You teach something every day without realizing it, and even though it is challenging, you can homeschool with no formal education if you can get books for your children to read. Libraries are everywhere.

Is homeschooling legal?

Home schooling is legal in all states and provinces, but each state has different laws and requirements. You can get this information from your local homeschool support group. These groups exist everywhere there are homeschoolers, and often have support, information, newsletters, field trips, book conventions, and more. You can find these on the Internet, in your local churches, or at the public library.

What are the considerations I need to think about?

1. How many students do you have?
2. How long do you want to spend homeschooling each day? (the average is 4–5 hours)
3. Discuss with your partner the division of teaching responsibilities.

What steps do I need to take?

1. Find out the learning styles of your children.
2. Find a curriculum method that fits those styles.
3. Buy a three-ring binder and write everything down that answers the above questions.
4. Assess the books you already own and find out what you need to get.
5. Find out if your library has some of the books you need.

6. Check garage sales, thrift stores, antique stores, book sales, and the classifieds for used books.
7. Get your children materials they will need and to create a homeschooling environment (see below).
8. Decide on a day to start in the fall—the day your local public school starts is a good day because then your children and their friends will not have jealousy issues.

How do I plan my school year?

Every year you as a parent will have to plan out the curriculum and projects you are going to do during the school year. If you are doing unschooling, you may decide to buy certain books or kits or other materials. If you are using some curriculum that you are buying from a supplier or catalog, you will need to start ordering those books. If you are doing the eclectic method you will have to do much more planning. Basically you will have to use your three-ring binder and map out week-by-week the books, writing projects, and other activities that your children will be doing. This is why the eclectic method is more time-consuming, but also gives you more freedom.

How do I create a homeschool environment?

Many (if not most) homeschoolers do school at the kitchen table. Often various chore charts, history timelines, alphabet posters, and other wall hangings are stapled all over the house because your home becomes a full-time learning environment and many homeschoolers are not lucky enough to have a dedicated schoolroom. If you do have a schoolroom, you can have desks or you can use a sofa or tables. It is important that the place you study is conducive to concentration and has room for messy projects. For this reason most homeschoolers have a large table somewhere, many bookshelves, and have either limited television or have gotten rid of it completely. It depends on your children how structured your school time is. If they have no problem studying in their pajamas on the sofa then that's great, but many children do need a dedicated time each day and have to feel a bit more awake and ready for study by getting dressed and sitting at the table.

How do I know if my child is at the right level?

Some states and provinces require testing to see if your students are learning. If your student can do the things outlined in these grade levels, they will be above what the average public schooled child is doing. You should also add some history, art, PE, science, and other activities too—which will come naturally as a homeschooler anyway.

Elementary School

Grade 1: Be able to read a short sentence aloud, write simple sentences, listen to and understand stories and instructions, speak well, compare, and do simple math.

Grade 2: Be able to understand the way things are written, read for personal enjoyment, find out things through reading, write short paragraphs, and do basic math.

Grade 3: Be able to tell and write stories, use a dictionary, and perform more advanced math calculations.

Grade 4: Be able to know a word's roots, read with skill and expression, do simple research and writing projects, read for literary value, discuss ideas, use writing for practical uses, and do more advanced math.

Grade 5: Listen to adults reading aloud to improve diction, think about and judge information, create original writings, understand parts of sentences, and use math for practical and thinking purposes.

Grade 6: Analyze reading for content and style, write all kinds of writings, develop speech, organize information for writing, and begin preparation for pre-algebra.

What do I do at the end of the year?

At the end of the school year it is time to make an evaluation of all that you did and how they worked. First look over the curriculum that you used and see how it worked for each child. Did it provide enough structure? Or did it have too much structure and do you need to loosen up a bit? Also make sure that each child is meeting the educational requirements that you want to fill . . . . what areas need to be worked on? What methods of learning worked best? Does hands-on work better, or a written approach? Does a child need to slow down, and does another child get bored and need a faster pace?

Then it is time to find new curriculum for next year. Look each book or series over before you buy and make sure you have the time to do it. Think twice about curriculum that requires lots of time from you, when you can just as easily get a book that allows your child to figure it out for themselves. Each year you will get better and better at doing these things. The first years you will spend hours agonizing over what to use, but after a while you will find that less is more and simple is better.

Helpful Rules to Prevent Burnout:

1. Know the last day of school. Either follow the public school schedule or your own (for example, how many days to finish all the books).
2. Have some projects that all your kids work on together. Unit studies work well for this.
3. Don't be afraid to take days off and have fun. Playing is learning too.

How many styles of homeschooling are there?

There are many approaches to learning in home schooling, all designed to fit the individual family, and especially individual students. None of them costs a lot of money (unless you decide to use an expensive correspondence school) and anyone can do them.

What do most homeschoolers do?

Most homeschoolers try all kinds of methods and curriculum as they go along, and their oldest child is the guinea pig. They have to change methods and/or curriculum as their children change and grow. Most of the time the younger children do an unstructured schooling, while older children have structured school that they do with almost no help or guidance from you.

What are the goals of homeschooling methods?

The different styles of homeschooling are either structured or unstructured. Structured is closer to the public school style, and unstructured focuses more

▲ **Montessori method of learning to differentiate dimensions**

on doing real-life projects to learn. Most homeschoolers use an in-between method. There are many homeschool methods with different names and methodology, when in fact they are all basically the same. They have these goals and principles:

1. Read the great literature that shaped our society.
2. Do not distinguish between classes in society.
3. Enhance the reasoning powers of the student.
4. Focus on principles and values.
5. Let the student guide themselves to some extent.
6. Begin real-life work early.

What are the major homeschooling methods?

Charlotte Mason Method: Charlotte Mason was a British Educator in the 1800s who developed a combination of assignments that included narration, nature notebooks, fine arts, language study, and focused on literature. It uses few textbooks or workbooks, as all information comes from literature. It is a more structured method.

Classical Education: Adopted by many Christian homeschoolers, it is based on ancient Greek and Roman educational methods. The focus is on language, logic, and fact, and children learn Latin. It is very structured and was used in America before the 1850s in private schools. Students read very difficult classical literature such as Cicero and Virgil and study many sciences.

Eclectic Method: Most home-schoolers are actually eclectic, combining many methods and books with a few workbooks, a lot of literature, many kinds of projects, and real-life experience, all focused around each individual child. Some other methods may be utilized and mixed in as well. It is somewhat structured because the children sit down in the morning to "do school" but usually in a relaxed manner (many times in their pajamas).

Montessori Method: Montessori works better for younger children, and is not good for high school. It is very hands-on and sensorial, and the children are put into an environment with no junk food, television, or computer. Books and toys are very high quality, and the toys are made of natural materials such as wood. The child is allowed to explore the controlled environment, with careful guidance to help increase concentration and other skills.

Resource Centers/Cottage Schools: Some communities of homeschoolers have created small homeschool-like schools by pooling their resources. Thus the children have the benefit of homeschooling, with more group learning opportunities and more kids to socialize with.

Umbrella Schools: Unlike a straight correspondence school, an umbrella school allows the family freedom to complete schoolwork at home, but also has teachers in a physical school that the child can meet with either in person or on the Internet. Charter schools often have a focus such as technology or the arts, and have facilities that a homeschooling family might not have at home. However, you will be accountable to the umbrella school, which gives you a little less freedom.

Unit Study Method: The family picks a topic and incorporates many disciplines (such as math, science, geography, art, music, language, history, etc.) and exhaustively studies that topic until they can go no further. Some unit studies use character traits as a topic, others might use historical time periods. There are now many different kinds of unit studies, some better than others. The student will remember much more after doing a unit study than from most other methods.

Unschooling: Also called Natural Learning or Child-led Learning, basically the parent finds out their child's interests and provides lots of materials for them to explore these interests. These materials might include books, movies, building projects, journaling, Internet, field trips, art projects, writing, kits, puzzles, music, puppets, collections, etc. Unschoolers spend lots of time on field trips, in the library, and at garage sales.

Waldorf Method: Developed by Rudolf Steiner (who also created biodynamic gardening), there are hundreds of Waldorf schools around the world, but it can be done at home as well. Young children don't do much academic study, simply focusing on writing as they get older. Art, music, gardening, and foreign languages are essential subjects, and the youngest children only do artistic projects. All children learn the recorder and how to knit. Young children have workbooks and older children have a few textbooks. There are no grades, and electronic media is discouraged.

What are children's basic learning styles?

Children (and you) have in-born learning styles, or ways in which they learn the best. Usually children use a combination of styles based on their personality. Most children change their styles as they age also—for instance young children tend to be type #1, while older children develop into types #2 and #4. No style is any better than any other:

1. Is disorganized, has a short attention span, has a hard time in groups, has trouble finishing things, doesn't like to work, impulsive, looks for more efficient ways to do things, would rather do activities than read, like arts, PE, and hands-on projects. This child would do well with unit studies, Montessori, and unschooling.
2. Likes plans and schedules, is responsible, is not very spontaneous, follows rules, likes organized school curriculum, isn't extremely creative,

does well with memorizing and drills, worries about requirements, likes facts and clear answers, needs approval. This child would do well with Charlotte Mason method, classical learning, or an umbrella or correspondence school.

3. Likes to be in control, is self-motivated, values wisdom, thinks and acts logically, likes to understand reasons behind ideas, organized, works alone, impatient with slow people, avoids social situations, makes long-term plans, likes math and science. This child would do well with eclectic, Waldorf, Charlotte Mason or maybe classical learning.
4. Enjoys socializing, likes being in a group, has a warm personality, looks for meaning and significance, likes to be recognized, worries about what others thinks, idealistic, can be organized, interested in general ideas and not details, like language arts, social studies, and fine arts. This child would enjoy an umbrella school, unschooling, unit studies, or Waldorf.

What are the goals of unschooling?

The theory (or non-theory) of unschooling is that children have no need to be taught to think and learn, and when provided with the resources and opportunities your child will educate herself much better than some "system" or "curriculum." Unschoolers often fight much opposition and even personal doubts as to whether their kids are getting educated properly, but standardized tests show that most unschoolers are doing just as well as other homeschoolers. There are several modern writers who have theorized why this method works, and their books are listed below.

The unschooled child's day is largely self-directed. For example, a child might have a passion for comic books. After reading about Charles Schulz, the child studies the whole Peanuts history, the role of comics in history and culture, and invents her own comic strip. Then the child might read a World War II comic and do a bit of study on airplanes used during the 1940s. Some of the projects include art, science, geography, language arts, history, and even math (to graph the panel transitions in Tintin comic books).

Another common idea among unschoolers is that children do not have to be hurried into education. If the interest is there then a child may be introduced into some formal training, but unschoolers would rather that young children have a very relaxed and informal education period even up to second or third grade. Traditional educational achievement is not really the goal of unschooling, but instead developing the whole child and their natural talents and abilities. By doing this, the rest of the "facts" that kids need to know will follow naturally. Work and play are combined and children really love "doing school" because it doesn't feel like school at all.

Another outcome that unschooling parents like to see is their children learning to know themselves. Many kids leave school completely in the dark on what to do with the rest of their life, and the goal of unschooling is to hopefully allow kids to know about the world around them and themselves well enough that making decisions about the future is the easy part.

How much freedom do unschooled kids get?

Most unschoolers would say complete freedom. This does not mean that their kids live wild and barefoot and draw on the walls with crayons. There is no classroom, no formal curriculum, lesson plans, or calendars. Often school goes on all year round. Some choose to add a math book, but that's usually it. Critics believe that kids who have total freedom won't learn the stuff that they just don't like and will end up stunted. Unschoolers believe that it doesn't matter when the learning happens, and by the time their kids reach adulthood, they will have watched their parents enough to learn what needs to be known.

What about state laws?

You need to contact your local homeschool group and find out. If a standardized test is required, or submitted written work, you should know so you can be prepared. Some states require that you do school a certain number of hours a day, but unschooling never stops. Think about astronomy classes in the middle of the night. A good idea is to document everything your kids do and to make a portfolio of work. This is especially important in high school years if your kids plan to enter college.

How do the kids learn anything?

For people who are not used to motivating themselves to study and who do not have access to learning resources, unschooling can seem completely insane. The truth is that most homeschoolers of any style will find that if there is something educational in their house, their children will eventually take up research and study outside of school hours, just for fun. Kids who love to learn will start to learn just out of enjoyment. Unschoolers (and most homeschoolers in general) should invest in a well-stocked library with a multitude of nonfiction and fiction subjects, art supplies and drawing tablets, board games, software, math manipulatives and puzzles, globes and maps, journals and notebooks, dress-up clothes and costumes; building toys such as Lego, Duplo, wood blocks, sand boxes and wood for building; musical instruments; and any science supplies or supplies for interests your kids might have. Unschoolers also keep informed about local events and performing arts, take classes in sports and arts, and visit art and science museums.

What is my role as an unschooling parent?

Besides being the provider of supplies and resources, and driving people around to various locations, unschooling parents often find themselves trying to catch up with the research that their kids are doing. The public library will become your best friend, and you will be reading many of the books your kids are reading in order to answer questions and partake of any discussion that might be happening. In its ideal form, unschooling is a partnership between child and parent in a learning adventure.

What about math?

There are many, many mathematical resources and learning methods besides the traditional textbook until they reach algebra. There are a plethora of games, puzzles, and books that are for people who don't find math fun, as well as books that give young children fun games to learn math.

Books loved by unschoolers:

Freedom and Beyond by John Holt
Miseducation: Preschoolers at Risk by David Elkind
Experience and Education by John Dewey
The Conspiracy Against Childhood by Eda LeShan
The Unschooling Handbook by Mary Griffith
I Learn Better by Teaching Myself/Still Teaching Ourselves by Agnes Leistico
Learning All the Time by John Holt
How Children Learn by John Holt
How Children Fail by John Holt
And What About College? by Cafi Cohen

Dyslexia and Dyspraxia:

Dyslexia is a learning disability that some children are born with. These kids often have trouble reading and spelling. It is generally limited to words and letters, and it often seems that kids will read a letter as something else, or write letters backwards. There are also related disabilities such as dyscalculia (mathematical trouble) and dysgraphia (writing trouble). Sometimes kids will have forms of all three. About 15–20 percent of people have some language-based learning disability, and dyslexia is the most common. Dyspraxia is when a child has poor coordination to the point that very simple things such as walking a straight path can be virtually impossible. The child may often not know his body position in space.

What can I do about it?

First of all, dyslexic children can learn to read and write, although it may be a struggle—and it is well worth the effort. Second, while dyspraxia may cause difficulties and delays, it is also possible for these children to learn things such as riding bikes, tying shoes, etc. Homeschooling is often a last resort for families with children with these learning disabilities, because of public school's inability to give the individual attention necessary for their proper development. Homeschooling families have had so much success that it should be the first choice for dyslexic children. Kids have the option of NOT sitting still during school, they can take breaks often during very frustrating learning sessions, and they can learn at their own pace.

◂ Unschooling allows children to enjoy life experiences as learning

What activities work best?

Reading games and tapes, especially interactive computer games, are the best method of learning for these children. For math there are also card games and dice and strategy games such as chess and Battleship. There are also many math board games and software available as well. For physical development swimming is a skill that should be learned and will definitely help with motor skills. Eventually you should also get your kids to learn to ride a bike and play traditional outdoor games with other children. Many of these kids also like thick pencils and crayons, and they should try doing manual activities like Lego and weaving.

Aspergers Syndrome:

Aspergers syndrome (or disorder) is a neurobiological disorder that is characterized by children having normal intelligence and language development, but who have autistic-like behavior (and difficulties in social situations and communication) ranging from mild to severe. Generally AS children will have trouble with social skills, won't like changes or transitions, and want things to always be the same. They often have obsessive routines (somewhat like OCD) and may be preoccupied with a particular subject of interest. They may have trouble reading body language and often have trouble figuring out where their body is in space so they may seem awkward. These children also are often really sensitive to sounds, tastes, smells, and sights so they will probably prefer soft clothing, certain foods, dim light, and quiet. They have a normal IQ and often have exceptional talent or skill in a very specific area, and often have very good vocabularies without being able to effectively communicate with others. AS is essentially like having just a tiny little bit of autism, or being high-functioning autistic. It is often misdiagnosed and many famous, eccentric, and brilliant people have it.

What to do about AS.

First of all, you will find that your child with AS will have some behavior problems, especially when young. Actually, homeschooling the AS child is the best possible solution. While full-blown autism may require additional help and therapy, Aspergers syndrome may cause difficulties but is much easier to handle when your child is at home. Public schools are not equipped to handle the "eccentricities" of this disorder, and rather than treating your child normally (which is the best way), the conventional educational methods and socializing will only make your child feel more isolated. Another benefit of homeschooling is your ability to control your child's diet. Autistic children can find much improvement by changing their diet, and some have found that eliminating wheat or dairy, or all sugar and food dyes has really helped. There is a link between autistic behavior and diet and it is because many autistic children can't break down certain proteins. The best thing you can do is become completely educated about Aspergers syndrome and get connected with other homeschoolers dealing with the same thing—and there are many!

How can I find used books?

Books are the single greatest resource for education at home, whether for your kids or for yourself. The best place to get books on any subject is your local thrift stores and secondhand shops. It is a necessity for any serious study to have your own home library, since you can't always rely on your public library to have everything you need. The secret to finding the books you need is to focus on classic literature, avoid revised editions, and pick up old schoolbooks for English, science and math. Library book sales and school auctions are also good places to find books.

What are publishers?

Publishers are companies that publish whole or partial education programs for homeschoolers. They also sometimes run correspondence schools. It is common for new homeschoolers with elementary-age children to buy their entire curriculum from a publisher because it's so easy. They also have high school programs, but it is more common to only get one or two books from a publisher for that age.

What is a unit study?

Unit studies are programs that use a theme to study many subjects. They are usually very hands-on, and study different topics under a category. Unit studies take a lot of research and preparation time, but the child retains more knowledge. A unit study can also be a one-subject study lasting for a limited amount of time. You can create your own full or partial unit study by reading a historical fiction book and then using a variety of writing, art, sculpture, and even dress-up, theater, and other projects to learn different parts of history or science.

What is book curriculum?

Book curriculum is the usual homeschool method. This is when a family buys different books for different subjects and assigns their own writing and hands-on projects around these books. Literature and writing are a big part of this kind of study. For example, you could read *Moby-Dick* and then use history books to study whaling life during that time period. Then you could have writing assignments and art projects . . . this is very similar to a unit study, except that a unit study does more pretend. Instead of building an actual ship, unit studies might build models out of cardboard. Book curriculum tends to be more real-life and focuses on writing projects.

What is a correspondence school?

Correspondence schools are schools that publish their own curriculum and send it to homeschoolers. The homeschoolers complete the work and return it to the school, and it is officially graded and recorded. These are much easier than putting together your own curriculum, and don't require much time investment except for actual schoolwork. When choosing a school, it is best to choose one that makes the curriculum specifically for homeschoolers—getting the books from a public school can cause a great deal of problems because they tend to be very critical of homeschoolers. It is also very wise to choose a school that has been around for a while—don't choose one of those "Get Your High School Diploma" television commercial schools.

What are the ways to do high school?

1. College preparatory—Do the required amount of study for each subject to create your own high school transcript, grading the work as you go. Minimum requirements for college preparation are:
   Religious: 4 years
   English: 4 years—emphasize writing skills
   Mathematics: algebra, geometry, and advanced math
   Social sciences: 3 years–1 year country and world history, 1 year government and economics
   Science: biology and chemistry with lab work
   Foreign language: some colleges require 2 years of one language
   Fine arts: 1 year
   Electives: computer
2. Correspondence school—If you want an accredited high school diploma, you can do a correspondence program. You can find these programs by talking to your local homeschool group, through a homeschool magazine, or online. Make sure you choose a well-recommended school that has been around for a while.
3. General education—If you are not planning on college, you can do

a minimum of requirements and supplement it with electives, grading the work so you can create a transcript. Minimum requirements for a general high school education are:
Religious: 4 years
English: 3 years—emphasize writing skills
Mathematics: 2 years—consumer or pre-algebra
Social sciences: 3 years–1 year country & world history, 1 semester government and economics
Science: 2 years of biology & chemistry, lab optional
Foreign language: 1 year language or fine art
Fine arts: can do 1 year language instead
Electives: any future job interest training

4. Unschooling - The student directs their study, with any degree of freedom and direction that the parent decides. This way is the most fun, you learn the most, and you can still take college entrance exams to get into most colleges.

Will I be able to get into college without a diploma?

Many homeschoolers are worried that they will not be accepted into a good university without having a diploma from an accredited high school. This is absolutely not true anymore, but it does depend on the college and what country you are in. As a teenager I could not decide what I wanted to do, and applied to several public universities including Central Washington and Penn State. Even though I had an accredited high school diploma from a correspondence school, I only used the printed diploma from my parents and my homeschool transcript. I was accepted easily into both of these schools. One of the keys to creating an excellent transcript is to make sure you cover every subject, and give a GPA on a 4.0 scale for each one. Be realistic with the GPA—no one will believe your child has a perfect 4.0 unless there are some awards or competitions to back it up. The second most important thing is to make the transcript look very professional—use a good word processor, make a logo for your "school," and sign it. My parents also typed a letter that said I had graduated high school on such-and-such date, and both of them signed it at the bank and it was notarized. Get several of those made if you plan to go to college. One thing that many homeschoolers forget to do is to enter contests, competitions, and other academic events. Winning or even simply participating creates an impressive résumé that most scholarship and school administrators will recognize. Take the SAT or ACT, or both, put that together with your diploma, transcript, and a portfolio of community involvement and achievements, and any university should open its door to you. In Canada and other countries it may be more of a challenge to get in as there is more competition, but even tougher universities will accept you under conditional status, meaning you are on probation until you prove yourself through your grades. Wherever you are, homeschooling is more and more recognized and accepted, and more universities are providing a more open-door policy, which benefits the untypical student.

7 | Food, Field, and Garden

GARDEN PLANNING

Last night, there came a frost, which has done great damage to my garden. . . . It is sad that Nature will play such tricks on us poor mortals, inviting us with sunny smiles to confide in her, and then, when we are entirely within her power, striking us to the heart.

~*Nathaniel Hawthorne,* The American Notebooks

What are the difficulties in living solely off your garden?

Planning is extremely important when you plan to live completely off your garden. If you have read about the food pyramid, you will need to grow a variety of things in order to maintain your family's health. It is much easier for a vegetarian to grow all she needs, because meat takes an incredible amount of work and space to grow. With a small amount of space, goats and chickens are more feasible but most of your area will be devoted to the garden. Make sure to follow these rules of thumb:

1. Use-space saving techniques (see below).
2. Save seed packets in a box inside the house and keep track of what you grow.
3. Have a scarecrow, a deer fence, a dog, a cat, and any other garden protection you need.

How much food will I need?

The first thing you need to do is plan out how much grain you will need, and how much space it will take to grow it and work around that. You need to know how much food your family eats, and how many plants or area will be need to produce it. You will need to plan for winter, and grow extra in case of shortages. You will also have to spend an amazing amount of time storing and preserving your food—canning, freezing, and drying all take some effort. You can use the guidelines for planning a food storage to estimate what each person eats in a year. When you know how much your whole family will eat, you will need to know how much seed is needed to grow that amount. Add together:

1. The amount your family eats and how much seed it will take to grow that amount.
2. The amount of seed will be needed to grow plants next year. You need to grow extra plants that will go to seed to grow more plants for the year after (if you are not buying them).
3. Overcompensate for shortages: pests, drought, storms and disease all shrink your crop, so add ⅓ more to the seeds (an organic crop loses about ⅓).
4. If you are feeding animals, make sure to add in enough for them.

How much can I get per acre?

Measurement per bushel:
Alfalfa: 60 pounds
Barley: 48 pounds
Buckwheat: 50 pounds
Clover: 60 pounds
Field corn: 56 pounds
Flax: 56 pounds
Oats: 32 pounds
Red clover: 50 pounds
Rye: 56 pounds
Sweet sorghum: 50 pounds
Spelt: 40 pounds
Wheat: 60 pounds
Apples: 42 pounds

The square footage to grow 1 bushel is:
Field corn: 500 square feet
Grain sorghum: 600 square feet
Oats: 600 square feet
Barley: 900 square feet
Wheat: 1000 square feet
Buckwheat: 1300 square feet
Rye: 1500 square feet

How can I maximize the yield of my garden?

1. Harvest your plants every day. Broccoli, cucumber, summer squash, beans, and chard will grow more if you regularly pick them.
2. Plant your vegetables close together using companion planting and careful planning.
3. Use transplants to start plants in March so you can start harvesting in June.
4. Build a greenhouse so you can grow during the winter.
5. Make cold frames or plastic covers to put on top of plants in early spring.
6. Use succession planting, double-cropping (planting a new crop as soon as one is harvested), or plant every two weeks.

▲ Transplanting seedlings

How do I transplant sprouts from inside the house to my garden?

Start seeds in the house or greenhouse, planted in any small container such as egg cartons, cans, boxes, etc. Don't use peat moss because it doesn't break down well enough, and if you have no container, simply fold newspaper in a circle or use eggshells and fill them with regular dirt. Water them whenever they look dry. When it's near to planting time, set them outside in the sun for longer and longer times each day so that they can adjust to the outdoors. Plant outside after the frost danger is over, at the beginning of wet weather, so they can get a good start.

What is a cold frame or cover?

You can lay thin plastic which you throw over your plants and hold down by rocks, or you can use a cloche, which is a light frame holding the plastic into a dome shape. A cold frame is a wooden box with

Elaborate cold frames ▸

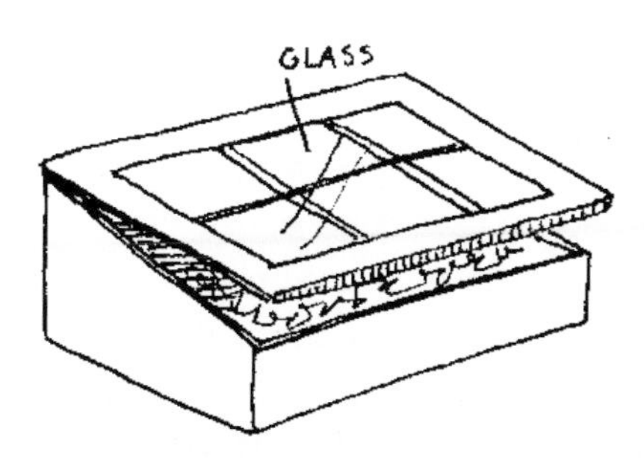

▴ Simple cold frame made of a window

a glass lid that can be opened and closed. To use a cold frame, you simply cover the plants when you know there will be frost. Generally this means close them at night, and open them during the day. Cold frames are harder to build and difficult to move but last longer and protect your plants from the weather. Cloches and plastic are easy to make but get wrecked easily. An easy cold frame is to just use straw bales to hold up an old window—be resourceful. To make a sturdier cold frame, build a 6x12-foot box on the ground with no bottom, facing south. Angle the top so that the higher end is on the northern side. Fasten two windows with hinges on the top as a lid. Use a stick to prop the windows open. This is like a miniature greenhouse.

▾ Basket row cover that can be used as shade or covered in plastic

▴ Row cover with plastic hoops

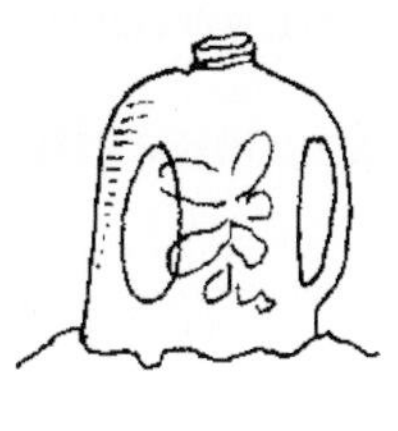

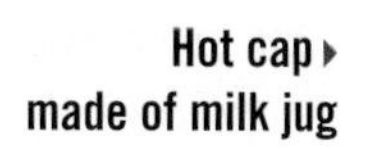

Hot cap ▸ made of milk jug

How do I grow plants when there is still frost?

If you are extending your season by planting several gardens during the year, the early and late gardens will need special care. Some plants are hardy to frost, others aren't. For a garden planted in August you will especially need frost covers over the plants. This means that you will have to be extra vigilant in watching the weather and be dedicated to covering them up.

What is interplanting?

Every plant has an ideal number of inches that it should be from other plants. You can grow two kinds of plants together in the same space if you carefully plan. A tall plant should be placed next to a relatively small plant that doesn't need much sun. To determine the spacing, add the spacing inches together then divide by 2. For instance, to grow cabbage and turnips, add 15 plus 4 = 19. Divide 19 by 2 = 8. Therefore, you can plant cabbage 8 inches from turnips. Also remember the companion planting rules for each species.

How do I deal with plants that take up too much space?

If you have very little space, summer and winter squash, cucumber, watermelon, muskmelon, cantaloupe, or corn all take up lots of room, although you could try using the "bush" variety of some of these plants. If you must have squash and corn, plant the corn and the squash together in the same place. Also you can plant pole beans in the corn and it will climb it, so you won't need poles.

Space-saving ideas:

1. Use plant stacking—plants that are of different heights are planted close together utilizing the direction of the sun. For instance, the tallest at the end, with graduating sizes down to the front: corn, then pole beans, then kohlrabi, then onions, then carrots.
2. Use containers on areas that have no soil (pack those plants in wherever you can), and especially use square containers because they utilize space better.
3. For climbing plants, use mesh for tendrils (grapes, etc.), poles for twining plants (pole beans, etc.), and lattices for other kinds of climbers.

What is intensive gardening?

The goal of intensive gardening is to have as many plants growing in as small an area as possible. An intensive garden also has as many plants harvesting throughout the year as possible, so succession planting is always used. You have to hand weed intensive gardens because there are no rows. You create a raised bed no wider than 3–4 feet so you can reach all the plants and the soil must be rich, with no rocks or clumps. Any climbing plants are given trellises, nets, strings, cages, or poles so they can grow up instead of out, and all plants are interplanted, hopefully with their companion plants.

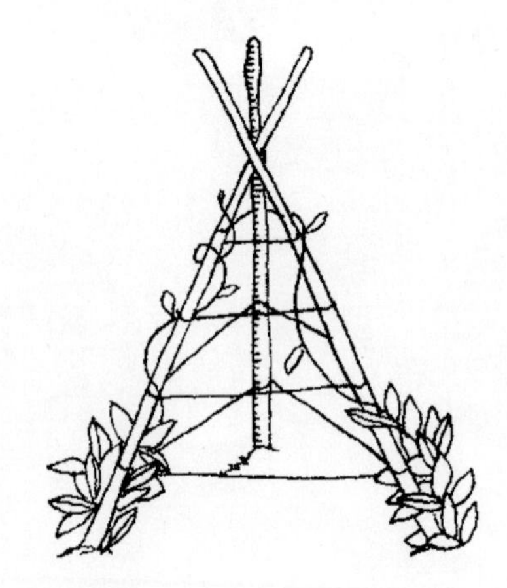

▲ **Tripod trellis**

What location considerations do I need to think about?

Sun and shade: Shade is sometimes just as important as sun for some plants. In very hot sunny regions, your plants may burn and die, and may need shade. But if you live in a shady and cool area, you need to put your garden in the sunniest place you have. If you use rows, they should run north to south so the plants will get equal sunlight.

Water: It is best to be near your water source, even if you are hand-watering. It is tiring to lug buckets, and difficult to pump

◂ **Intensive garden**

irrigation ditches from far away. And it is hard to find an extra-long hose to reach to your tap.

Slope: Plant on flat ground to avoid soil erosion. If you don't have flat ground, you will need to create terraces, or steps up the slope with access for you and water.

Traffic: Don't put your garden directly near a highway. The most lead-filled soil is within 100 feet of a street, and the lead will enter your vegetables and you can get lead poisoning. If you can't do it far away, make an entire new layer of topsoil full of compost in order to be safe.

Size: It takes about 50 square feet per person (planting in rows), unless you are using intensive gardening. An intensive garden can theoretically grow everything in a very tiny space.

How do I choose plant varieties?

It is important to know your own climate and the frost dates. Decide what your family wants to eat during the year, and then decide what kind of varieties will work best in your temperature zone. If you are going to save your own seed, buy open pollinated varieties instead of hybrid varieties. It can actually be illegal to allow a hybrid (an engineered plant) to go to seed if it is a patented variety.

Where can I get seeds?

It is best to find a small, organic company if possible, instead of the big dealers. Don't buy the seeds from a store where they were in direct sunlight—they need to be kept dark and cool. Some companies specialize in heirloom seeds,

plants that are old rare varieties that are often well-suited to the small organic farmer who wants to save seed. Heirloom plants also tend to be hardier and more pest-resistant than modern varieties.

What is the simplest way to do succession gardening?

Divide your garden space into four parts. The parts will be used for early spring, spring, and summer and fall gardens. In each of the four parts you can also delay some of the planting by two weeks, so you can continuously harvest. However, for most people this is not necessary.

Another way to do succession gardening:

1. Divide your season into two parts—cool season and warm season. The cool season is when the night temperature is between 25–60°F. The warm season is when the night temperature is 15°F cooler than the day.
2. Plant cool season plants when you know that the night temperature is not below 25°F: arugula, broad beans, beets, broccoli, Brussels sprouts, cabbage, Chinese cabbage, cauliflower, collards, endive, fennel, kale, kohlrabi, lettuce, parsley, parsnip, peas, radish, spinach, and turnips. You may need to use the greenhouse or cold frames.
3. Plant another garden of warm season plants at the normal planting dates: artichokes, asparagus, bush beans, carrots, celery, sweet corn, cucumber, eggplant, garlic, leek, melon, onion, peanuts, sweet peppers, pole beans, potato, pumpkin, rhubarb, squash, sweet potato, Swiss chard, tomato, and watermelon.

What are Latin names?

Plants are identified with both a popular name and a Latin name, for example: cabbage is also called *Brassica oleracea*. Latin names are the scientific way of identifying each species, and are much more accurate than popular names. When buying seeds, always make sure that the Latin name is specified so you know what you are getting.

Latin classification system:

Kingdoms: All living things are classified into five kingdoms. Plants are part of kingdom Plantae, fungus is in kingdom Fungi, and animals are part of kingdom Animalia. There are also kingdom Monera, made up of microscopic single-celled creatures such as blue-green algae, and kingdom Protista, which are similar but have complex cells, such as seaweed.

Phyla: Each kingdom has several phyla that further categorize things. The mushroom group is phylum Basidiomycota, the yeast group is called phylum Ascomycota. In the Plantae species there are two groups: One is for conifer, moss, and fern-type plants called gymnosperms. The other is the group of plants which animals (including us) eat, called angiosperms.

Classes: Angiosperms come in two classes, Monocoyledoneae and Dicotyledoneae, or monocots and dicots. Monocots have seeds with one leaf, the plant has narrow leaves, and the flower has parts in multiples of three. Dicots have seeds with two leaves, the plant has broad leaves, and the flower has parts in multiples of four or five with large colorful petals.

Within each monocot and dicot family there are genera. The genus of a plant identifies exactly which species it is.

Monocot families you should know:

Arecaceae: palm family—coconuts and palms (equivalent of wheat in a tropical area)

Graminae: grass family—wheat, bamboo, corn, rice

Liliaceae: lily family—onion, lily, tulips

Musaceae: banana family

Dicot families you should know:

Apiaceae: carrot family—carrot, parsley

Asteraceae: sunflower family—dandelion, sunflower

Brassicaceae: cabbage family—cabbage, cauliflower, kale, turnip

Cucurbitaceae: melon family—cucumber, melons, squash, pumpkin

Fabaceae: pea family—peas, peanuts

Lamiaceae: mint family—lavender, mint

Leguminoseae: legume family—alfalfa, bean, peanut, pea, soybean

Poaceae: grass family—wheat, barley

Rosaceae: rose family—rose, apple

Solanaceae: nightshade family—pepper, potato, tomato

How do I understand Latin names?

The Latin name of the common foxglove is *Digitalis purpurea*. In Latin, the genus is most often capitalized, and the species name is not. Often in books and seed catalogs, they will use abbreviations, so it could also be listed as *D. purpurea*. It is important to get the right variety since this wild variety is used to make a drug that treats heart disease—other foxgloves (*D. mertonensis, D. grandiflora*, etc.) do not have the same properties.

Kingdom: Plantae

Phylum: Angiospermophyta

Class: Dicotyledoneae

Family: Scrophulariaceae (popular name: figwort, related to foxgloves)

Genus: Digitalis

Species: Purpurea

How do I draw out my garden on paper?

Measure how many square feet your garden space is, which way the ground slopes, the direction of the sun (or light source) and where it hits, and where there is shade. Also find out how far away and what direction the water source is and use graph paper to draw a picture of all these things. Then you can draw your garden after you have decided your plants, varieties, your calendar, etc.

What other things do I need to write down?

In your record you also need to write down all your plant varieties, what seed brand (or source) that you got your seeds from and how many you used, any problems per plant, when you started harvesting and how much you got, and notes for yourself for next year.

Example garden record:

Orchard

Apple—got caterpillars eating worms. Canned some apples, dried others. Graft on other branch from neighbor this fall.

Plum—plums did not grow enough to can, so dried them.

Herb Garden

Fennel—harvested all through the summer.

Garlic—some of the garlic was small, but it did make my roses better.

Yarrow—made into stored infusion for medicine.

Grain

Soybean—grew some soybean, although I did not water enough and could have done better.

Kitchen Garden

Sprouts—did bean sprouts which did well, but were sour tasting.

Vegetable Garden

Corn—got tall but did not produce big ears, need more sun and more water.

Tomato—tomatoes did well, I had enough to can and freeze and dry.

CULTIVATION

Many things grow in the garden that were never sown there.

~*Thomas Fuller,* Gnomologia, *1732*

How do I clear large stones from a field?

Get a draft animal and a chain. Dig around the stone, and then put a chain tightly around it so that it won't slip off. Have the animal pull the chain and roll the stone right onto a stoneboat. You can make the stones into a fence.

How do I clear away stumps?

Hitch a draft animal to a mallet lever. Wrap a root on the far side of the stump to the mallet lever "handle" with strong rope. When the animal pulls, it will turn the mallet lever, forcing the stump out of the ground. What is cultivation? Cultivating (also called tilling) is the process of turning over soil with a plow or disc, or using hand methods such as a shovel to turn over the earth. There is an alternative to cultivating called natural farming (see below). The goal in cultivation is to loosen up the dirt and remove any rocks so that plants will have the optimum environment. When cultivating with hand tools, people most often make a raised bed. A raised bed has the same amount of dirt, but it has been stirred and loosened so that it is fluffier, making it higher than the surrounding soil. Many raised beds are "double dug," a process of digging that makes it simpler to raise the bed.

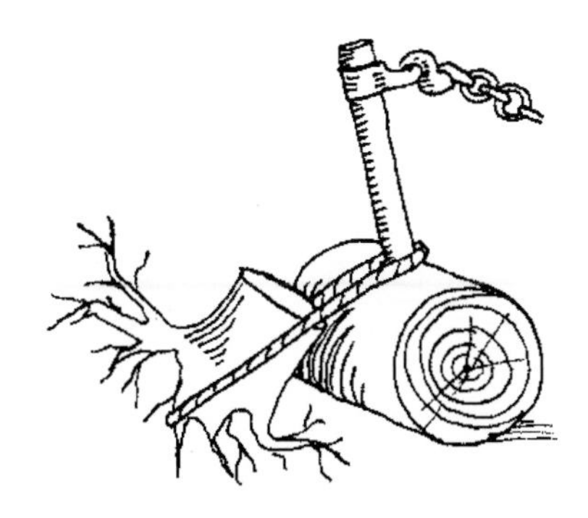

▲ Mallet lever

Cultivation calendar:

Fall: Cultivate all your ground, including the new areas you will plant next year.

Spring: Cultivate all the ground you will plant after the soil is dry enough. To see if it is, make a ball of soil and drop it from 3 feet up. If it doesn't break apart, don't till. Cultivate as early as you can.

How do I double dig with hand tools?

The first time you dig will be the hardest. Cover the plot 6 inches deep with your fertilizer or compost. Dig a trench as deep as you need, then dig another trench next to it. Put the second trench's dirt into

Double digging

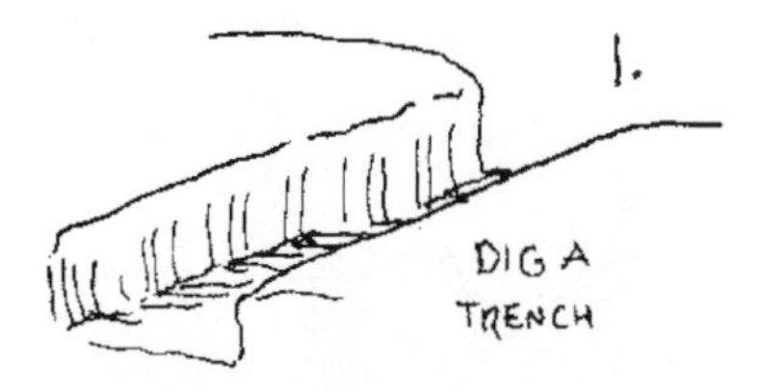

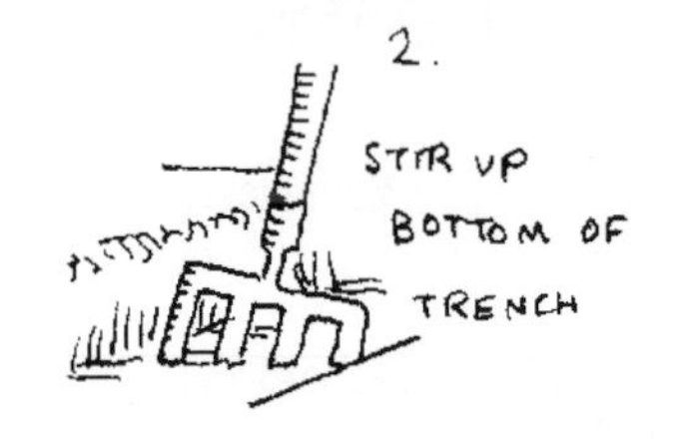

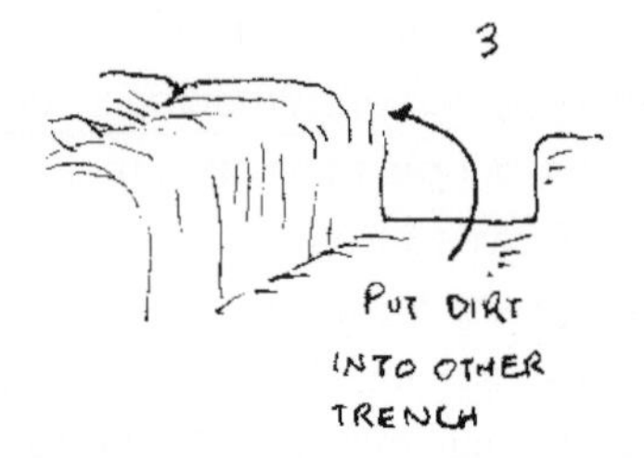

the first trench, dig a third trench and put that dirt into the second trench, and so on throughout the plot. If there is a lot of grass in the dirt, shake apart the clumps before throwing them in the trench. Let the plot sit a few days, and then break up the clods with a hoe. Water the area and see if it is higher than the ground around it. If you did it right then it should be at least a few inches taller than the rest of the area.

What is a hill?

Corn, watermelon, pumpkin, and other similar types of plants can be planted in hills. In the desert, your "hill" will actually be a hole that can catch water. In wet areas, make a hill 6 inches tall and plant your seeds in the top of it. This method uses a lot of space, but works especially well in clay-type soil.

Why would I make a raised bed rather than rows?

Rows can be very closely spaced together, or they can be far apart. A raised bed is either a double-dug plot or a box of wood that has been filled with soil. Raised beds generally work well in a small area, but rows are easier to weed. Generally you just decide how much space you have, what equipment you have and how much time you have. With a lot of time and no

Raised bed

Row planting ▲

space, do a raised bed. For no time and big space, use rows which you can plow up.

What is plowing?

Plowing is when you use a tool with a blade that turns over the soil. It is usually pulled by a tractor or horses, and sometimes has more than one blade. The rows must be wide and spaced far apart, but much more area can be covered and many more plants planted because of the additional horsepower. It must be done at certain times during the year.

How do I cultivate with farm implements?

1. First the plow is used to cut furrows (or long turned-over ditches of earth) in a circular pattern or in long rows, depending on the type of plow used.
2. A harrow is used to break up the earth further. A plow leaves clods of earth and grass roots, and a harrow breaks these up. The type of harrow depends on the type of soil.
3. If further work is needed to break up the clumps, a roller is drawn over the soil. A roller flattens the ground which will make your work during harvest much easier. A roller can also be used after planting to push seeds into the soil further.

▼ Plowing a furrow

How do I plow new land?

When you plow a field regularly in one direction, each year you lose the amount of topsoil that your plow turns. Plowing also brings up stones, so that you constantly have to clear them out. When preparing a field for planting, plow it once, then go through with a spade and stone boat and get rid of all the rocks. Plow around the field in a spiral from one corner, and make sure to avoid sloughs (very wet places) or sudden hills. Don't plow soggy ground, because your plow will get stuck and plants won't grow there anyway. For green (or new) fields, use the plow to turn the ground over completely (turning the grass upside down) or lap-furrow it where it is not completely turned over. A lap-furrow is when you overlap the furrows instead of making them right next to each other, and they allow the cultivator or clodcrusher to more thoroughly break up the soil. If you will be using a horse to cultivate the weeds out after plants have grown, space the rows at least 40 inches apart.

How do I plow cultivated land?

If the soil has been plowed before, and it doesn't have very much clay, you can plow a furrow that twists the soil and turns it completely over. Or you can use a double Michigan plow which makes two twists. Both of these methods break up the soil enough that it is no longer necessary to use a cultivator or disc to break up further, which eliminates the need to lap-furrow.

How does a plow work?

The moldboard determines the amount of twist in a furrow, and the cutter or coulter determines how the furrow lies. A coulter that is horizontal to the ground and perfectly vertical creates flat furrow

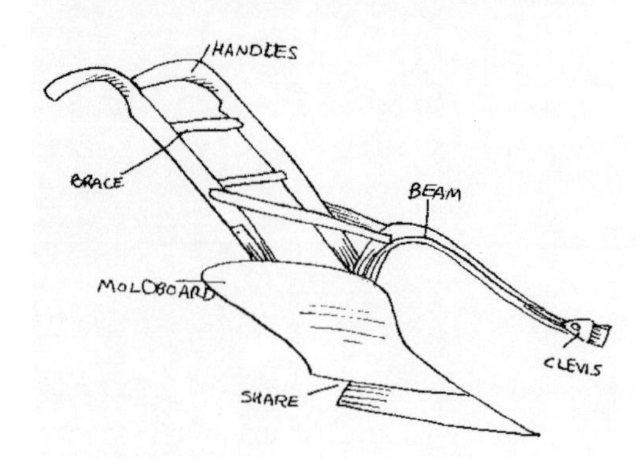

▲ Parts of a plow

slices, while a coulter inclined will make better furrow slices. For horse farming, the average length of a trace-chain is 10 feet, including all the material from the clevis to the horse's shoulders. This is why using a plow and horse is only feasible when you are planting acreage. Plows come in two categories: walking plows and riding plows. With walking plows you have to do what the name says—walk behind and hold the handles. With the riding plow, also called a sulky plow, you can sit and ride while your horse does most of the work.

What if I am having problems plowing?

If the plow has a tendency to rise at the heel, the ring is too high in the clevis, and must be put 1–2 holes downward. The plow should run evenly and flat along the ground. If the plow rises at the point of the share, the ring is too low in the clevis and should be raised 1–2 holes. The factors that cause these problems would be the height of the

▼ Plow angles

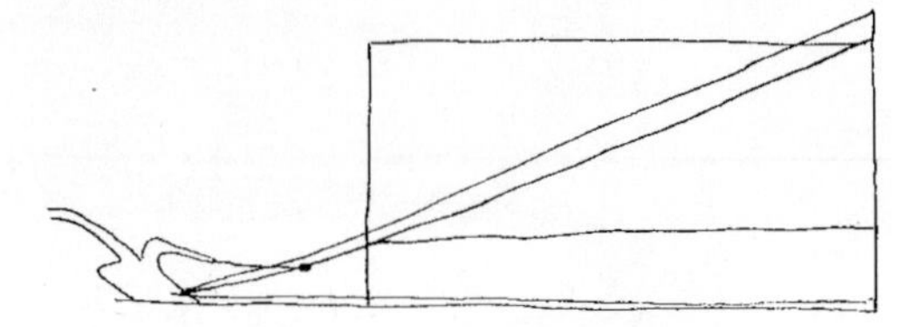

horses, the length of the draught-chains, the depth of the furrow and the type of soil being plowed. The more evenly the plow works, the easier it will be for you to plow.

What is natural farming?

Natural farming is a technique of growing plants without plowing, cultivation, or sometimes even fertilizer. After the initial transition from cultivation to natural (which can take a few years), the promoters of this method say that the plants are healthier and have fewer pests and disease than cultivated plants.

How do I plant a field with no cultivation?

1. In the fall put clover, vetch, or alfalfa seeds in a tray, pour fine powdered clay over them, and spray with a thin mist of water. This makes pellets ½ inch in diameter.
2. Spread the pellets (also called seed balls) over the field. Cover the field with straw by throwing it down in all directions. This is your ground cover.
3. Let 10 ducks per ¼ acre forage loose in the field.
4. Plant your grain crop (with clay-covered seeds) when the preceding crop is ripening. After harvest, sow white clover with the grain as a ground cover. Cover with straw.

How do I maintain an orchard with no cultivation?

Treat the soil the same as for fields. Cut weeds and tree sprouts with a scythe.

How do I grow vegetables with no cultivation?

The techniques differ according to climate. Wait for rain that will fall for several days in early spring. Try to plant so sprouts will grow before the weeds. Cut some of the weeds, throw down clay-covered seeds, and lay the weeds on top as your ground cover instead of straw. You may need to cut the weeds back 2–3 times. Let chickens scratch through the garden—they won't eat the seed balls because they look and taste like dirt, and they'll eat much of the insect pests. Plant clover in late summer or fall to keep back weeds.

What is the sandwich method?

Before planting, layer barn litter, compost, grass clippings, chopped leaves, and wood ashes between layers of peat moss. Water it well, until it is damp and spongy. Then plant your seeds in it.

How can I double dig and use natural farming?

You can use a cross between totally natural and a little bit of cultivation. Double dig a well-developed and fertilized bed to make a raised bed (see Cultivation). After this, it must never be dug or walked on again. Put earthworms in the bed to do natural tilling, and then use the seed balls to plant and regular ground cover to control weeds.

SOIL CARE

There is a great pleasure in working in the soil, apart from the ownership of it. The man who has planted a garden feels that he has done something for the good of the world.

~Charles Dudley Warner, 1870

What are the parts of the soil?

Humus: Organic matter in the final stage of decomposition, unrecognizable as plant material.

▲ Thin layer topsoil and rocky subsoil

Loam: The ideal soil, made of sand, silt, and clay.

Subsoil: The deeper layer of soil, usually lighter in color, which stores water.

Topsoil: The top layer of soil, darker and more crumbly, where most nutrients exist.

Weed indicators of soil type:

Alkali soil:

Coast - saltgrass (*Distichlis spicata*)

Desert - chickweed (*Stellaria media*)

Shepherd's purse (*Capsella bursa-pastor*)

Blueweed (*Echium vulgare*)

Gromwell/Puccoon (*Lithospermum*)

Field peppergrass (*Lepidium campestre*)

True chamomile (*Chamomilla matricaria*)

Bellflower (*Campanula*)

Salad burnet (*Poterium sanguisorba*)

Scarlet pimpernel (*Anagallis arvensis*)

Bladder campion (*Silene latifolia*)

Poor drainage:

Dwarf St. John's Wort (*Hypericum*)

Horsetail (*Equisetum*)

Silverwood (*Potentila anserina*)

Creeping buttercup (*Ranunculus repens*)

▼ Chamomile

▲ Mayapple

Mosses
Sumac (*Rhus integrifolia*)
Curly dock (*Rumex crispus*)
Sorrel (*Rumex acetosella*)
Hedge nettle (*Stachys palustris*)
Mayapple (*Podophyllum peltatum*)
Thyme-leaved speedwell (*Veronica serpyllifolia*)
American hellebore (*Veratrum viride*)
White avens (*Geum album*)

Acid soil:

Bracken fern (*Pteridium aquifolium, aquilinum*)
Spurrey (*Spergula arvensis*)
Corn marigold (*Chrysanthemum segetum*)
Sow thistle (*Sonchus arvensis*)
Scentless mayweed (*Matricaria inodora*)
Plantain (*Plantago*)
Lady's thumb (*Polygonum persicaria*)
Goose tansy, rough cinquefoil (*Potentilla monspeliensis*)
Wild strawberry (*Fragaria*) Rabbit-foot clover (*Trifolium arvense*)
Horsetail (*Equisetum*)
Dock (Rumex)

Slightly acid soil:

English daisy (*Bellis perennis*)
Sorrel (*Rumex*)
Prostrate knotweed (*Polygonum aviculare*)

Very acid soil:

Knapweed (*Centaurea nigra*)
Hawkweed (*Hieracium*)
Silvery cinquefoil (*Potentill argentea*)
Horsetail (*Equisetum*) if swampy type

Heavy soil:

Coltsfoot (*Tussilage farfara*)
Creeping buttercup (*Ranunculus repens*)
Dandelion (*Taraxacum officinale*)
Plantain (*Plantago*)
English daisy (*Bellis perennis*)
Broadleaf dock (*Rumex obtusifolius*)

Light/sandy soil:

Spurry (*Spergula arvensis*)

▼ Wild stawberry

▲ Daisy

▲ Hawkweed

▼ Coltsfoot

Corn marigold (*Chrysanthemum segetum*)

Sheep's sorrel (*Rumex*)

Cornflower (*Centaurea cyanus*) especially when flowers are pink

Small nettle (*Urtica urens*)

Shepherd's purse (*Capsella bursa-pastoris*)

White campion (*Lychnis alba*)

Maltese thistle (*Centaurea melitensis*)

St. Barnaby's thistle (*Centaurea solstitialis*)

Hard/crusted soil:

All chamomiles

Mustards

Morning glory

Quack grass (*Agropyron repens*)

▲ Morning glory

▼ Campion

Goosefoot (*Chenopodiums*) no matter how bad the soil is

Salty soil:

Russian thistle (*Salsola kali*)

Sea aster (*Aster tripolium*)

Asparagus (*Asparagus officinalis*)

Beet (*Beta*)

Shepherd's purse (*Capsella bursa-pastoris*)

Mustards

What are subsoil indicators?

If you dig down 2–3 feet deep and look at the subsoil, it will indicate the health of the topsoil.

Red/yellow subsoil: Lots of iron oxides, good drainage, acidic soil, common in warm climates.

Blue/blue-gray subsoil: Lack of oxygen and poor drainage, common in thick layers of clay.

White to ash-gray subsoil: Nutrients and humus have leached away, acidic and/or sandy, common under pine trees.

Even medium brown subsoil: Good drainage.

Pale subsoil very similar to topsoil: Poorly developed soil, topsoil may have been removed.

Dark brown subsoil: Abundant decomposed organic matter, usually where wetlands have been.

Patches or streaks of colors: Pockets of poor drainage, different soils—plants may have difficulty.

Roots end at same depth: Layer of compacted soil, cemented layer, and poor drainage.

What makes up the best possible

▾ **Mustard**

soil?

Matter and organisms: 5 percent of soil—add soybean meal, garbage, cottonseed meal, plant residue, sludge manure and compost.

Minerals: 45 percent of soil—add lime, green sand, granite, dust, nitrates in organic matter.

Water: 25 percent, sprinkle or irrigate, but use the water table when possible.

Air: 25 percent of soil—make the soil loose with organic material and cultivation.

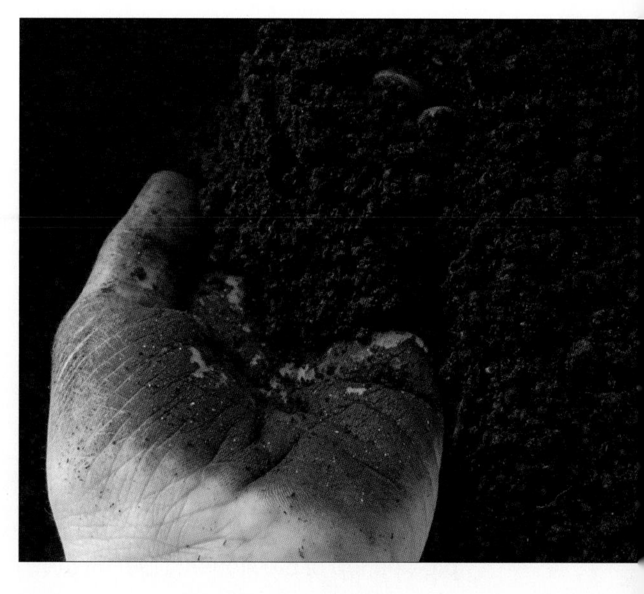

How do I recognize good soil?

The best soil looks dark, crumbly (like chocolate cake), and has things living in it. It's true that some plants like sandy soil, and others like clay, and some like acidic soil, but any plant will grow in fertile soil. To improve your soil, plant legumes, fertilize by either letting animals live there or adding some well-composted manure, and mix in compost and green manure. Your soil will either have good drainage or bad drainage. For bad drainage, you will need to help your soil drain. Dig a ditch, and/or build raised beds that keep plant roots above the water table.

How do I make an irrigation/ drainage ditch?

1. Carefully analyze the ground you want to drain, and figure out the terrain. Make a drawing of the highest point, the lowest point, and sketch the best layout for a drainage system.
2. Dig the trench about 5 inches wide and 2 feet deep. Test the trench by running water down it—the water should flow smoothly.
3. Fill the bottom of the trench with small 1-inch stones. At this point you may also put in 1-inch perforated drainage pipe on top of the stones. Then fill the rest of the trench with stones up to about 1 inch from the top.
4. Fill the last inch with some of the dirt that was originally dug from the trench.

How do I irrigate my garden?

The basic rule of watering a garden is to put enough water to reach the roots, and when it dries out, water again. Generally this means that you need it to be moist about 6 inches down (you can dig to find out), and you will have to water in the morning and evening (usually—unless you have wet weather). Soak them good and deep when the sprouts are well started.

Ways to water:

Bucket: The cheapest, most tiring method, but quite effective and doesn't waste much. Many people use a bucket or watering can with great results.

Upside-down bottle: Used mainly for trees, get a jar or jug with a small opening and fill it with water. Stick it in the ground upside down next to the plant.

Drip hose: A hose with small holes that lets water drip out that is laid by all the plants.

Soaker hose: A hose with small holes on one side and a cap on one end which forces all the water out like a sprinkler 3–5 feet.

Ditch irrigation: A ditch around a field through which water from a well, spring, or community irrigation ditch flows. It is diverted into the field by contour ditches 6–12 inches deep through the field downhill. You don't want to wash things out, but you don't want it too dry. To make the water go into the contour ditches, dam the big ditch.

Traditional irrigation: Big farms use water pressure and large rotating sprinklers, either with wheels or simply rotating heads. These are quick and easy, but expensive and wasteful.

Rules of watering plants:

1. Try to keep the soil damp all the time, though not necessarily soaking wet.
2. The most arid place can grow things as long as it has topsoil and water. A desert is simply dry.
3. Water in the morning faithfully, or vegetables will be stunted, and you may also have to water at night.
4. Mulch the ground when the soil is warm, usually late spring.

▾ Ditch irrigation

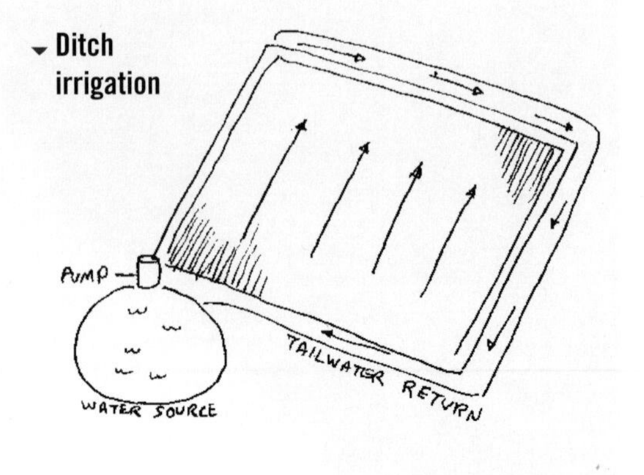

5. Know how much water each plant needs, so that you don't overwater the tolerant plants.
6. Prevent erosion by using cover crops, mulching, and wind breaks.
7. Most plants need about an inch of water per week, but check each individual plant for its preference.
8. Don't lose water to surface runoff, puddles, or evaporation. But don't scrimp on water either.

▲ Compost bin

How do I make compost?

To compost is to purposely rot organic materials to make fertile soil. Build a long wooden crate at least 8 feet long and 4 feet wide and 4 feet tall, with inch-wide spaces in between the slats, to make a compost bin. An average person makes 2 pounds of garbage a week so this should hold enough for a 4-person family. Put a brick or stone floor on the bottom. Dump in any garden waste, except weeds with seeds and diseased plants. Put in any kitchen scraps, except animal products. If you don't have enough garbage to make compost for your garden, get some from neighbors and grocery stores. If you put grass in, add straw also to help it decompose.

Can I use human waste for compost?

You can design a composting toilet that can be connected to the bin for your regular food and garden waste. This compost bin can be used to heat water and warm a greenhouse in winter.

How do I care for my compost?

1. Spread the material evenly, and mix the soft materials together.
2. Mix in manure and dirt every now and then when you can.
3. Introduce worms into the bin to help speed it along—this is called vermicompost (see Worms).
4. In dry weather water it and cover.
5. Stir it once after a few weeks.
6. It will be done rotting after about 3 months. Cover it until you use it.

What if my compost doesn't seem right?

Note: All garbage will compost but troubleshooting these things will make it happen quicker.

Pile is cool and dry: Add water until the center is evenly moist, and in dry places water and cover.

Pile is cool and wet: Make sure pile is at least 3 feet square and 4 feet tall.

Large pile is cool and wet: Add alfalfa meal, manure, or fresh grass clippings, and stir.

Pile cools before composting: Turn and stir.

Smells bad and is wet: Add shredded newspaper or straw and stir. Cover to keep off rain.

Not completely composted: Add newspaper or straw and stir.

What is green manure?

Green manure is an easy way to fertilize, but is more practical for a big field. It can also be done on small plot though—basically you plant something, cover it with dirt, and wait for it to decay. Plant rye grass, buckwheat, barley, pearl millet, oats, or alfalfa in your field, then wait until it is mature to cut it down with a scythe. Plow it under and wait until it is composted before planting seeds. It will look like your other soil when it has composted.

Types of green manure:

Legumes: Sow in fall so flowers will attract beneficial insects.

Alfalfa (*Medicago sativa*): Sow in spring or summer.

Bell or fava bean (*Vicia faba*): Sow in fall or very early spring—they can tolerate as low as 15°F.

Clovers: Berseem clover (*Trifolium alexandrinium*) is a summer or winter annual. Crimson clover (*T. incarnaturn*) is winter-hardy and easily tilled. Dutch white

▾ **Alfalfa**

clover (*T. repens*) is easily cultivated. Red clover (*T. pratense*) is a quick-growing biennial that can be planted from spring through fall. Subterranean clover (*T. subterraneum*) is a cool-season reseeding annual, best for sowing under taller crops. New Zealand white clover (*T. repens*) is a hardy, long-lived perennial and is heat resistant.

Peas: Field peas (*Pisum sativum*) and Austrian field peas (*Lathyrus hirsutus*) sow in fall or very early spring. Cowpeas or Southern peas (*Vigna sinensis*) grow as a summer annual.

Vetch: Common vetch (*Vicia sativa*) grows in most soils. Hairy vetch (*V. villosa*) is tolerant of very cold weather. Purple vetch (*V. atropurpurea*) is less cold-tolerant, making good winter-kill mulch.

How do I compost manure?

Pile it up outside in a pile at least 6 feet deep and no more than 6 feet wide and as long as you want. Cover it with dirt, and let it sit all winter. If you use it before then, it may kill your plants. If you keep your animals inside, also gather the bedding which traps the urine. The best animal beddings for your garden are chopped straw, corncobs, wood shavings, or sawdust because they will decompose well. Don't let the pile get too dry.

Manure not to use:

Bat guano, dog and cat manure, sewage, fresh manure, or horse manure that hasn't been thoroughly composted (it contains weed seeds). Human waste directly from your toilet is different from sewage because sewage is any material being taken away by sewer pipes. Once it is in the sewer it is contaminated, so only use waste that you save with your composting toilet.

How do I use fertilizer?

Green manure, manure, and compost in various mixtures make fertilizer. Fertilizer adds needed nutrients to the soil to make healthier plants. Mix together your compost and manure, then spread it on top of your green manure. Plow, or turn the soil over to mix them into the dirt. Always put the manure into the soil immediately after you uncover it from the dirt, to save the gases. Even the gases released by the fertilizers are good for your soil.

How important are worms in fertilizing?

Another important step to fertilizing and building your soil is worms. It is necessary to cultivate worms into your garden, which means you can't Rototill. A regular plow will kill a few worms, and double-digging and raised beds kill hardly any. Not only do worms help put nutrients in the soil by chewing up organic material, their manure, called castings, are the best manure for soil there is.

How do I use mulch?

Mulch is any cover put over the soil to prevent weeds from growing and/or to insulate the ground (not all soil needs this). Put mulch on after the ground has thawed, unless you are growing strawberries. Strawberries like cold, wet ground, so you can put it on before the ground has thawed for them. Use organic material or newspaper to cover all around your plants. Mulch keeps water in the soil, smothers weeds, and if you use organic material, fertilizes the ground. Some people swear by it because you may never have to weed.

GROWING ENVIRONMENTS

Who loves a garden loves a greenhouse too.

~William Cowper

How do I read a seed packet for plant variety temperature requirements?

Frost-tender: Can't survive even a light frost. Don't plant early, and make sure the soil temperature is warm enough.

Semi hardy: Can live through a light frost. Won't survive a heavy frost, but you can plant right around the last frost date for your area.

Hardy: Will survive early spring and late fall frosts, and can be planted a couple weeks before the last frost date.

What is a greenhouse?

A greenhouse is a structure that allows heat and light from the sun to penetrate but doesn't allow all of the heat to escape. Greenhouses are used mainly to grow plants that need the warmth, to extend the growing season, and to protect from frost. They can also be used as an attachment to your home so you can utilize solar heat.

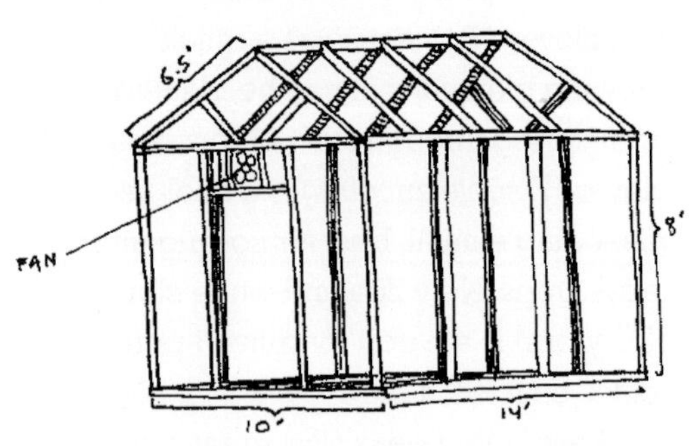

▲ Wood-framed greenhouse that can be covered in plastic

What is the simplest greenhouse I can build?

The easiest greenhouse is a frame made of plywood or PVC piping that is fastened firmly to the ground, which is then covered in heavy-duty plastic. The

◂ Simple PVC greenhouse

bottom edge can be a wood frame or a square made of bales of hay, and the plastic stapled to it. The plastic can be fastened to the frame with leftover 1-inch PVC piping cut into 2-inch lengths and sawed along the edge to from a "C". This is then snapped over the pipe and the plastic.

What is a more complicated greenhouse to build?

A more expensive and difficult greenhouse is a wood frame that is covered in glass or Plexiglas or other hard plastic sheeting. The frame would be similar to a

Glass greenhouse

▲ Large steel-frame plastic greenhouse

barn frame, with gussets (or reinforcements to the joints) to strengthen it.

Growing plants in the house:

If you live in a place where there is no natural light and you don't have a backyard or greenhouse, you might need to grow plants inside. Some people do this hydroponically, where they use factory-made fertilizer in water and grow under lights. For a situation in which you couldn't get the special fertilizer your plants would quickly die, so the best method is really to use soil and electric light.

How do I grow plants indoors?

Basically you would need to find soil and make your own compost using your food scraps and perhaps a composting toilet. Your plants would need to be put in

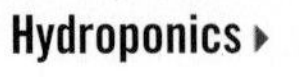

Hydroponics ▸

containers—the cheapest way to get them is from thrift stores or to use materials you have. Square containers save space and hold more soil. You will also need water trays for under the plants, tables for the containers, and electric lights. Plants need certain colors of light in order to optimize their growth. Red and blue are the most important, and these must be balanced or plants will get too short or too long and skinny.

Types of light bulbs:

Incandescent bulbs: The normal, house light bulb. These work well but do not have enough blue light. They can also be somewhat hot so if you put the plant too close it can scorch.

Fluorescent bulbs: Produce 3–4 times as much light as incandescent with the same amount of energy and are relatively inexpensive. They are sold in cool white (with more blue and yellow-green colors), or warm white (with more orange and red colors), so the best way is to put the two together in a fixture so the plants get all the necessary colors. Full-spectrum fluorescent bulbs are also sold which last twice as long and contain all the colors—these would be the best kind for the small homesteader.

High-intensity discharge lamps (HID): Used by professionals, use a lot of energy and are more expensive, and some produce so much light that protection must be worn around them.

How do I regulate night and day?

Garden plants and vegetables require 14–18 hours of light each day. Then the lights must be shut off so that the plants can rest. If you notice the plants seem to be reaching for the light or have huge leaves then they may need more light.

How do I care for my indoor garden?

Rotate the plants each week to get an even distribution of light, and use white trays or foil reflectors. Clean fluorescent bulbs once a month to clear them of dust.

Feel your plants—if they are warm, the light is too close. And just like any other garden you must water the plants and fertilize the soil with your compost.

BASIC PLANT CARE

The Sun, with all the planets revolving around it, and depending on it, can still ripen a bunch of grapes as though it had nothing else in the Universe to do.

~Galileo Galilei

Propagation:

Propagation is the process by which some plants reproduce, instead of making seeds. Potatoes, bananas, and many kinds of herbs use this process. They may grow just from their own roots, or even by planting a piece of the fruit.

How do I propagate a plant with lots of roots?

Plants with many stems, or a root clump (a large mass of smaller roots), can be propagated this way. Take a plant that is at least 2 years old, dig it up, and soak the roots overnight in water. This will soften them so you can gently pull the plant apart without tearing it. Plant pieces separately.

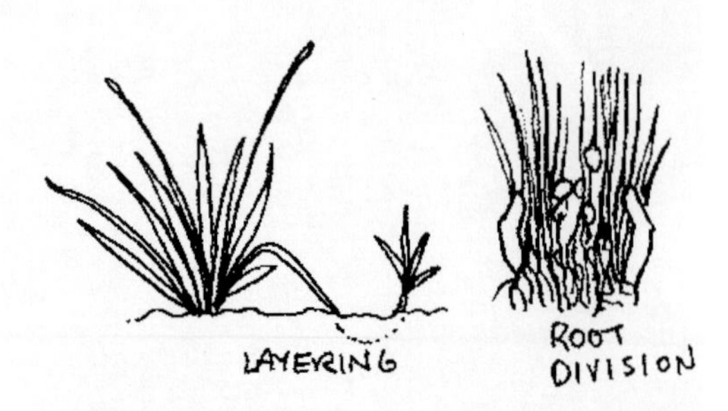

▲ Propagation through layering and root division

How do I propagate by cutting a root?

Lemon balm, comfrey, mint, horseradish, and others can be propagated by taking a plant at least 2 years old, and cutting off a piece of root at least 2 inches long. These plants have a single, fat, almost woody root, and you need to find a big, healthy-looking one. Plant it in the ground like you would a seed.

How do I propagate by layering?

For some shrubs, just bend a branch over and cover the middle of it with dirt (leaving the end out). The middle will grow roots into the ground and you can cut the plant away from the old one.

How do I propagate a tuber?

Get a nice potato, hopefully one that represents your ideal potato, and cut it into pieces. Each piece should have an eye. On a potato the eyes are darker brown spots which indent into the tuber making it lumpy—on other tubers they are somewhat similar. When you cut be careful not to damage the eye. Plant each piece and a potato plant will grow from it.

How do I prevent tree injuries?

To find out if a tree has new growth, look at the bark. There will be new cracks and openings in it. Please cut down as few trees as possible, because trees are becoming precious. If you find an alternative to wood, use it. Never make a very large gash in a tree. The sap will run out of the gash and run down the side of the trunk, and diseases can enter the tree through the hole. Never cut a ring of bark around the tree. This is called girdling and will kill the tree.

How do I plant a tree?

1. A tree has a root ball—a ball of soil that is surrounding the mass of roots. Don't let this ball dry out before you plant it. If the ball is wrapped in something, take that wrapping off before you plant.
2. Dig a hole deep enough that the root ball will be just covered by the soil. A tree has a part at the base where it begins to flare out and turn into roots. This flare is where you will want your hole to come to.
3. Put the tree in the hole and fill it with dirt. Leave a circular compression or trench all around the tree a little thinner than the root ball so that water can settle in it.
4. Spread fully composted organic material on top of the soil all around the tree (don't mix it with the soil because it can actually cause harm), and water very thoroughly to soak the roots. Don't step on the dirt around the tree.
5. You shouldn't have to stake a tree unless the dirt around it is loose—trees need to learn to hold themselves up. Pruning also does not help the tree grow unless you are only removing broken branches.
6. When watering, soak down to the roots, but then wait a while until the soil dries out to water again. Only swamp trees like their roots wet all the time.

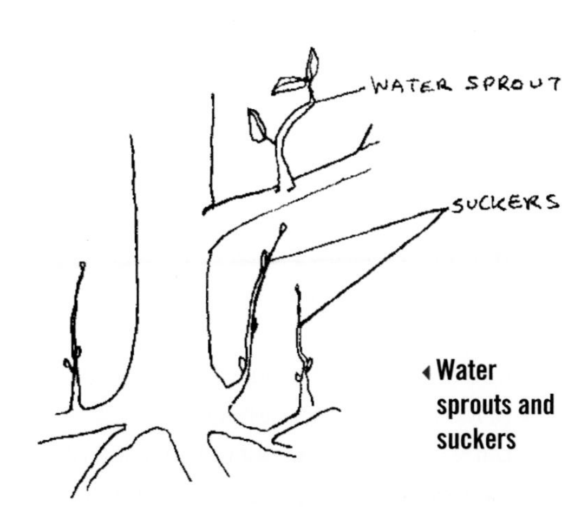

◂ Water sprouts and suckers

Tree pruning rules:

1. Each plant is different and requires different pruning techniques and schedules, so research first.
2. Careful, balanced pruning makes your plants grow better and fruit grow bigger. Excessive pruning kills trees and shrubs.
3. If cutting a diseased plant, clean your pruning shears with rubbing alcohol afterwards.
4. Never prune the first year, and never during a drought.
5. Prune as little as possible. It is better to prune too little than too much.

How do I prune a tree?

1. Each plant is different, but in general you can prune in early spring and late fall when all the leaves are off the tree.

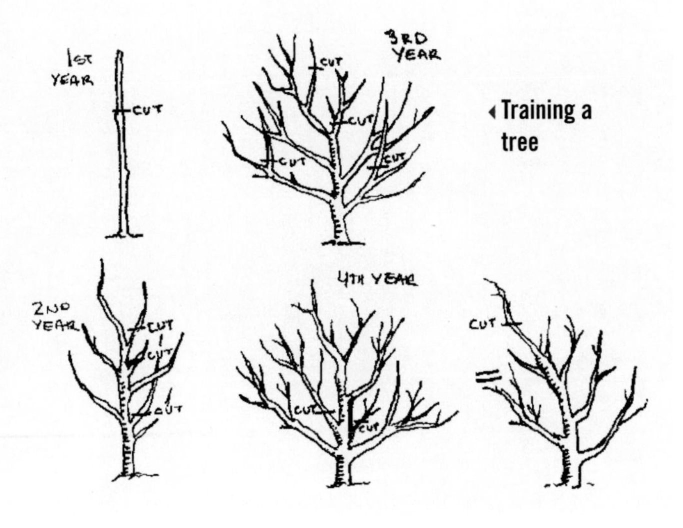

◂ Training a tree

2. Prune off the dead wood, and the branches that are rubbing against other branches (you should only need to do the worst ones).
3. Pinch off some of the suckers and watersprouts, but not all of them, leave a few to flower. A sucker is a small sprout that grows out of the roots of the tree and comes up out of the ground (it looks like a completely different baby tree), or right out of the trunk of the tree. A watersprout looks a lot like a sucker but is a very green shoot that comes out of branches after too much pruning or an injury. If you take them all, they will all come back, but take a few and only a few will come back.
4. You can now attempt to train the branches of the tree to be more conducive to an optimum tree shape. Fruit trees are supposed to have more horizontal branches, while other trees are supposed to have branches at an angle of 45° to 90°. Mature fruit and nut trees need more light within the tree and so they might be thinned, and they can also be pruned smaller than another tree so that the branches will grow short and fat for strength and easy picking.

What is grafting?

Grafting is the process in which a branch from one type of fruit tree is grafted onto another. The purpose of this is varied, but the most common reason is by putting a difficult-to-grow fruit species on an easy-to-grow tree, the fruit will grow faster and easier. Grafting makes it possible to grow several kinds of apples on the same tree, saving you space.

What is one way to graft trees?

1. In late winter (February), cut the scions (fruit-tree shoots with buds) that you want to graft at both ends so they are about two inches long and as thick as a pencil. Do this by pressing them

Well-trained trees

1 CUT A 'T' IN THE TREE

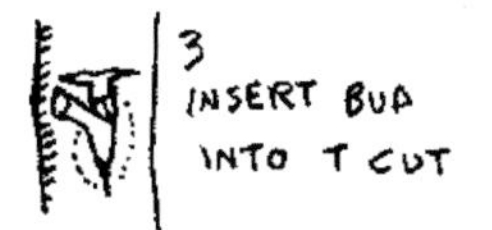

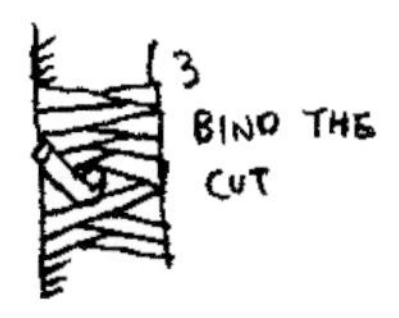

▲ **A method of grafting**

against the knife with your thumb and rolling them back and forth.

2. Start grafting in late March or April when the buds on the trees you are grafting start to swell. Bring grafting wax with you in a dish of warm water.
3. On the tree you are grafting to (you can only graft a cherry onto a cherry, an apple to an apple, etc.), saw the top off a branch and split the branch down the middle a little way with a chisel or knife. Wedge the split open a little bit with a small piece of wood.
4. Shape the end of two scions so that they will wedge into the branch better.
5. Carefully put two scions into the cleft, one on each side, so that the end of the branch appears to be forked. Take out the wedge.
6. Take a ball of grafting wax, lay it on the end of the stalk between the two scions, and press it down so that it seals up all the openings in the wood and is firmly in place. Grafting wax is any sealant to be used for plugging plant wounds. Some people use simple Elmer's glue, but beeswax or other sealants will also work. Bind it with tape or cloth.

▼ **A tree with grafted scions**

How can I make my own grafting wax?

Either mix together 10 ounces rosin, 2 ounces beeswax, 1 ounce of charcoal powder, and 1 tablespoon of linseed oil,

▼ **Another grafting method**

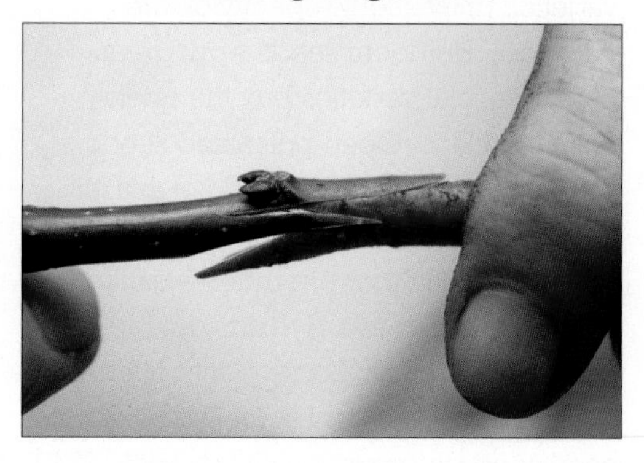

or mix equal parts of tallow, rosin, and beeswax.

What are the benefits of weeding?

Some people don't weed at all, and others weed fanatically. Most people weed sometimes and do a medium job. The less weeds the better because they don't have to compete for water, sunlight, and nourishment. After a garden has been in the same spot a while you will have less weeds to pull, and this is where natural farming starts to really work. But no matter what, the first few years you will have to weed at least a little bit.

How should I weed?

Weeding is hard work, although some people find it therapeutic. Weed when the soil is damp, such as in the morning after the dew, or after rain. During hot, dry days the weeds hold on very tight to the soil. It is uncomfortable to weed bending over, so the most comfortable way is on all fours, squatting, or by sitting on something such as newspaper.

Plants that you can save the seeds from:

Don't buy hybrid seeds or patented varieties (they should specify on the package). Heirloom seeds work great, but for normal varieties buy the open-pollinated kind. Open-pollinated seeds were pollinated naturally, and not in a lab. When you have diverse seeds that were grown naturally, they are more resistant to pests and disease. Hybrids get bigger but die more easily. Some plants, such as bananas, potatoes, and some herbs don't need seeds to grow; you simply stick a piece of the plant into the ground (a process called propagation).

Types of plants:

Annuals: Makes a flower and then makes seed for each flower in 1 year.

Biennials: Most root vegetables are biennial. They flower and make seed after 2 years. The first year they just store up food in a root.

Self-seeders: Self-seeding plants create volunteers, or plants that grow without you having to do anything; for example: a sunflower drops its seeds before you get to them.

How do I save seeds?

1. First make sure your plants get pollinated. This means either encouraging bees, or sticking your finger in one flower and rubbing the pollen in another.
2. Flowers that make a flower head will create a seedpod. Simply cut it off, dry it, and break it open when you need to plant. Fruit are another example of a pod, except you can eat it. Cut open the fully ripe fruit and scrape out the seeds, and let them dry. For any other plant, make sure there are no weed seeds in with your good seeds, by making a sieve. The mesh should only allow the seeds you want to keep to stay on top.
3. After putting them through a mesh and sifting, dry the seeds for at least 2 weeks before storing them, unless the seeds have beards or tufts.

How do I store seeds?

Let them dry well, and then store them in a cool, dry place. The fridge is good if you keep them in tightly sealed very dry jars and containers. A dry cellar or basement is also a good place.

What is companion planting?

Companion planting is when you plant complementary plants next to each other to counteract bugs. Many plants have insect repellant properties and by placing then near other plants, you can provide an effective bug control. Some plants pollinate, attract good bugs, and shelter other plants.

Type of Plant	Does Well With	Does Poorly With
Amaranth	Sweet corn	
Anise	Coriander (improves growth and flavor), cabbage family	
Apple	Chives Clover	
Asparagus	Parsley and basil (deters asparagus beetle) Tomato Nasturtium	Onion Garlic Gladiolus
Basil	Tomato Everything (improves flavor and growth) Asparagus Nasturtium Pepper	Rue

Cabbage companion planted with marigold

Type of Plant	Does Well With	Does Poorly With
Beans	Carrot Cabbage family Beets Cucumber Corn Grain Peas (improves growth) Spinach Eggplant Mustard Potato Rosemary	Pole beans Wormwood Marigold
Beets	Bush beans Cabbage Lettuce Onion Kohlrabi Lima beans Radish Celery	
Borage	Tomato Squash Strawberry	
Broccoli Cauliflower	Sage	
Bush beans	Beets Carrot Cucumber Marigold Potato Cabbage Celery Corn Eggplant Lettuce Peas	Fennel Garlic Onion
	Radish Strawberry	
Cabbage family	Sage (deters pests and improves growth) Celery Beets Onion family Chamomile Spinach Chard Peas (improves growth) Tansy (deters cutworm and cabbage worm) Anise Beans Bush beans Cucumber Hyssop Mint Cabbage Rosemary Thyme	Dill Pole beans Strawberry Tomato
Carrot	Leaf lettuce Parsley Tomato Sage (deters rust or carrot flies and improves growth)	Dill

Type of Plant	Does Well With	Does Poorly With
Carrot (continued)	Chervil (deters Japanese beetle) Pea (improves growth) Radish (deters Cucumber beetle, rust flies and disease) Beans Bush beans Chives Flax Onion Peas Rosemary	
Celery	Cabbage Bush beans Onion Spinach Tomato	
Chard	Cabbage	
Chervil	Radish Lettuce Carrot Grapes Roses Tomato	
Chives	Carrot Tomato Apple Grapes Roses	
Clover	Apple	
Collard		Tansy
Coriander	Anise	

Type of Plant	Does Well With	Does Poorly With
Corn	Snap or soybeans (improves corn growth) Cucumber Peas (improves growth) Potato Pumpkin Squash Amaranth Radish Sunflower Bush beans Pole beans Onion	
Cucumber	Beans Cabbage family Corn Peas (improves growth) Radish (deters cucumber beetle, rust flies, and disease) Bush beans Dill Lettuce	Sage Potato
Dill	Lettuce Onion Cucumber	Carrot Tomato Cabbage

Type of Plant	Does Well With	Does Poorly With
Eggplant	Beans Four o'clock Bush beans Pole beans Spinach	Potato Peppers
Flax	Carrot Potato	
Garlic	Roses Everything (deters aphids and beetles)	Asparagus Bush beans Peas Pole beans
Grapes	Chervil (deters Japanese beetles) Chives Hyssop Nasturtium Radish Tansy	
Hyssop	Cabbage Grapes	Radish
Kohlrabi	Beets	
Lettuce	Carrot Cucumber Onion Radish Beets Chervil Dill Peas Bush beans Pole beans Strawberry	Chrysan-themum
Mint	Cabbage Strawberry	
Mustard	Beans	

Type of Plant	Does Well With	Does Poorly With
Onion	Beets Cabbage family Lettuce Tomato Carrot (together deters rust flies and nematodes) Dill Celery Cucumber Pepper Squash Strawberry	Asparagus Bush beans Peas Pole beans
Parsley	Tomato Asparagus Roses Carrot	
Peas	Beans Carrot Corn Cucumber Potato Radish Turnip Lettuce Spinach Cabbage Bush beans Pole beans Peas Squash	Garlic Onion Wormwood Potato
Pepper	Four o'clock Onion	Potato Eggplant Tomato

Type of Plant	Does Well With	Does Poorly With
Pole beans	Marigold Radish Carrot Corn Cucumber Eggplant Lettuce Peas Radish	Garlic Onion Wormwood Beans Cabbage Beets
Potato	Beans Cabbage Corn Marigold Peas (improves growth) Horseradish or tansy (deters Colorado potato beetle) Bush beans Flax	Sunflower Peppers Cucumber Eggplant Pumpkin Tomato
Pumpkin	Corn	Potato
Radish	Beets Carrot Spinach Squash Corn Peas (improves growth) Chervil Cucumber Lettuce Pole beans Grapes Bush beans	Hyssop

Type of Plant	Does Well With	Does Poorly With
Roses	Chervil (deters Japanese beetles) Chives Garlic Parsley	
Rosemary	Cabbage Beans Carrots Sage	
Rutabaga Turnip	Peas (improves growth)	
Sage	Broccoli Cauliflower Rosemary Cabbage Carrot Strawberry Tomato Beets	Cucumber Rue
Savory	Beans Onion Cabbage	
Spinach	Peas (improves growth) Beans or tomato (improves growth with shade) Cabbage Radish Celery Eggplant Strawberry	

Type of Plant	Does Well With	Does Poorly With
Squash	Nasturtium Radish (deters cucumber beetle, rust flies, and disease) Tansy Borage Corn Onion	
Strawberry	Borage or sage (enhances flavor, deters rust flies, and disease) Mint (deters aphids and ants) Bush beans Lettuce Onion Spinach	Cabbage
Sunflower	Corn	Potato
Tansy	Cabbage Potato Squash Grapes	Collards
Thyme	Cabbage family Tomato (with cabbage deters flea beetles, cabbage maggot, cabbage butterflies, Colorado potato beetle and cabbage worms) Borage	
Tomato	Asparagus Basil Garlic Marigold Parsley Chervil (deters Japanese beetles) Spinach Carrot Chives Sage Thyme Celery Onion	Cabbage family Fennel Potato Pepper Dill Corn

Aphids

What can I plant to attract good insects?

Bees: bee balm, borage, summer savory

Hoverflies: buckwheat, German chamomile, dill, morning glory, parsley

Ladybugs: yarrow

Predatory ground beetles: amaranth, lovage

Predatory wasps: anise, borage, German chamomile, dill, yarrow

What can I plant to deter bad insects and pests?

Aphids: anise, catnip, chervil, coriander, dill, garlic, mint, yellow nasturtiums, peppermint, petunias, sunflowers

Ants: catnip, mint, tansy, pennyroyal

Asparagus beetles: parsley, petunias, pot marigold

Bean beetles: rosemary, tomato, summer savory

Black flea beetles: sage, catnip, wormwood

Blister beetles: horseradish

Cabbage moths: hyssop, peppermint, rosemary, sage, summer savory

Cabbage worms: borage, thyme

Carrot rust flies: rosemary, salsify, wormwood, sage

Codling moths: garlic, mint

Colorado potato bug: flax, horseradish, eggplant

Corn borers: radish

Cucumber beetles: nasturtiums, radish

Flea beetles: catnip, hyssop, mint, peppermint

Fleas: lavender, pennyroyal

Flies: basil, rue

Gophers: elderberry

Japanese beetles: catnip, chives, white flowering chrysanthemum, garlic, rue, tansy, white geraniums, larkspur

Leafhopper: petunias

Mexican bean beetles: petunias

Mosquitoes: basil

Moths: costmary, lavender

Potato beetle: coriander, horseradish

Root maggots: garlic

Root nematodes: chrysanthemum

Slugs: comfrey, wormwood

Snails: garlic

Spider mites: coriander, dill

Striped cucumber beetles: tansy

Striped pumpkin beetles: nasturtium

Squash vine borers: radish

Squash bugs: catnip, dill, nasturtiums, tansy

Thrips: basil

Tomato hornworms: borage, petunias, pot marigold

Weevils: catnip

White cabbage butterfly: peppermint, wormwood

White flies: nasturtiums, marigolds (calendula)

Wooly aphid: clover

Plant pest control rules:

1. Turn over your soil in the fall so the birds can eat the bugs.
2. Rotate where you plant your crops every year, so bugs can't get a residence.

3. Fertilize and help your soil.
4. If you can't weed, mulch.
5. Keep your garden clear of weeds and debris.
6. Cultivate the plants in the fall.
7. Destroy infected plants.
8. Keep the ground clear of everything around trees.

What are some organic pest control methods?

Note: Insects usually go through several life stages . . . egg, pupa, larva, and adult. Some of the photos include several stages. A pupa is like a cocoon, while larvae look more like grubs.

Alfalfa weevil: Encourage parasitic wasps.

Aphids: Spray with watered-down clay or soapy water, and bring in ladybugs.

Asparagus beetle: Shake beetles into can of soapy water, or spray tea on plants.

Black flea beetle: Dust with soot and ashes, plant near shade, and spray with garlic or hot pepper.

Blister beetle: Also toxic to horses or grazing animals. Cut hay with a sickle bar or rotary mower and don't crimp or crush hay.

Cabbage looper: Sprinkle worms with flour or salt.

Cabbage maggot: Apply wood ashes to the soil.

Cabbage worm: Put sour milk in the center of the cabbage head, and dust with 1 cup flour. Or use mint cuttings as mulch.

Carrot rust fly: Sprinkle wood ashes at plant base.

Codling moth: Spray with soapy water.

Colorado potato beetle: When plants are wet, spray with wheat bran, remove beetles and eggs by hand, and spray with mix of basil leaves and water. Cover ground with 1 inch of clean hay or straw mulch.

Corn earworm: Put a little mineral oil in the silk on the tip of each ear.

Cutworm: Sprinkle wood ashes around plant and press a tin can with the bottom cut out around the stem 3 inches deep.

Harlequin bug: Handpick bugs off.

Hessian fly: Encourage parasitic wasps.

Hornworm: Handpick off and sprinkle dried hot peppers on plant.

Leafhopper: Shelter plants.

Mexican bean beetle: Spray with garlic, destroy eggs, handpick beetles, and plant earlier.

Mosquito: Empty out all standing water.

Moth: Sprinkle around dried sprigs of lavender.

Onion maggot: Plant onions all over instead of in one place.

Slug: Handpick off, make borders of ashes or sand, and mulch with wood shavings or oak leaves.

Spider mite: Spray cold water or spray mix of wheat flour, buttermilk, and water on leaves, or spray soapy water, or spray coriander, and introduce predators.

Squash bug: Handpick, grow on trellises, and dust with wood ashes.

Squash vine borer: Pile up soil as high as the blossoms.

Striped cucumber beetle: Handpick, mulch heavily, dust with wood ashes, and grow with trellises.

Tarnished plant bug: Remove plant after harvest.

Thrips: Spray with oil and water mix, spray with soapy water.

Weevil: Hill up soil around sweet potato vines.

Wireworm: Plant green manure like clover.

What general pest control can I use to prevent pests?

Bay leaf: Sprinkle dried leaves on plants.

Catnip: Spread sprigs around and it will repel a variety of pests.

Chrysanthemum (pyrethrum): Mix with water and spray as a general insecticide.

Garlic: Concentrated garlic spray deters fungus and insects.

Horseradish: The root can be made into a spray, either raw or as a tea.

Hot peppers: Make a spray from the tea for most insects.

Kelp: Use as powder or tea as spray both kills bugs and fertilizes.

Lemon balm: Use powder and sprinkle throughout garden.

Marigold: Plant the scented varieties throughout the garden.

Petunia: The leaves can make a tea for a potent bug spray.

How can I encourage pest-eating creatures?

Note: These are friendly creatures . . .

Spiders: Do not kill the spiders you find in your garden.

Minute pirate bug: Eats spider mites.

Mite destroyer beetle: Eats spider mites, introduce into garden.

Parasitic wasp: Eats Hessian flies and others.

Six-spotted thrips: Eats spider mites, introduce into the garden.

Toad: Lay a flower pot upside-down in your garden.

Purple martin birds: Make a bird house with many apartments or a group of birdhouses clustered together, painted white, and put on a high pole away from trees and buildings. Remove other bird's nests except the purple martins, and take down the house in the winter.

What poisonous plants are used as a natural pesticide but are toxic to people?

Four o'clocks: Poisonous especially to children.

Larkspur: Poisonous to both humans and animals.

Marigold (Calendula): Unlike pot marigold, French and Mexican marigolds are poisonous.

Mole plants: Deters moles and mice and isn't good for humans.

Stinging nettles: Hairs on the leaves have formic acid which stings you. However, these can be ingested when properly prepared.

Tansy: Toxic to many animals, and don't let it go to seed or it will grow everywhere.

Wormwood: This medicinal herb is used to create a botanical poison that should not be used directly on food crops.

Rules for plant disease prevention:

1. Rotate crops—don't plant the same thing in the same spot two years in a row.
2. Don't work in the garden when it's wet.
3. If you have a disease, don't save seed.
4. Clean up old debris.
5. Maintain even soil moisture.
6. Fertilize properly—don't overdo it.
7. Don't put diseased plants in the compost pile.
8. Try not to get the tops of the plants wet when watering.

9. Wash your hands after touching a diseased plant.
10. Grow resistant varieties.
11. Prune and weed to improve air circulation.
12. Avoid injuring plants and trees.
13. Sterilize pruning shears after pruning diseased plant.

What should I plant the next year if I am not growing the same plant in that spot again?

Beans: plant cauliflower, carrots, broccoli, cabbage, corn. Don't plant onions or garlic.

Beets: plant spinach.

Carrots: plant lettuce or tomatoes. Don't plant dill.

Cucumbers: plant peas or radishes. Don't plant potatoes.

Kale: plant beans or peas.

Lettuce: plant carrots, cucumbers, or radishes.

Onions: plant radishes or lettuce. Don't plant beans.

Peas: plant carrots, beans, or corn.

Potatoes: plant beans, cabbage, corn, or turnips. Don't plant tomatoes, squash, or pumpkins.

Radish: plant beans.

Tomatoes: plant carrots or onion.

How do I identify a disease?

A disease often looks like a weird brown or rust spot on some part of the plant. Tan spot wheat disease makes small brown dots on the leaves, while alfalfa phytophthora root rot makes a big brown spot on the root. Another kind of disease is a caused by a nematode, or a worm that creates cysts on the plants. The cysts make the plant stunted and/or yellow. There is no cure for this disease except for prevention and removal. Every location has different diseases and the best possible identification is the local extension office.

What do I do if my plants are infected?

1. Destroy infected plants/trees or pruned-off parts.
2. Scrape off loose bark from trees.
3. Cut out diseased wood, then patch with tree-patching compound.

CULTIVATING WILD EDIBLE FOODS

Eat leeks in March and wild garlic in May
And all year after physicians may play.

~Old Welsh Rhyme

Obviously (and any wild food book will tell you this) if everyone ran into the woods and started grabbing up wild plants it would be devastating, especially since many of these are endangered. But wild plants are full of nutritional value and, once you identify what grows in your climate, can be simple to grow. Starting out cultivating native species can be tricky especially if you have a landscaped yard. You'll have to rip out any non-native plants that will compete, and carefully choose what you are going to put in to make sure it is appropriate for the soil and climate that you have. It's also going to be a challenge creating a small ecosystem, especially if you want to grow fish and water plants.

Getting the seeds and plants initially can also be a challenge. These need to be locally sourced plants and seeds, but not pulled directly from the wild. Your best bet is to search Google for "native plants" or "wild heritage plants" for your province or state. You could also ask your local nursery, or local county extension. Make a list of the plants you want, and then search for those specific species, because most places will carry mainly ornamental plants, with one or two edible varieties. But, once you have your first plants, most of these wild varieties simply spread on their own. Your goal is to create a garden that you never have to plant again, and everything is edible.

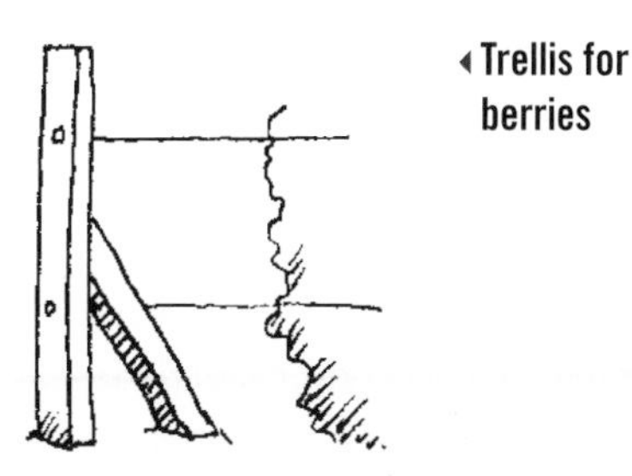

Trellis for berries

Fruits

Your best bets are to start a couple of trees from young saplings, such as a nut tree and, fruit tree. These will take a while to mature, so you could also start some berry bushes. Blackberries and raspberries are very easy to grow, and last a long time. You have to get the

Blackberries

wild varieties, and fortunately they are plentiful so you can get them for free if you want. You simply use root cuttings, or even cut the suckers (a shoot growing off the root) off and replant. In a small backyard you'll have to keep them pruned back, and you can decide if you want to trellis them for easy picking or keep it wild and natural. A good compromise is an old fashioned tipi trellis woven out of sticks. Another highly valuable plant is the wild strawberry, because it tastes good and the whole plant is edible. They also reproduce with runners, but the strawberries themselves can be planted as well.

▲ Duckweed

Fish and Water Plants

Building a backyard pond is a common practice, and you can raise fish in them. Many people raise channel fish, trout, striped bass, and tilapia. Some people raise a couple of these, although mostly the channel fish and tilapia. For a self-sufficient pond that feeds itself, the

▼ Backyard pond

tilapia is the easiest because they eat mostly plant material and plankton. You can introduce small, edible, cleaning plants to your pond, such as duckweed, which will feed the fish and clean the waste at the same time. Some people are experimenting with cleaning gray water with duckweed, which could be worked into the design as well. If you were to throw a decent number of breeding tilapia into your pond they would reproduce and choke the pond as they overpopulate, so regular fishing or trapping to keep a balance is the key. Extras can be dried or smoked.

Roots, Vegetables, and Herbs

One great example of an easy-to-obtain wild vegetable is any kind of wild onion. Alliums include garlic, onions, and leeks. They are an excellent companion plant for all the rest, and are incredibly healthy, but a little trickier to grow. You want all of your plants to be either self-propagating or perennial, and wild garlic is one of these. You simply don't harvest all of them, and they will continue to send out bulbs and seeds. Many herbs are also perennials, and if they go to seed you can save the seeds as well. For example, wild mint is easy to get, very hardy, likes damp soil, and just spreads on its own.

Wild onion ▸

Grains

The wild rice you buy at the store is farmed, but is still the wild variety, so the seeds (which is really a type of water grass) are fairly easy to find. It is harvested by hitting the heads over baskets with sticks so the rice falls in, and then it is spread out to dry in the sun which takes a few days. Wild rice is the main grain of most places in North America, supplemented by roots, nuts and in some areas, corn (or maize). You'll have to do some research for what is native to your area. Many times acorn powder was used as a flour instead of a grain flour. If you do grow rice, it is an annual so some of what you harvest needs to be saved for seed.

▲ **Chicken of the woods**

Mushrooms

Some people like to eat wild mushrooms, which should only be done by an expert. It is much better to grow them so you know what you are eating. It's actually quite simple because all mushrooms are wild mushrooms. The key to growing them outside in your little ecosystem is to replicate what they grow on in the wild, and then buy the fungus plugs from a mushroom supplier. For example, in the Pacific Northwest (and many other places) we have chicken of the woods, which likes to grow mostly on oak, though it also likes yew, cherrywood, sweet chestnut, and willow. If you get a log of that wood, preferably an old cheap log that is rotting a bit. Soak it for 48 hours, and drill holes all over, which you then put the plugs into. Then you seal them with melted cheese wax, and bury the log about ⅓ into the ground. Make sure the log stays moist, and by fall you should have mushrooms. Chicken of the woods is particularly valuable because it can be prepared like chicken and contains very high protein.

▼ **Wild rice**

GROWING AND HARVESTING GRAIN

If you can look into the seeds of time, and say which grain will grow and which will not, speak then unto me.

~William Shakespeare

Preparing the field:

The soil must be as free of weeds as possible. You can do this by planting in ground that has been cultivated for at least a year. If you start with new ground, plow it in late summer, then disc or till several times.

About winter grain:

Winter grain is planted in September or later, so the seeds will "hibernate" during the cold weather and grow as soon as the weather warms up. It's better to have good snow cover for insulation—really cold weather will kill it. But don't let it stool (grow a stalk) because that can also kill it. Avoid it by planting later and letting animal graze over it a little. Plant winter wheat, barley, oats, or rye in early fall to sprout the next spring.

Growing:

Rotate what your fields grow for best results. Grain doesn't like rich soil so don't add manure, just add some compost—if it does have it, the grain gets too tall and falls over (called "lodging" or "going down"). Put the seeds in a sack slung over your shoulder and scatter over the surface of the field as evenly as possible (called broadcasting). Then use a harrow or rake to cover them with dirt. Don't walk in your field at all after the grass comes up, even to weed. To prevent weeds, plant buckwheat or amaranth (they grow fast), or cultivate a lot before planting, or plant a lot of grain to offset losses. To water, wait till it rains, use flood irrigation, or set up a sprinkler beforehand.

The grain stages:

Heading out: the seed head forms—this happens within two months.

Milk stage: when you press it a milky juice comes out.

Dough stage: no milk, but soft enough to mash. After a little, you can just make a dent in it.

Dead ripe: kernel is totally dry and can't be dented with your fingernail.

Young wheat

▲ Ripe wheat

Tillers and suckers:

A grain plant has a main stalk and head, but it also has secondary stalks that also produce grain, called tillers. Corn also has these but they are called suckers because they don't produce anything. Sometimes planting less seed will cause more tillers, producing more grain than you would normally get, because there is less competition between plants.

Pounds per bushel for different varieties of grain:

Alfalfa: 60 pounds
Barley: 48 pounds
Buckwheat: 50 pounds
Clover: 60 pounds
Field corn: 56 pounds
Flax: 56 pounds
Oats: 32 pounds
Red Clover: 50 pounds
Rye: 56 pounds
Sweet Sorghum: 50 pounds
Spelt: 40 pounds
Wheat: 60 pounds

Yield per grain:

The average yield of an acre is 30 bushels (threshed and winnowed) or 1 ton. The square footage to grow 1 bushel is:

Field corn: 500 (10 x 50) square feet
Grain sorghum: 600 (10 x 60) square feet
Oats: 600 (10 x 60) square feet
Barley: 900 (10 x 90) square feet
Wheat: 1,000 (10 x 100) square feet
Buckwheat: 1,300 (10 x 130) square feet
Rye: 1,500 (10 x 150) square feet

An acre is 43,560 square feet. A family of five eats about 1,000 pounds of wheat in a year. That's 16 bushels of wheat, which could be produced on 16,000 square feet, or about ⅓ acre.

Note: If fungus has got your grain, don't eat it and don't give it to animals. The mold can cause all kinds of problems.

When to harvest:

Harvest when it is in the late dough stage, not milk but able to be dented. When you let it dry it will become dead ripe.

How to harvest:

1. Cut with a scythe or sickle or horse-drawn mower. Holding on to the nibs (handgrips) of the snath, and keeping your arms straight, twist at the waist while stepping forward. That way you move forward and cut at the same time. Keep the top of the snath down, so that the tip of your blade won't go into the ground. Try not to cut the weeds and grass. A windrow is a long row of wheat; a pile is a small pile of wheat. The mower will make windrows, a person with a scythe and cradle can make a pile or windrows. Make a pile if the grain is dead ripe, so it can be shocked. If not dead ripe, make windrows.

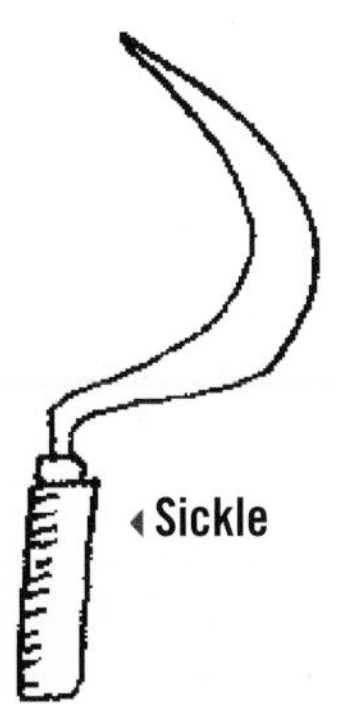

Sickle

Scythe

2. Sheave and shock them and let dry in the field in the fall. A sheave is a large handful of stalks, which is tied near the top with barren tillers (the secondary stalks that grow from a grain plant's base) into a binder's knot. To tie it, twist

Mower

the stalks together, then tuck under the band going around the sheave. Two to three sheaves are leaned against each other in the field, and another sheave is set on top of them, making a shock big enough for your hands to meet on the other side.

3. You may have to harvest early and let it dry. You can let it dry in the field as it is, or bring it into the barn, if the barn is large and dry. When there are no green stems at all, and when you tap corn stalks and they sound hollow, it is dry.
4. Oats and barley cannot be hulled by flailing. They must be steamed and ground in a mill. To thresh grain, use a hand flail to whack the grain while it is spread on the barn floor. Pile the straw separately. To use a flail, swing the flail back and forth quickly, hitting the grain on the barn floor. When you do all of it, flip it all over with pitchforks and flail it again. Do your flailing in the winter. When done, winnow it. For oats

▲ Shock of wheat

▼ Binder's knot

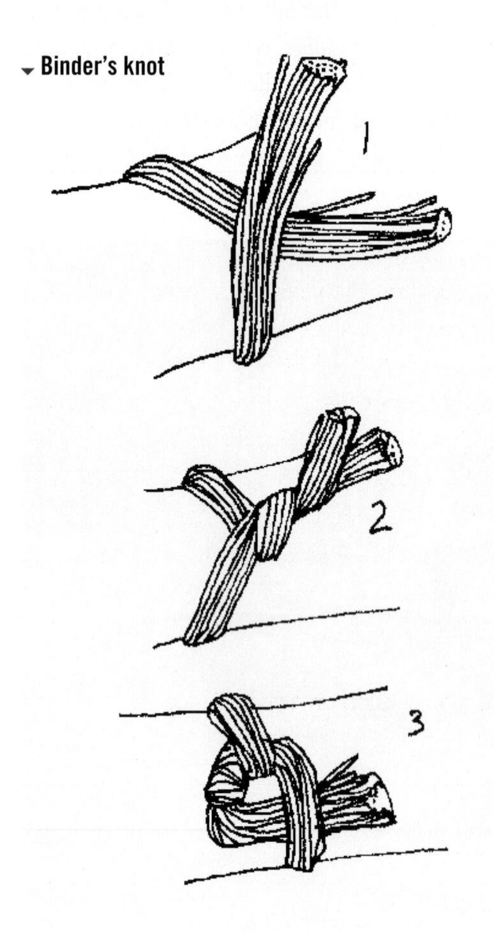

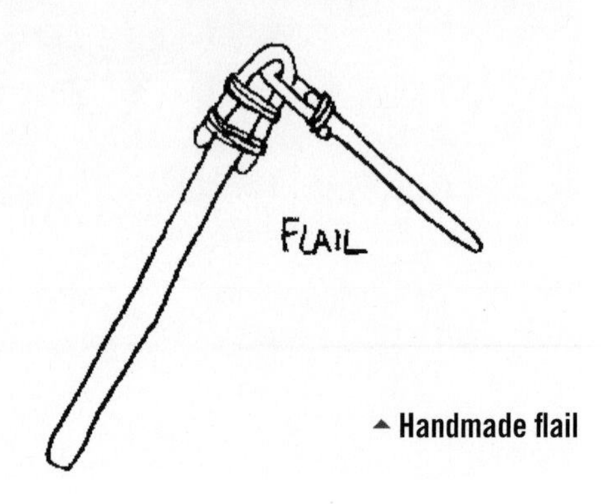

▲ Handmade flail

use a heavy swingle, for beans a light swingle.

5. To winnow (separate the chaff from the grain), toss the grain up with a basket and catch it again outside on a windy day. Or toss it in the air with a pitchfork, or use a winnowing tray. To winnow beans, wait for a windy day and pour them back and forth between two tubs. Save the chaff for livestock feed and save the straw or pile it in the field for the animals to eat.
6. Make sure the grain is as dry as possible. Sun dry it if you need to. Then store it.

To save grain seed:

You can save seed from your second crop of grain (unless you're using a hybrid). Your seed should be the best seed heads, unbroken and healthy. Let it dry in the shock for at least a month until it is totally dry. Then thresh it, and store the seed in a cool, dark place.

Where to store grain:

You have to store grain in a rodent-proof container (don't put it in a sack that has held seed grain because it can be highly poisonous) and keep hungry cats around. Keep the containers in a cool, dry place. It will keep for a year or more, until you grind it. Then use it immediately.

Preventing damage:

Don't store in an old shed since these are full of bugs.

The grain should be really dry.

Take out all twigs and leaves.

Check on it all the time to make sure there is no mold, bugs, or rodents. They will contaminate your food.

▲ **Traditional flail with a heavy swingle**

Mix lots of bay leaves in with your grain.

If you do get pests:

Carbon dioxide with dry ice: Get 1 tablespoon dry ice per 5 gallons of grain. Get an airtight container, put the ice on the bottom, pour the grain on top, wait an hour, then seal the container.

Carbon dioxide in a pit: Dig a pit in dirt that will stay dry, line it with woven straw, pour in the grain, and then seal it airtight with clay. The grain produces the carbon dioxide.

Heat: (Don't use for seed wheat). Spread ¼ inch on a pan and heat in the oven 140°F for 30 minutes.

Fungus:

Grain can get ergot fungus sometimes (rarely). Don't ever eat moldy grain, or feed it to your animals, because you can die and your animals can get sick. It turns the grain hard black and purple on the inside. It is caused when your grain gets damp. If your seed grain gets it, you should throw it out but if you are desperate, soak the grain in really salty water. The sclerotia (ergot masses) will float to the top where they can be skimmed off.

Ways to processing grains:

Cracking: breaking the kernel in two or more pieces, done especially on corn.

Crimping: flattening the kernel slightly, done especially with oats.

Flaking: treating with heat and/or moisture, then flattening it.

Grinding: forcing through rollers and screens.

Rolling: smashing between rollers at different speeds with or without steaming.

Saving: save the wheat to grow as seed next year.

Grain grinding methods:

Use a mortar and pestle, a hand grinder, or electric mill. For large jobs you can either buy a big electric mill or a water-powered stone mill like the old 1800s type, which is beyond complicated. The best for a family is the electric mill because you only have to put it through once. A hand grinder you must crank for a long time, sift it, and put it through several times depending on the texture you want.

Grits to cake:

The most coarse "setting" on an electric mill makes grits, which can be used to make cereal and animal feed. The settings then get finer to make cake flour,

▾ Grinding grain in a mortar

▲ Haystacks

which is very fine flour. Generally 1 cup of grain makes 1½ cups of flour.

Kinds of hay:

Alfalfa: makes the best hay, especially when mixed with bromegrass. Produces at least 5 tons of hay per acre for three cuttings a year. Susceptible to alfalfa weevil.

Red Clover & Timothy: mixed together this is the second best kind. Produces 2½ tons per yearly harvest (can't do three cuttings, but doesn't get alfalfa weevil).

Millet: grows very fast, can be fed 30 days after planting, and gets 5 tons per acre. Feed with buckwheat for full nutrition.

Oats: straw makes the most nutritious hay of all, and likes cooler weather.

Growing hay:

Plant the year before you first harvest your hay. You can broadcast (toss the seeds evenly by hand) on snow in February or on frozen ground in March. Cut the clovers with a scythe or mower just as they begin to blossom.

After cutting:

You have to let the hay dry in the field before putting it in the barn. Rain really hurts the hay, and a lot of rain ruins it. It will probably take two days after cutting for it to dry, but if it is hot, dry, and windy it could take only one. When it's almost dry, rake it into windrows with a hand rake or a horse-drawn hay rake. Test to see if it's dry enough to put in the barn by taking a bunch about 2 inches thick and twist it. If it breaks on the third twist, it's ready. Use a pitchfork to load into a hay wagon, and take it to the barn.

FOOD PRESERVATION

And the Quangle Wangle said
To himself on the Crumpetty Tree,—
"Jam; and Jelly; and bread;
Are the best of food for me!"

~Edward Lear (1812-1888)

What is blanching?

Before freezing vegetables, you must blanch them. There are two ways to do this: boiling and steaming. Blanching slows or stops the enzymes that make vegetables lose their flavor and color. If you blanch too much then they will lose nutritional value, but blanching too little will speed up the enzymes.

Boiling method:

1. Wash, drain, sort, trim, cut vegetables.
2. Put one gallon of water per pound of prepared vegetables, or two gallons per pound of leafy greens, into a pot and bring to a boil.
3. Put vegetables into a wire basket, coarse mesh bag or metal strainer and lower into the pot.
4. Keep boiling for the specified time for that kind of vegetable.
5. Cool the vegetables in ice water for the same time that you boiled them (except corn on the cob), stirring occasionally.
6. Drain thoroughly and pack into a container and freeze.

Blanching peas in a steamer

Steam method:

Asparagus: small stalk, 2 minutes; medium stalk, 3 minutes; large or all stalk, 4 minutes.

Broccoli: 5 minutes.

Brussels sprouts: small heads, 3 minutes; medium heads, 4 minutes; large or all heads, 5 minutes.

Butter beans: small, 2 minutes; medium, 3 minutes; large or all, 4 minutes.

Cabbage: shredded, 1½ minutes; wedges, 3 minutes.

Carrots: whole, 5 minutes; diced or sliced, 2 minutes.

Cauliflower: 3 minutes.

Celery: 3 minutes.

Collard greens: 3 minutes.

Corn on the cob (Note: cooling is twice the time): small ears, 7 minutes; medium ears, 9 minutes; large or all ears, 11 minutes.

Corn whole kernel/cream style: 4 minutes.

Eggplant: 4 minutes.

Globe hearts artichoke: 7 minutes.

Green beans: 3 minutes.

Green peas: 2 minutes.

Greens: 2 minutes.

Irish potato: 3–5 minutes.

Jerusalem artichoke: 3–5 minutes.

Kohlrabi: whole, 3 minutes; cubes, 1 minute.

Lima beans: small, 2 minutes; medium, 3 minutes; larger or all, 4 minutes.

Mushrooms: whole steamed, 5 minutes; slices steamed, 5 minutes; buttons/quarters steamed, 3½ minutes.

Okra: small pods, 3 minutes; medium pods, 4 minutes.

Onions: until center is heated, 3–7 minutes.

Parsnips: 2 minutes.

Peas: edible pod, 1½–3 minutes.

Pinto beans: small, 2 minutes; medium, 3 minutes; large or all, 4 minutes.

Rutabagas: 3 minutes.

Snap beans: 3 minutes.

Soybeans: green, 3 minutes.

Summer squash: 2 minutes.

Sweet peppers: halves, 3 minutes; strips/rings, 2 minutes.

Turnips: 2 minutes.

Wax beans: 3 minutes.

Freezing:

1. Gather freezing equipment: bags, steamers, and pots. Turn the freezer temperature to -10° F a day ahead of time.
2. Boil vegetables (or blanch if necessary), then chill in ice water. Use only the freshest and best food, and smaller is better—use not-quite-full-grown vegetables such as baby carrots and half-grown beans. Fruit can be treated by adding a bit of lemon juice, or adding ½ cup of sugar per 1 pound of fruit.
3. Cut as needed and put in plastic freezer baggies or wide-mouthed jars (leave 1 inch to spare at the top). Label them with type of food and the date.
4. Put bags in the freezer, and once they are frozen, turn the temperature back to 0° F. Frozen fruits and vegetables last about year (except onions), and baked foods last 6 months. Animal products and meat only last 3–6 months, so be careful.

Running a freezer:

The best temperature is -5° F, but to save energy you can go as high as 0° (but no higher!). Put cartons or buckets full of water into the bottom of your freezer, so if the electricity goes out, food will last longer, and you will have a small water supply. Keep the freezer in the coolest room of the house, but not where it freezes since it can withstand hot temperatures but not cold.

If the electricity goes out:

Keep the door closed, and cover the freezer with blankets except for the motor vent, if you know the power is coming back on soon. If it isn't, you will have to pull everything out and use non-electric preservation.

1. The first thing to do is to build an outside refrigerator using a cooler. Dig a big whole in the ground, stick the cooler in and insulate it with materials like straw and bricks and then cover it up with something very heavy so animals don't get in. I would also move lots of stuff from the fridge into the cold storage. If you have a running stream you can try to create a waterproof container for food, which would keep it even colder.
2. However, meat won't keep long in a cooler. Use a fire or barbecue to cook some of the meat that you plan to eat in the next week. Cooked meat will stay good just being refrigerated much longer than raw meat, probably 5–6 days. The rest of the meat you need to salt and dry. You could smoke the meat, but to do this properly takes a smokehouse and several weeks of time and constant vigilance. If you suddenly have take care of all your meat, salting is much more practical.
3. Clean the meat, and cut off anything you don't like, but you might want to leave the fat because that can be valuable later. Dry it off thoroughly and you ***can*** leave it whole, but it is easier to cut it into smaller strips to make it more likely to preserve in the middle. Rub spices into them, and then rub tons and tons of salt into them.
4. When you've rubbed in as much salt as you can, then cover it in a layer of salt to coat it. Hang it up somewhere that is about 59°F for at least 3 weeks, checking often for spoilage. A basement or cold storage is ideal. When you are ready to cook it, wash off the salt. The way this works is that the salt dissolves into the water in the meat and prevents bacteria from growing if that balance is greater than 3.5 percent salt to water. You want it to be over 10 percent, which you can't really control but if you rub tons of salt in there you can be pretty sure you've got it.

Food that does better if you freeze it than drying or canning it:

Asparagus, sweet green peas, snow peas, whole berries, melons, spinach, kale, broccoli, cauliflower, freshwater fish.

Food that you can't refreeze after it's been thawed:

Any kind of meat, ice cream, and vegetables. Fruit and bread you can refreeze without any problems.

Note: If canning isn't done properly, you can end up with botulism and die. Follow the instructions exactly, and use proper jars and seals.

Equipment needed for canning:

Pressure canner, enameled canner, glass canning jars, a sieve, canning jar lids and seals, wide-mouthed funnel, rubber gloves, jar tongs or lifters, a loud timer.

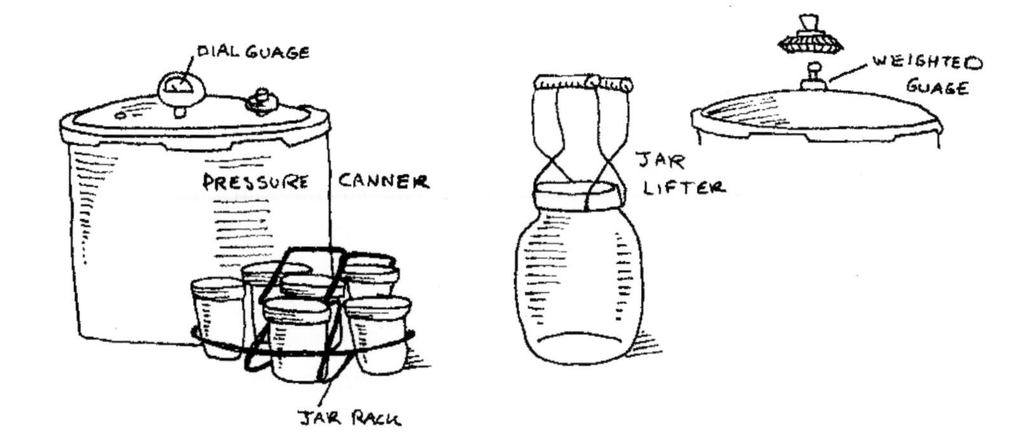

The canners:

A pressure canner has a dial-type temperature gauge, pressure regulator, and lock down clamps. It is not recommended to can without an accurate pressure regulator. In the old days it was done but today, and for beginners, always use one. This is because water boils at 212°F, not hot enough to kill all bacteria, so pressure is used to raise the temperature to 250°F.

An enameled canner is blue or black with white speckles, has a lid and canning rack, and is used for sterilizing and boiling things. It is also used to can things such as fruit that has lots of acid so it's safe to can—which is called water bath canning.

The jars:

The jars must be heavy mason or mason-type jars made for canning. Using old mayonnaise jars doesn't work because they break very easily under pressure (although it is possible in an enameled canner). Some companies are now selling classic mason jars with the wire bail lids, but the cheapest method is to get reusable jars and get new lid and screw bands. This type has a small lid with a rubber seal, and a metal band or ring that tightens it (the rubber seal can be used once, but the ring can be re-used). Then you throw out just the lid after one use. Never use cracked, chipped jars, or jars with a worn rim.

Pressure canning:

1. Prepare the food for canning by blanching, skinning, pitting, slicing, and poaching as needed (use a canning recipe for each mixture). Use only the best fruit, and make sure to remove all the bruised or infected parts.
2. Boil all your plastic and stainless steel equipment for 30 minutes, then wrap in a clean towel. Wipe down counters with chlorine scouring powder (like Comet) or other antibacterial solution, and rinse with boiling water. Dip knives with wooden handles in boiling water. Wash canning jars and put in simmering water (180°F) for 10 minutes. Heat canning lids in hot water but don't boil.
3. Put the food into the canning jar and add liquid to within a ½ inch from the top (for air space). Put on a lid snugly, but not so tight that air can't escape.
4. Put the jars in a rack in the pressure canner with 2–3 inches of boiling water in the bottom. Make sure the jars don't touch the sides, the bottom of the canner, or other jars. Fasten the lid and open the petcock. For fruit: Put jars in the rack, and cover with 2 inches of briskly boiling water. Put the lid on, but don't fasten it down. Leave the petcock open so that steam can escape.
5. Turn on the heat until steam comes out of the petcock in a steady stream (about 10 minutes). When the steam is nearly invisible 1 or 2 inches from the petcock, close the petcock.
6. Raise pressure rapidly to 2 pounds less than you need, then lower the heat and bring up the pressure the remaining 2 pounds slowly. This process should last for the amount of time specified for that particular food. Turn off the heat and let the pressure drop to zero. Wait 2 minutes, and then slowly open the petcock. Open the lid away from you so you won't get burned by steam, or just wait a while until it cools.
7. Pull the jars out with tongs, holding them straight upright (don't tip them!); let them cool on a dry, non-metal surface for at least 20 hours. When they are cool, test the seals, wash, dry and label them, remove the jar rings, and store in a cool, dry, and dark place.

Pressure canning altitude adjustments:

To use: find out your altitude, and for every thousand feet use the appropriate pounds. The poundage raises the temperature (which takes longer the higher you are).

Altitude in feet	*Weighted gauge*	*Dial gauge*
under 1,000	10 pounds	11 pounds
1,000 to 2,000	15 pounds	11 pounds
2,000 to 4,000	15 pounds	12 pounds
4,000 to 6,000	15 pounds	13 pounds
6,000 to 8,000	15 pounds	14 pounds
8,000 or more	15 pounds	15 pounds

Water-bath canning:

1. Use only highly acidic foods such as high-acid tomatoes and tomato sauce (no mushrooms or meat), jam, jelly, juices, barbecue sauce, chili sauce, relish, pickles, etc., or use a recipe which calls for an acidic additive specifically for water-bath canning in an enamel canner. If you are in doubt of the acidity of something, add 2 tablespoons lemon juice or ½ teaspoon citric acid (vitamin C) per quart.

2. Clean and sterilize the jars by boiling and heat lids (same as step 2 for pressure canning), then fill the jars with the hot food (as called for in the recipe). Put a new lid on the jar and screw on a ring firmly.
3. Fill the canner with water so that when the jars are put in the water will reach 1–2 inches above the top of the tallest jars. Bring to boil, and put the jars into the wire rack in the canner. This is the water bath. Put the cover on and let it reach a full rolling boil.
4. Processing time starts when the water reaches full boil. At the end of the processing time (check the recipe), lift out each jar carefully, and place on dry folded towels. Each jar will seal and you will usually hear a ping as the lid is sucked in. Don't touch them until they are cool.
5. The next day (usually takes a long time to cool), remove the rings, check for a good seal, wash the jar, label it with contents and date, and put it in a cool, dark, dry place.

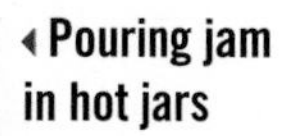

◂ **Pouring jam in hot jars**

Water-bath canning altitude adjustments:

Altitude in feet	increase processing time
1,000 to 3,000	5 minutes
3,000 to 6,000	10 minutes
6,000 to 8,000	15 minutes
8,000 to 10,000	20 minutes

Food safety rules:

Never eat from a jar that has lost its seal. It won't make a suction noise when you open it, and the lid won't be sucked in.

When canning, make sure the temperature is high enough and that you do it for a long enough time.

Keep the jars stored below 40°F.

Cook canned food for at least 10 minutes at boiling or 350°F in the oven before eating.

Don't use recipes or methods from before the mid-1980s—always use modern recipes.

Throw out anything with mold—don't try to scrape it off. Store jars loosely to prevent mold.

Don't use jars larger than a quart because they can't be heated enough.

Check the top rim of the jars for nicks before you use them.

If your boil drops or your temperature drops at any time, start over completely from the beginning.

A boil means a super-hot-really bubbling boil, not tiny bubbles.

Put hot food in hot jars and cold food in cold jars as temperature changes will break the glass.

Don't put hot jars on a cold surface or in cold air.

Check the processing time for altitude and adjust as needed.

If your water bath is not covering the jars by at least an inch, cover with more boiling water.

Add a piece of tomato to everything you can. Tomato has enough acid in it to prevent botulism from occurring.

How long canned food lasts:

Government authorities say that canned food lasts 1 year, and that's when you should throw it out. However, most home canners use their food far after that time period without problems. If you follow all the safety rules your food should be fine for a very long time, even

◂ **Onions can also be hung to dry**

10–20 years. However, it will not have the same nutritional value and the longer you wait the bigger the risk.

About food dehydrators:

A food dehydrator can dry food of any kind (including chips, jerky, and fruit leather). It can make stale chips and crackers taste better, de-crystalize honey, dry bread sticks, pasta, flowers, dyed wool, start seedlings, grow sprouts and make yogurt. A low-wattage dehydrator can run well on a home power source.

How to dry herbs by hanging:

Bunch the plant and tie up with a string. Then hang them upside down from the ceiling. It should take about 2 weeks to completely dry, and they will also hold their flavor longer if you keep them this way. When you are ready to use them, remove

the leaves and crumble them up. Make sure the room is cool, dry and airy.

How to screen dry (fruits, vegetables, herbs and seed pods) in the sun:

1. Use ripe fruits and vegetables. For fruit leather they can be overripe. Wash them thoroughly, peel, and slice very thin unless you are doing peas or corn. For seed pods, harvest before they burst.
2. Dip fruits in a gallon of water that contains 6 tablespoons of pickling salt and soak for no more than 5 minutes. Blanch vegetables, and chill in cold water, then soak up the water with a cloth. Prepare fruit leather.
3. Spread one layer on a drying tray (remember to line with plastic when doing leather). Keep moist foods away from dry foods—in fact, it is best if each type of food is dried separately. Label everything so you know what it is.
4. Put the trays in the hot sun, or in your oven or electric dryer. Dry at a low enough temperature that it doesn't cook, but high enough that it will dry before it spoils.

Dried apples

▲ Sun-drying tomatoes

5. Turn big chunks of food three times a day, and small foods once or twice a day. In a dehydrator, move almost dry food to the top, and moist food to the bottom.
6. Vegetables are dry when they are brittle and break when bent. Fruits are leathery or brittle, and should produce no moisture drops when squeezed. Pods will become dry and brittle.
7. Put the food into a wide-mouthed bowl for a week, stirring it 2–3 times per day. Keep it covered with a screen or porous cloth. This conditions the food to resist mold. Then repack it more tightly somewhere else.
8. If you want to pasteurize, put it in the oven for 30 minutes at 175° F. Store dried foods in the dark, or in a dark container or it will lose nutrients. Remember to label, and check it in the first 2 weeks for moisture—if there is some, dry some more.
9. Food should stay good for at least 6 months and more. If you have bugs, remove the bugs and roast food item for 300° F for 30 minutes.

Preparing fruit and vegetable leather:

1. Even fruit that has already started fermenting can still make good leather. Wash, peel, remove seeds and pits, then grind the fruit up by mashing or blending. Vegetables must be precooked and spiced and sweetened to taste (although it will be sweeter when it is dried so do it very lightly).
2. The puree must be thin enough to pour but not too thin to be watery. If it is too thick, add fruit juice or water. If it is too thin, add another kind of fruit puree.

3. Line the tray with plastic wrap, then pour the puree onto it. Tilt the tray to make it the same depth all over. The best drying temperature is 120°F.
4. Dry until the leather is still slightly sticky but peels easily from the plastic. Cut into strips and roll up with the plastic wrap.

Sun-dried jerky:

Cut meat into a lean, trimmed strips 1½ inches by ½ inch and as long as you want, cut along the grain. Trim off tendons, gristle, and fat, and sprinkle with ground pepper and salt. Dry in the sun 4 feet above a slow fire. The firewood should be non-resinous hardwood (green is ok) and flames kept very low because the fire is just to keep away birds and flies. Dry in the middle of the day, not at time when dew may cover it.

Storing eggs:

Cold storage:

Don't store the eggs near anything smelly like onions. Pack the eggs in a wood, plastic, or ceramic container in sawdust or oatmeal with the small end down. If you don't have cold storage, use the fridge, basement, or root cellar. Use eggs that you gather as soon after laying as possible. Eggs in the fridge will last 6 weeks, but if you seal fresh eggs in plastic bags they will last 2 months. If they are

stored at 30–40° F in fairly high humidity they will store about 3 months.

Pickled eggs:

Hard boil eggs, cool, and remove the shells. Soak the eggs in a brine of ½ cup of salt per 2 cups water for 2 days. Pour off brine and heat 1 quart vinegar, ¼ cup pickling spice, 2 cloves garlic, and 1 tablespoon sugar to boiling, pour it over the eggs and leave for 7 days to cure.

Freezing:

Use only very fresh, clean eggs (not ones that you had to clean). Crack the eggs and put the contents into the freezer container. Only freeze as many eggs per container that you will use at one time because you can't refreeze the eggs once you thaw them. Stir together without whipping in air, and add 1 tablespoon of sugar or ½ teaspoon of salt per cup of egg. They will store 8 months.

Drying:

Beat very fresh eggs well and pour into a layer ⅛ inch thick onto a drying surface that is lined with plastic or foil. Plates work for outside, and pans work for the oven, or you can use a dehydrator. In an oven or dryer, dry at 120° for 24–36 hours, then turn the egg over, remove the plastic or foil, break it up and dry for 12–24 more hours. In the sun, it will take 5 days until they are dry enough to break easily when touched. Grind the egg into a powder and use in baking, or to reconstitute add an equal amount of water (½ cup egg powder with ½ cup water). Dried eggs will last 3–4 months.

Lard pack:

Use very fresh clean eggs and dip in melted lard. Then pack the eggs in salt in a large bucket so that no eggs touch each other. Put the eggs in a cool place such as a root cellar, and they may stay good for a year.

What is live storage?

Live storage is a way of keeping certain vegetables and fruits in a cellar over a long period of time without any processing. The best plants for this are pumpkins, potatoes, dry beans and peas, onions, parsnips, apples, oranges, pears, grapes, tomatoes, and most other root vegetables.

Preparation:

When preparing to store foods in a live storage, leave the dirt on, which will

◂ Well-stocked cold storage

help preserve the food from decay. Use untreated wood bins, plastic buckets, or enamel cans, and keep fruits away from vegetables because the gas produced by apples can cause vegetables to sprout. Pack root vegetables in damp sawdust, sand, or moss. Keep potatoes out of any light or they will turn green and be poisonous.

Maintaining the storage:

Put a thermometer on the inside and outside of the cellar and monitor the temperature every day. Use doors and windows to maintain a temperature of 32°F, opening the door in cold weather and closing the door in really cold weather and really hot weather. The food needs humidity so it doesn't dry out, usually 60–75 percent, so if necessary put out pans of water, sprinkle the floor with water, or cover the floor with damp sawdust. If it is too damp, take pumpkins, squash, and onions to a dryer area or they will rot. Remove all spoiled food, and if something is about to spoil, dry it or can it quickly before it rots. Throw it out if something is rotting, molding, and treat for insects right away.

Saving honey:

Honey does not need to be frozen, canned, or refrigerated and keeps well in any type of container. Don't refrigerate it or it will crystallize sooner, and when storing for a long time don't let it get warmer than 75°F or it will lose flavor. Use a container that has a wide mouth because eventually the honey will crystallize and then you won't have to be pouring it, you can scoop. When you keep a small amount in your kitchen keep it in a warm place.

Types of preservation:

Freezing: There is no need to freeze except that it doesn't crystallize. Just warm it up and it will liquefy.

Crystallizing: This happens naturally to honey, and it simply dries. It can be used exactly the same in a recipe and to make it liquid you simply warm it up to 130°F as quickly as you can and then cool it as quickly you can. If it is in a can, put it on your woodstove. If in a jar, put it in a double boiler. Don't let it get hotter than 130°F.

Homestead Dictionary

Above the bit: When a horse raises his head above the level of the rider's hands in order to ignore the rider's commands, therefore reducing the control the rider has over the horse.

Abscess: A pocket of pus-surrounded, inflamed tissue.

AC: Alternating Current, an electric current provided by electric companies.

Achilles tendon: The large tendon that connects the heel bone to the calf muscle on the leg.

Action: In horse terms, the movement of the horse's legs.

Adze: A cutting tool with a curved blade at a right angle from the handle, used to hollow out, hew, scrape, and roughly shape large pieces of wood, especially logs.

Aerial roots: Roots that come out of the ground as additional support for the plant, to enable climbing, or to increase the amount of nutrients and gases the plant is absorbing.

Aerobic: A process that depends on free oxygen air, especially for fermentation or decomposition (such as for human waste).

Afterbirth: The placenta and blood delivered from the uterus after giving birth to offspring.

Agalactia: When a lactating mother does not produce milk right after giving birth.

Aids: In horse terms, the signals or cues that a rider gives to the horse to tell it what to do. Natural aids are voice; legs, hands, and weight, and artificial aids are whips and spurs.

Air compressor: A device that takes air and compresses it, and then delivers it at a high pressure.

Airs above the ground: When a highly trained horse performs complicated movements (called high school movements) where either the front legs or all four legs are off the ground.

Albino: Lacking pigment in the skin. A true albino has very white or pink skin, white hair or fur, and pink eyes.

Alkali: Any water-soluble salts found in soils that turn litmus paper blue, and in too large quantities can be harmful to agriculture.

Amble: In horse terms, a slower gait than pacing.

Amperage: The strength of a current of electricity (i.e. this has 30 amps).

Anaerobic: A process of being alive or active without free oxygen (the opposite of aerobic) such as for anaerobic bacteria that breaks down human waste.

Anesthetic: A partial or total loss of sensation, induced by injecting a drug or with a pill.

Anhidrosis: The limited or complete inability to sweat.

Annual: A plant that completes its life cycle within a year.

Anther: The primary male reproductive structure at the tip of a flower's stamen. The anther produces pollen that fertilizes the ovules to make seeds.

Anthropomorphize: To ascribe human characteristics to something that is not human, such as an animal or object.

Antifreeze: A liquid added to water in a cooling system in order to lower the temperature that it freezes at.

Anvil: An anvil is a heavy iron table used by blacksmiths to hammer metal. It has different odd-shaped parts for different kinds of work, including a face, pritchel hole, table, and horn.

Areola: A small ring of color around the nipple of the breast.

Array: A single solar module is a glass sheet enclosing either single-crystal or poly-crystal solar cells on top of a waterproof backing material and edged with an aluminum mounting frame. Several modules make a solar array.

Artery: Any tube in the body that carries blood from the heart to the cells, tissues, and organs.

Artesian: A type of well in which water fills it through internal hydrostatic pressure, or simply the pressure the water has just sitting still. When digging such a well, it will fill with water quickly without further effort.

Artificial aids: Tools used to help command the horse, including spurs and whips.

Ascarids: Internal parasites that plague animals, also called roundworms.

At grass: A horse that has been turned out in a pasture, paddock, or field.

Auger: A metal hand tool for drilling holes, especially in wood.

Awl: A tool with a handle and a sharp, straight, metal end which is used to poke holes in tough material such as heavy canvas and leather.

Ax: A tool with a wooden handle called a helve, and a heavy metal head with a sharp edge. It is used in a chopping motion to cut wood.

Azoturia: The technical term for tying-up or Monday morning sickness. When a horse has prolonged muscle contractions during exercise.

Back at the knee: A conformational fault in a horse where the upper leg is set back further than the lower leg. More serious than over at the knee because there is more strain on the tendons.

Back breeding: In horse terms, breeding back to a certain stallion to preserve a certain genetic trait.

Back cut: Also called a felling cut, the second cut which causes a tree to fall.

Bacteria: A single-celled organism, some of which are good for humans (breaking down waste material) and some of which are bad (causing infections and disease). A single bacteria is called a bacterium.

Bail: The handle of a bucket, a half-circle made of wire or wood.

Balustrade: A railing used on the edge of balconies and stairs.

Bandy-legged: When a horse's hock turns outward, also called bow-hocks. The opposite of cow-hocks.

Banged tail: A horse's tail that has been trimmed flat at the bottom, only seen on dressage and hunter horses.

Barn sour: A horse that does not like being ridden away from the barn or leaving their pasture mates.

Barrel: The area of the horse's body between the forelegs and the loins.

Barren tillers: Secondary stalks that grow from the base of a wheat plant which don't grow grain seeds.

Bars: In horse terms, the fleshy area between the front and back teeth where the bit rests.

Bascule: In horse terms, the arc a horse makes when jumping a fence.

Bast: The fibers of a hemp plant used in making hemp paper.

Bat: Also called a whip or crop, an artificial aid used to encourage a reluctant or lazy horse to move forward or to punish. In making pottery, a wooden or plastic circular board which is put on the wheel so that the pot can be easily removed once it is formed.

Bay: In horse terms, a deep reddish brown coat color with a black mane and tail.

Bay window: Any window space that projects outward from the walls of a building.

Beam: A structural member of a building made of wood or metal that transversely supports a load.

Bedding: Absorbent material put on the ground in animal housing to soak up urine and make cleanup easier. Also provides softness for hooves and feet.

Beetle: A wooden mallet, or hammer, made of ironwood, and bound with metal bands on the head to strengthen it. Used to drive in gluts or make mortise-and-tenon joints.

Bellows: Two triangular pieces of wood, hinged together. The top piece has a handle and a hole in the middle of the board. Between the boards is a folding frame covered in airtight material. On the end of the bellows is a round hole or spout to let air out. The top board is pushed up and down, forging air out of the spout. Bellows are used to heat a forge fire.

Behind the bit: When a horse holds his head close in order to ignore the rider's commands. Also called overbent.

Bent: A section of a timber frame house, or one "frame" which consists of two vertical sides and a peak which supports the roof of the house.

Biennial: A plant that takes two seasons to complete its life cycle.

Billets/billet straps: Straps used to attach the girth to a saddle.

Billy: A male goat.

Bind: When cutting down a tree, when the saw gets stuck in the tree as you're cutting, as in, "My saw bound in the tree."

Binder's knot: A knot tied with wheat stalks around a handful of stalks to make a sheave.

Biodiesel: A mixture of ethanol, vegetable oil, and lye, which is processed through transesterification to produce a substitute for regular diesel fuel.

Biodynamic: A method of agriculture created by Rudolf Steiner that seeks to work with the health-giving forces of nature. It attempts to follow the rhythms of the earth and the cosmos, and has special formulas of compost for the soil.

Biofilter: Another name for biological filtration, which is a process of filtration and purification using natural biomass, and can be used to remove odors, purify water, clean fish tanks, and treat home waste.

Biointensive: A method of agriculture that utilizes small spaces to create a sustainable garden that can grow enough calories to feed people.

Uses compost, intensive planting, companion planting, and open-pollinated seeds.

Biomass: The total mass of plant material and animal waste in a given area that is being converted into a fuel. Usually the result of this process is heat.

Bit: A mouthpiece made of metal, rubber, or some other material that is held on by the bridle. Used to convey instructions to a horse.

Bitumastic: A protective coating over metal structures that are exposed to weathering, usually made partly of asphalt.

Blade: The metal or wooden part of the plow that turns over the soil. It does not cut the soil.

Blanch: To cook something briefly, especially for vegetables.

Block plane: Has a shallow-pitched blade, used to shrink a stick that fits too tight.

Bog spavin: Soft, liquid-filled swelling on the inside of a horse's hock that does not usually cause lameness (unlike regular spavin).

Boll: A ball formed of leaves on a cotton plant that form after the flower falls off. The boll holds the ball of cotton that holds the seeds, sort of like a seedpod.

Bone: In horse terms, the measurement around the leg just below the knee or hock. The measurement determines the horse's ability to carry weight (light-boned means a limited weight capacity).

Bosal: A braided noseband used in western riding that is the equivalent of a bitless bridle.

Bot: A parasite that bothers horses.

Bottom: The blade of the plow that turns over the soil.

Bovine: The family of mammals that include bulls, cows, steers, and oxen.

Bow-hocks: In horse terms, also called bandy-legged, when the hock turns outward. The opposite of cow-hocks.

Bowed tendon: An injury to the tendon that runs down the back of a horse's lower leg.

Boxy hooves: In horse terms, describes a narrow, upright hoof with a small frog and closed heel, which is also called clubfoot.

Breaking/breaking-in: When a young horse is taught basic skills prior to learning riding and driving.

Breaking loose: When removing a tire, loosening all the lug nuts before removing them.

Breed: A certain type of animal that is bred for certain characteristics over a long period of time. Also the word describing the act of mating two animals together.

Bridle: A piece of equipment worn on the horse's head which bits and reins can be attached to communicate with the horse.

Bridging: A network of boards that connect floor joists together to help support the floor.

Bridoon: A snaffle bit used with a curb bit in a double bridle.

Broadax: An ax with a 12–14" blade. The eye (the hole where the blade fits into the handle) is at an angle so that it will swing out of the way for your knuckles.

Broadcasting: A method of planting seeds in which you scatter them randomly rather than planting rows, sometimes utilizing the wind to sow them farther.

Brood: A group of young animals hatched at one time, especially for birds but also applicable to bees. To brood is to sit on or hatch eggs, or the mothering protection of hovering near and defending one's young.

Brood cell: A cell in a honeycomb that holds a growing baby bee.

Brood mare: In horse terms, a mare used for breeding.

Broken-in/broke to ride: A horse that is used to tack and rider and has begun basic training, also called greenbroke.

Broken winded: A horse that has an abnormal breathing pattern from Chronic Obstructive Pulmonary Disease (COPD). Also called the heaves.

Brushing: In horse terms, when the hoof or shoe hits the inside of the leg near the fetlock, usually due to poor conformation or action.

Brushing boots: Boots made especially for horses that have problems with brushing to avoid injury.

Buck: In horse terms, when the horse leaps in the air with the head lowered and the back arched.

Bucking: Cutting a log into the lengths of wood you need to build with.

Bucksaw: The best tool for sawing wood. It has a wood frame, and a wire tightener.

Bull: A male bovine animal.

Bungee: An elastic cord with J-or S-shaped hooks on each end, which are used to do anything from tying down things to holding a clutch together on a truck.

Burr mill: Two millstones set a certain distance apart so that it can scrap the hulls of oats off without crumbling the groats (the grain) inside. The distance is 1⁄16 inches.

Butcher: To kill and prepare animals for food, or a person with the knowledge to do so.

Butter mold: A round dish with sides, sometimes with a lid, that has a carved picture on the inside. It is greased, and then butter is pressed in to make a picture in the butter.

Butt-up: A type of roof ridge made in thatching which forces the straw together from both sides of the roof to form a peak.

Cable hoist puller: Also called a come-along, a tool attached to a logchain that ratchets the chains tight and helps to control a tree's fall as it is being cut down.

Calf: A young cattle usually under 1 ½ years old.

Calorie: A unit of how much energy a food has the potential to make when it is oxidized in the body. This energy is created by heat.

Calve: When a cow gives birth to a calf.

Calyx: The outer ring of sepals that protect the unopened forming flower bud.

Cannon bone: In horses, the bone in the lower foreleg between the knee and the fetlock. Also called a shinbone.

Canter: A horse's gait with three beats, where the leading hind leg strides first, then the opposite diagonal pair, then the opposite foreleg. In western riding called the "lope."

Cantle: The back ridge of an English saddle.

Capped hocks: In horse terms, swelling on the point of the hock caused by an injury or lying down without enough bedding.

Capstone: A flat stone set on top of a rock wall to finish it off.

Carburetor: The part of an engine that mixes gasoline vapor with air before it is ignited.

Cast: A horse that rolls and gets stuck against some wall or fence is cast.

Castings: Worm manure, the best manure for the soil.

Castrate: To remove the reproductive organs, either the testicles or ovaries. An animal which has had this done is castrated, and is most often done to male mammals.

Catgut: A thin strip of sheep intestine sterilized and used like thread to suture or stitch up a wound.

Caulking: A material used to stop up joints and seams to make them waterproof. To caulk is to install this material.

Cavalletti: A low wooden jump used to teach a horse and rider how to jump.

Cavesson: A noseband that fits to the bridle, or a leather or nylon headgear with attachments for side reins and lunge line for lunging.

CCA: Stands for Cold Cranking Amps. A number rating how well an engine battery starts in cold temperatures. A higher CCA number means it is better starting in cold weather.

Cellar: A dark, cold room usually dug underground used to store food. A storm cellar is used to protect people from tornadoes and hurricanes.

Cellular respiration: The process which plants use to extract energy through a chemical breakdown of stored food molecules.

Cetane: The cetane number is a measure of how well a fuel ignites when it is compressed. The higher the cetane number the cooler it burns, and the better the fuel.

Chaff: Dry outer parts over the husk of mature wheat and other grains that can be removed by threshing or flailing.

Chainsaw: A cutting tool with a gas engine that turns a chain at high speeds. It is used to cut large pieces of wood and logs.

Chainsaw file: A special round file for sharpening the hooks in the chain of a chainsaw.

Charkha: A tiny spinning wheel designed by Gandhi to spin cotton, which unfolds from a small box and sits on the floor.

Checking: When a board that is drying shrinks unevenly, causing it to warp and crack.

Chestnut: In horse terms, a small rubbery bump on the inside of the legs.

Chicken wire: A type of fencing which is made of thin wire woven into small hexagons, which chickens can't get through.

Chinking: Waterproof material such as clay or mortar that is stuffed into the spaces between the logs in a cabin. To chink is to perform this task.

Chip/chip-in: When a horse adds a small step right in front of a fence before jumping.

Churn: A container that uses either paddles run by a handle and crank mechanism or a pole that is dashed up and down into milk in the container. Churns are used to make cream into butter.

Cider press: A mill and press that crushes apples and squashes them for juice, which is turned by a hand crank.

Cinch: In horse terms, a strap attaching the saddle to the horse. Called a girth in English riding.

Cistern: Any kind of underground tank used to store water, but especially for storing rainwater.

Class: In Latin classification, a category within the phyla of a kingdom. For instance, the monocot class in the angiosperm phyla in the plant kingdom.

Clean-legged: A horse that does not have feathering on the lower legs.

Clevis: See clevy.

Clevy: On a plow, where the draught is attached. Also called a clevis.

Cloaca: A cavity in birds and fish that has intestinal, genital, and urinary tracts in it.

Cloche: Like a miniature greenhouse, a small flexible frame with a plastic cover in a dome shape that sits over plants to protect them from frost.

Clod crusher: A machine drawn by horses that breaks up tough clods after plowing. There are 24 cast iron disks with teeth that revolve separately. The wheels alternately rise and fall to prevent clogging.

Cob: A horse that is stocky and well built for carrying heavy riders. Also can refer to a form of natural building material based on mud.

Coffin bone: In horse terms, a small bone within the hoof.

Cohousing: Usually built in a city, housing that is created to help build community, usually to lower housing costs. However, most cohousing is now more expensive because of community benefits.

Coldblood: Any heavy European breed of horses descended from the Forest horse.

Cold frame: A wooden box with a glass lid on a hinge so it can be opened and closed. Plants are grown in the frame so that it can act like a small greenhouse, protecting from cold weather.

Collection: In horse terms, when the rider causes the horse's frame to be compacted, the horse light and supple, the baseline shortened, the croup lowered, the shoulder raised, and the head vertical.

Colic: In any animal or human, severe abdominal pain.

Colostrum: The first milk produced by a mammal that is completely essential for the survival of a creature. Colostrum protects babies from disease and gives complete immunity.

Colt: An uncastrated male horse up to four years old. A male foal is called a colt foal.

Come-along: Also called a cable hoist puller, a tool attached to a logchain that ratchets the chains tight and helps to control a tree's fall as it is being cut down.

Coming: In horse terms, when a horse is approaching a certain age, he is said to be coming to it (i.e. "coming two").

Commune: A community in which all the members have given up personal property rights, and instead the community as a whole owns the property. These people often share common interests, work, and income.

Companion planting: Growing a plant next to another plant because they compliment and benefit each other.

Compost: The act of converting organic material into its basic state—rich dirt and soil. Any mixture of decaying plant waste and manure is compost.

Condensation: With liquids, the process of changing from a gas to a liquid or solid, such as turning steam back to water.

Conformation: The overall way in which a horse is put together; the proportions of its anatomy.

Coolant: A liquid that produces a cooling effect, especially if it is used to transfer heat away from one part to another.

COPD: Abbreviation for chronic obstructive pulmonary disease, also called being broken winded.

Cord: A stack of wood 4 feet high, 4 feet wide, and 8 feet long.

Corn cultivator: A cultivator made specifically to cultivate corn.

Corn knife: A corn knife has a straight blade, is about 1 ½ feet long and is used to cut corn stalks. It is kept very sharp. The older type of corn knife straps to your leg so you can don't have to bend over.

Corolla: The part of the plant known as the flower. Layers of petals, which ring the receptacle, and are pretty to attract bugs and birds.

Corrosion: Any kind of decay or erosion caused by a chemical action.

Cotton cultivator: A cultivator made specifically to cultivate cotton.

Cotyledon: The part of the seed where the seedling gets food. In some cases it is a storage area, and in other seeds it absorbs the food from the endosperm.

Coulter: On a plow, it cuts the furrow slice in front of the share.

Counter canter: The horse canters in a circle with the outside leg leading instead of the common inside leg.

Cow: Any bovine female animal.

Cow-hocks: In horse terms, the hocks turned in like a cow. The opposite of bow-hocks.

CPR: Cardiopulmonary Resuscitation, a process of restoring a person's heartbeat and respiration when they have lost consciousness.

Cracked heels: In horse terms, when the heels become inflamed, the skin cracks and pus comes out.

Cracking: Breaking the kernel of a grain into two or more pieces, especially for corn.

Cradle: A rack connected to a snath for catching the grain as it is cut.

Crawl space: A low space beneath the floor to give people access to plumbing and wiring.

Cream separator: A device that uses centrifugal force to remove cream from milk. To remove cream from goat's milk the separator must be adjusted for it or be specifically designed for it. They have a maximum of 18 disks that milk is forced through.

Crib-biting/cribbing: When a horse hooks his teeth into something solid and sucks air in through the mouth. If a horse does this without latching on to something it is called wind sucking.

Crimping: Flattening the kernel of a grain slightly, especially for oats.

Crop: An artificial aid also called a whip or bat, used to encourage reluctant or lazy horses, or to punish. Also, a group of a specific agriculture produce grown in a year, such as a crop of corn. In a bird, a pouch in the gullet where food is partially digested so it can be regurgitated for its young.

Crossbreeding: Mating two horses of different breeds or types.

Crosscut saw: A saw for cutting across the grain of the wood. It can have one handle or two for two people to use it at the same time.

Cross peen: A blacksmith's hammer, shaped with a ball instead of a flat circle like on a claw hammer.

Cross-pollinator: A plant that has to have help from wind or insects to pollinate.

Crossties: Using two ropes or ties to tether a horse by putting one on each side and connected them to a post or wall.

Croup: In horse terms, the top of the hindquarters from the point of the hip to the tail.

Crow hopping: When a horse leaps in the air with all four feet off the ground at the same time.

Cues: Also called aids, signals used by a rider to communicate to a horse.

Cultivator: The name properly means any implement that is used to cultivate plants after they have started growing, but it has been improperly used for tools that prepare the ground, such as the harrow. It is mostly used to destroy weeds that have grown between plants that have been planted in rows only. Most cultivators can be adjusted for the width of the rows using screws or iron keys. They are usually shaped with several bars with curved teeth coming out of them, but a few are shaped like plows, which throws dirt on either side.

Curb: When the tendon or ligament below the point of the hock thickens from strain.

Curb bit: A bit that has cheeks and a curb chain which lies in a chin groove. In a double bridle it is used with a bridoon or snaffle bit.

Curb chain: A chain used with a curb bit.

Curriculum: All the courses of study offered by an educational institution, or a group of related courses in a field of study.

Cutting down ax: Has a handle the perfect length for the user, used for cutting down trees.

Cutting up ax: An ax with a good length helve (handle) for the user, used to cut up cordwood.

Cyst: A pouch or sac without an opening that contains fluid and develops abnormally in a body.

Daisy clipper: A horse that seems to be hugging the ground when it moves.

Dam: A horse's or calf's biological mother. Or, an obstruction to running water (such as a stream or river) that slows the flow of water to create a pond or lake.

Damping off: Death of seedlings by fungal disease before or after they emerge from the pod.

Debeak: Clipping the tip off of a chicken's beak so that it can't hurt itself or others.

Decant: To pour off a liquid without disturbing the sediment at the bottom.

Deckle: A frame with a screen that is used in the making of paper.

Deadhead: To remove dead flowers from a plant by pinching off the flower and ovary from the stem, in order to encourage more blooming.

Depth of girth: The measurement from the top of the withers to the elbow. If a horse has a large measurement it has good depth of girth.

Detangler: Something put into hair or fur which prevents it from tangling, usually a conditioner.

Dewormer: A medicine for curing an animal of worms.

Diagonals: In horse terms, when the legs move in pairs. When the left foreleg and right hindleg move together it is left diagonal, and when the right foreleg and left hindleg move it is right diagonal.

Dibber: A tool used to make holes in the soil to plant bulbs or seedlings in.

Dicot: A plant that has two seed leaves or "cotyledons."

Dimension lumber: Wood cut into pieces of a specific size.

Dipped back: Also called sway back, when the horse's back is unusually hollow, or dipped in.

Dipstick: A metal rod that is dipped into a container in order to find out the fluid level.

Disc: Also called disk or wheel harrow. Used for land with many roots or stiff clay, instead of teeth. The shaft that the disks are fixed (called the wheel gang) is attached to the pole and draft bars by a ball joint. It cuts and separates, rather than scratches.

Dished face: In horse terms, when the nose of the horse is concave instead of straight.

Dishing: An undesirable action, where the toe of a horse's foreleg is moved outward in a circular motion with each step.

Distemper: In horse terms, a highly contagious disease also called strangles.

Distillation: A process of purifying a liquid by boiling it and condensing the vapors, using a still.

Disunited: In horse terms, when the horse's legs are out of sequence during a canter.

Dock: The bony part of a horse's tail where the hair grows from.

Docking tool: A bloody, dangerous way of removing an animal's testicles. It used to be used to remove an animal's tail, which is now considered inhumane.

Dormant: A state of suspended activity in which plants stop growing to protect themselves against extreme cold or heat.

Dormer: For rooms that are located in the roof of a house, the windows are put into the roof. Usually they are perpendicular to the walls of the house and have their own roofs.

Double bridle: A traditional English bridle with two bits, a snaffle, and curb, which gives the rider greater control.

Double cultivator: Cultivates three rows at a time.

Double dug: A type of raised bed garden in which the soil is double dug, or dug twice to make it fluffy and light, also called French intensive gardening.

Double Michigan plow: Also called sod and subsoil plow. It has two blades, which creates two furrows that are twisted into one. Completely breaks up soil, and plows a deep furrow, but takes more work.

Dovetail: A type of joint in woodworking that likes somewhat like the tail of a bird, with a flared shape that fits into a notch.

Dowsing: To find water using the mysterious skill of using sticks or wire to locate underground pipes or groundwater. A person who has this skill is a dowser.

Draft horse: Any horse used for hauling heavy loads but usually used for large breed horses.

Draught: Another term for a harness for a animal team, or an animal team that is specifically bred for pulling.

Draw rein: A rein that attaches to the girth at one end, goes through the rings of the bit and then back to the rider's hands. Gives the rider more control but is easily used incorrectly.

Dressage: The art of training a horse so that he is completely obedient, or the competitive sport in which the horse's dressage is judged.

Drone: A bee whose only purpose is to compete to mate with the queen bee. Drones have no stingers and are very fat.

Dropped/drop noseband: A noseband buckled beneath the bit, which prevents the horse grabbing the bit in his teeth.

Drop spindle: A wooden disc usually rounded on the bottom, with a small dowel going through it. The end of the dowel protrudes down through the disc so that when it hits the floor it can spin. The end of unspun wool is attached to the dowel and then it is spun down to the floor, making thread.

Dryland distemper: Also called pigeon fever, causes abscesses in the chest and belly.

Dutch oven: A cast iron pot with legs and a lid used directly in the fire like an oven.

Easement: A legal right-of-way on another person's property, such as road access.

Eaves: The lower border of a roof that overhangs a wall.

EIA: Equine Infectious Anemia, a virus without a cure. Also called swamp fever.

Elastrator: A tool used to castrate an animal, with the use of an elastic band. Also called an emasculator.

Electrolyte: In the human body, ionized salts in body fluids that the body needs to live. The mixture of electrolytes conducts electricity.

Elizabethan collar: A special collar made for goats that prevent a female goat from sucking from her own udder.

Emasculator: A tool used to castrate an animal with the use of an elastic band. Also called an elastrator.

Embryo: An organism that is in the process of growing before it is born or hatched. An egg is not an embryo unless it has been fertilized. A plant also has an embryo that is inside the seed.

Embankment: A mound of earth or stone built to hold back water or support a road.

Enamel: A paint that dries to a hard glossy finish, or a colored glassy compound that is fused to metal, glass or pottery for decoration or protection.

Encumbrances: On a property, when the former owner sold or reserved the right to mine, cut trees or take the water, even though it is your land. If you cut trees and there was an encumbrance you would be stealing the former owner's trees.

Endemic: A species that is only found in a certain location and nowhere else.

Endosperm: Food storage that surrounds the plant embryo inside the seed. The endosperm nourishes the seedling during germination.

Engagement: In horse terms, when the hind legs are well under the body.

Entire: Also called a stallion, an uncastrated male horse.

Epiphyte: Plants that grow on other plants but aren't parasites because they get their nutrients from the air. Orchids, mosses, lichens, and some cacti are epiphytes.

EPM: Equine Protozoal Myeloencephalitis, a neurological disorder.

EPSM: Equine Polysaccharide Storage Myopathy, when the muscles waste away, usually in draft horses.

Equine Infectious Anemia: A virus with no cure, also called EIA or swamp fever.

Equitation: The art of riding a horse.

Ergot: A horny growth on the back of a horse's fetlock joint.

Erosion: When something is worn away over time, but especially for soil. Erosion can be caused by wind and water.

Ethanol: Another term for ethyl alcohol, a pure alcohol which is used as a fuel and in perfumes, medicines, and as a solvent.

Euthanization: A method of purposely killing an animal in the least painful way possible (usually with gas or injection) in order to prevent further suffering.

Ewe: A female sheep.

Ewe-neck: In horse terms, a conformation fault in which the neck is concave along its top edge, instead of straight or curved outward.

Evaporator tray: A tray 5 feet wide and 16 feet long, used to evaporate sweet sorghum green juice to make syrup. It has 3–10 compartments with small openings, and a spigot at the end to pour the syrup out. It should be made of black iron or stainless steel.

Eventing: An equestrian competition where riders demonstrate dressage, cross-country, and show jumping; also called combined training.

Extension: In horse terms, the lengthening of the horse's frame and step, the opposite of collection.

Extravagant action: Hackneys and Saddlebreds are often trained to have high knee and hock action that is called "extravagant."

Eyering: The edges of a bird's eyelids.

Face: The top of an anvil.

Farrier: A person who shoes horses and has knowledge of hoof care.

Farrier's anvil: A smaller anvil used by a farrier to create horseshoes, and weighing between 50–150 pounds.

Fault: A crack in the surface of the earth. Basically the space between two plates of the earth's crust, which often shift, causing an earthquake.

Feathering: In horse terms, long hair on the lower legs and fetlocks, usually seen on heavy breeds.

Feed cutter: A grinder specifically made for cutting straw, corn, and corncobs into feed for cattle and horses. A straw cutter has self-feeding spiked rollers turned by a hand crank. It turns it into a revolving knife grinding against a fixed knife and through a chute.

Felling: When cutting down a tree, a second cut above first cut that causes the tree to fall. Also, the act of cutting a tree.

Fermenter: A device used to ferment alcohol, usually with an airlock and a gauge.

Fertile: Anything that is capable of starting or supporting reproduction.

Fertilization: To add fertilizer to the soil. Or the act of starting reproduction either through insemination or pollination.

Fertilizer: A large amount of materials that when mixed together and worked into the soil increase its nutrient value for growing plants.

Fetlock (joint): The lowest joint on a animal's leg.

Figure-eight noseband: In horse terms, also called a grackle noseband, it has thin leather straps that cross over at the front and buckle above and below the bit.

File: Files are metal tools with rough sides used to sharpen and grind things by hand. There are wood files, metal files, chainsaw files, and many other files in a variety of shapes.

Filly: A female horse under four years old. A female foal is called a filly foal.

Firing: In horse terms, when the skin over a leg injury is burned with a hot iron to make scar tissue.

Fistulous withers: Inflammation and infection at the top of a horse's withers.

Five gaited: A horse shown at a walk, trot, canter, slow gait, and rack, instead of just the first three.

Flail: A tool for threshing, or beating grain to remove hulls and other unwanted material called chaff. The handle is called a stave, and the swinging part is the swingle. The swingle is about 2 feet long. It is all made of wood, and the swingle is attached to the stave with wire or something so that it swings right and left.

Flashing: Sheet metal put around the chimney and dormers on a house where they connect to the roof, in order to keep out moisture.

Flax breaker: A machine with two big timbers fastened together that are hinged on one end so that they can go up and down, shaped into triangular blades. They drop down into triangular slots on special table. It looks kind of like a mouth with two teeth on the top, and two dents on the bottom to fit them in.

Flax card: Wooden paddles with wires poking out, similar to wool cards, used to comb out flax.

Flax wheel: A flax wheel is the smallest spinning wheel, and runs with a foot treadle (all the others are run by hand). It is used for flax because it spins it tighter than a big wheel.

Flexion: When a horse bends his head in to loosen the pressure of the bit on the lower jaw. Also describes the full bending of the hock joints, and vets perform flexion tests when diagnosing lameness.

Flexor tendon: The tendon at the back of an animal's legs.

Flint: A hard stone that when struck can create a spark, which is useful for starting fires.

Floating: Rasping a horse's teeth as part of normal dental care.

Flush/flushing: Putting a ewe on the richest feed or pasture in order to increase the chances of having twins or triplets.

Flying change: In horse terms, changing the canter lead to gain balance during turns.

Foal: Any horse up to one year old.

Foaling: The act of a horse giving birth. A dam is said to be foaling.

Fodder: Food for domestic animals.

Footing drain: Found mainly on older homes with a basement and no sump pumps. The footing drain is a pipe underground around the basement that diverts floodwater away or into the sewer.

Forage: Food for domestic animals that they usually find in the field or pasture.

Also the act of looking or searching for food, such as foraging for wild strawberries.

Force: To make a plant growing earlier than the normal season by tricking it to think it's warm out. Bulb plants are often forced.

Forearm: In animals, the part of the foreleg above the knee.

Forehand: In horse terms, the head, neck, shoulder, withers and forelegs. Horses with little training that pull forward on the bridle are on the forehand.

Forelock: In horse terms, the bit of mane between the ear which lies on the forehead.

Forge: A forge is a building housing a large fireplace for heating the metal, large bellows to heat the fire, a tub of water, an anvil, forging tools, and a woodpile. Also the apparatus or fire that heats the metal is called a forge.

Forge hammer: A hammer with a large, rounded metal head, used to pound hot metal.

Forging anvil: A very heavy anvil used specifically by a blacksmith, weighing between 75–500 pounds.

Foundation: The concrete or rock that is put into a hole the shape of the house, which supports the entire house. In a beehive, a sheet of beeswax with a honeycomb imprint which encourages bees to build combs where you want them.

Founder: In animals, when the coffin bone in the foot becomes detached and rotates during a severe case of laminitis. A horse with this problem has foundered.

Four-in-hand: A team of four harness horses.

Furrow: A long strip of earth that has been turned over by a plow.

Framing chisel: A thick chisel for making the holes in mortises square.

French intensive: Also called "double dug" raised bed gardening, which produces the most produce for the least area. It helped inspire biointensive gardening.

Friedman harrows: An angular harrow that is shaped like two wings, or a triangle. It is pulled from the peak of the triangle.

Frog: In horse terms, a triangular rubber pad on the sole of the foot that acts as a shock absorber.

Frying pan: An iron pan with an iron lid and a 3-foot iron handle.

Full mouth: A horse that has all his permanent teeth after 6 years old.

Fungus: A member of the fungi family, which are organisms that live as parasites, and include mushrooms, mold, toadstools, mildew, and rust.

Fuse: A safety device that melts and interrupts an electrical circuit.

Gable: The vertical triangular end of a building from the level of the eaves to the ridge of roof.

Gad: A 4–5-foot long hardwood sapling with 2–3 feet of leather used as a whip for ox teams.

Gaggle: A group of geese.

Gait: In horse terms, the different paces at which it moves, usually walking, trotting, cantering and galloping.

Gaited horse: A horse that moves at paces other than the regular walk, trot, and canter.

Gallop: The four-beat gait of the horse where each foot touches the ground separately.

Galvayne's groove: A dark line appearing on the upper corner incisor of horses between the ages of 8–10, used to estimate a horse's age.

Gambrel: A large metal tool for hanging an animal during butchering, usually used with a hoist. It is actually two hooks that are inserted between the Achilles tendon and ankles of the animal.

Gander: A male goose.

Gauge wheel: On a plow, it runs along the surface of the ground and determines the depth of a furrow.

Gaskin: In animals, the second thigh extending from above the hock upwards to the stifle.

Gelatin: A yellowish, transparent substance taken from boiling down animal parts, which is used in different foods.

Gelding: A castrated male horse.

Generator: An engine, either diesel or gasoline powered, which converts mechanical energy into electrical energy.

Gestation: The period of development from conception until birth inside the uterus. Also called pregnancy.

Girth: The circumference of the body measured around the barrel right behind the withers, or the strap on an English saddle that secures the saddle to the horse. Called a cinch in western riding.

Glut: A round wedge with a thin, flat end made of ironwood that is used to wedge into logs to split boards.

Glycerin: Sweet, syrupy by-product of a chemical reaction of fats and oils. In a homestead situation it would be a byproduct of biodiesel and useful for texture in soaps and herbal remedies.

Going: In horse terms, the texture of the ground, deep going, good going, rough going, etc.

Good doer: An easy-to-keep horse that keeps healthy on a minimal amount of food.

Goose: A female goose.

Goose-rumped: A muscle developed by jumping horses on the croup, also called jumper's bump.

Gosling: A baby goose.

Grackle noseband: Also called a figure-eight noseband, it has thin leather straps that cross over at the front and buckle above and below the bit.

Grade: A horse that is not registered by any breed association.

Grafting: To attach a branch of one variety of fruit to a tree of another variety.

Grain mill: A machine with two grinders that grain is runs between, grinding into smaller and smaller pieces. A grain mill can be either cranked by hand, or electric.

Graywater: Waste water that is not sewage or toilet waste, such as what goes down your sink or shower drain.

Grease: In horse terms, inflamed skin at the back of the fetlock and pasterns from being in the wet all the time.

Green: A horse still in the early stages of training is a green horse.

Greenbroke: A horse that is used to tack and rider and knows basic training, also called broken-in or broke to ride.

Greenhouse: A building made of panels that let in sunlight, used to grow plants, especially if the plants require warmer temperatures than are currently available.

Green manure: A crop that is planted, grown to maturity, then plowed into the soil to provide fertilizer.

Grindstone: A flat, circular stone set so it can spin when it is turned with a crank or pedal. It has a small bucket hanging over the stone to be filled with water, and a trough under the stone filled with water. When the wheel is in use it runs through the trough to stay wet.

Groats: Hulled buckwheat.

Grog: When making pottery, bits of dried clay particles in wet clay that gives it a grainy or sandy feeling.

Grooming kit: A collection of brushes, combs and equipment for cleaning an animal's coat, hair, and hooves.

Ground: To fasten electrical equipment to the earth to safely discharge an electrical flow.

Ground line: A pole put on the ground right in front of a fence so the horse and rider to judge the take off point for jumping.

Ground manners: The behavior of a horse while being handled during grooming, saddling, etc.

Grout: A thin mortar used to fill cracks in masonry or brickwork, but especially for laying tile. Also the action of filling cracks with grout.

Gullet: The throat and the esophagus.

Gymnastic: Fence placed at certain distances from each other for training a jumping horse.

Habit: Riding clothes made for riding sidesaddle.

Hack: A type of horse that are elegant riding horses, or to hack or go for a ride.

Half halt: An exercise that tells the horse to pay attention before changing direction, gait or some other movement.

Halter-broke: A young horse that only knows how to wear a halter and not much more.

Hames: In horse terms, metal arms fitted into the harness collar and linked to the traces.

Hand: In horse terms, a unit of measure to gauge the height. One hand equals 4 inches.

Hand ax: A small ax similar to a small broadax, with a handle about 1 ½ feet long, used for hewing and cutting small things.

Hardwood: Some hardwoods may be softer than some softwoods. The difference is in the way they produce seeds. Hardwoods make seeds with some covering, such as apples or acorns.

Hardy: A pointed tool used in blacksmithing used to make straight edges and cut off metal.

Harness: Horse equipment for driving a horse instead of riding.

Harness horse: A horse used for harness that has harness conformation: straight shoulders, etc.

Harrow: Also called a spring-tooth. A tool used to break up dirt into finer pieces (than a disc) and cover up planted seeds that are drawn by a draft animal that works something like a large rake. A spike harrow consists of a lattice frame with teeth coming straight down which are pulled through the soil, a wheel harrow consists of a row of wheels, and a shares harrow consists of sharp, flat blades.

Hay: Grass or other plants such as clover or alfalfa that has been cut and dried for animal feed.

Hay rake: Gathers up the hay before it is removed from the field. A sulky hay rake has a seat to ride on, while the spring-teeth gather the hay until a lever is pulled which lifts the teeth. The hay passes under the rake, and then the teeth are lowered to gather more.

Hay wagon: A wagon with tall, fence-like sides so that a great amount of hay can be tossed in.

Header: A board at the top of the wall of a house to which the roof is attached.

Head: The measure of the pressure of falling water, such as in a stream.

Head loss: In calculating how much power is available from a stream, the amount the stream slows down from obstructions and bends.

Heartgirth: On a horse, the girth of the barrel just behind the withers, or where the heart might be.

Heart room: A horse's barrel, or the space within it. A horse with a deep chest has lots of heart room.

Heaves: Another term for the abnormal breathing pattern Chronic Obstructive Pulmonary Disease (COPD), also called broken winded.

Heavy horse: Any large draft horse.

Heavyweight: A horse capable of carrying more than 196 pounds.

Heddle: The part of a loom that holds the warp. The heddle may be stationary or it may be movable with hand levers or foot peddles in order to move the warp up and down for easier weaving.

Heel: The back of an anvil.

Heifer: A female cow before she has had her first or second calf.

Heirloom: A strain of seeds, which has been preserved for its unique traits, usually varieties that are very old. They are non-hybrid, organic, and open-pollinated.

Helve: A helve is an ax handle.

Hemorrhage: Too much bleeding, which starts rapidly.

Herniated disc: Also called a ruptured disc, when the cartilage between the vertebrae of the spine ruptures painfully.

Hetchel: A wooden board with long, sharp, metal spikes sticking out of it, used to rip flax into shreds prior to weaving.

Hew: To wear away wood, usually by carving, whittling or sanding, when shaping a tool or other object.

Hindquarters: On animals, the area from the rear of the flank to the top of the tail, down to the top of the gaskin. Also called quarters.

Hinney: The child of a male horse and a female donkey.

Hives: An allergic reaction with bumps or red spots on the skin. The technical term is urticaria.

Hock: On animals, the joint midway up the hind leg that provides most of the forward motion of the animal.

Hocks well let down: In horse terms, a horse with short cannon bones (or shanks) which is desirable because it gives strength. Long cannons are a conformational fault.

Hoe: A garden tool with a long wooden handle and a metal blade at right angles to the handle that is used to cut and dig up weeds.

Hogged mane: A horse's mane that has been shaved close.

Homeschool: Also called home education, when parents remove their children from an outside educational institution and educate their own children at home.

Homestead: A place where all the necessities of life are produced. In the 1800s, a homestead was free government land given to people to promote the expansion of the United States.

Horn: In horse terms, either the hard outer covering of the hoof (the toenail), or the prominent pommel on the front of a western saddle, also called a saddle horn. On an anvil, the pointed cone on the back.

Horsemanship: The art of riding or equitation.

Horsepower: In the old days, how many horses it would take to pull a load. In modern times, a unit of power equal to 746 watts.

Hot: In horse terms, a horse that is overly excited, or a horse that becomes easily excited.

Hotblood: Any horse from Arabian or Thoroughbred blood.

Huller: A grinder not unlike a grain mill, which is specially made to remove hulls of grain. Barley requires a huller to remove the tough outer hull. It is either hand cranked, or electric.

Humanure: Human waste that is prepared to be used as manure for growing plants.

Humus: The part of dirt that comes from organic material, such as compost. It is in the final stage of decomposition, unrecognizable as plant material.

Hunter: A type of horse suitable for being ridden with dogs for fox hunting, or any well-mannered, smooth-gaited jumping horse.

Hurd: The pulp made from a hemp plant that is used in making hemp paper.

Husk: The tough outer cover of a grain such as corn or barley, which is not edible. To husk is to remove this outer cover.

Husking peg: A round pointed peg 3–4 inches long with a thong to attach it to the fingers. It is used to tear the cornhusk loose with one quick motion

Hybrid: A cross between a horse and any other horse-relation such as an ass or zebra. Also a plant that is the result of crossbreeding two related species in very controlled conditions.

Hydrometer: A device used to find out the proof of your alcohol.

Hydroponic: A method of growing plants indoors, using water and special growing chemicals under ultraviolet lights.

Implement: Any farm equipment that is dragged behind a horse or tractor, such as a cultivator, harrow, plow, etc. used to work the ground and grow plants.

Impulsion: When a horse is moving forward in a controlled and strong fashion (not the speed).

Inbreeding: The mating of relatives and siblings. In horses it is done to promote a certain family trait.

Incisor: The teeth located at the front of the mouth that are made for cutting and gnawing.

Incubate: To provide heat to eggs so that they can complete their development in the embryo and then can hatch. Birds incubate eggs by sitting on them, and humans can do it with an incubator.

Indirect rein: Pressing the rein opposite the direction the horse is moving, against the horse's neck.

In front of the bit: When a horse pulls or hangs heavily on the rider's hand.

In hand: Controlling a horse from the ground instead of from the horse's back.

Insemination: To put semen from a male into the reproductive tract of a female.

Inside leg: In horse terms, the legs of the horse and rider which are on the inside of the circle being traveled.

Insulation: Any material that insulates something. The material stops loss of heat or cold from a container or building.

Interplanting: With careful planning, some plants can be grown very close together so that they can benefit each other.

Intramuscular: Within a muscle, or into a muscle.

Irons: Also called stirrups, metal pieces attached by leather straps to the saddle used for the rider's feet.

Jack: A male donkey. Or, in automotive lingo, a tool used to raise a vehicle off the ground in order to remove a tire or get under the car.

Jennet/jenny: A female donkey.

Jig: A flat piece of board with a roller on the bottom, used to sharpen chisels. The chisel is fastened to the board so that the blade hangs over the end a little. The roller is set on the grindstone so that it spins and supports the board while the chisel is sharpened. There is an adjuster attached to the board so that different angles can be made for different chisels.

Jog: In western riding, the term for trot. In English riding it describes a slower, shortened pace.

Joint III: A disease in foals caused by bacteria in the navel.

Jointer: A long plane with block 2 feet long and a 2-inch wide blade. A jointer is used to level off joints, but mostly to make the edge of a board smooth. It is used until you can take a long shaving all the way down the board.

Joist: Supports for the floor (floor joists) or for the ceiling (ceiling joists) in a building.

Jump-and-coulter plow: All wood, with a square blade, and a stick before the blade. The stuck cuts a furrow (or small ditch for seeds), the bottom makes it deeper.

Jumper: A horse bred for jumping which competes in jumping classes.

Junction box: A box that houses electrical circuits that are connected.

Kettle: An iron kettle has legs to stand right in the fire, and a handle to hang from a kettle crane.

Kettle crane: A crane in the fireplace for hanging a kettle over the fire.

Kid: A young goat.

Kingdom: In Latin classification, all living things are categorized into five kingdoms.

Knob-and-tube: A type of electrical wiring used before 1940.

Knock-kneed: In horse terms, a conformation fault in which the knees point towards each other.

Lagoon: A septic system that is a large pool into which human waste is stirred. Air is mixed with the waste to create aerobic treatment.

Laminitis: In animals, a condition in which the laminae inside the hoof get inflamed and painful, which can lead to founder.

Lampas: When the hard palate inside an animal's mouth swells.

Lampblack: A black substance made mostly of carbon (such as soot) and used to make pigments and ink.

Lateral cartilages: In horse terms, the wings of cartilage attached to the coffin bone with the foot.

Lath: Thin strips of wood made in a lattice as a backing for plaster on a wall.

Laryngeal hemiplegia: Partial paralysis of the larynx causing breathing difficulty called roaring.

Lead: In horse terms, the leading leg during a canter (i.e. right lead canter, left lead canter).

Leader: In horse terms, either two leading horses in a team of four or a horse harnessed in front of one or more horses. The near leader is on the left; the off leader is on the right. When spinning, any previously spun yarn that is used to start off a new bunch of wool or other fiber for spinning.

Legume: Any erect or climbing bean or pea plant of the ***Leguminosae*** family.

Leg up: Helping someone mount a horse by holding his or her leg, or some other assistance in climbing up something tall.

Level: A long board with a square cavity in the center. A half circle of wood, arching over the center of the board horizontally, has a notch in the top center of it. Tied on the notch is a horsehair string with a metal weight on the end that hangs into the cavity. The weight stays inside the cavity by a small gate of wood. The board is set on the thing to be measured. When the weight is off center, the thing is not level.

Lien: The right to take someone's property if they don't pay off a debt or loan. When buying property it can have a lien, or someone else can have the right to take it away if the person selling it to you does not repay a debt.

Light horse: Any horse other than a heavy horse or pony that is used for riding or carriages.

Light of bone: A conformation fault in which there is not enough bone below the knee to support horse and rider without strain.

Limb: Any of the main branches coming from the trunk of a tree.

Limbing: Removing any of the main branches from the trunk of the tree, usually when preparing to cut up a tree that has been felled.

Line breeding: Mating horses with a common ancestor several generations back in order to promote a particular trait.

Loam: The ideal mixture of sand, silt, clay, and organic material in soil to make the best dirt for growing plants.

Loft: A floor in a building that is closer to the roof than a normal upstairs, and usually only covers a small portion of the house space.

Logchain: A chain used to anchor a tree as it is being chopped down, in order to control its fall.

Loins: The area on either side of the vertebrae just behind the middle of the back. In horses, this area is just behind the saddle and is the weakest part of a horse's back.

Longe/lunge: Training a horse by working through paces in a circle with a lunge rein attached to a cavesson.

Longitudinal: When drying wood, shrinkage that causes the wood to shrink along its length.

Loom: An apparatus with several frames that are used to weave thread or yarn into fabric and rugs.

Lope: In western riding the same as a slow canter.

Lymphangitis: The lymphatic system becomes swollen and painful, usually in the hind legs.

Mallet lever: A mallet-shaped wood tool, made out of a section of a log with one branch sticking out. On the end of the "handle" is a metal loop for hitching an animal. It works like a lever to pull stumps out of the ground, by putting it next to the stump with the "handle" sticking up.

Malt: Grain that has been prepared through soaking and drying before using it to create mash, in the process of brewing alcohol.

Manure: Barnyard and stable dung, often mixed with animal bedding that is gathered and used to fertilize the soil.

Mare: A female horse 4 years old and older.

Martingale: In horse terms, a neck strap which buckles around the horse's neck, and another strap which attaches to the girth, goes through the neck strap and attaches to the noseband or reins. The noseband attachment is a standing martingale, the reins is a running martingale. Prevents the horse raising his head.

Mash: A mixture of mashed malted grains and hot water used in brewing alcohol.

Maul: A wooden tool used to pound things, looking somewhat like a potato masher.

Meat grinder: A kitchen tool that either has a cutter blade on the outside of the grinder body or a blade or knife on the inside of the grinder body. A meat grinder can be used to grind anything, but primarily to grind meat.

Meconium: The dark brown or black feces passed by babies shortly after birth (and sometimes during birth). If the baby passes meconium before birth it means it is under stress. The meconium can cause severe brain damage.

Membrane: A thin, flexible layer of tissue covering a surface or separating regions or organs in an animal or plant. When a mammal gives birth, there may be a membrane around the face of the new offspring.

Menstruation: A woman's menstrual flow, or menses, when a woman's uterus goes through a monthly cycle of releasing blood and debris.

Middleweight: A horse that is capable of carrying up to 196 pounds.

Mitbah: The angle of the neck of the Arabian horse that gives it the characteristic arch.

Molt: Many birds lose their feathers once a year, usually after egg-laying season is over. A bird doing this is said to be molting.

Monday morning disease: The technical term is azoturia, and it is also called tying up. See Azoturia.

Monocot: A plant that has one seed leaf or "cotyledon."

Mortar: A building material made of lime, cement or plaster of Paris used in laying bricks, tile, stone, etc.

Mortar and pestle: A mortar is a hollow block of wood or a stone bowl, and the pestle is a heavy rounded tool with a handle. The grain or herb is put into the hollow or bowl, and the pestle is used to grind it.

Mortgage: When the person lending you the money to buy your property holds the title on the property until you pay it off. If you default on your loan then they take the land away.

Mother: Mother of vinegar is a bacteria drawn from the air that makes vinegar.

Mower: A horse-drawn mower is pulled through a grain field and cuts the grain down. It leaves the grain in windrows, long narrow rows of grass 8–10 feet apart. They are made to be ridden on and have many levers for adjustability. The driving wheels are usually about 30 inches in diameter. They have two kinds of knives, a toothed knife for cutting grain, and a regular knife for cutting grass.

Mucking/mucking out: Removing wet and soiled bedding and cleaning a stable.

Mulch: A cover over the soil after plants have come up which prevents weeds from growing and insulates the ground. It can be plastic, organic material or newspaper.

Mule: The child of a male donkey and female horse.

Mustang: A wild horse from the American West.

Mutton withers: A horse with wide, flat withers, such as on a quarter horse.

Nail header: A wood tool, all one piece, with a handle and a square on the end of the handle. In the square is a hole through which unfinished nails are put. The nail header holds the nail while a head is pounded on it.

Nail rod: A long metal rod which is about the width of the desired nail, which is used during forging to cut into pieces to shape into nails.

Nanny: A female goat.

Natural farming: A technique of growing plants without plowing, cultivation or fertilizer.

Naturalist: A person who studies the workings of nature and its workings as a whole, as well as creatures and organic things that live in it.

Navicular bone: A small bone inside a horse's hoof that fits horizontally between the short pastern and coffin bone.

Navicular disease: When the navicular bone degenerates causing pain and lameness.

Near side: The-left hand side of a horse.

Neck reining: Turning a horse by using the opposite rein against the neck.

Neck strap: In horse terms, a leather strap buckled around the horse's neck to help new riders, or a martingale.

Negative: Something with a negative electric charge, or that the electrons are negative.

Nematode: A worm without segments that is pointed at both ends. Some are parasites.

Net head: In calculating the power available from a stream, the Total Gross Head minus the Head Loss.

New York plow: Designed for deep tilling, used for sugar cultivation. Instead of the width being adjusted by the clevis, it is adjusted at the end of the beam where it connects to the handle-frame.

Nibs: The handgrips on a snath.

Off side: The right-hand side of a horse.

On the bit: When the horse is correctly holding his head near vertical and accepting rein commands.

Open-pollinated: The opposite of hybrid, the seeds were pollinated by a whole crop of seeds that have been protected from cross pollination.

Organic: According to the National Organic Standards Board, agriculture that promotes biodiversity, natural cycles, and sustainability. Pollution from air, soil, and water is minimized.

Outhouse: A small house with a pit under it that is used as a toilet, instead of plumbing.

Ovary: A flower's female reproductive system, located at the base of the pistil. The ovary contains ovules that require fertilization by sperm from pollen.

Oven rake: A tool that looks like a hoe but with a shorter handle, used to rake ashes out of the oven.

Overbent: Also called behind the bit, when the horse tucks in his head to avoid the bit.

Overface: Trying to make a young horse jump a fence when it is beyond his capability.

Overreaching: In horse terms, when the toe of the hind foot catches and injures the back of the pastern of the front foot, which happens when galloping or jumping.

Ovule: The little eggs inside the flower's ovary. The ovules contain an "egg" to be fertilized by one sperm from a pollen tube.

Oxen: A team of bovine draft animals, usually steers which are used to pull heavy loads.

Pacer: A horse that moves its legs laterally instead of diagonally.

Packhorse: A horse that carries goods in packs on either side of its back.

Paddock: A small enclosure for putting animals out to graze.

Paddy: An irrigated or flooded field where rice is grown.

Parasite: An animal or plant that lives off a host animal or plant without benefiting or killing the host.

Parch: To dry or roast by exposing to heat, but especially for making things very dry.

Parietal bones: The bones on top of the skull.

Parrot mouth: In horses, an overbite where the top jaw extends forward over the bottom.

Part-bred: A horse that is part Thoroughbred and part something else.

Pastern: In animals, the sloping bone in the lower leg connecting the hoof to the fetlock.

Pasteurization: Heating a liquid to a specific temperature for a period of time to kill harmful microorganisms.

Pasture: A fenced field free of harmful objects and plants that is used as a grazing area for animals. The grass is often cultivated like any other crop.

Peel: A wooden shovel used to shovel food in and out of a fireplace bread oven.

Perineum: On a female, the region between the anus and the back of the vulva, and during childbirth the part the stretches the most.

Percolation: When a liquid passes through a porous substance. For instance, when water passes through soil at a certain rate, that speed is called its percolation rate.

Permaculture: A patented word created by Bill Mollison that combines ***permanent*** and ***agriculture***. It is a method of agriculture and living that looks at the whole landscape and lifestyle to design the most efficient and caring methods for all the species living there.

PETE: Polyethylene terephthalate, which can be identified by its recycling symbol, which will have a 1 in the center, with the letters PET or PETE under it.

pH: Stands for the potential of hydrogen, and measures on a scale from 0 to 14 the acidity or alkalinity of a solution. Seven is neutral, while above 7 is alkaline and less than 7 is acidic.

Phosphorus: A non-metallic element that occurs in phosphates and is used in safety matches, pyrotechnics, incendiary shells, etc.

Photosensitive: Being light sensitive.

Photovoltaic: Anything that produces voltage when exposed to radiant energy, especially light. When several solar arrays are connected together, it is a photovoltaic array.

Phyla: In Latin classification, a category with a kingdom, such as the gymnosperm phyla in the plant kingdom.

Pier: A column of masonry that supports other structural parts of a building.

Pigeon-toed: A conformation fault in horses where the feet turn inward.

Pipping: When a baby bird starts to peck a hole from the inside of the egg.

Pitch: A substance that is related to tar, and is made from pine wood charcoal and resin. It is used to make wood waterproof. Also, the amount of slope of a roof.

Pitchfork: A metal-pronged tool with a wooden handle used to toss hay, manure, grain, and anything else.

Plane: A wood block made of beech that holds a blade for the plane body. The blade is held in with little shims (wedges). A plane is used to smooth wood.

Plaster: A white powder (usually a form of calcium sulphate) that when mixed with water forms a paste that later hardens into a solid. The action of applying the paste is called plastering.

Plat logs: The top wall logs of a cabin, which support the lower end of the rafters and ceiling joists.

Plow: Also spelled plough. A plow is used to turn the earth over in preparation for planting. The blade (or bottom) turns over the soil, the frame is the overall structure, the gauge wheel determines the depth of the furrow, the coulter cuts the furrow slice from the land in front of the share, and the clevis is where the draught (harness) is attached.

Plumb bob: A string with a weight at the end that is used to make sure things are vertically straight.

Pointing up: Fixing problems with drywall after it is installed.

Points: External features of a horse making up its conformation, or when identifying color combinations refers to the lower legs, mane and tail (i.e. a bay with black points has black legs).

Poll: The highest point on the top of the animal's head.

Pollen: A fine powder produced by seed-bearing plants, containing the male parts, which fly through the air and fertilize the female parts of another plant.

Pollination: When pollen from the stamen goes into the stigma of a plant, causing reproduction.

Pommel: The center front of an English saddle.

Pony: Any small horse that is 14.2 hands or less.

Port: A raised section in the center of the mouthpiece on some curb bits. A low port is a mild bit.

Positive: Something with a positive electric charge, or that the protons are positive.

Posting trot: Also called a riding trot, when the rider rises in the saddle in rhythm with the horse's trot.

Potato digger: Similar to a plow, it has two wheels which are followed by a blade which turns up the earth and the potatoes with it. It pushes the plants to one side so they do not get tangled.

Poult: A baby turkey.

Prairie-breaking plow: Makes a furrow 4 inches deep, very heavy and long. Uses a wheel coulter, and the clevis is adjusted both side-to-side (for width of the furrow) and vertically (for depth).

Prepotency: A horse's ability to consistently pass on character traits to offspring.

Pressure canner: A pot with a pressure gauge that brings the contents to a pressure hot enough to kill bacteria.

Prime: When painting, to prime is to cover the surface with a white (or brown) flat paint before covering it with color. Priming a hand pump is to add water to the pump to create a suction in order to bring water to the surface.

Pritchel hole: A hole in an anvil for making specific things.

Prolapsed: When an organ or part of an organ has fallen down or slipped out of place.

Proof: The amount of ethanol in your alcohol. All alcohol has a little water, and 100 proof alcohol has 50 percent water and 50 percent ethanol, which is drinkable. 198 proof alcohol has very little water and can kill you.

Propagation: In growing plants, methods of reproduction without using seeds, such as root cutting.

Protein: A group of complex tiny molecules that contain the basic components of all living cells, and are essential in the diet for the growth and repair of tissue.

Pruning: To trim a tree or shrub in order to stimulate growth and train its shape.

Pruning shears: What are ***not*** pruning shears: scissors, grass trimmers, power tools, and hedge clippers. A pruning shear is a curved cutter with short blades and small handle that is used with one hand to prune branches of trees and shrubs.

Purebred: A horse that had both parents from the same breed.

Purlin: On a timber frame house, a long beam which sits on top of the bents and connects them together. Usually a house will have four on each side of the peak of the roof. On a log cabin the purlins are horizontal logs in a roof that are supported at each end, and hold up the rafters.

Quarter round: A 3–6 inch pole that has been cut into quarters and is used for chinking, window and door trim and molding.

Quarters: The part of an animal's body from the rear of the flank to the top of the tail down to the top of the gaskin, also called the hindquarters.

Quenching bucket: A large tub full of water kept near a forge to cool metal.

Quicksand: A pit filled with loose wet sand, found in nature, which often traps heavy objects that sink below the surface.

Quidding: When a horse drops half-chewed food from his mouth from age or dental problems, fixed by floating the teeth.

Quilting: Stitching in a quilt that goes through all the layers in order to hold the filler in place, often with a decorative design.

Quilting frame: A rectangular frame that holds the quilt tight while you stitch it.

Racehorse: A horse bred for racing, usually Thoroughbred, Quarter Horse, Arabian, or Standardbred.

Radial: Shrinkage when drying wood that causes a board to become more skinny.

Rafter: A sloping beam running from the peak of a roof to the bottom which helps support the boards and shingles (or other roofing material). They are usually spaced 16–24 inches apart.

Rake: A large wooden rake with a half circle of wood to support the raking part, used to rake hay. These can only be made by hand.

Rainrot: A painful skin inflammation on horses, causes raised hair, hair loss, and crustiness.

Rangy: A horse with a larger size and scope of movement.

Ratchet: A mechanical device with a toothed wheel that only allows movement in one direction.

RC: A number rating the reserve of an engine battery. A higher RC number means that your battery has a bigger reserve in case the engine fails.

Reining: A type of western riding where spins and slides are done in patterns.

Resuscitation: To restore someone to consciousness; to revive someone. Usually this refers to a process such as CPR.

Retting: A process of rotting the stalks of flax, hemp and nettle before it is broken down into fibers so it can be made into cloth and paper.

Reverse osmosis: The finest filtration known, in which water is forced through a membrane. It can filter out salts, sugars, proteins, particles, dyes, etc.

Ridgepole: A long pole put at the peak of the house that will support the rafters, usually on log cabins.

Ridgling/rig: A male horse that has not dropped one testicle, which can cause stallion-like behavior.

Riding horse: A horse bred specifically to be comfortable for riding.

Ringbone: Any bony changes in the pastern or coffin joints that may cause temporary lameness.

Ringworm: A contagious fungal disease that has small circular patches where hair falls out.

Ripple: A plank with wood or wire teeth through which flax is pulled in order to remove the seeds.

Rising trot: Also called a posting trot, when the rider rises in the saddle in time to the horse's trot.

Roach back: The opposite of a hollow back, when the curvature of the spine goes outward.

Roached mane: Also called a hogged mane. A mane that has been shaved close for its whole length.

Road plow: Cuts a furrow 7-9 inches deep. A very durable plow with no wheel for making roads.

Roll roofing: A roofing material saturated with asphalt that comes in rolls.

Roughing in: Installing sewer lines or water pipes under the concrete of the foundation. The lines and pipes are called rough-ins.

Roaring: A noise made by horses when breathing because they have *Laryngeal hemiplegia*.

Rolag: An even roll of unspun wool that is prepared for spinning.

Roller: Pulled by horses or by hand, a tool that crushes sod on top of the ground after using a harrow. It forces small stones level with the surface, makes the ground smooth for using a scythe and rake, and presses soil around seeds. A roller is especially useful for compacting manure gases into light soil, and in clay soils prevents winterkilling.

Rolling: Smashing grain between rollers going at different speeds.

Roman nose: When the nose curves outward instead of being flat or concave.

Root ball: The root system of a tree mixed with a ball of dirt clinging to the roots, making a large, heavy ball.

Root bound: When a plant's roots have grown so tightly and tangled that they completely fill the container they are in.

Rosin: A fine-powdered by-product left from distilling pine resin. It is sometimes added to soap to help it create a lather.

Rototill: To till your ground with a machine that can be pushed by hand. It has several sharp blades turned by a motor that cut and turn the earth rapidly. Although quick and produces light soil, it kills worms.

Roundworm: Common name of ascarids, an internal parasite.

RPM: Stands for Revolutions Per Minute. The number of times an engine revolves in one minute.

Ruptured disc: Also called a herniated disc, when the cartilage between the vertebrae of the spine ruptures painfully.

Saddle horn: A very prominent pommel on a Western saddle, also called just a horn.

Saddle horse: A riding horse.

Saddle marks: White hairs in the saddle area on a horse.

Sap: A mixture of sugars, salts, and minerals circulating through a plant, somewhat like the blood of a person.

Sap yoke: A piece of wood hollowed out to fit your neck, with a hole on each end for a rope or chain to attach a bucket.

Saw: A tool used to cut wood, with a serrated edge, usually for big cuts. Usually it has as long, sharp blade with teeth, made of metal, and a wooden handle.

Scion: A fruit tree shoot with buds, used in grafting on another tree.

Scope: A horse has scope when it has potential and capability of movement. A special horse.

Scoring ax: An ax with a long handle, used to score (cut a line into) a tree.

Scotch harrow: Made of two rectangular pieces that are chained together and pulled from the ends.

Scotch sub-soil plow: Used right after a turning plow in the same furrow, used to break up and pulverize the soil further.

Scours: Diarrhea in baby animals.

Scoville unit: A unit used to measure how hot a food is, invented for chili peppers, and devised by Wilbur Scoville in 1912.

Scratches: In horse terms, a scabby, oozing inflammation on the back of the pasterns just above the heel.

Scythe: A large blade used to cut hay and grain, attached at right angles to a snath (the handle).

Seed drill: A wheeled machine that is drawn behind a team of horses which has a bin, or hopper which pulls seeds down into tubes which feed into drills. The drills drop the seeds into the earth, and can be set to any depth. The soil naturally drops over the seeds as the drill pulls out.

Seedy toe: Separation of the hoof wall from the laminae, sometimes accompanied by laminitis.

Self-pollinator: A plant whose individual flowers contain all the parts to successfully pollinated themselves.

Self-seeder: A plant that grows from seed that it drops itself, and is spread naturally without any help from you.

Selvage: The edge of a fabric that is woven so that it won't unravel or fray.

Septic tank: A tank in the ground which human waste is piped into where it can decompose safely using anaerobic treatment and leech into the soil.

Setting: When a fowl (such as a chicken or duck) is incubating her eggs by sitting on them and keeping them warm with down from her body.

Sewage: Any material taken away by sewer drains.

Shake: A shingle that has been split from a piece of log, 2–4 feet long.

Shank bone: The hind cannon of an animal.

Shares harrow: An effect harrow that digs deeper (2–3 inches) than a regular harrow. Has sharp, flat blades shaped like a sled runner.

Shaving horse: A shaving horse is like a vise, used to clamp your wood while you use a drawshave on it. It is a four-legged bench with a heavy plank attached at one end going halfway to the seat, a bench with a stick that clamps your work and a bottom bar to rest your feet. You sit on the seat end and push the bar with your feet to clamp your work tight, so that the harder you pull the drawshave, the harder you have to push with your feet.

Sheath: A protective outer covering around a stallion's penis.

Sheave: A large handful of wheat stalks, tied together.

Shingle frow: A thick blade with a handle 18 inches long set at right angles from the blade, which is used to make shakes (handmade shingles).

Shivers: In draft horses, an abnormally high leg gait where the horse flexes one or both hind legs and you can see tremors. It is thought to be caused by EPSM.

Shock: A stack of sheaves of wheat, made to shed water.

Screwband: A metal band used to fasten the sealed canning lid to the jar.

Shovel: A wooden shovel has a handle shaped like a D, and an iron shoe on the shoveling part. It is used for any digging except postholes, and for shoveling snow.

Shoulder-in: In horse terms, a movement where the horse is evenly bent along the length of the spin away from the direction it is moving.

Shuttle: A handheld tool used in weaving which thread or other material is wrapped around to be woven through the warp, creating the weft.

Shy: In horse terms, when the horse jumps suddenly to one side when startled.

Sickle hocks: In horse terms, when the hocks are bent giving the hind legs a sickle shape, making the legs too far under the body.

Sidebone: When the lateral cartilage on either side of a horse's coffin bone gets rigid within the hoof.

Side-hill plow: Also called a swivel plow, it throws the furrow-slice down hill. It is pivoted so that it can move from side to side when at the end of a furrow, so that you can plow straight lines instead of in a circular pattern.

Side reigns: Reins used in training to help position the horse's head, attached to the bit and to the girth or a training surcingle.

Side-stick collar: A collar made for goats that prevent a female from sucking from her own udder.

Silk: The tassel on sweet corn located at the top of the ear.

Sill: The wood under the house that rests directly on the foundation.

Single cultivator: Cultivates two rows at a time.

Sippy bottle: A bottle with a special spout that only releases its liquid when it is sucked on.

Sire: A horse's male parent.

Skein: A length of spun fibers (such as flax or wool) that is looped into a loose coil.

Skep: A man-made beehive shaped like a dome, and often made of a coil of twisted grasses.

Slab-sided: In horse terms, having narrow ribs.

Slaked: Also called slacked, lime that has been treated with water to cause it to heat and crumble. A chemical reaction.

Slub: An irregularity in raw wool such as thin spots or thick spots, which are pinched out prior to spinning.

Slurry: The plant pulp that is layered onto a deckle in the making of paper.

Smoker: A tool used in keeping bees which allows a controlled flow of smoke to be put in with bees in order to stupefy them.

Smoothing plane: A plane 8–10 inches long used to finish a rough surface.

Snaffle (bit): A type of bit that acts at the corners or bars of a horse's mouth and uses only one rein.

Snath: The handle of a scythe, with two nibs (handgrips). For harvesting grain, a cradle (a rack) is attached to it. The best snath is a steam-bent, black cherry snath.

Sock darner: A small rounded tool that is put in a sock to support it while you are darning.

Softwood: Some softwoods may be harder than some hardwoods.ntent The difference is in the way they produce seeds. Softwoods let seeds fall to the ground without any covering, such as pinecones.

Solar: Anything relating to the sun or utilizing the energy of the sun.

Solvent: A liquid that can dissolve other substances, often used for removing paint.

Sorghum press: A large mill that is powered by a horse or mule walking around it has three rollers that are turned by reduction gears attached to a pole, which is attached to a horse. The juice goes through several strainers, and into a holding tank.

Sound: Free from lameness and injury.

Spacer board: Also called a sticker, a piece of dried wood put between layers of newly cut wood that is being dried outdoors.

Spark plug: An electrical device that fits into the cylinder head of an engine and ignites the gas with an electric spark.

Spavin: In animals, a degenerative arthritis in the lower joints of the hock which can be seen as bony swelling, also called bog spavin.

Spillway: A runoff from a dam that allows overflow to be directed to a place that you designate, rather than flooding.

Spindle: A stick or pin used to twist the yarn or thread when spinning.

Spinning wheel: There are several kinds of wheels. The small wheels with foot pedals are for wool, the big wheels are for fine fibers such as flax, silk and cotton. Small wheels come in two styles, Saxony and Norwegian. The Saxony style is the most common.

Spline: Peeled-off strips of wood prepared for making baskets, which are woven together. Also can refer to a wood strip that fits into any groove or slot between parts, such as on a cabin.

Splints: In animals, an injury to one or both the metacarpal or splint bones running up the back of the cannon bone.

Split ax: An ax for splitting logs, with a wedge-shaped head and a long handle.

Splitting maul: A combination between an axe and a maul. It has a wedge on one side and a hammer on the other so that the wedge can be used like an axe and the hammer to dislodge things.

Spoke shave: A tool with handles and blades that are used to shave wood. It is used as the final shaving tool when making a helve, and is also used in making wooden boats.

Spraddle legs: When a baby bird's legs turn outwards abnormally because of slippery footing.

Spurs: A small metal device worn on the rider's boot to enforce leg aids during riding.

Stoneboat: A large sled with curved front curved up so that it would be 3 feet from the ground. Use to haul stones from a field.

Square harrow: A square shaped harrow that is pulled from one corner.

Stallion: Also called an entire, an uncastrated male horse.

Starch: A substance found in seeds, fruit, tubers, roots, wheat and rice which is a complex carbohydrate. It is used as food and also as a stiffener in papers and fabrics.

Starter: Animal feed specially made for chicks, ducklings and other baby fowl.

Stave: In basket-making, a long piece of material used to weave the walls of the basket. Also the handle of a flail is called a stave.

Steer: A castrated male bovine animal.

Sticker: Also called a spacer board, a piece of dried wood put between layers of newly cut wood that is being dried outdoors.

Still: An apparatus used in the process of the distillation of a liquid, such as the purification of water or in the making of ethanol.

Stock horse: A horse used in ranch work, driving and cutting cattle, etc.

Stoneboat: A wooden sled for collecting rocks from a field, which is pulled by an draft animal.

Strangles: A highly contagious disease, also called distemper.

Straw: After grain has been threshed off, the remaining stalk is dried and used for bedding for animals, thatching, weaving or braiding.

Stringhalt: In animals, the over-flexion of the hind legs in which the leg is jerked up to the belly.

Stripper: A long blade on the end of a straight handle used to strip the leaves off sweet sorghum plants. The tip is curved like a sickle, and the rest is straight, and the curved part is used to cut stalks and heads.

Strongyles: Also called bloodworms, an internal parasite.

Stud: A male animal used almost exclusively for breeding.

Subfloor: The layer of flooring under your hardwood or carpet made of floor joists, bridging, plywood and/or concrete.

Subsoil: The deeper layer of soil that is lighter colored, where water is stored.

Substance: A horse that has quality build has substance.

Sucker: A small sprout that grows out of the roots of a tree and up out of the ground (it looks like a whole new baby tree) or comes out of the trunk of the tree.

Sulky cultivator: A cultivator with a seat for the driver, unlike a regular cultivator.

Sulky plow: A single plow that is mounted on a frame with a seat. A brake allows the plow to be pulled out of the ground without having to do it by hand, and the beam relieves pressure to the horse's necks.

Super: A section of a beehive used to store excess honey, normally put on top of a brood chamber.

Surcingle: In horse terms, webbed strap which goes under the barrel to which side reins can be attached, blankets can be secured, etc.

Suspensory ligament: In animals, ligaments that support the fetlock and spread around the fetlock joint.

Swaddling: To wrap an infant tightly so that they feel as if they are in a similar place as the womb, which they find comforting.

Swingle: A tool that looks like a large knife about two feet long, used for winnowing. It can also be attached to a stave as part of flail.

Swivel plow: Also called a turn-twist plow. Two plows are attached to one beam so that you can switch between to plow either right or left. The forward plow turns a depth of 3 inches, and the rear plow reaches a depth of 5–7 inches.

Table: The front lower section of an anvil.

Tack: Short for tackle, riding equipment such as saddle, bridle, etc.

Tag/tagging: Clipping the wool away from a ewe's vagina in order to make breeding easier and cleaner.

Tallow: Hard fat taken from the bodies of cattle, sheep or horses, which is used for candles, leather dressing, soap and lubricants.

Tangential: Shrinkage when drying wood that causes the wood to shrink along its width.

Tapeworm: Internal parasites.

Tarpaper: Heavy paper full of tar which is used on a roof for waterproofing.

Teaser: A stallion used to test a mare's readiness for breeding, but who does not actually breed with her.

Tetanus: Also called lockjaw, a serious bacterial infection that enters the body through a puncture wound.

Thinning: To remove excess seedlings to make more space for other plants to grow larger and healthier.

Thresh: To beat with a flail or other tool to remove the chaff from the grain.

Thrifty: In horses, a horse that maintains health on a small ration, also called good-doer.

Throat: The inside of the bottom of the hook on the chain of a chainsaw.

Thrush: In horses, a fungal or bacterial infection in the frog that discharges a bad-smelling liquid.

Ticking: Any strong fabric used for mattress or pillow covers.

Tied in below the knee: In horses, a conformation fault in which the measurement below the knee is much less that the measurement above the fetlock.

Tie log: A beam or one or more logs that connect and provide lateral support for two opposite walls.

Tilling: Another word for cultivating, but usually for the initial plowing of a field.

Tipi: A house shaped like a cone made of several wooden poles and a half-circle of fabric or leather, designed by the Plains Indians.

Tire chain: A special grid of chains made to fit around a tire to improve an automobile's traction on an icy or snowy road.

Tire iron: A metal tool used to pry a tire from the rim on a vehicle.

Title: A legal deed to a property, or a document that shows the legal ownership of a property.

Titration: A chemical process used to find out how concentrated a substance is.

Tofu frame: A wooden frame 7 inches by 7 inches that has a wire mesh or screen in the bottom. The screen is covered in cheesecloth or unbleached muslin, and a board 6 ½ inches by 6 ½ inches is pressed down on the tofu when it is in the frame to squeeze out the whey.

Tom: A male turkey.

Tong: Metal tool for holding onto hot metal. A farrier's tong has two round disks on the end. You can shape the tong to hold whatever it is you want.

Top line: In animals, the line from the back of the withers to the end of the croup.

Top heavy: An animal with a heavy body in relation to the substance of the legs.

Topsoil: The top, darker layer of soil, which is more crumbly and where most of the nutrients are.

Total Gross Head: The distance that a stream is moving or dropping, used in calculating the power available from the stream.

Trailer: A transportation vehicle that is towed behind another vehicle that is used for animals and equipment.

Transcendentalism: A philosophy of the 1800s begun by Margaret Fuller, Ralph Waldo Emerson, Henry David Thoreau, Bronson Alcott, and Emily Dickinson which advocated simplicity, truth, intuition and other ideals.

Transesterification: A chemical reaction used in the making of biodiesel in which ethanol and lye are mixed to make sodium methoxide.

Transition: In horse terms, changing from one pace to another. Changing to a faster pace is an upward transition, and changing to a slower pace is a downward transition.

Tread: The bumps on a tire that improve traction on the road.

Tree-pruning hook: A long pole with a hook at right angles from the pole on the end of it, for cutting branches.

Trimming out: Finishing the plumbing in a house by installing the fixtures.

Trivet: A small piece of metal used to elevate pots and pans off the top of a wood cook stove so that they don't get so hot.

Trot: In horse terms, a moderately fast gait where the horse moves from one diagonal pair of legs to the other with a period of not touching the ground in between.

Trunnel: A wooden peg that is used instead of metal nails, and are hammered in with a mallet.

Truss: An assembly of wood or metal formed into a triangle framework, usually to support a roof.

Tuber: The fleshy underground stem or root of a plant for reproduction and food storage, such as a potato.

Turning plow: A simple plow with a curved blade, a rotating disc in the middle, and a wheel in the front. The disc cuts a furrow, and then the bottom turns it over.

Turnout: Turning out horses in a field for a length of time during the day, or a standard of dress for a horse and rider. When making a road, a shallow notch in the outside slope so that water can flow out.

Tuyere pipe: A metal pipe running through a forge directly to the fire through which air can be blasted with a bellows to make the fire hotter.

Twill: A heavy cloth with parallel lines or ribs, often woven in rather than died, and used for making beds or ticking.

Type: A horse that fulfills a certain purpose, such as a cob, hack or hunter, but of no particular breed.

Udder: An organ resembling a bag that has mammary glands and produces milk in female mammals such as cows, sheep and goats.

Undercut: When cutting down a tree, a cut made on a tree at least 8 inches in diameter, on the same side as the direction you want the tree to fall.

Underlay: In a house, a layer of plywood or hardboard sheets on top of the subfloor. This plywood lies directly under the hardwood of a wood floor.

Undershot: A horse deformity in which the lower jaw sticks out further than the upper jaw.

Unschool: A form of homeschooling in which the homeschooled children are allowed to educate themselves in a more natural way. It is meant to be the opposite of institutionalized education.

Uterus: An organ in a female mammal in which a fertilized egg implants and develops. It is also called the womb.

Vent: The anus of an animal, especially of fowl such as chicken and ducks.

Vermicompost: Compost to which worms are added in order to break up the compost, decay it faster, and add worm castings to increase the nutrients.

Vermiculite: A mineral added to a variety of materials including brake linings, as a filler, in textured paint, in concrete, in plaster and as fireproofing. It has also been used in stormwater biofilters and has several gardening uses.

Vise: A clamping device with two jaws that close with a screw.

Volunteer: A plant that grew spontaneously from a self-seeder.

Vulva: The external genital organs of a female.

Warp: The stationary, taut threads held vertically by a loom.

Washtub: A wooden or metal tub used to wash clothes in.

Water bath canner: A pot which is big enough to can fruit, but does not get hot enough for other foods to kill the bacteria.

Watersprout: Like a sucker, but is a very green shoot that comes out of a tree branch after too much pruning or an injury.

Watt: The standard unit of electrical power. 746 watts makes 1 horsepower.

Weanling: An animal or child that has been recently weaned from its mother's milk.

Weather vane: An arrow or flat silhouette (such as a rooster) that rotates on a pole. On the pole are visible indicators of the four directions. Use a compass to place the vane facing the right direction. The pointing end tells which way the wind is coming from.

Weft: The thread that has been wound on a bobbin and is woven through the warp on a loom.

Wheel: When making pottery, an electric or foot-powered rotating circular table on which pots are formed.

Whittle: To shape wood with a whittling knife or pocketknife, usually for fine carving.

Winch: A tool that is sometimes motorized, that has a coiled cord and a hook, which is attached to things too heavy for a human to pull. It assists in hoisting or pulling heavy objects.

Windbreak: A barrier that slows or stops the wind, often a row of trees or shrubbery.

Windrow: A long row of wheat piled up before making into sheaves.

Winnow: To separate the chaff from the grain.

Winnowing tray: A wooden box with handholds on the sides, no end, and no top, kind of like a large scoop.

Womb: An organ in a female mammal in which a fertilized egg implants and develops. It is also called the uterus.

Woodlot: A patch of forest on a person's property that is used specifically for firewood. Maintaining a woodlot takes expertise in order to not use up all the wood.

Wort: When brewing alcohol, the unfermented malt, or the malt in the process of fermenting.

Wrap-over: A type of roof ridge when thatching that folds a thick layer of straw over the edge of the roof-peak and is fastened on both sides.

Wrench: A tool that is used to twist a nut or bolt.

Yealm: When thatching a roof, a bundle of longstraw that has been made into a tight layer of straw which has been cleaned up and is level at both ends. To make these is called yealming.

Yoke: A piece of wood shaped to a person's or animal's shoulders to help in carrying or pulling a great weight.

Yurt: A traditional Mongolian portable house with a round shape that provides excellent shelter in extreme weather.

Bibliography

"A basic guide to raising sheep." *Countryside & Small Stock Journal* (July/August 1999): 83.

"A beekeeper's year." *Countryside & Small Stock Journal* (March/April 1999): 80.

Ackland, Tony. "Home Distillation of Alcohol" *Home Distiller*. http://homedistiller.org (Jan. 2004).

Aguilar, Francisco X. "How to Install a Polyethylene Biogas Plant" United Nations University. http://www.ias.unu.edu/proceedings/icibs/ibs/info/ecuador/install-polydig.htm (Mar. 2004).

"Alternatives to bottles" AskDrSears.com. http://www.askdrsears.com/html/2/T026000.asp (Aug. 2003).

Andel, Cynthia. "For health and comfort, try wild herbal teas" *Backwoods Home Magazine*. http://www.backwoodshome.com/articles2/andal76.html (Sept. 2003).

Anderson, Karen. "Dear Karen: Topic: Food Storage" Y2K Women. http://www.y2kwomen.com/archives/DK5.html (Sept. 2003).

Anderson, Kathy, PhD. "Creep Feeding Growing Horses" NebGuide. http://www.ianr.unl.edu/pubs/animals/g1276.htm (Sept. 2003).

——"Feeding and Care of Orphaned Foals" NebGuide. http://www.ianr.unl.edu/pubs/animals/g1291.htm (Sept. 2003).

——"Housing for Horses Flooring for Stalls" NebGuide. http://www.ianr.unl.edu/pubs/animals/g1237.htm (Sept. 2003).

——"Nutrition Management of Pregnant and Lactating Mares" University of Nebraska Cooperative Extension. http://www.ianr.unl.edu/pubs/animals/ec272.pdf (Sept. 2003).

——"Winter Care for Horses" NebGuide. http://www.ianr.unl.edu/pubs/animals/g1292.htm (Sept. 2003).

Änderung, Letzte. "Die Webseite fuer Spinner: Spinnen mit dem Charkha" Handspinnen.de. http://www.handspinnen.de/charka/ (Apr. 2004).

"Are mordants used for natural dyeing safe?" Aurora Silk. http://www.aurorasilk.com/info/are_mordants_safe.shtml (Aug. 2003).

Arnet, Gary, DDS. "No Dentist? Oh, no!" *Backwoods Home Magazine*. http://www.backwoodshome.com/articles2/arnet75b.html (Sept. 2003).

—— "What to do when there's no doctor" *Backwoods Home Magazine*. http://www.backwoodshome.com/articles2/arnet75a.html (Sept. 2003).

"A to Z Home's Cool" A to Z Home's Cool. http://homeschooling.gomilpitas.com/index.htm (Apr. 2004).

"Automotive Fluids—Checking Fluids" DIY. http://www.diynetwork.com/diy/en_maintenance/article/0,2021,DIY_13808_2270917,00.html (Feb 2004).

Avery, Barbara and Ann Lemley. "Cryptosporidium: A Waterborne Pathogen" U.S. Department of Agriculture. http://www.nal.usda.gov/wqic/cornell.html (Apr. 2004).

Ayush, Tseeliin. "Daily Life of the Mongols" http://www.indiana.edu/~mongsoc/mong/daily.htm (Jan. 2006).

"Baby Sling Instructions" Maya Wrap. http://www.mayawrap.com/sewSling.shtml (Sept. 2003).

"Babywearing" Wears the Baby. http://www.wearsthebaby.com/babywearing/podegi.html (Sept. 2003).

Barakat, Christine. "Top Ten Trailering Tips" EquiSearch. http://www.equisearch.com/care/trailering/trailering1560/ (Oct. 2003).

Barringer, Sam, DVM. "What Will You Do When Injury Strikes?" *Mother Earth News* May 1999: 60.

Bartel, Marvin. "Learning to Throw" Learning to Throw on Potter's Wheel. http://www.goshen.edu/~marvinpb/throw/cover39.html (Oct. 2003).

"Basic Beeswax Formulas" Lumina Candles. http://www.luminacandles.com/formulas.htm (Sept. 2003).

"Basic Skin Care" Holistic-online.com. http://1stholistic.com/Beauty/skin/skin_basic_skin_care.htm (Aug. 2003).

"Basic soapmaking" *Countryside & Small Stock Journal*. July/August 1999: 117.

Bayley, Lesley and Richard Maxwell. *Understanding Your Horse: How to Overcome Common Behavior Problems*. North Pomfret: Trafalgar Square, 1996.

Bear, John B., PhD and Mariah P. Bear, MA. Bear's Guide to Earning College Degrees Nontraditionally. Benicia: C & B, 1994.

Beard, Dan. "Billy Bow-Leg Moccasins" The Inquiry Net! http://www.inquiry.net/outdoor/winter/gear/moccasins/billy_bowleg.htm (Mar. 2004).

Beard, Daniel C. *The American Boy's Handybook*. Boston: David Godine, 1994.

Beatson, Peter. "Experiments With Early Medieval Pottery" New Varangian Guard Inc. http://users.bigpond.net.au/quarfwa/miklagard/Articles/Pottery.htm (Apr. 2004).

Bell, Amy. "Unschooling FAQ's" Unschooling—Delight-Driven Learning. http://home.rmci.net/abell/page7.htm (Apr. 2005).

Benfatto, Elaine. "Introduction to Charkha Spinning" Urban Spinner. http://www.urbanspinner.com/charkha/charkha.pdf (Apr. 2004).

Bewely, Sara. "Be Your Own Biosecurity Guard" *Hobby Farm* (March/April 2005).

Bigbee, Daniel and Glenn Froning. "Egg Cleaning Procedures for the Household Flock" NebGuide. http://www.ianr.unl.edu/pubs/poultry/g466.htm (Sept. 2003).

"Botany Pages" Botany Pages. http://mung.pittsfieldrr.net/ (Mar. 2005).

Boyles, Denis. *The Lost Lore of Man's Life*. New York: Harper Collins, 1997.

"Breed Information" Hoof.com. http://www.hoof.com/draft/index.html (Oct. 2003).

Breningstall, F. Thomas. "Hoof Trimming Tools and How to Use Them" *Rural Heritage*. http://www.ruralheritage.com/village_smithy/tools_trim.htm (Oct. 2003).

Brewer, Thomas. "Ambidextrous chainsaw filing" *Backwoods Home Magazine*. http://www.backwoodshome.com/articles/brewer57.html (Sept. 2003).

Brill, Steve. "Wild Plants" Wildman Steve Brill. http://www.wildmanstevebrill.com/ (Mar. 2004).

Britt, Deborah M. *Horse Training Basics: An Indispensable Guide for Beginning Trainers*. Loveland: Alpine, 1994.

"Bronchitis" Mamashealth.com. http://www.mamashealth.com/infect/bronchitis.asp (Mar. 2004).

Brown, Chris, MD. "Veggie v. meat diet" Handbag.com. http://www.handbag.com/healthfit/diet/veggievsmeat/ (Aug. 2005).

Brusie, Bob, DVM. "Preventing Choke in Horses" EquiSearch. http://www.equisearch.com/care/injuries/eqchoke396/ (Oct. 2003).

Bubel, Mike and Nancy. *Root Cellaring; the Simple No-Processing Way to Store Fruits and Vegetables.* Emmaus: Rodale, 1979.

"Build a Home Distillation Apparatus" Bureau of Alcohol, Tobacco and Firearms. http://64.91.224.91/arch/Moonshine/still.pdf (Jan. 2004).

"Built-In Units" Carnegie Mellon Mechanical Engineering. http://www.andrew.cmu.edu/course/24-311/mathcad/built-in_units.htm (Oct. 2003).

Burris, Marjorie. "How to maintain a dirt road" *Backwoods Home Magazine.* http://www.backwoodshome.com/articles/burris48.html (Sept. 2003).

"Business Development Centre—CROPS" Department of Agriculture, Malaysia. http://agrolink.moa.my/doa/bdc/ (Sept. 2003).

Butler, Jean. "Home butchering – The versatile meat grinder" *Countryside Magazine.* http://www.countrysidemag.com/issues/5_2001.htm (Sept. 2003).

"Can Sizes and Equivalents" Home Cooking. http://homecooking.about.com/library/archive/blhelp7.htm (Sept. 2003).

"Cardiopulmonary resuscitation" Wikipedia. http://en.wikipedia.org/wiki/Cardiopulmonary_resuscitation (May 2004).

"Catgut" The 1911 Encyclopedia. http://20.1911encyclopedia.org/C/CA/CATGUT.htm (Mar. 2004).

"Chinese Oolong Tea" No-Occident. http://www.no-occident.com/noool.htm (Sept. 2003).

Clay, Jackie. "Ask Jackie: issue #70" *Backwoods Home Magazine.* http://www.backwoodshome.com/advice/aj71.html (Sept. 2003).

—— "Ask Jackie: issue #71" *Backwoods Home Magazine.* http://www.backwoodshome.com/advice/aj71.html (Sept. 2003).

—— "Ask Jackie: issue #73" *Backwoods Home Magazine.* http://www.backwoodshome.com/advice/aj73.html (Sept. 2003).

—— "Ask Jackie: issue #74" *Backwoods Home Magazine.* http://www.backwoodshome.com/advice/aj74.html (Sept. 2003).

—— "Ask Jackie: issue #78" Backwoods Home Magazine. http://www.backwoodshome.com/advice/aj78.html (Sept. 2003).

—— "Build your own log home in the woods" *Backwoods Home Magazine.* http://www.backwoodshome.com/articles2/clay72.html (Sept. 2003).

——"Canning 101—pickles, fruits, jams, jellies, etc." *Backwoods Home Magazine.* http://www.backwoodshome.com/articles/clay53.html (Sept. 2003).

—— "Medical kits for self-reliant families" *Backwoods Home Magazine.* http://www.backwoodshome.com/articles/clay60.html (Sept. 2003).

—— "Tips and Handy Hints for 4x4 Living" *Backwoods Home Magazine.* http://www.backwoodshome.com/articles2/clay71.html (Sept. 2003).

Clayton, Tommy. "Syrup Cooking at Southern Cross Farm" Syrupmakers. http://www.syrupmakers.com/clayton/index.htm (Aug. 2003).

"Companion Planting" Golden Harvest Organics. http://www.ghorganics.com/page2.html (Aug. 2003).

"Construct a garden pond" Garden Advice. http://www.gardenadvice.co.uk/howto/water/construct/index.html (Sept. 2003).

"Construction Tips" Building a Log Cabin. http://www.maqs.net/~cabin/ (Aug. 2003).

"Cooking Abbreviations" Cooking Basics. http://merlin.capcollege.bc.ca/comp101/Assignments/Assignment4/Dayna/Cooking%20Abbreviations.htm (Aug. 2003).

"Corn" *Countryside & Small Stock Journal.* March/April 1999: 50.

Corrie, Roger. "Clod-Crushers" Appletons' Cyclopaedia of Applied Mechanics. http://www.history.rochester.edu/appleton/a/agmac-1.html (Sept. 2003).

—— "Cultivators" Appletons' Cyclopaedia of Applied Mechanics. http://www.history.rochester.edu/appleton/a/agmac-3.html (Sept. 2003).

—— "Harrows" Appletons' Cyclopaedia of Applied Mechanics. http://www.history.rochester.edu/appleton/a/potatodi.html (Sept. 2003).

—— "Hay Rake" Appletons' Cyclopaedia of Applied Mechanics. http://www.history.rochester.edu/appleton/a/hayrake.html (Sept. 2003).

—— "Potato-Digger" Appletons' Cyclopaedia of Applied Mechanics. http://www.history.rochester.edu/appleton/a/harrows.html (Sept. 2003).

——"Ploughs" Appletons' Cyclopaedia of Applied Mechanics. http://www.history.rochester.edu/appleton/a/ploughs.html (Sept. 2003).

—— "Rollers" Appletons' Cyclopaedia of Applied Mechanics. http://www.history.rochester.edu/appleton/a/rollers.html (Sept. 2003).

—— "Sulky-Plough" Appletons' Cyclopaedia of Applied Mechanics. http://www.history.rochester.edu/appleton/a/sulkyplo.html (Sept. 2003).

"Cough and Cold Remedies" Make-Stuff.com. http://www.make-stuff.com/formulas/cough_cold.html (Sept. 2003).

Crabbe, Barb, DVM. "Cleaning a Horse's Sheath" EquiSearch. http://www.equisearch.com/care/grooming/eqsavvy386/ (Oct. 2003).

—— "How Much Does My Baby Weigh?" EquiSearch. http://www.equisearch.com/care/breeding/eqweigh28/ (Oct. 2003).

—— "How to Take Your Horse's Digital Pulse" EquiSearch. http://www.equisearch.com/care/firstaid/eqpulse178/ (Oct. 2003).

—— "Trail Terrors: Part 5: Tying Up, How to Recognize it and What to Do" EquiSearch. http://www.equisearch.com/sports/trailriding/eqtrailtr51582/ (Oct. 2003).

—— "Vaccination & Deworming Primer" EquiSearch. http://www.equisearch.com/care/vaccinations/eqvaccine2183/ (Oct. 2003).

Crook, Catherine. "Thatched Roofs—An Introduction" Building Conservation. http://www.buildingconservation.com/articles/thatchrf/thatchrf.htm (Feb. 2004).

Damerow, Gail. "Horse Power vs. Horsepower" Rural Heritage. http://www.ruralheritage.com/horse_paddock/horsepower.htm (Oct. 2003).

Danielson, Steve, Robert Wright, Gary Hein, Leroy Peters, Jim Kalisch. "Insects that Attack Seeds and Seedlings of Field Crops" NebGuide. http://www.ianr.unl.edu/pubs/insects/g1023.htm (Oct. 2003).

Dempsey, Jock. "How Do I Get Started in Blacksmithing?" Anvilfire.com. http://www.anvilfire.com/FAQs/getstart/index.htm (Mar. 2004).

Dempsey, John. "Selecting an Anvil" Anvilfire.com. http://www.anvilfire.com/FAQs/getstart/index.htm (Mar. 2004).

Densmore, Frances. *How Indians Use Wild Plants.* Toronto: Dover, 1974.

De Saulles, Denys. *Home Grown.* Boston: Houghton Mifflin, 1988.

Deyo, Holly. "Is It Soap Yet?" Millennium-Ark. http://www.millennium-ark.net/News_Files/Soap/Soapmaking_Holly.html (Aug. 2003).

Dickey, Elbert, Robert Pharris, Phillip Harlan and Gary Hosek. "Home Sewage Treatment Systems" NebGuide. http://www.ianr.unl.edu/pubs/wastemgt/g512.htm (Oct. 2003).

Dickson, Charles, Ph.D. "Cayenne: The Burning Balm." *Mother Earth News* (September 1999): 20.

"Digging a Well by Hand" Walton Feed. http://waltonfeed.com/old/well.html (Sept. 2003).

Doan, Ted. "Creep Feeding Lambs" NebGuide. http://www.ianr.unl.edu/pubs/sheep/g432.htm (Oct. 2003).

Dorling Kindersley Limited. *Eyewitness Books: Weather.* New York: Alfred A. Knopf, 1991.

Doty, Walter L. *All about Vegetables.* San Ramon: Chevron-Ortho, 1980.

"Dovetail Joints" Integrated Publishing. http://www.tpub.com/builder2n3/32.htm (Aug. 2003).

"Drying Firewood" End Times Report. http://www.endtimesreport.com/storing_firewood.html (Sept. 2003).

Duffy, Cathy. *Christian Home Educator's Curriculum Manual. 1997 –'98 Elementary Grades.* Westminster: Home Run-Grove, 1997.

—— *Christian Home Educator's Curriculum Manual. 1997–'98 Junior/Senior High.* Westminster: Grove, 1997.

Durtschi, Al. "Making a Cistern" Walton Feed. http://waltonfeed.com/old/cistern.html (Sept. 2003).

Dusault, Allen. "Methane Digesters for Dairies: New Opportunities for Industry and the Environment" Eco-Farm. http://www.eco-farm.org/sa/sa_dairy_synopsis_digester.html (Mar. 2004).

Eborn, Doug. "The Grains" Walton Feed. http://waltonfeed.net/self/grains.html (Mar. 2005).

Eastman, Charles A. *Indian Boyhood.* New York: Dover, 1971.

"Edible Motor Oil" Looksmart. http://www.findarticles.com/cf_dls/m1511/n8_v19/20979809/p1/article.jhtml (May 2004).

Eisenberg, Arlene, Heidi E. Murkhoff, and Sandee E. Hathaway, BSN. *What to Expect the First Year.* New York: Workman, 1989.

Elliot, Larry. "For truly independent energy system, your choices are solar, wind, and water" *Backwoods Home Magazine.* http://www.backwoodshome.com/articles/elliott28.html (Sept. 2003).

"Emergency Childbirth" Bagelhole.org. http://www.bagelhole.org/article.php/Survival/80/ (Jan. 2004).

Emery, Carla. *The Encyclopedia of Country Living: An Old Fashioned Recipe Book.* Seattle: Sasquatch, 1994.

"Encyclopedia by Common Names" Botany.com. http://www.botany.com/Common%7E1.html (Sept. 2003).

Epps, Bill. "3 Versions of Tongs" iForge. http://www.anvilfire.com/iForge/ (Mar. 2004).
"Euthanasia" University of Iowa Institutional Animal Care and Use Committee. http://www.uiowa.edu/~ancare/EUTH0001.HTM (Mar. 2005).
Evangelista, Anita. "Finding the best dog for the country life" *Backwoods Home Magazine*. http://www.backwoodshome.com/articles/evangelista63.html (Sept. 2003).
Everett, Erin. "Yurt Sweet Yurt: The Value of Simple Living." http://www.newlifejournal.com/junjul04/everett_0704.shtml (Jan. 2006).
Ewer, Cynthia Townley. OrganizedHome.com. http://organizedhome.com/clean/cleansol.html (Aug. 2003).
"Farm Equipment – Mechanics" EquiSearch. http://www.equisearch.com/farm/tools/mechanics1973/ (Apr. 2004).
Fears, J. Wayne. *How to Build Your Dream Cabin in the Woods: the Ultimate Guide to Building and Maintaining a Backcountry Getaway*. Guilford: Lyons, 2002.
"Fever: 26 Coping Tactics" MotherNature.com. http://www.mothernature.com/Library/Bookshelf/Books/47/58.cfm (Apr. 2004).
"Figuring Your Due Date" Plus Size Pregnancy. http://www.plus-size-pregnancy.org/figuring.htm (Sept. 2003).
"Fitting Harness and Collar" Hoof.com. http://www.hoof.com/draft/index.html (Oct. 2003).
"Fly Spray Formulas" EquiSearch. http://www.equisearch.com/care/pest_control/eqflyspray2169/ (Oct. 2003).
"Folk Remedies" Health 911. http://www.health911.com/remedies/rem_indx.htm (Apr. 2004).
"Food Storage and Emergency Preparation" Church of Jesus Christ of Latter-day Saints. http://www.providentliving.org/emergencyprep/calculator/0,11242,2008-1,00.html (Sept. 2003).
"Food Storage Calculator" The SurvivalRing Homepage. http://www.survivalring.org/foodcalc.htm (Sept. 2003).
"Formulas to get rid of insects in your home" Make-Stuff.com. http://www.make-stuff.com/formulas/insects.html (Sept. 2003).
"Frugal Baby Patter…ONLINE!!!!" Mothers Helping Hands. http://www.angelfire.com/biz/mothershelpinghands/mypatt.html (Sept. 2003).
"Frugal Baby Tips—Diapers" Born to Love. http://webhome.idirect.com/~born2luv/frugal-diapers.html#February10 (Sept. 2003).
Fulhage, Charles D., Dennis Sievers, James R. Fischer. "Generating Methane Gas From Manure" University of Missouri Extension. http://muextension.missouri.edu/explore/agguides/agengin/g01881.htm (Mar. 2004).
Garden Way Publishing. Just the Facts!. Pownal: Storey, 1993.
Geissal, Dynah. "Slaughtering and Butchering" *Backwoods Home Magazine*. http://www.backwoodshome.com/articles/geissal23.html (Sept. 2003).
Geller, Jon, DVM. "Poison on the Farm" *Mother Earth News* (August September 1999): 66.

"Give Yourself Stitches" *Maxim.* http://www.maximonline.com/maximwear/articles/article_4444.html (Oct. 2003).

Glad, Bill, MD "Caring for wounds in the field" *Backwoods Home Magazine.* http://www.backwoodshome.com/articles2/glade75.html (Sept. 2003).

Gleaves, Earl. "The Home Laying Flock, Part II: Management" NebGuide. http://www.ianr.unl.edu/pubs/poultry/g542.htm (Oct. 2003).

—— "Managing the Home Goose Breeder Flock" NebGuide. http://www.ianr.unl.edu/pubs/poultry/g711.htm (Oct. 2003).

Groleau, Rick. "Build a Rice Paddy" Japan's Secret Garden. http://www.pbs.org/wgbh/nova/satoyama/hillside.html (Sept. 2003).

Grotelueschen, Dale, DVM; Duane Rice, DVM. "Enterotoxemia in Lambs" NebGuide. http://www.ianr.unl.edu/pubs/animaldisease/g794.htm (Sept. 2003).

—— "Vaccinations in Sheep Flocks" NebGuide. http://www.ianr.unl.edu/pubs/animaldisease/g849.htm (Sept. 2003).

"Growing Artichokes" The Tasteful Garden. http://www.tastefulgarden.com/growingartichokes.htm (Mar. 2005).

"Growing Avocado Seeds into Trees" The Garden Helper. http://www.thegardenhelper.com/avocado.html (Mar. 2005).

"Growing & using Grains." *Countryside & Small Stock Journal* (March/April 1999): 34.

"Growing Under Lights: Grow Plants Even When the Sun Doesn't Shine" Gardener's Company. http://www.gardeners.com/gardening/BGB_underlightsb.asp (Oct. 2003).

"Grow Your Own Tea Plants: Camellia sinensis" Live Herb Nursery. http://liveherbnursery.com/teacamelliasinensis.html (Sept. 2003).

Hale, Rich. "Horse Shoe Demonstration" iForge. http://www.anvilfire.com/iForge/ (Mar. 2004).

"Hand, Face and Body Lotions You Can Make" Make-Stuff.com. http://www.make-stuff.com/formulas/lotions.html (Sept. 2003).

"Hand Laundry and Clothes Wringers" Rhema Publishing, Inc. http://members.aol.com/keninga/washing.htm (Nov. 2004).

"Hand (manual) expression of breast milk" Sutter Health. http://babies.sutterhealth.org/health/healthinfo/index.cfm?section=healthinfo&page=article&sgml_id=tp16330 (Aug. 2003).

"Hand-Milled Soap Additives" Millennium-Ark. http://www.millennium-ark.net/News_Files/Soap/Hand_Milled_Ing.html (Aug. 2003).

Hanson, Chris. *The Cohousing Handbook.* Point Roberts: Hartley & Marks, 1996.

"Harness Maintenance" Hoof.com. http://www.hoof.com/draft/index.html (Oct. 2003).

Harris, Martin. "Trusses—low cost marvels to roof over most large spaces" *Backwoods Home Magazine.* http://www.backwoodshome.com/articles/harris23.html (Sept. 2003).

Hawken, Paul and Fred Rohe. "The History of Vegetable Oil" Greentrust. http://ww2.green-trust.org:8383/2000/biofuel/oilhistory.htm (Jan. 2004).

Hayes, Karen, DVM, MS. "Busted! 12 Stallion-Care Myths" EquiSearch. http://www.equisearch.com/care/breeding/eqmyth642/ (Oct. 2003).

—— "Foaling Checklist" EquiSearch. http://www.equisearch.com/care/breeding/eqfoaling1073/ (Oct. 2003).

—— "How to Pull a Loose Shoe" EquiSearch. http://www.equisearch.com/care/hoofcare/eqloosesho379/ (Oct. 2003).

Henry, Chris. "Disposal Methods of Livestock Mortality" NebGuide. http://www.ianr.unl.edu/pubs/animals/g1421.htm (Sept. 2003).

"Herbal Remedies for Increasing Milk Supply" Kellymom.com. http://www.kellymom.com/herbal/milksupply/herbal-rem_a.html (Apr. 2004).

"Herbs to Avoid in Pregnancy" Snowbound Herbals. http://www.sbherbals.com/AvoidInPregnancy.html (Sept. 2003).

"Herd Handbook I & II" Canadian Meat Goat Journal. http://www.arjenboergoats.com/ (Sept. 2003).

"Hide Tanning" Homestead.org. http://www.homestead.org/tanning.htm (Sept. 2003).

Holling, Holling C. *The Book of Indians*. New York: Platt & Munk, 1962.

"Homemade Hand Lotion" Stretcher.com. http://www.stretcher.com/stories/980108b.cfm (Apr. 2004).

"Home Made Milk Paint Recipe" Real Milk Paint. http://www.realmilkpaint.com/recipe.html (Aug. 2003).

"Home Remedies for Minor Burns and Sunburn" Make-Stuff.com. http://www.make-stuff.com/formulas/burns.html (Sept. 2003).

"Hospital Corpsman 3 & 2" Integrated Publishing. http://www.tpub.com/content/medical/ (Mar. 2004).

Hough, Champ and Elizabeth Lliff. "Build a Barn that Works" EquiSearch. http://www.equisearch.com/farm/stable/eqbarn915/ (Oct. 2003).

"How to do the Heimlich Maneuver" Heimlich Institute. http://www.heimlichinstitute.org/howtodo.html (May 2004).

"How to Fillet a Fish" Fishing Cairns. http://www.fishingcairns.com.au/page17-1.html (Sept. 2003).

"How to Fillet a Fish" What You Need to Know about Tampa Bay, FL. http://tampa.about.com/library/howto/htfillet.htm (Sept. 2003).

"How to grow cotton" Cotton Australia. http://www.cottonaustralia.com.au/KD_howtogrow.html (Apr. 2004).

"How to measure the wind" *Countryside & Small Stock Journal* (March/April 1999): 119.

"How to Substitute for Wheat Flour" Food Resource. http://food.oregonstate.edu/g/subflour.html (Mar. 2004).

"How You Can Grow a Peanut Plant" Virginia-Carolina Peanuts. http://www.aboutpeanuts.com/infougro.html (Aug. 2005).

Hunt, W. Ben. *Indian Crafts and Lore*. New York: Golden Press, 1954.

Hunter, Jim. "Drying apples." *Countryside & Small Stock Journal* (July/August 1999): 53.

Hunter, Julie. "Patching clothes." *Countryside & Small Stock Journal* (July/August 1999): 123.

—— "Poison ivy preventative: fight fire with fire" *Countryside & Small Stock Journal* July/August 1999: 95.

"Install a drainage system in your lawn" Garden Advice. http://www.gardenadvice.co.uk/howto/lawns/drainage/ (Sept. 2003).

"Instructions for one version of the knit or crochet soakers from The Frugal Baby Pattern..." Mothers Helping Hands. http://www.angelfire.com/biz/mothershelpinghands/crochetedsoaker.html (Sept. 2003).

"Intensive Gardening Methods" Sunny Boy Gardens. http://www.sunnyboygardens.com/herb-information/essays/intensive.htm (Apr. 2004).

"Intensive Spacing Guide" Virginia Cooperative Extension. http://www.ext.vt.edu/pubs/envirohort/426-335/426-335.html (Sept. 2003).

"Introduction to Timber Framing" Timber Framer's Guild. http://tfguild.org/intropg.html (Mar. 2004).

"Is a Micro-Hydroelectric System Feasible for You?" U.S. Department of Energy. http://www.eere.energy.gov/consumerinfo/refbriefs/ab2.html (Mar. 2004).

Jackson, Jackie Lee. "How to Measure Cinch Size" EquiSearch. http://www.equisearch.com/tack/eqcinchgau120/ (Oct. 2003).

Jaworski, Joe. "The Care and Feeding of Sourdough Starters" JoeJaworski.com. http://www.joejaworski.com/bread3.htm (Aug. 2003).

Jenkins, Joseph. The Humanure Handbook. http://www.weblife.org/humanure/default.html (Sept. 2003).

Johnson, Duane, Blaine Rhodes and Robert Allen. "Canola-based Motor Oils" Purdue University Center for New Crops & Plant Products. http://www.hort.purdue.edu/newcrop/ncnu02/v5-029.html (May 2004).

Julin, Brian S. "Cannabis-Marijuana FAQ" Cannabis-Marijuana FAQ. http://www-unix.oit.umass.edu/~verdant/Marijuana_FAQ/Index.html (Mar. 2004).

King, P. R. "Build Your Own Yurt." http://www.woodlandyurts.co.uk/Yurt_Facts/Build_Your_Own.html (Jan. 2006).

Kirby, Barbara. "What is Asperger Syndrome?" O.A.S.I.S. http://www.udel.edu/bkirby/asperger/ (Apr. 2005).

Kuhns, Mike; Tom Schmidt. "Heating With Wood: I. Species Characteristics and Volumes" NebGuide. http://www.ianr.unl.edu/pubs/forestry/g881.htm (Sept. 2003).

Kloss, Jethro. *Back to Eden.* Lotus Light, 1989.

Koppedrayer, K. I. "Cleaning Your Feathers" Primitive Archer. http://www.primitivearcher.com/articles/cleaning.html (Aug. 2003).

Kovach, Tom R. "Blanching vegetables" *Backwoods Home Magazine.* September/October 1999: 53.

Keupper, George, Mardi Dodson. "Companion Planting: Basic Concepts and Resources" ATTRA. http://attra.ncat.org/attra-pub/complant.html (Aug. 2003).

Laing, Ken. "Horse Power for Organic Farms" Rural Heritage. http://www.ruralheritage.com/horse_paddock/horse_power.htm (Oct. 2003).

Lanza, Patricia. "Lasagna Gardening" *Mother Earth News.* May 1999: 51.

Leitch, William. *Hand-Hewn: The Art of Building Your Own Cabin.* San Francisco: Chronicle, 1976.

Ligda, David J. "The Water Buffalo" The Water Buffalo. http://ww2.netnitco.net/users/djligda/waterbuf.htm (Mar. 2004).

Lindstrom, Carl. "Greywater irrigation" Greyewater.com. http://www.greywater.com/ (Oct. 2003).

Lingren, Herbert. "Creating Sustainable Families" NebGuide. http://www.ianr.unl.edu/pubs/family/g1269.htm (Sept. 2003).

—— "Daily Activities for Family Time Together" NebGuide. http://www.ianr.unl.edu/pubs/family/nf320.htm (Sept. 2003).

"Listen to your weeds!" *Countryside & Small Stock Journal* July/August 1999: 120.

Lotven, Melynda. "Growing Gourds FAQ" Just Gourds. http://www.justgourds.com/GourdInfo.htm (Feb. 2005).

"Maintaining water quality in the pool" Garden Advice. http://www.gardenadvice.co.uk/howto/water/maintaining/index.html (Sept. 2003).

"Make Your Own Aftershave" Make-Stuff.com. http://www.make-stuff.com/formulas/aftershave.html (Sept. 2003).

"Make Your Own Biodiesel" Journey to Forever. http://journeytoforever.org/biodiesel_make.html (Jan. 2004).

"Make Your Own Powder" Make-Stuff.com. http://www.make-stuff.com/formulas/powder.html (Sept. 2003).

"Making a Drop Spindle" The Joy of Handspinning. http://www.joyofhandspinning.com/make-dropspin.html (Aug. 2003).

"Making Biodiesel Fuel at Home" Bagelhole.org. http://www.bagelhole.org/article.php/Transportation/149/ (Jan. 2004).

"Making Herbal Infusions, Decoctions and Ointments" Healthy New Age. http://www.healthynewage.com/makeherb2.html (Jan. 2004)

"Making Jam Without Added Pectin" National Center for Home Food Preservation. http://www.uga.edu/nchfp/how/can_07/jam_without_pectin.html (Feb. 2005).

"Making lump charcoal" California BBQ Association. http://www.cbbqa.com/faq/18.html (Mar. 2004).

"Making lye from wood ash" Journey to Forever. http://journeytoforever.org/biodiesel_ashlye.html (Oct. 2003).

"Making Lye Water" End Times Report. http://www.endtimesreport.com/making_lye.html (Sept. 2003).

"Making Natural Dyes From Plants" Pioneer Thinking. http://www.pioneerthinking.com/naturaldyes.html (Apr. 2004).

"Making Yarn with a Drop Spindle" The Joy of Handspinning. http://www.joyofhandspinning.com/HowToDropspin.html (Aug. 2003).

Marusek, James. "Tutorial on Cutting Firewood" Impact. http://personals.galaxyinternet.net/tunga/N15.htm (Oct. 2003).

Masanoba, Fukuoka. *The One Straw Revolution: An Introduction to Natural Farming*. n.p.: Rodale, 1978.

Matlack, Pamela: "Build a warp weighted loom" Essortment.com. http://vt.essortment.com/warpweighted_rkmy.htm (Apr. 2004).

—— "Warping and weaving on a warp-weighted loom" Essortment.com. http://vt.essortment.com/warpingweaving_rkpp.htm (Apr. 2004).

Maxted-Frost, Tanyia "The Benefits of Organic Food: Our right to eat uncontaminated food" Positive Health. http://www.positivehealth.com/permit/Articles/Organic%20and%20Vegetarian/frost47.htm (Aug. 2003).

McCutcheon, Marc. *Roget's Super Thesaurus.* Cincinnati: Writer's Digest F&W, 1995.
McLeod, Judyth. *Botanica's Organic Gardening: The Healthy Way to Live and Grow.* San Diego: Laurel Glen, 2002.
"Measurements and Conversions" Millennium-Ark. http://www.millennium-ark.net/News_Files/INFO_Files/Metric_Imp_Charts.html (Aug. 2003).
"Menstrual Odor" Museum of Menstruation and Women's Health. http://www.mum.org/Odor.htm (May 2004).
Mercer, Steve. "A root beer FAQ" Jason's Bistro. http://www.jagaimo.com/bistro/rootbeerfaq.html (Dec. 2004).
"Metric prefixes and unit conversions" All about Circuits. http://www.allaboutcircuits.com/vol_5/chpt_1/10.html (Aug. 2003).
Mettot, Beverly. "Companion Planting" *Backwoods Home Magazine.* http://www.backwoodshome.com/articles2/mettot81.html (Sept. 2003).
"Mix your own mortar" Garden Advice. http://www.gardenadvice.co.uk/howto/garden-build/mortar/index.html (Sept. 2003).
Mollison, Bill. *Introduction to Permaculture.* Tyalgum: Tagari, 1991.
Moore, Deborah. "Learning to Cook on a Wood Stove" *Countryside Magazine.* November/December 1998. http://countrysidemag.com/issues/6_1998.htm (Sept. 2003).
Morley, G.M., MB, ChB, FACOG. "Clamp the Umbilical Cord" Dr. Joseph Mercola. http://www.mercola.com/2002/jan/2/umbilical_cord.htm (Aug. 2003).
Muir, Cynthia. "Bee sting treatment" http://papa.essortment.com/beestingtreatm_rzdn.htm (Apr. 2004).
—— "Make your own cooking spices" PageWise. http://inin.essortment.com/cookingspices_rxux.htm (Oct. 2003).
National Council Boy Scouts of America. *Boy Scout Handbook.* New Brunswick: SA, 1962.
"Natural Treatment for Diabetes" Shirley's Wellness Café. http://www.shirleys-wellness-cafe.com/diabetes.htm (Mar. 2004).
Needham, Walter and Barrows Mussey. *A Book of Country Things.* Brattleboro: Stephen Greene, 1965.
"Nosebleed" MedlinePlus. http://www.nlm.nih.gov/medlineplus/ency/article/003106.htm (Apr. 2004).
Nuñes, Angel. "No Need for Shampoo!" 25 Years Ago on Ambergris Caye. http://www.ambergriscaye.com/25years/noneedshampoo.html (Sept. 2003).
—— "Life without Toilet Paper!!" 25 Years Ago on Ambergris Caye. http://www.ambergriscaye.com/25years/notoiletpaper.html (Sept. 2003).
—— "Roasting Peppers" 25 Years Ago on Ambergris Caye. http://www.ambergriscaye.com/25years/roastingpeppers.html (Sept. 2003).
Nyerges, Christopher. "Make bread from weeds" *Countryside & Small Stock Journal* (March/April 1999): 58.
Ody, Penelope. *The Complete Medicinal Herbal.* New York: Dorling Kindersley, 1993.
Ong, Samuel, MD. "Bronchitis" eMedicine. http://www.emedicine.com/EMERG/topic69.htm (Mar. 2004).

"On the Farm" The Norse Storm. http://www.internet-at-work.com/hos_mcgrane/viking/eg_viking_menu1.html (Mar. 2004).

"Our Tooth Sense Worth" Word of Mouth. http://www.massdental.org/public/wordofmouth.cfm?doc_id=532 (Apr. 2004).

"Owenlea Farm greenhouse" Owenlea Farm. http://www.bright.net/~fwo/greenhouse/greenhouse.html (Sept. 2003).

Pearsall, Kendra. "Naturopathic Recommendations for Diabetics" Mercola. http://www.mercola.com/2003/aug/20/diabetes_naturopathic.htm (Mar. 2004).

Pedigo, Jayne. "Glossary of Equine Terms" EquiSearch. http://www.equisearch.com/glossary/ (Oct. 2003).

——"Grooming for Health" EquiSearch. http://www.equisearch.com/care/grooming/grooming082497/ (Oct. 2003).

—— "How to Clean a Horse's Hooves" EquiSearch. http://www.equisearch.com/care/howto/hooves/ (Oct. 2003).

—— "How to Put a Halter on a Horse" EquiSearch. http://www.equisearch.com/care/howto/halter/ (Oct. 2003).

——"Obvious or Suspected Fracture" EquiSearch. http://www.equisearch.com/care/firstaid/fracture102801c/ (Oct. 2003).

—— "Riding Bareback" EquiSearch. http://www.equisearch.com/novice/learntoride/bareback070100a/ (Oct. 2003).

—— "Tack Safety" EquiSearch. http://www.equisearch.com/tack/tacksafety081097/ (Oct. 2003).

—— "Weaning Time" EquiSearch. http://www.equisearch.com/care/breeding/weaning072500a/ (Oct. 2003).

"Periodontal (Gum) Disease" American Academy of Periodontology. http://www.perio.org/consumer/2a.html (Apr. 2004).

Peterson, Dorene, "Making Essential Oil" Henriette's Herbal Homepage. http://www.ibiblio.org/herbmed/faqs/medi-4-1-distilling.html (Jan. 2004).

"Pit Fired Pottery" Mathin.com. http://mathin.com/pottery/pitfire/index.html (Jan. 2004).

Piven, Joshua and David Borgenicht. *The Worst-Case Scenario Survival Handbook*. San Francisco: Chronicle, 1999.

"Plants Database" Natural Resources Conservation Service. http://plants.usda.gov/index.html (Jan. 2004).

"Plaster and plastering" Technosolutions. http://www.technosolution.co.uk/diy/building/Plastering/plastering.htm (Sept. 2003).

"Plastering, Stuccoing, and Ceramic Tile" Integrated Publishing. http://www.tpub.com/content/construction/14044/css/14044_211.htm (Mar. 2004).

Pratt, Danette. "Build a Warp Weighted Loom" Build a Warp Weighted Loom. http://www.geocities.com/attedragon/ (Apr. 2004).

"Primitive Cooking Methods" Bagelhole.org. http://www.bagelhole.org/article.php/Survival/93/ (Jan. 2004).

Probst, Sarah. "Can Imprinting Go Too Far?" Rural Heritage. http://www.ruralheritage.com/horse_paddock/horse_imprint.htm (Oct. 2003).

"Proper Canning Jars and Lids" Garden and Hearth. http://gardenandhearth.com/Canning/canning_jars.htm (Aug. 2003).

Quilt Designs: Old Favorites—and New. North Kansas City: Aunt Martha's Creations, n.d.

"Quince Growing" NSW Department of Primary Industries: Agriculture. http://www.agric.nsw.gov.au/reader/deciduous-fruits/h413.htm (Aug. 2005).

"Rainwater Harvesting and Purification System" Experiments in Sustainable Urban Living. http://users.easystreet.com/ersson/rainwatr.htm (Sept. 2003).

Ramey, David W. DVM.; Stephen E. Duren, PhD. *Concise Guide to Nutrition in the Horse*. New York: Howell Book House, 1998.

"Recipes for Whole Foods Baby Formula" Weston A. Price Foundation. http://www.westonaprice.org/children/recipes.html (Oct. 2003).

"Recycling agricultural wastes to produce hot water" Experiments in Sustainable Urban Living. http://users.easystreet.com/ersson/composti.htm (Sept. 2003).

Repensky, Lisa. "Raising Paint Horses" A Lil' Bit of Color. http://geocities.com/Heartland/Hills/3619/raising.html (Oct. 2003).

——"Training Paint Horses" A Lil' Bit of Color. http://geocities.com/Heartland/Hills/3619/training1.html (Oct. 2003).

"Replacing a Car Battery" DIY. http://www.diynetwork.com/diy/ab_auto_electrical_system/article/0,2021,DIY_13677_2276056,00.html (Feb. 2004).

Reynolds, Biz Fairchild. "What the Angels Eat" *Mother Earth News* July 1999: 34–38.

Richards, Matt. "Bark Tanning" Braintan.com. http://www.braintan.com/barktan/1basics.htm (Mar. 2004).

Riggs, Kathleen. "Using Boiling Water bath Canners" Utah State University. http://extension.usu.edu/files/foodpubs/canfs02.htm (Sept. 2003).

Rivero, Lisa. "Unschooling Mambo" Unschooling.com. http://www.unschooling.com/library/index.shtml (Apr. 2004).

"Roadside Emergencies—Servicing a Flat Tire" DIY. http://www.diynetwork.com/diy/ab_tires/article/0,2021,DIY_13697_2270907,00.html (Feb. 2004).

Roberts, Monty. *The Man Who Listens to Horses*. New York: Random House, 1997.

"Rock Construction" Bagelhole.org. http://www.bagelhole.org/article.php/Housing/21/ (Jan. 2004).

Rodale, Maria. Organic Gardening. Emmaus: Rodale, 1998.

Rood, Mary Ann. "Tips on Growing Gourds" North Carolina Gourd Society. http://www.twincreek.com/gourds/growing.htm (Mar. 2005).

"Rug Braiding" Craftown. http://www.craftown.com/instruction/rugs.htm (Oct. 2003).

Rutherford, Kim, MD. "What is the Apgar Score?" KidsHealth. http://kidshealth.org/parent/pregnancy_newborn/pregnancy/apgar.html (Aug. 2003).

Rymer, Eric. "European Farming During Middle Ages to 1800s" Historylink101.http://www.historylink101.com/lessons/farm-city/middle-ages.htm (Mar. 2004).

Salisbury, D.L., DVM. "Doc Salisbury makes sourdough" *Countryside & Small Stock Journal*. July/August 1999: 100.

Sadanandan, Smitha. "Good earth" The Hindu. http://www.hinduonnet.com/thehindu/mp/2003/08/04/stories/2003080401900100.htm (Sept. 2003).
Savino, Kelly. "Check out my $50 greenhouse!" Earthworks Studio. http://www.primalmommy.com/greenhouse.html (Sept. 2003).
—— "Recipes for a small and crowded planet" Earthworks Studio. http://pages.ivillage.com/primalmommy/recipes.html (Sept. 2003).
Schmidt, Vicki. "Horseshoe Savvy" Rural Heritage. http://www.ruralheritage.com/village_smithy/savvy.htm (Oct. 2003).
Schwalbaum, Martin. "(Apple Butter) and More Jam Recipes" Fourpeaks. http://www.4peaks.com/fjam.htm (Feb. 2005).
"Sea Sponges instead of Tampons" Valley Café. http://www.valleycafe.com/resources/sea-sponges.html (Sept. 2003).
Sears, William, MD, Martha Sears RN. *The Baby Book.* Boston: Little, Brown and Company, 1993.
Sebesta, Judith Lynn. "Warp-Weighted Loom Photos" Judith Lynn Sebesta. http://www.usd.edu/~jsebesta/Headingband.html (Apr. 2004).
Seton, Ernest Thompson. "Teepee Plans 10'" Old School Scouting: What to Do, and How to Do It. http://www.inquiry.net/outdoor/native/skills/teepee.htm (Oct. 2003).
"Sewage Treatment and Leaching Beds" Thunder Bay District Health Unit. http://www.tbdhu.com/inspection/Sewage.htm (Aug. 2003).
"Shampoos, Rinses and Conditioners" Make-Stuff.com. http://www.make-stuff.com/formulas/hair_care.html#trulyhomemade (Sept. 2003).
Shannon, Ron. "Water Wheel Engineering" IPC-VI Designing for Sustainable Future. http://www.rosneath.com.au/ipc6/ch08/shannon/ (Mar. 2004).
Shepard, Mark. "Charkha Tips: How to Spin Cotton on Mahatma Gandhi's Spinning Wheel." http://www.markshep.com/nonviolence/Charkha.html (Apr. 2004).
Shover, James G. "Reduce losses from drone comb." *Countryside & Small Stock Journal* (May/June 1999): 85.
Sloane, Eric. *Diary of an Early American Boy: Noah Blake 1805.* New York: Ballantine/Random House, 1965.
Smith, Lucy. *Improve Your Survival Skills.* London: Usborne, 1987.
"Snakes" Carolinas Health System. http://carolinas.org/services/poison/snakes.cfm (Aug. 2003).
"Something in the Air: Airborne Allergens" National Institute of Allergy and Infectious Diseases. http://www.niaid.nih.gov/publications/allergens/title.htm (Oct. 2003).
Stedham, Glen. *Bush Basics: A Common Sense Guide to Backwoods Adventure.* Victoria: Orca, 1997.
Storey, John and Martha. *Storey's Basic Country Living.* Canada: Storey, 1999.
"Survival Meat Preserving—Part 2, Jerky" End Times Report. http://www.endtimesreport.com/storing_meat_2.html (Sept. 2003).
"Symptoms of type 1 diabetes" LillyDiabetes.com. http://www.lillydiabetes.com/Education/DiabetesSymptoms.cfm (Mar. 2004).

"Tallow Candles" Fort Clatsop National Memorial. http://www.nps.gov/focl/candles.htm (Sept. 2003).

Tatchell, Judy and Dilys Wells. *Food Fitness and Health.* London: Usborne, 1991.

"Tending to Mortise-and-Tenon Joints" Dummies.com. http://www.dummies.com/WileyCDA/DummiesArticle/id-2325.html (Mar. 2004).

The American National Red Cross. *Standard First Aid.* St. Louis: Mosby-Year Book, 1993.

"Testing vinegar." *Countryside & Small Stock Journal* July/August 1999: 47.

"Thatching: The traditional British craft" Britain Express. http://www.britainexpress.com/History/thatching.htm (Feb. 2004).

"The basics of basic bread." *Countryside & Small Stock Journal* (March/April 1999): 37.

"The fab four." *Countryside & Small Stock Journal* March/April 1999: 42.

The Guys Six Feet Under. "Making Alcohol" Temple of the Screaming Electron. http://www.totse.com/en/drugs/booze_the_legal_drug/alcohol.html (Jan. 2004).

"The Heimlich Maneuver" The Heimlich Institute. http://www.heimlichinstitute.org/correcting.htm (Apr. 2004).

"The homesteader's guide to vinegar." *Countryside & Small Stock Journal* (July/August 1999): 44.

"The Household Cyclopedia" Haunted Planet. http://www.mspong.org/cyclopedia/about.html (Mar. 2004).

The Mother Earth News. *The Mother Earth News Almanac: A Guide through the Seasons.* New York: Bantam, 1973.

The Reader's Digest Association. *Back to Basics.* Pleasantville: Reader's Digest, 1981.

The Reader's Digest Association. *Complete Do It Yourself Manual.* Pleasantville: Reader's Digest, 1973.

The Reader's Digest Association. *Great Health Hints & Handy Tips.* Pleasantville: Reader's Digest, 1994.

"The Teachings of Lao Tzu" Taoism for Teenagers. http://www.angelfire.com/wa/tao4teens/What.html (Aug. 2003).

"The Timber Framing Process" Vermont Timber Works. http://www.vermonttimberworks.com/framing.html (Mar. 2004).

Thomsen, Skip. "Diesel generator power is a sensible choice especially when integrated into the total system" *Backwoods Home Magazine.* http://www.backwoodshome.com/articles/thomsen43.html (Sept. 2003).

Tracy, Paul, Paul,Barry D. Sims, Steven G. Hefner and John P. Cairns. "Guidelines for Producing Rice Using Furrow Irrigation" University of Missouri Extension. http://muextension.missouri.edu/xplor/agguides/crops/g04361.htm (Apr. 2005).

"Treating Shock" The Church of Jesus Christ of Latter-day Saints. http://www.lds.org/library/display/0,4945,31-1-15-51,00.html (Apr. 2004).

Trebor, Leonard. "Growing the Eternal Tomato" *Backwoods Home Magazine.* (May/June 1999): 82

"Typhus and Cholera" End Times Report. http://www.endtimesreport.com/cholera.html (Sept. 2003).

"Unconventional Therapies—Sassafrass Tea" BC Cancer Agency. http://www.bccancer.bc.ca/PPI/UnconventionalTherapies/SassafrasTea.htm (Dec. 2004).

United States Department of the Army. *Survival.* Washington: Army, 1970.

USDA. "Harvesting, Retting and Fiber Separation" Industrial Hemp in the United States. http://www.globalhemp.com/Archives/Government_Research/USDA/ages001Ee.pdf (Mar. 2004).

"USDA Plant Hardiness Zone Map" United States National Arboretum. http://www.usna.usda.gov/Hardzone/ushzmap.html (Aug. 2003).

Vandermark, Traci. "Homemade cosmetics recipe" Pagewise. http://arar.essortment.com/cosmeticsrecipe_rfdk.htm (Apr. 2004).

"Vegetable Information" ENVEG Project. http://www.hri.ac.uk/enveg/ (Sept. 2003).

"Vegetable Oils" Plants for a Future. http://www.scs.leeds.ac.uk/pfaf/vegoils.html (Jan. 2004).

Vitruvius. De Architectura. The Water Screw. http://www.mcs.drexel.edu/~crorres/Archimedes/Screw/SourcesScrew.html (Mar. 2004).

Vivian, John. "Growing Grains" *Mother Earth News* July 1999: 50.

— "The New Science of Freezing & Canning" August/September 1999: 38.

Wagner, Dale. "Loads" Hoof.com. http://www.hoof.com/draft/html/range.html (Oct. 2003).

Walker, Tas B. "The Pitch for Noah's Ark" Answers in Genesis. http://www.answersingenesis.org/Home/Area/Magazines/docs/v7n1_ark.asp (Mar. 2004).

Wallner-Pendleton, Eva. "Vaccination Guide for the Small Poultry Flock" NebGuide. http://www.ianr.unl.edu/pubs/animaldisease/g1202.htm (Sept. 2003).

"Water" End Times Report. http://www.endtimesreport.com/waterarticle.html (Sept. 2003).

"Water Distillation Principles" Bagelhole.org. http://www.bagelhole.org/article.php/Water/101/ (Jan. 2004).

"Ways to get rid of head lice" Make-Stuff.com. http://www.make-stuff.com/formulas/lice.html (Sept. 2003).

Weaver, Sue. "Wild on the Farm: Maple Syrup" *Hobby Farm* March/April 2005.

Webster, Noah. *The New Universities Webster's Dictionary*. Ed. Joseph Devlin, M.A. New York: World Syndicate, 1938.

Wendelken, David & Rebecca. "Warping Your Loom" Early Period. http://www.housebarra.com/EP/ep03/14warping.html (Apr. 2004).

Wetherbee, Kris. "Striking Gold with Green Manure" *Mother Earth News* May 2000: 64.

"What is a Scoville unit?" wiseGeek. http://www.wisegeek.com/what-is-a-scoville-unit.htm (Feb. 2004).

"What Is It and Where Can I Find It?" Make-Stuff.com. http://www.make-stuff.com/home_business/resource_list.html (Sept. 2003).

"What to Do About Aggressive Moose" Wildlife Conservation. http://www.wildlife.alaska.gov/aawildlife/agmoose.cfm (May 2004).

White, Elaine C. "Soap Recipes: Seventy Tried-and-True Ways to Make Modern Soap" http://members.aol.com/oelaineo/directions.html (Nov. 2003).

Whitefeather, Willy. *Outdoor Survival Handbook for Kids*. Tucson: Harbinger House, 1990.

Wilson, Janet. "Electrical Appliances and the Energy Dollar" NebGuide. http://ianrpubs.unl.edu/consumered/heg94.htm (Apr. 2004).

Wilson, Melanie. "Calcium: getting enough without milk" Vegetarian Baby and Toddler. http://www.vegetarianbaby.com/magazine/articles/calcium.html (Aug. 2003).

"Wind chill is a guide to winter danger" *USA Today*. http://www.usatoday.com/weather/resources/basics/windchill/wind-chill-chart.htm (May 2004).

Wiseman, Lisa. "Buying Your First Horse" Homestead.org. http://www.homestead.org/Equis/BuyingYourFirstHorse.htm (Mar. 2005).

"Wounds" Wilderness Survival. http://www.wilderness-survival.net/medicine-6.php (Oct. 2003).

Yago, Jeffrey, PE, CEM. "A solar primer: How it works, how it's made, what it costs" *Backwoods Home Magazine*. http://www.backwoodshome.com/articles2/yago72.html (Sept. 2003).

Yeager, Alice Brantley. "Cucumbers: Cool is the word!" *Backwoods Home Magazine* May/June 1999: 8.

Index

pregnancy, 179, 186-7

About the Author

▲ The author and two of her daughters

Nicole Faires is an adventurer, self-proclaimed eccentric, wife, and mother of three girls.

She grew up in a semi-nomadic homeschooling family and spent her early years in rural Montana on a hobby farm raising chickens and growing her own food, learning to crochet, reading out-of-print books by Masanobu Fukuoka, and dreaming of the Amish.

She has earned her living as a piano accompanist, Internet technical support agent, greeting card designer, portrait artist, hospital housekeeper, life coach, and web administrator, but every spare minute away from work was spent in taking notes on the great thinkers and gardeners of the past.

She now lives in beautiful British Columbia on Vancouver Island as a champion of renaissance sustainable lifestyles.

www.nicolefaires.com
www.twitter.com/nicolefaires